6th International Conference on Mathematics and Natural Sciences (ICMNS 2016)

On the Road Towards Sustainable Development

Journal of Physics: Conference Series Volume 1127

Bandung, Indonesia
2-3 November 2016

Editors:

Dessy Natalia
Oki Neswan
Didin Mujahidin
Khairul Basar
M. Ikbal Arifyanto

ISBN: 978-1-5108-8100-6
ISSN: 1742-6588

Printed from e-media with permission by:

Curran Associates, Inc.
57 Morehouse Lane
Red Hook, NY 12571

Some format issues inherent in the e-media version may also appear in this print version.

Printed by Curran Associates, Inc. (2019)

For permission requests, please contact the Institute of Physics
at the address below.

Institute of Physics
Dirac House, Temple Back
Bristol BS1 6BE UK

Phone: 44 1 17 929 7481
Fax: 44 1 17 920 0979

techtracking@iop.org

Additional copies of this publication are available from:

Curran Associates, Inc.
57 Morehouse Lane
Red Hook, NY 12571 USA
Phone: 845-758-0400
Fax: 845-758-2633
Email: curran@proceedings.com
Web: www.proceedings.com

TABLE OF CONTENTS

IOP Conf. Series: Journal of Physics: Conf. Series **1127** (2019) 011001 doi:10.1088/1742-6596/1127/1/011001

PREFACE

The 6th International Conference on Mathematics and Natural Sciences

"On the Road Towards Sustainable Development"

The 6th ICMNS was organized by Faculty of Mathematics and Natural Sciences ITB and supported by School of Pharmacy and School of Life Sciences and Technology ITB. This conference aimed to disseminate recent research results in the field of sciences and mathematics, and to promote multi- and inter-disciplinary researches in sciences and their applications. The two days intensive and warm interaction between researchers from across the world are the driving force for the advancement of science, and in turn for future sustainable development of humanity. The scope of ICMNS covered several fields of studies, namely chemistry, physics, mathematics, material science, computational science, medical and pharmaceutical sciences, life sciences, environmental sciences, science of renewable energy, earth and space science.

It was wonderful to have over 400 speakers and participants representing various universities, research institutes, government agencies, and industries from Indonesia, Malaysia, Singapore, Korea, Japan, Australia, The Netherlands, Germany, and Kazakhstan. We were very honored to have Prof. Gerard 't Hooft, the 1999 Nobel Laureate in Physics, in this conference. The conference program consisted of a Nobel Laureate Lecture, plenary sessions, parallel sessions, and poster sessions. We would like to express our gratitude to all distinguished speakers, presenters and participants. We also sincerely thank the Steering Committee, the Scientific Committee, the Organizing Committee and all sponsors for their great efforts and contributions to make this conference possible.

We are very pleased to introduce the Proceedings of the 6th ICMNS which contain a selection of papers from contributed presentation in the conference. The Proceedings features recent results and contemporary research in Mathematics, Physics, Astronomy and the related fields. We hope that this Proceedings will foster the advancement of science and technology which is essential to support sustainable development.

Bandung, November 2 2018

Editors,
Dessy Natalia
Khairul Basar
Oki Neswan
M. Ikbal Arifyanto
Didin Mujahidin

ICMNS 2016

IOP Conf. Series: Journal of Physics: Conf. Series **1127** (2019) 011002 doi:10.1088/1742-6596/1127/1/011002

COMMITTEES

STEERING COMMITTEE

- Prof. Abdul Waris (ITB, Indonesia)
- Prof. Daryono T. Hadi (ITB, Indonesia)
- Prof. Edy Tri Baskoro (ITB, Indonesia)
- Dr. Indra Noviandri (ITB, Indonesia)
- Dr. I Nyoman P. Aryanta (ITB, Indonesia)
- Prof. Kajta Loos (University of Groningen, The Netherlands)
- Dr. Lily Euwilaichtr (Biotech, Thailand)
- Prof. Martin Baca (Technical University in Kosice, Slovak Republic)
- Prof. Mikio Kurita (Kyoto University, Japan)
- Prof. F.P. Zen (ITB, Indonesia)
- Prof. Taufik Hidayat (ITB, Indonesia)
- Prof. Yong-Hoon Cho (KAIST, South Korea)

SCIENTIFIC COMMITTEE

- Dr. Agus Suroso (ITB, Indonesia)
- Dr. Anggraini Berlian (ITB, Indonesia)
- Dr. Akfiny H. Aimon (ITB, Indonesia)
- Dr. Azzania Fibriani (ITB, Indonesia)
- Dr. Biswarup Mukhopadhyay (Virginia Tech, USA)
- Dr. Debbie S. Retnoningrum (ITB, Indonesia)
- Dr. Dian Rosleine (ITB, Indonesia)
- Dr. Enny Ratnanigsih (ITB, Indonesia)
- Dr. Fourier Dzar E.L. (ITB, Indonesia)
- Prof. Hidemi Nagao (Kanazawa University, Japan)
- Dr. Ihsanawati (ITB, Indonesia)
- Dr. Ilma Nugrahani (ITB, Indonesia)
- Dr. Khairul Basar (ITB, Indonesia)
- Dr. Kiki Vierdayanti (ITB, Indonesia)
- Prof. I. Made Arcana (ITB, Indonesia)
- Prof. Katja Loos (University of Groningen, The Netherlands)
- Dr. Lily Euwilaichtr (Biotech, Thailand)
- Prof. M. Bachri Amran (ITB, Indonesia)
- Dr. M. Ikbal Arifyanto (ITB, Indonesia)
- Dr. Muhamad Abdulkadir Martoprawiro (ITB, Indonesia)
- Dr. Rachmawati (ITB, Indonesia)
- Prof. Yana M. Syah (ITB, Indonesia)
- Dr. Oki Neswan (ITB, Indonesia)
- Dr. Sri Harjati Suhardi (ITB, Indonesia)
- Dr. Sparisoma Viridi (ITB, Indonesia)
- Dr. Yudi Suharyadi (ITB, Indonesia)
- Prof. Zeily Nurrachman (ITB, Indonesia)

ORGANIZING COMMITTEE

- Dr. Dessy Natalia (Chairperson)
- Dr. Didin Mujahidin (Co-Chairperson)
- Dr. Fifi F. Masduki (Secretary)
- Dr. Agus Suroso
- Dr. Akfiny Hasdi Aimon
- Alif Sussardi, M.Eng
- Anindyajati Anin, M.Si
- Dr. Anita A. Artarini
- Dr. Azzania Fibriani
- Dede Enan, S.Si
- Dr. Dian Rosleine
- Elvira Hermawati, M.Si
- Evan Irawan Akbar, M.Si
- Dr. Fiki Taufik Akbar
- Firmansyah Heru B.
- Dr. Fourier Dzar Eljabbar L.
- Dr. Gantina Rachmaputri
- Dr. Handajaya Rusli
- Dr. Inge Magdalena S.
- Dr. Ihsanawati
- Dr. Intan Timur
- Isti Istiqomah, M.Si.
- Dr. Khairul Basar
- Dr. Kiki Vierdayanti
- Dr. Mochamad Apri
- Dr. M. Ikbal Arifyanto
- Dr. Muhammad Iqbal
- Muhammad Reza, M.Si.
- Dr. M Yudhistira Azis
- Dr. M. Yusuf
- Dr. Nizar Happyana
- Noviyanti
- Dr. Rachmawati
- Reza Aditama, M.Si
- Dr. Robby Roswanda
- Dr. Syailendra
- Sulaeman
- Dr. Utriweni Mukhaiyar
- Untung Triadhi, M.Si

IOP Conf. Series: Journal of Physics: Conf. Series **1127** (2019) 011003 doi:10.1088/1742-6596/1127/1/011003

Peer review statement

All papers published in this volume of *Journal of Physics: Conference Series* have been peer reviewed through processes administered by the proceedings Editors. Reviews were conducted by expert referees to the professional and scientific standards expected of a proceedings journal published by IOP Publishing.

IOP Conf. Series: Journal of Physics: Conf. Series **1127** (2019) 012001 doi:10.1088/1742-6596/1127/1/012001

The Standard Model and Beyond

Gerard 't Hooft

(*The 1999 Nobel Laureate for Physics*)

University of Utrecht, The Netherlands

g.thooft@uu.nl

Abstract. The 20th century brought us enormous advances in understanding the most basic constituents of space, time, and matter. All known particles and forces are assembled and described in what is known as "The Standard Model". This allows us to calculate precisely what happens when particles collide with energies up to the equivalent of a trillion Volt for every charged particle. But there still is unexplored terrain: what exactly happens beyond those trillion Volts, and how exactly does the gravitational force enter into the picture? Gravity can turn space and time into "black holes", and when these black holes interact with particles, our mathematical tools fall short.

IOP Conf. Series: Journal of Physics: Conf. Series **1127** (2019) 012002 doi:10.1088/1742-6596/1127/1/012002

Near Horizon Extreme Kerr Geometry

M F A R Sakti[1], P Y D Sagita[1], A Suroso[1,2], F P Zen[1,2]

[1]Theoretical Physics Lab., THEPI Devision, and
[2]Indonesia Center for Theoretical and Mathematical Physics (ICTMP), Institut Teknologi Bandung, Jl. Ganesha 10 Bandung 40132, Indonesia

E-mail: m.fitrah@students.itb.ac.id

Abstract. The conjectured Kerr/CFT correspondence states that the quantum theory of gravity in the near horizon of the Near Horizon Extreme Kerr (NHEK) black holes, is holographic dual to a two-dimensional chiral conformal field theory. To obtain NHEK geometry, we take the extreme limit $a = M$ of the Kerr metric and continue by transforming the coordinates. Consequently, we can obtain the new Ernst potentials for that geometry by using the same coordinates transformation. Finally, we can find the central charge as a function of the Papapetrou-Ernst potentials from stretched horizon formalism.

1. Introduction

In mathematical physics, black hole solutions are related to mathematical equations which are equivalent to the Einstein field equation called by Ernst equations in stationary case. The solutions of these equations are named as Ernst potentials [1, 2]. Ernst equations are also related to the general rotating black hole solutions proposed in [3] which correspond to the killing vector $\hat{\xi}_t$ and $\hat{\xi}_\phi$. This metric can be applied to the more general black holes, Kerr-Newman black holes, so by assuming some cases it will reduce to Schwarzschild, Reissner-Nordström, and also Kerr solutions. So in our case, we take the intrinsic charge of the Kerr-Newman spacetimes to zero in order to obtain the Kerr one.

In our universe, we believe that there so many physical phenomena that have been proved indirectly from the observation such as the cosmological model of Dark Energy, that have been modeled in many theories such in [4, 5, 6], and theoretically appear but still less of observational evidence such as holographic duality in black holes theory. Furthermore, physical theory that is recently discussed is about the extreme black holes, especially about the discoveries of the holographic dualities, which tell the relation between the quantum theory of gravity with the quantum field theory or we can call it by AdS/CFT correspondence [7]. AdS/CFT correspondence leads the emergence of the Kerr/CFT correspondence. The conjectured Kerr/CFT correspondence, that states the quantum theory of gravity in the near horizon of extreme black holes is holographic dual to the conformal field theory (CFT), also has been quite successful as the tools to find the microscopic origin of the Bekenstein-Hawking entropy which is proportional to the central charge (that appears from the Cardy formula) times the extremal Frolov-Thorne temperature. To obtain the corresponding central charge, the method proposed by Brown and Henneaux [8] which is associated to AdS_3 spacetime can be adopted. Besides the application for the black holes, AdS/CFT correspondence also has been used on the other gravitational objects such as Neutron stars [9] and Boson stars [10] that have cosmological constant in the action. In [10], there also a study about the counterpart of the boson stars, i.e. Q-balls with the non-zero cosmological constant. We can also find the others study of the Boson stars and Q-balls in [11, 12, 13].

In the holographic dualities, the central charge of the Virasoro algebra that appears from extremal Kerr which generates the asymptotic symmetries of the near horizon geometry is also found [14, 15]. In addition, the central charge of the extremal magnetized Kerr-Newman metric has been found by Astorino [16]. He also shows the case when the intrinsic charge vanishes and it reduces to magnetized Kerr one such in [17], independently. By taking B to zero, the central charge that appears from extremal Kerr-Newman metric also can be obtained. However, the central charge that is associated with the Reisser-Nordström metric cannot be obtained except we extend to 5 dimensions such in [18] because in 4 dimensions there are no off-diagonal terms of the metric.

The Near Horizon Extreme Kerr (NHEK) geometry of the Kerr black holes has been obtained taking the extreme limit of the event horizon, i.e. $M = a$. Finally, the central charge can also be obtained by using stretched horizon formalism with four-dimensional stationary black hole metric in an ADM form [19]. This method can be applied for all stationary black hole metric in N-dimension.

In this following, we show the Ernst potential of the extremal Kerr by using the coordinate transformation that has never been shown yet. To find that that Ernst potential, we use the same coordinate transformation like we use to find the NHEK metric. We can see that the potential just depends on the angular coordinate because we have taken the radial coordinate on the horizon of the black holes. Note that we know the central charge of extremal geometry, which corresponds to the entropy of the black holes. We also find the central charge as a function of the Papapetrou and Ernst potentials. This result will help us to find the extreme Kerr metric from the Ernst equation and relates to our unpublished work [20].

We organize the remaining parts of the paper as follows. In section 2, we review the Ernst equations and its solution is related to the Papapetrou metric. In section 3, the central charge of the stretched horizon formalism is obtained in the function of the Papapetrou and Ernst potentials. Here, we investigate the NHEK geometry and find the new Ernst potential. In the last section, we end with the summary and conclude the rest of the paper.

2. Review on the Papapetrou metric and Ernst Equations

We begin with the general stationary metric in Papapetrou form [3] such that

$$ds^2 = f^{-1}\left[e^{2\gamma}(d\rho^2 + dz^2) + \rho^2 d\hat{\phi}^2\right] - f(d\hat{t} - \omega d\hat{\phi})^2 \tag{1}$$

which is related to the killing vectors $\hat{\xi}_{\hat{t}}$ and $\hat{\xi}_{\hat{\phi}}$. By defining $e^{2\gamma} = -2P^{-2}$, $d\rho^2 + dz^2 = d\zeta . d\zeta^*$, and transforming $\hat{t} \to \hat{\phi}$ and $\hat{\phi} \to \hat{t}$, we obtain

$$ds^2 = f^{-1}\left(-2P^{-2}d\zeta . d\zeta^* + \rho^2 d\hat{t}^2\right) - f\left(d\hat{\phi} - \omega d\hat{t}\right)^2 . \tag{2}$$

For the case of the metric which consists of radial and angular coordinates, we can choose

$$d\zeta = \frac{1}{\sqrt{2}}\left(\frac{d\hat{r}}{\sqrt{\Delta}} + i d\theta\right) . \tag{3}$$

Hence, equation (1) becomes

$$ds^2 = f^{-1}\left[-P^{-2}\left(\frac{d\hat{r}^2}{\Delta} + d\theta^2\right) + \rho^2 d\hat{t}^2\right] - f\left(d\hat{\phi} - \omega d\hat{t}\right)^2 , \tag{4}$$

and we have chosen that all potentials in the Papapetrou metric to depend on r, θ coordinates.
These equations,(1) and (4), are the solutions of Ernst equations that are defined by

$$(\text{Re }\varepsilon + |\Phi|^2)\nabla^2\varepsilon = (\nabla\varepsilon + 2\Phi^*\nabla\Phi).\nabla\varepsilon, \quad (5)$$
$$(\text{Re }\varepsilon + |\Phi|^2)\nabla^2\Phi = (\nabla\varepsilon + 2\Phi^*\nabla\Phi).\nabla\Phi, \quad (6)$$

where $\varepsilon = f - |\Phi|^2 + i\hat{\varphi}$ and Φ are the complex function solutions. For example, we show the solution of Kerr-Newman metric with the corresponding Ernst potentials which is found in [2]. The Ernst potentials are

$$\Phi = -iq\cos\theta + \frac{qa\sin^2\theta}{\hat{r} + ia\cos\theta}, \quad (7)$$
$$\varepsilon = -(\hat{r}^2 + a^2)\sin^2\theta - q^2\cos^2\theta + 2Mai\cos\theta(3 - \cos^2\theta) - \frac{2a\sin^2\theta(Ma\sin^2\theta + iq^2\cos\theta)}{\hat{r} + ia\cos\theta}. \quad (8)$$

By taking these Papapetrou potentials

$$\rho = \sqrt{\Delta}\sin\theta,\ P = (\sqrt{A}\sin\theta)^{-1},\ f = -\frac{A\sin^2\theta}{\Sigma},\ \text{and}\ \omega = \frac{(2M\hat{r} - q^2)a}{A}, \quad (9)$$

the Kerr-Newman metric is obtained to be ($G = c = \hbar = 1$)

$$ds^2 = \Sigma\left(-\frac{\Delta}{A}d\hat{t}^2 + \frac{d\hat{r}^2}{\Delta} + d\theta^2\right) + \frac{A\sin^2\theta}{\Sigma}\left(d\hat{\phi} - \frac{(2M\hat{r} - q^2)a}{A}d\hat{t}^2\right)^2, \quad (10)$$

where

$$\Delta \equiv \hat{r}^2 - 2M\hat{r} + a^2 + q^2,\ \Sigma \equiv \hat{r}^2 + a^2\cos^2\theta,\ A \equiv (\hat{r}^2 + a^2)^2 - \Delta a^2\sin^2\theta.$$

By setting $q = 0$, Kerr-Newman metric reduces to Kerr or $a = 0$ reduces to Reissner-Nordström, or setting both of $a = q = 0$ reduces to Schwarzschild.

3. NHEK Black Holes and The Corresponding Ernst Potentials

Here, we will study the extremal Kerr geometry by using new coordinates [14, 15, 21] on Kerr metric in Boyer-Lindquist coordinates defined by

$$ds^2 = \Sigma\left(-\frac{\Delta}{A}d\hat{t}^2 + \frac{d\hat{r}^2}{\Delta} + d\theta^2\right) + \frac{A\sin^2\theta}{\Sigma}\left(d\hat{\phi} - \frac{2M\hat{r}a}{A}d\hat{t}^2\right)^2,. \quad (11)$$

Potential ω is the angular velocity of the Kerr black hole which has an important role in finding the central charge in the extremal case. There is an event horizon at

$$r_{\pm} = M \pm \sqrt{M^2 - a^2}. \quad (12)$$

The surface gravity, Hawking temperature, and the angular velocity of the event horizon are

$$\kappa = \frac{r_+ - M}{2Mr_+},\ T_H = \frac{\kappa}{2\pi},\ \Omega_H = \frac{a}{2Mr_+}. \quad (13)$$

For the extremal case $a = M$, the event horizon is at $r_+ = M$ and the near horizon is at $\hat{r} = r_+$. We see that the Hawking temperature vanishes on the extremal case but later we will see that the temperature on the stretched horizon has a non-zero value.

Before showing the Ernst potential of the extreme Kerr, we introduce this coordinates transformation

$$\hat{t} = \frac{2Mt}{\lambda},\ \hat{\phi} = \phi + \frac{t}{\lambda},\ \hat{r} = \frac{\lambda M}{y} + M, \tag{14}$$

on the extremal Kerr metric and take the limit $\lambda \to 0$. Finally, by using (14), we can obtain

$$ds^2 = 2J\Omega^2 \left[\frac{-dt^2 + dy^2}{y^2} + d\theta^2 + \Lambda^2 \left(\frac{dt}{y} + d\phi\right)^2\right], \tag{15}$$

which is in the Poincaré-type coordinates and where

$$J = M^2,\ \Omega^2 = \frac{1 + \cos^2\theta}{2},\ \Lambda = \frac{2\sin\theta}{1 + \cos^2\theta}.$$

By using $d\hat{t} \to y^2 d\hat{t}$, we have

$$ds^2 = 2J\Omega^2 \left[-y^2 dt^2 + \frac{dy^2}{y^2} + d\theta^2 + \Lambda^2 \left(d\phi + y d\hat{t}\right)^2\right], \tag{16}$$

which is NHEK in the form of general near horizon geometry that is still a vacuum solution. We also find the new Ernst potential along with using this transformation (14) that only depends on the angular coordinate because we have take $\hat{r} = M$. The corresponding Ernst potential is

$$\varepsilon = -2M^2\sin^2\theta - \frac{2M^2\sin^4\theta}{1 + \cos^2\theta} + 2iM^2\cos\theta \left[(3 - \cos^2\theta) + \frac{\sin^4\theta}{1 + \cos^2\theta}\right]. \tag{17}$$

Then global NHEK geometry [22] also can be obtained from coordinates transformation (14) on the metric (15), which are defined by

$$y = \frac{1}{r + \sqrt{1 + r^2}\cos\tau},\ t = y\ \sin\tau\sqrt{1 + r^2},\ \text{and}\ \phi = \varphi + \ln\left(\frac{\cos\tau + r\ \sin\tau}{1 + \sin\tau\sqrt{1 + r^2}}\right). \tag{18}$$

Then we find

$$ds^2 = 2J\Omega^2 \left[-(1 + r^2)d\tau^2 + \frac{dr^2}{1 + r^2} + d\theta^2 + \Lambda^2 \left(d\varphi + r d\tau\right)^2\right]. \tag{19}$$

Now, the central charge that is associated with NHEK geometry can be calculated in the stretched horizon formalism. In this formalism, the metric which is used is four-dimensional stationary black hole in ADM form [19]

$$ds^2 = -N^2 d\hat{t}^2 + h_{\hat{r}\hat{r}} d\hat{r}^2 + h_{\hat{\theta}\hat{\theta}} d\hat{\theta}^2 + h_{\hat{\phi}\hat{\phi}}(d\hat{\phi} + N^{\hat{\phi}} d\hat{t})^2, \tag{20}$$

The central charge which corresponds to the (20) has general form as

$$c = \frac{3f A_{BH}}{2\pi}, \tag{21}$$

where A_{BH} is the horizon area of the black hole and $\hat{f}$ is a function which is defined by

$$\hat{f} = \left.\frac{f_2 f_3}{f_1}\right|_{\hat{r} = \hat{r}_+} \tag{22}$$

where

$$f_1 = \frac{N}{\hat{r} - \hat{r}_+}, \; f_2 = (\hat{r} - \hat{r}_+)\sqrt{h_{\hat{r}\hat{r}}}, \text{ and } f_3 = \frac{N^{\hat{\phi}} + \Omega_H}{\hat{r} - \hat{r}_+}. \tag{23}$$

The Hawking temperature is defined by

$$T_H = \frac{\partial_r N}{2\pi\sqrt{h_{\hat{r}\hat{r}}}} = \frac{f_1(r - r_+)}{2\pi f_2}, \tag{24}$$

and the extremal Frolov-Thorne temperature in the stretched horizon formalism [19] is

$$T = \frac{T_H}{\epsilon}, \tag{25}$$

where $\epsilon = (r - r_+)f_3|_{r=r_+}$. Then we have

$$T = \frac{1}{2\pi\hat{f}}. \tag{26}$$

We will see the central charge of the NHEK geometry. The event horizon of the (16) is in $r_+ = 0$ [19, 23, 24]. So, from the metric (19), we see that

$$f_1 = f_2 = \sqrt{2M^2\Omega^2}, \quad f_3 = 1, \tag{27}$$

and it causes $\hat{f} = 1$. In addition, the horizon area is

$$A_{BH} = \int \sqrt{|\gamma|} d\theta d\phi, \tag{28}$$

where γ is the determinant of the Kerr metric only on the components g_{22} and g_{33}. Then we find A_{BH} is $8\pi M^2$ for extreme Kerr. Then we end with the central charge (21) and the temperature (26) to be

$$c = 12M^2, \tag{29}$$

$$T = \frac{1}{2\pi}, \tag{30}$$

where it is equal to the Kerr metric by taking $r_+ = M$. Actually, we can find the central charge directly from the Papapetrou and Ernst potentials in the definition of function $\hat{f}$. So, from (23) we can find

$$f_1 = \frac{\rho}{(\hat{r} - \hat{r}_+)\sqrt{-\text{Re}\ \varepsilon}}, \; f_2 = (\hat{r} - \hat{r}_+)(\text{Re}\ \varepsilon\ \Delta P^2)^{-1/2}, \; f_3 = \frac{\Omega_H - \omega}{\hat{r} - \hat{r}_+}. \tag{31}$$

4. Summary and Conclusion

Black holes solutions are related to mathematical equations which are equivalent to the Einstein field equation in stationary case called by Ernst equations. Ernst equations are also related to the general rotating black hole metric. This metric can be applied to the more general black holes, Kerr-Newman black holes. In our case, we take the intrinsic charge of the Kerr-Newman spacetimes to be zero in order to obtain Kerr spacetimes. One of the great discoveries recently in theoretical physics is holographic dualities which happen in extremal black holes such as Kerr one and are called by Kerr/CFT correspondence. The conjectured Kerr/CFT correspondence has been quite successful as the tools to find the microscopic origin of the Bekenstein-Hawking entropy. For calculating the entropy of the extremal black holes, one also need to find the central charge that has appeared from the Cardy formula.

Here, we reconstruct the Near Horizon Extreme Kerr geometry. Then by using the same coordinate transformation on the Kerr metric, we obtain the Ernst potential for this geometry. We find that the Ernst potential only depends on θ

$$\varepsilon = -2M^2\sin^2\theta - \frac{2M^2\sin^4\theta}{1+\cos^2\theta} + 2iM^2\cos\theta\left[(3-\cos^2\theta) + \frac{\sin^4\theta}{1+\cos^2\theta}\right],$$

because of the near horizon condition. This Ernst potential is the solution of the Ernst equation represents the vacuum Einstein field equation. We show that central charge of the metric can be defined in the function of Papapetrou and Ernst potentials. So, we can find directly the central charge and the extremal Frolov-Thorne temperature from those potentials. This result will help us to find the extreme Kerr metric from the Ernst equation.

Acknowledgments
We gratefully acknowledge support by Riset PMDSU 2016 from Ministry of Research, Technology, and Higher Education of the Republic of Indonesia. M.F.A.R.S would like to thank Haryanto M. Siahaan for giving us the topic and the stimulating discussion. M.F.A.R.S also thanks all members of Theoretical Physics Laboratory, Institut Teknologi Bandung for the valuable support.

References

[1] Ernst FJ 1967 *Phys. Rev.* **167** 1175
[2] Ernst FJ 1967 *Phys. Rev.* **168** 1415
[3] Reina C and Treves A 1976 *Gen. Relativ. Grav.* **7** 817
[4] Arianto, Zen FP, Gunara B E, Triyanta and Supardi 2007 *J. High Energy Phys.* **0709** 048
[5] Arianto, Zen FP, Triyanta and Gunara BE 2008 *Phys. Rev.* D **77** 123517
[6] Zen FP, Arianto, Gunara BE, Triyanta and Purwanto A 2009 *Eur. Phys. J.* C **63** 477
[7] Maldacena J 1998 *Adv. Theor. Math. Phys.*, **2**, 231-252
[8] Brown JD and Henneaux M 1986 *Commun. Math. Phys.* **104** 207
[9] de Boer J, Papadodimas K and Verlinde E 2010 *J. High Energy Phys.* **1010** 020.
[10] Hartmann B and Riedel J 2013 *Phys. Rev.* D **87** 044003
[11] Sakti MFAR and Sulaksono A 2016 *J. Phys.: Conf. Ser.* **739** 012075
[12] Sakti MFAR and Sulaksono A 2016 *AIP Conf. Proc.* Eds. Terry Mart and Djoko Triyono **1729** 020035
[13] Sakti MFAR, Sagita PYD, Suroso A and Zen FP 2017 *J. Phys.: Conf. Ser.* **856** 012009
[14] Hartman T, Murata K, Nishioka T and Strominger A 2009 *J. High Energy Phys.* **0904** 019
[15] Compere G 2012 *Living Rev. Rel.* **15** 11
[16] Astorino M 2015 *Phys. Lett.* B **751** 96
[17] Siahaan HM 2016 *Class. Quantum Grav.* **33** 155013
[18] Garousi MR and Ghodsi A 2010 *Phys. Lett.* B **687** 79
[19] Chen B and Zhang JJ 2012 *Nucl. Phys.* B **856** 449
[20] Sakti MFAR, Sagita PYD, Suroso A, and Zen FP 2016 Near Horizon Extreme Magnetized Kerr Geometry. *arXiv preprint arXiv:1612.00701*
[21] Bardeen J and Horowitz GT 1999 *Phys. Rev.* D **60** 104030
[22] Guica M, Hartman T, Song W and Strominger A 2008 *Phys. Rev* D **80** 124008
[23] Chen B and Chu CS 2010 *J. High Energy Phys.* **1005** 004
[24] Chen B, Long J and Zhang JJ 2010 *Phys. Rev.* D **82** 104017

IOP Conf. Series: Journal of Physics: Conf. Series **1127** (2019) 012003 doi:10.1088/1742-6596/1127/1/012003

Automation of data acquisition Surface Plasmon Resonance (SPR) based on LabVIEW

Hendro [1], Jerfi [1] and Viridi S[2]

[1] Physic Instrumentation Laboratory, Bandung Institute of Technology
[2] Nuclear Physics and Biophysics Research Group, Faculty of Mathematics and Natural Science, Bandung Institute of Technology

E-mail : fjerfi@gmail.com

Abstract. Evanescent wave is the result of Attenuated Total Reflection (ATR), Surface Plasmon Resonance (SPR) occurs at a specific angle on condition ATR. When a beam is passed through a prism, the beam will experience a reflection on the part of the prism which is in contact with the sample. The result of this reflection to be captured by the detector, On the condition of SPR weaken the intensity of the reflected beam of laser light that occurs because the refracted exponentially away from the boundary of the medium. The LabVIEW-based program has been created that can be used as a tool for experimental SPR, using Motor SHOT 602 and Lock-In Amplifier (SR510). Programs that have been made already equipped with the condition automation starts of movement Motor 2 (Prism), Motor 1 (Detectors), Readings Voltage detectors by Lock-In Amplifier (SR510) and storage of results (time, a position of the motor 1 and 2, the voltage read Lock in Amplifier). Automation program is made by studying the way of the light at Surface Plasmon Resonance (SPR), to determine the relationship between the rotational angle of motor prism and the rotational angle of motor, it can be used for control the position of the sample on the prism and the detector in a program based on LabVIEW.

1. Introduction

Total Internal Reflection (TIR) is an optical phenomenon that occurs when there is the light that penetrates the boundary of a medium with the angle of incidence greater than the critical angle, it could lead to total reflection events that occur at the boundary of the medium [9], but in this case, does not mean that all light will be reflected - there is a bias wave which decays exponentially with distance from the interface known as evanescent waves. Evanescent wave is the result of Attenuated Total Reflection (ATR), Surface Plasmon Resonance (SPR) occurs at a specific angle on condition ATR [5].

Surface Plasmon Resonance (SPR) is an optical phenomenon when the electromagnetic (EM) with a certain energy shot at a certain angle, resulting in a resonance that occurs in the electron on the surface of metal - dielectric interfaces. Surface Plasmon Resonance (SPR) is an optical technique that can be used to monitor molecular interactions on substrates in situ. Surface Plasmon Resonance (SPR) allows to study and determine the kinetic parameters of interactions between molecules in real time without the need for labeling [2, 4, 8].

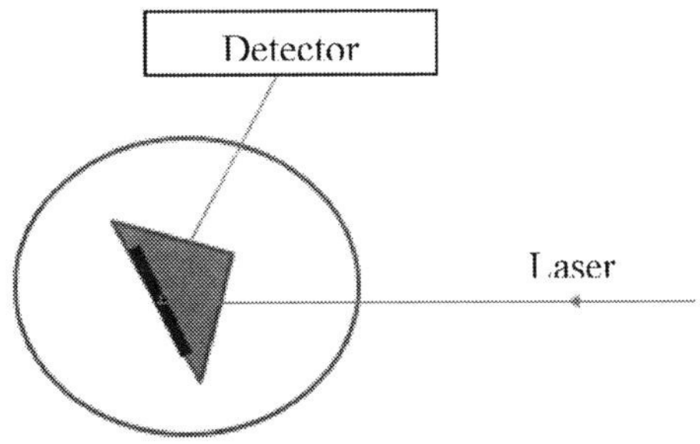

Figure 1. Configuration Experiment Surface Plasmon Resonance

An already polarized laser beam passed in a specific direction toward a metal sample which has a smaller refractive index than the refractive index Prism. In particular, the Prism angle swivel when the angle Prism at an angle greater than the critical reflection, the light will be reflected occur on the inside of the Prism which comes into contact with the sample. The resulting reflection rays then want to be captured by detectors. The phenomenon of Surface Plasmon Resonance (SPR) occurs at a certain angle, when the angle of incident light on the prism is greater than critical angle prism. This occurs when the momentum of a laser beam of the same value with momentum of electrons in the sample, the result is the electrons separated from the sample and interact with laser light until it reaches the resonance condition, the result of this condition is a decrease in the intensity of the laser beam which wants to be captured by detectors [5]. On this configuration, data retrieval is done by changing the angle of motor Prism. With the change in motor position prism, then the motor position detector should also be changed.

SHOT 602 is one tool that can be used as a support of this experiment because SHOT 602 features 2 motors, adjustable swivel angles manually on his tools, or through the computer using specific applications. The company's production of Motor SHOT 602 is Sigma Koki. In terms of usage, Sigma Koki issued a sample application that can be used to control the motor SHOT 602. One of the software being used is LabVIEW (Laboratory Virtual Instrumentation Engineering Workbench). Problems occurred is a program issued by the company is less efficient because of limitations in each function, one program only has one function. Therefore, the complexity of the software program style needs to be reduced to make it more efficient and effective

Based on these problems, this research aims to create a program based on LabVEW that can be used to control the Motor SHOT 602 with style program that is simpler and improves the level of efficiency of the program, by means of combining multiple functions into one program. The new things that want to add is a process automation from the program between motor 1 and motor 2 with knowing the relationship between the rotational angle of Prism (Motor 2) and rotational angle of detector (Motor 1)

2. Theory

Motor SHOT 602 connected to the computer using the RS232 interface and commands given are then sent through Hyper Terminal application. LabVIEW is used for processing and visualization of data in the areas of data acquisition, control and instrumentation, as well as the Automation Industry [1]. Process automation program, created by knowing the relationship between the rotational angle of Prism (Motor 2) and rotational angle of detector (Motor 1), can be seen in the picture below:

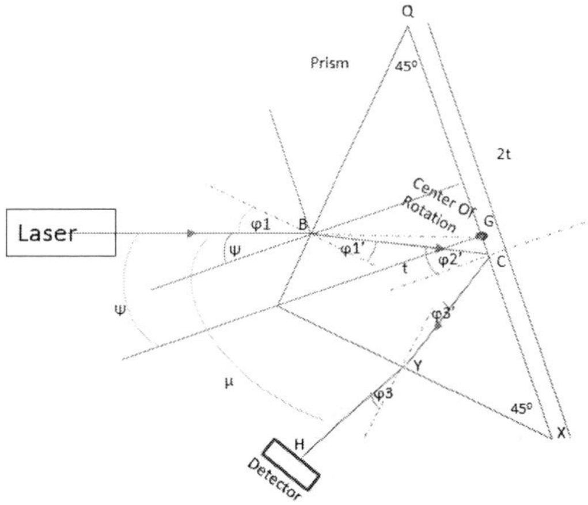

Figure 2. Relationship between ψ and μ [3,9]

Figure 2 shows the relationship between rotational angle of prism ψ (Motor 1) and rotational angle of detector μ (Motor 2), can be calculated using the following equation [3,9]:

$$\mu = \psi + (90 - \angle YGX - \angle HGY \quad (1)$$

The Angel $\angle YGX, \angle HGY$ can be searched by using Snell's Law. Snell's Law is a formula used to describe the relationship between the angles of incidence and refraction, when referring to light or other waves passing through a boundary between two different isotropicmedia, such as water, glass, or air. Snell's law states that the ratio of the sine of the angles of incidence and refraction is equivalent to the ratio of phase velocities in the two media, or equivalent to the reciprocal of the ratio of the indices of refraction.

$$n_1 \sin\theta_1 = n_2 \sin\theta_2 \quad (2)$$

Law of sines $$\frac{a}{\sin\alpha} = \frac{b}{\sin\beta} = \frac{c}{\sin\gamma} \quad (3)$$

And find the length of the side of triangle

$$ab^2 = bc^2 + ac^2 - 2(bc)(ac)\cos(\angle acb) \quad (4)$$

By using the three equations, the relationship between rotational angle of Prism (Motor 2) and rotational angle of Detector (Motor 1) can be determined. Based on the results of the calculation. The relationship between the rotational angle of the Prism (μ) and rotational angle of detector (ψ) can be written as follows

$$\angle YGX = \sin^{-1}\left\{\frac{XY}{YG}.\frac{\sqrt{2}}{2}\right\} \quad (5)$$

$$\angle HGY = 180^0 - \angle GYH - \angle YHG \quad (6)$$

$$\mu = \psi + \left[90^0 - \sin^{-1}\left\{\frac{XY}{YG}.\frac{\sqrt{2}}{2}\right\} - \left(180^0 - \angle GYH - \angle YHG\right)\right] \quad (7)$$

With

$$XY = \frac{\sin\left(45^0 + \varphi 1'\right)}{\sin\left(90^0 - \varphi 1'\right)}\left(2t - \left\{\left[\sqrt{2t} - \left(t\frac{\sin\psi}{\sin\left(135^0 - \psi\right)}\right)\right] \cdot \left[\frac{\sin\left(90^0 - \varphi 1'\right)}{\sin\left(45^0 + \varphi 1'\right)}\right]\right\}\right) \quad (8)$$

$$YG = \sqrt{t^2 + XY^2 - 2t.XY.\cos 45^0} \quad (9)$$

$$\angle GYH = 135^0 - \varphi 1 + \sin^{-1}\left\{\frac{CY}{YG}\sin\left(135^0 - \varphi 1'\right)\right\} \quad (10)$$

$$\angle YHG = \sin^{-1}\left\{\frac{YG}{R\ \text{detektor}}.\sin\left(\angle GYH\right)\right\} \quad (11)$$

With

$$CY = \frac{\sqrt{2}}{2.\sin\left(90^0 - \varphi 1'\right)}\left(2t - \left\{\left[\sqrt{2t} - \left(t\frac{\sin\psi}{\sin\left(135^0 - \psi\right)}\right)\right].\left[\frac{\sin\left(90^0 - \varphi 1'\right)}{\sin\left(45^0 + \varphi 1'\right)}\right]\right\}\right) \quad (12)$$

$$\varphi 1' = \sin^{-1}\left(\frac{n}{n'}\sin\varphi 1\right) \quad (13)$$

$$\varphi 1 = 45^0 - \psi \quad (14)$$

t is the high of prism.

3. Method

The laser beam is connected to the circuit chopper will produce a light that clipped in accordance with a frequency that is set up on a Signal Generator. The value of the frequency of the Signal Generator is what will go into Lock In Amplifiers as a frequency reference, so Lock In Amplifiers will only read the voltage value on the detector corresponds to the frequency of the incoming laser beam reference. The scheme and the conditions that need to be considered before the data capture process is running, it can be seen in figure 3 and figure 4.

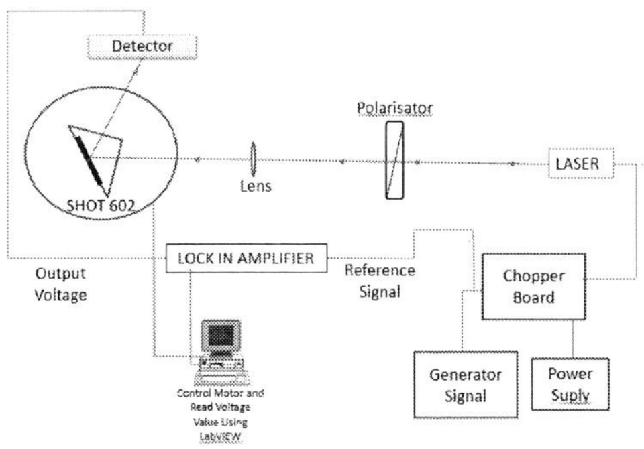

Figure 3. Scheme on Experiment Surface Plasmon Resonance

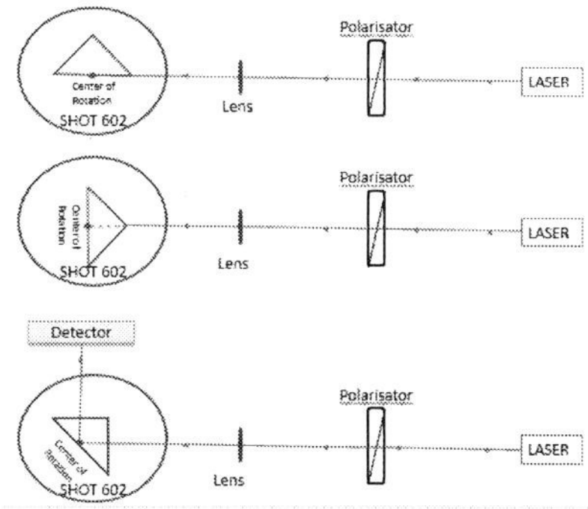

Figure 4. Set up Prism

4. Result

Overall Program control and Data acquisition Surface Plasmons Resonance (SPR) that was made using the function of the tab control that divides some of the functions into specific parts. As for the parts comprising the initial settings, program control and data acquisition of the SPR, and read the results.

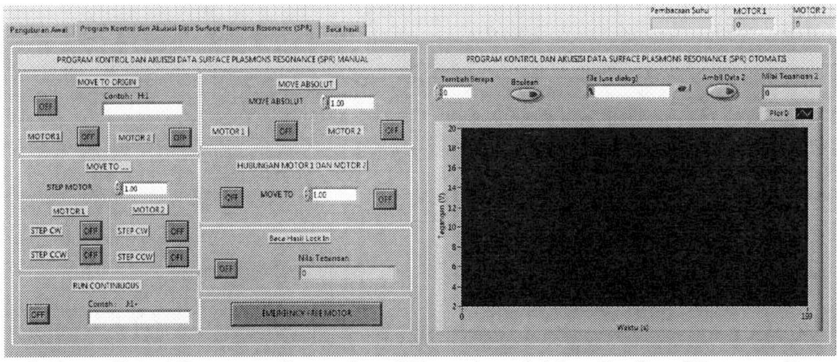

Figure 5. Front Panel Program Control and Data Acquisition Surface Plasmon Resonance

Figure 5 Shows that, the program is further divided into two parts. The first part is related to the Data acquisition and control Program SPR manually, and the second part is related to the Data acquisition and control Program SPR automatically.

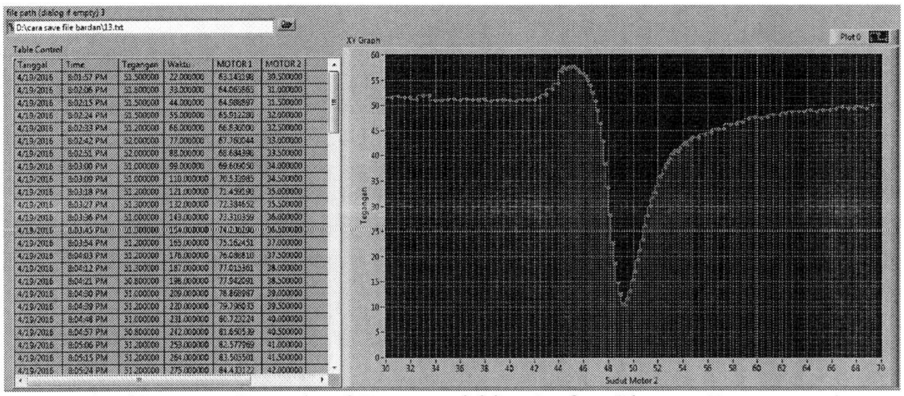

Figure 6. Result of Program Control and Data acquisition Surface Plasmon Resonance Automatic

Figure 6 Shows one of the results obtained by using data acquisition and control Program SPR automatic. Those results are the result of data storage that has already been done. Stored data associated with the necessary quantities such as time, the voltage on the detector, the position of motor 1 and motor 2 position. The data already obtained automatically in a plot by the program by taking the angle Motor 2 as X axis and voltage on the detector as Y-axis. During the process of automation of data retrieval, the magnitude of change in the angle of turn of Motor 2 (Prism), can be changed during the process of data capture is running, it can be seen in Figure 6. To move the angle with a range of between 30.5 -44, with a transition each time is 0.50. While the rest of the corners with a 34.1-44 range, has a transition each time was 0.20, and for range, 34.1-43.1 has a transition each time of 0.50. This is done to simplify and speed up data retrieval in the process. In this 1-time data retrieval by using a sample of gold, all of the take-ups from the position of Motor 2 (Prism) 30.50 up to 84 with the taking of the data points as many as 110 were off 7 minutes 30 seconds. This automation program runs for 1 time point data retrieval takes 9 seconds, with parts, 3 seconds pause time to move the Motor 2 (Prism) in accordance with the added value of how much filled, 1 second pause time to move the Motor 1 (Detector) in accordance with the calculation Relationship of Motor 1 and Motor 2, 4.5 seconds pause time for 3 times the reading of Lock-in. and 0.5 seconds pause time to average his point and display the value in the form of graphs of 3 times the reading Lock-ins. The overall results obtained by running the Program Data acquisition and Control these SPR can be seen in Figure 7.

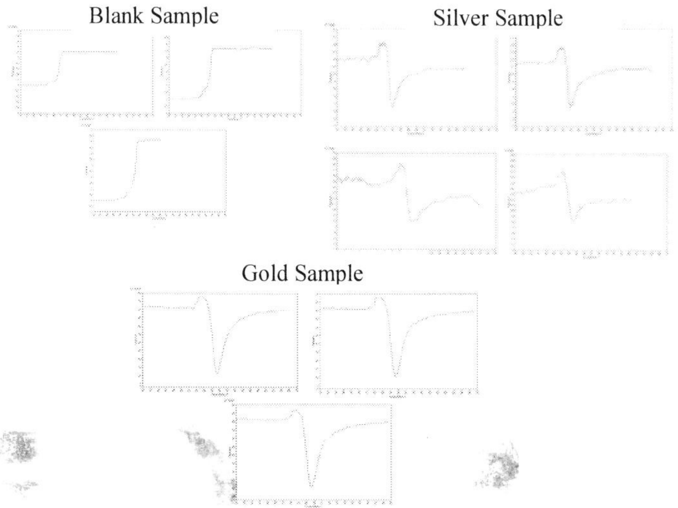

Figure 7. Result with blank sample, silver sample, and gold sample

5. Conclusion

Has created a program control Motor SHOT 602 that can be used as an experimental tool for Surface Plasmon Resonance (SPR), by the condition of automation of the data collection process starts from the movement of the motor prism , the movement of the motor detector , reading the voltage value at the detector and data storage experimental results (date , time, position motor prism, position motor detector , the result of reading lock -in amplifier).

References

[1] Artanto,D. (2012): Interaksi Arduino dan LabVIEW, PT Alex Media Komputindo Jakarta.
[2] Resonance Platform for Detection of Biological Molecules, Journal Sensors Volume 13, 201-210.
[3] Hendro, S. Viridi, dan Y. Pratama (2012): Determination of detector rotation angle in the experiment based on the total internal reflection using an equilateral right angle prism, The 5th Asian Physics Symposium, 040012-1–040012-5.
[4] Marco Rusnati, Antonella Bugatti, Stefania Mitola, Daria Leali, Paolo Bergese, Laura E. Depero and Marco Presta (2009): Exploiting Surface Plasmon Resonance (SPR) Technology for the Identification of Fibroblast Growth Factor-2 (FGF2) Antagonists Endowed with Antiangiogenic Activity, Journal Sensors Volume 9, 6471-6503.
[5] Sholihah,A., Rufaida,R. dan Abraha,K. (2015): Pengamatan fenomena Surface Plasmon Resonance (SPR) pada permukaan lapisan tipis perak menggunakan laser dengan panjang gelombang berbeda dalam konfigurasi Kretschmann, Prosiding 25 Jateng & DIY, 119-202.
[6] Sigma-Koki CO (2010): User Manual SHOT-602, Japan.
[7] Stoke.R (2015): Quantum Information Processing with 1010 electrocns?, IQIS seminar.

[8] Vesna Hodnik dan Gregor Anderluh (2009): Toxin Detection by Surface Plasmon Resonance, Journal Sensors Volume 9, 1339-1354.
[9] Y. Pratama (2011): Setup Prisma Siku untuk Studi Pemantulan Total Teredam, Skripsi S.Si., Institut Teknologi Bandung.

IOP Conf. Series: Journal of Physics: Conf. Series **1127** (2019) 012004 doi:10.1088/1742-6596/1127/1/012004

Design and Development of Radiated Heater with Temperature Control Program for Surface Plasmon Resonance Experiment

Utomo D K[1], Kurniawati A[1], Hendro[1] and Viridi S[2]

[1] *Theoretical High Energy Physics and Instrumentation Research Group, Faculty of Mathematics and Natural Science, Bandung Institute of Technology, Jl. Ganesha no. 10 Bandung, Indonesia, 40132.*
[2] *Nuclear Physics and Biophysics Research Group, Faculty of Mathematics and Natural Science, Bandung Institute of Technology, Jl. Ganesha no. 10 Bandung, Indonesia, 40132.*

Abstract. This research aims to design and to develop of radiated heater with temperature control program to be used in surface plasmon resonance experiments (SPR). The heater is made of NiCr solenoid with a value of 1 ohm resistance and 5 volt DC of voltage source. The control device in this experiment used the Arduino microcontroller which was connected to the LabVIEW software with an interface. The result that heating control program has been functioning in good condition with a deviation value at 3% from setpoint value at a distance of 0.8 cm, 1.0 cm and 1.2 cm from the heater surface.

1. Introduction

Heat energy can be transferred from an object that has a high temperature (heat source) to another object which has a low temperature. Thermal energy is required in surface plasmon resonance experiments with controlled processes. The design of the devices of surface plasmon resonance has been made by assessing the effect of temperature on the detection of materials such as volatile organic compound [1]. Another research has been finished to study about temperature effect and refraction index in surface plasmon resonance [2].

Furthermore, the research about an application in surface plasmon resonance experiment has been done concerning sensitivity and light intensity in an electrochemical process [3]. Further and even more importantly, though, the research about study of thickness gold and silver sample has been completed in surface plasmon resonance experiment [4]. It is also interesting to note that the application in biosensor has been worked on food characteristic of a chemical compound by surface plasmon resonance experiment [5]. Equally interestingly, surface plasmon resonance experiment in medical research has been done for initial study of avian influenza virus subtype H6 characterization [6]. Because of that, it is important to know application about surface plasmon resonance experiment that used of the temperature effect.

Generally, the heater is made from solenoid with AC voltage source [7]. In this research, the heater is made from solenoid with DC source voltage. The designed plant is a low cost application with components easy to access. This heater design is made to fulfill of surface plasmon resonance experiment that needs temperature effect to study such as biomolecules [8]. The heater is designed by specification with small area 1 x 1.5 cm^2 and the position is near from the experiment sample with range of mm. Therefore, in this research aims to design and to develop of radiated heater which has ability to control temperature value in surface plasmon resonance experiment.

2. Theory

The assumption in this study is based on the law of energy conservation that the power radiated from the core solenoid to the surface solenoid with r_s distance that is described by red line. This solenoid heating with a certain distance r_x that is described by blue line. The assumption of wave front is solid cylinder as shown in Figure 1.

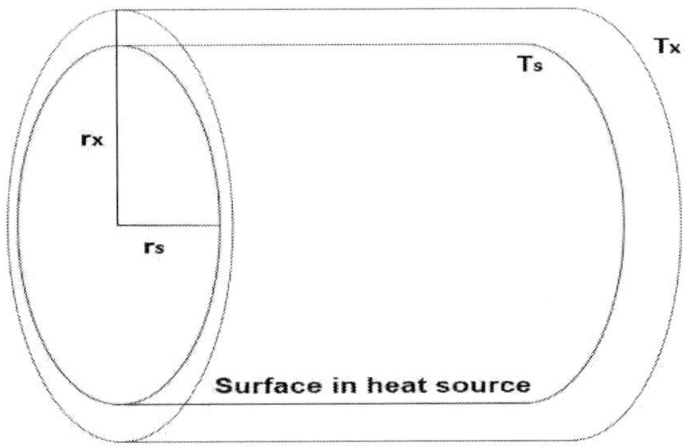

Figure 1. Wave front of solid cylinder of heat transfer scheme.

Based on the heat transfer scheme can be determined intensity of the heat radiation on the surface in heat source of heater as shown in equation 1.

$$I_s = \frac{e\sigma T_s^4}{2\pi r_s l} \tag{1}$$

I_s is the intensity of the radiation on the surface of solenoid, r_s is the distance from the core of solenoid to the surface of solenoid and T_s is the surface of solenoid temperature. Then, the determination of temperature at a distance from the core of solenoid can be determined by equation 2.

$$I_x = \frac{e\sigma T_x^4}{2\pi r_x l} \tag{2}$$

I_x is the intensity of radiation at a distance of r_x from the core of solenoid and T_x is the temperature at a distance of r_x from the core of solenoid. Assuming that the power radiated from the core solenoid is same, then the equation 1 and equation 2 is the same, so that the intensity of the heat radiation only depends on the distance from the core solenoid as shown in equation 3.

$$T_x = T_s\left(\frac{r_s}{r_x}\right)^{\frac{1}{4}} \tag{3}$$

3. Experiment Method

The design of heater in this study is made of NiCr solenoid wire with 53 coils, the length is 0.469 m, the diameter is 2.5 x 10^{-4} m, the mass is 2.5 x 10^{-4} kg and the area is 3.68 x 10^{-4} m^2. This solenoid has a resistance value of 1 ohm. This is connected by power supply 5V DC, relay SSR 25-DA, microcontroller Arduino UNO R3, and a temperature sensor used is the LM35 sensor. The method used in this study is an experiment with including the heaters design, microcontroller program design and LabVIEW software program design and test tools with a heating control program. In this testing phase to vary the distance from the heat source (solenoid) in one dimension at the distance of 0.8 cm, 1.0 cm, and 1.2 cm.

4. Experiment Result

4.1. Temperature control display result in LabVIEW

Display result of temperature measurement in this research can be shown in the front panel of LabVIEW as shown in Figure 2. Data retrieval begins when the program is started by clicking the Run button and is finished at a specified time by clicking the stop button in the front panel in LabVIEW software. Data graph is automatically stored on computer drives with format .txt files that it can be analyzed using Microsoft Excel.

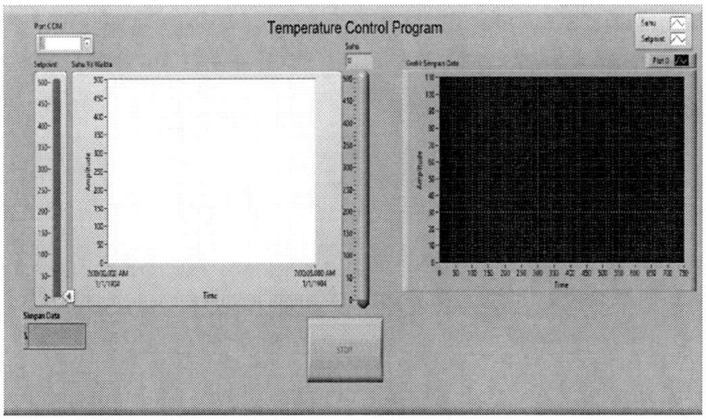

Figure 2. Display of Temperature Control Program in LabVIEW.

4.2. Determination of temperature value at the certain distance

Measurements were taken at the distance 0.8 cm, 1 cm, and 1.2 cm from the surface solenoid heaters. The results of temperature measurements are made during 1000 seconds from the room temperature which is shown in Figure 3.

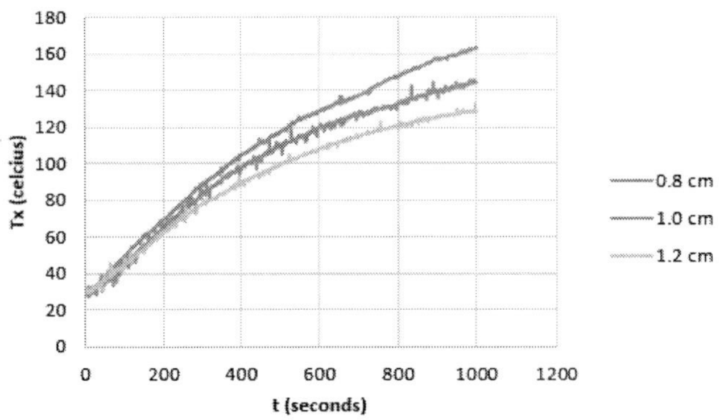

Figure 3. Temperature measurements during 1000 seconds based on distance variation.

Based on the measurement results in Figure 3, the line graph reveals increase dramatically rate at the distance 0.8 cm, 1 cm, and 1.2 cm from the surface solenoid heaters, the experiment result shows that the temperatures explain the value at 163 °C, 145 °C, and 129 °C respectively. The measurement data can be further analyzed to determine the relationship between the temperature values against certain distance experimentally by comparing with theory based on equation 3. Based on theory, the value of T_s used is 422 °C with r_s cm corresponding experimental results and measurements. This equation can determine the relationship between the temperature values against a certain distance as shown in Figure 4.

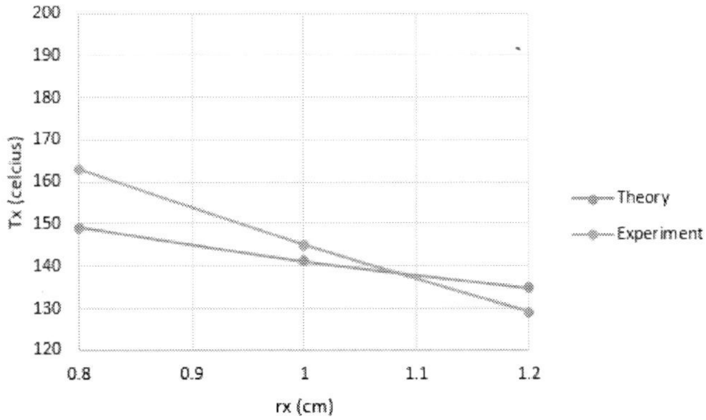

Figure 4. Temperature measurement based on theory and experiment at 1000 seconds.

A glance at the Figure 4 reveals the temperature values slide away at 1000 seconds. Figure 4 describes the temperature measurements against the distance based on theory and experiment. According to the theory, the model can predict at a certain distance which has assumption wave front as solid cylindrical. The heater is radiated heat from the surface solenoid which can be determined by thermal radiation model. At a distance of 0.8 cm from the surface solenoid, the temperature shows 149 °C theoretically, meanwhile, based on experiment at 1000 seconds, the temperature at a distance of 0.8 cm, shows at 163 °C. The difference value also occurs at a distance of 1 cm from the surface of solenoid, the temperature shows 141 °C theoretically. On the other hand, experiment result at a distance of 1 cm at 145 °C for 1000 seconds. At the last measurement at a distance of 1.2 cm from the surface of solenoid, the temperature shows 135 °C theoretically and explains 129 °C experimentally at 1000 seconds. Based on these results, the intensity of the heat radiation only depends on the distance from the heat source (solenoid). Therefore, the relationship of temperature and distance explains inverse relationship between the values of temperature versus distance from the heat source. The result of this analysis can be used as a reference analysis temperature determination of the distance that it needs on surface plasmon resonance experiments.

4.3. Temperature Control Result

The line graph at the Figure 5 describes the temperature rate increase and levelled off at the setpoint value from the room temperature. The experiment has done at a distance of 0.8 cm with setpoint value at 120 °C. This is followed by two measurements at a distance of 1.0 cm and 1.2 cm with setpoint value at 100 °C and 80 °C respectively. The heater temperature control program can control the temperature value in good condition to somewhere in the vicinity of 3 % error from the setpoint value.

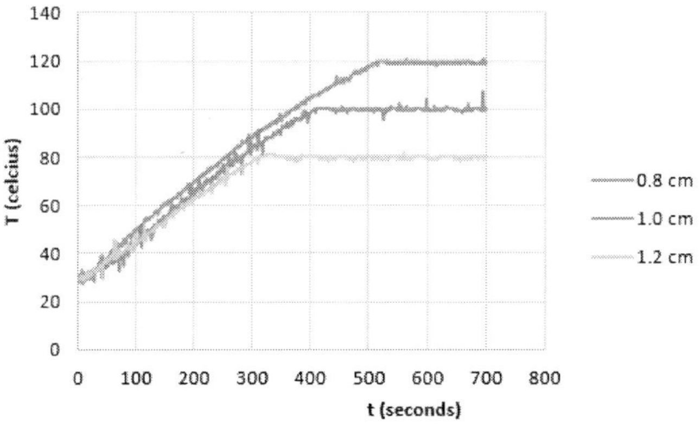

Figure 5. Temperature control result at the distance of 0.8 cm, 1.0 cm and 1.2 cm.

Based on the results, temperature control program has been functioning well and can be used to set a certain temperature with a certain distance in accordance with the needs of surface plasmon resonance experiments. But the results of these tests indicate that the heat rate still needs long time to reach setpoint so it needs novelty design with isolator material around the heater to produce transfer heat energy more quickly towards to the setpoint. Alternative way, to get the maximum heat energy can be done by setting the smallest possible distance between the heating sources and sample. The reason is that the smallest distance from the heat source, the sample experiment will receive the largest heat intensity based on the law of energy conservation with assumption of the radiated power is constant. Another interesting point, the maximum intensity of heat radiation can also be obtained using a small resistance value (under 1 ohm) in heater which is used to accelerate current on the heating process.

5. Conclusion

The design of heater has been created for surface plasmon resonance experiment which can be controlled at certain temperature value. The heater specification is made of NiCr solenoid with 53 coils, the length is 0.469 m, the diameter is 2.5×10^{-4} m, the mass is 2.5×10^{-4} kg and the area is 3.68×10^{-4} m^2. The result that temperature control program of heater has been functioning in good condition with a deviation value approximately 3% from the setpoint value at a distance of 0.8 cm, 1.0 cm and 1.2 cm.

6. Acknowledgments

This research is supported by student activity funds from Department of Physics Bandung Institute of Technology Indonesia 2016 and PUPT RistekDikti 2016.

7. References

[1] Choi S R, Si H, Kim D H, Oh S J, Kim J H, Kim, Si S and Kim W K 2013 *Patent* No. US 8,502.983 B2.
[2] Xiao F, Michel D, Li G, Xu A, Alameh K 2013 *J. Lightwave Technology* **32** p 3567-3571.
[3] Abayzeed S A, Smith R J, Webb K F, Somekh M G, See C W 2016 *J. Sensor and Actuators B:Chemical 235(2016)*627-635.
[4] Chen, Shujing, Lin, Chengyou 2016 *J. Light and Electron Optics* **127** p 7514-7519.
[5] Yakes B J, Buijs J, Elliott C T, Campbell K 2016 *J. Talanta* 156-157(2016)55-63.
[6] Zhao X, Tsao Y C, Lee F J, Tsai W H, Wang C H, Chuang T L, Wu M S, Lin W L 2016 *J. Virological Methods* **233** p 15-22.
[7] Levărdă B. and Budaciu C 2010 *Proc. ICSTC* p 282 – 286.
[8] Wagner C E, Macedo L J A, and Opdahl A 2015 *Proc. Anal. Chem* p 7825-7832.

Exact Solutions of Schrodinger Equation in Cylindrical Coordinates for Double Ring-Shaped Coulomb Oscillator Potential Using SUSY QM Method

Miftahul Ma'arif, M Naufal I, A Suparmi, C Cari

Physics Departement, Faculty of Mathematics and Natural Sciences, Sebelas Maret University, Surakarta, Indonesia
E-mail: miftahulmaarif15@student.uns.ac.id

Abstract. Schrodinger equation for a doubled ring-shaped coulomb oscillator potential is investigated using supersymmetric quantum mechanics approach. The three dimensional Schrodinger equation for a doubled ring-shaped coulomb oscillator potential in cylindrical coordinates is separated into three parts that contains one dimensional Schrodinger type equation which are solved using supersymmetric operator and the idea of shape invariant potential. The energy spectrum and the total wave functions are obtained.
Keywords : Schrodinger equation, Doubled Ring-Shaped Oscillator, Coulomb, Supersymmetry

1. Introduction

The Schrodinger equation is the most fundamental equation in non-relativistic quantum mechanics and the closely related Heisenberg equation, playing the same role as Hamilton's laws of motion in non-relativistic classical mechanics. In pure mathematics, the Schrodinger equation and its variants are one of the basic equations studied in the field of partial differential equations, and has application to geometry, spectral and scattering theory, and to integrable systems [1]. Exact analytical solutions of the Schrodinger equation for some physical potential are important to obtain the wave function and the energy spectrum.

One of exactly solvable quantum systems that has been applied to various fields is the spherical oscillator [2,3]. Many authors have carried out significant investigations on non-spherical oscillator [4], ring-shaped oscillator [5-8], and ring-shaped non-spherical oscillator [9-12]. These studies became interesting in mathematics itself and have potential applications in theoretical chemistry. DRSO potential that proposed by Carpio-Bernido in 1989 has been investigated using path integral method [13-14], algebraic method [15], AIM method [16], SUSY QM and shape invariance approach [17]. The potential can be described as follows

$$V(r,\theta)=\frac{1}{2}M\omega^2r^2+\frac{\hbar^2}{2m}\left(\frac{b}{r^2\sin^2\theta}+\frac{c}{r^2\cos^2\theta}\right) \quad (1)$$

In this paper, a doubled ring-shaped oscillator potential have modified by adding coulomb potential in axial part for cartesian coordinates system becomes

$$V(x,y,z)=\frac{1}{2}M\omega^2(x^2+y^2)+\frac{\hbar^2}{2m}\left(\frac{b}{x^2}+\frac{c}{y^2}\right)+\frac{\alpha}{z^2}+\frac{\beta}{z} \quad (2)$$

Equation (2) is called doubled ring-shaped oscillator potential by applying cylindrical coordinates system in Schrodinger equation and using separation variables method, so each part can be solved to get the wave functions and energy spectra by using supersymmetric approach.

2. Review of Supersymmetric Quantum Mechanics Approach Using Operator

The exact analytical solutions of Schrodinger equations for some physical potentials are very important. According to the definition proposed by Witten, the Hamiltonian in a SUSY QM is expressed as [18]

$$H_{ss}=\begin{pmatrix} -\frac{\hbar^2}{2m}\frac{d^2}{dx^2}+\frac{\hbar}{\sqrt{2m}}\frac{d\varphi(x)}{dx}+\varphi^2(x) & 0 \\ 0 & -\frac{\hbar^2}{2m}\frac{d^2}{dx^2}-\frac{\hbar}{\sqrt{2m}}\frac{d\varphi(x)}{dx}+\varphi^2(x) \end{pmatrix};H_{ss}=\begin{pmatrix} H_+ & 0 \\ 0 & H_- \end{pmatrix} \tag{3}$$

With

$$V_-(x,a_0)=\varphi^2(x)-\frac{\hbar}{\sqrt{2m}}\varphi'(x); \qquad V_+(x,a_0)=\varphi^2(x)+\frac{\hbar}{\sqrt{2m}}\varphi'(x) \tag{4}$$

Here H_- and H_+, are supersymmetry partner of the Hamiltonian, $V_-(x)$ and $V_+(x)$ are the supersymmetry partner potential. To simplify the determination of the energy spectrum and the wave functions, the new operators, raising and lowering operators, are introduced as

$$A^+=-\frac{\hbar}{\sqrt{2m}}\frac{d}{dx}+\phi(x) \quad \text{and} \quad A=\frac{\hbar}{\sqrt{2m}}\frac{d}{dx}+\phi(x) \tag{5}$$

By inserting equation (5) into equation (3) we get the SUSY Hamiltonian as

$$H_-(x)=A^+A \quad \text{and} \quad H_+(x)=AA^+ \tag{6}$$

It is always possible to factorize the usual Hamiltonian as

$$H=H_-+E_0=-\frac{\hbar^2}{2m}\frac{d^2}{dx^2}+V_-(x;a_0)+E_0 \tag{7}$$

By using equations (4) and (7) it is obtained that

$$V(x)=V_-(x;a_0)+E_0=\phi^2(x;a_0)-\frac{\hbar}{\sqrt{2m}}\phi'(x;a_0)+E_0 \tag{8}$$

where $V(x)$ is the effective potential, while $\phi(x)$ is determined hypothetically from equation (8) based on the shape of effective potential from the associated system. In addition to the supersymmetry that gives the relationship between the eigenvalues and eigenfunctions between the two Hamiltonian partners but does not provide the actual spectrum [21], Gendenshtein [19] proposed the condition of shape invariant to determine the energy spectrum. A pair of potentials $V_\pm(x)$ in equations (5) are said to be shape invariant if they are similar in shape but different in the parameters and this condition is given as,

$$V_+(x;a_j)=V_-(x;a_{j+1})+R(a_{j+1}) \tag{9}$$

where $j = 0,1,2,..$, and a is a parameter in our original potential, V-, whose ground state energy is zero, $a_j=f_j(a_0)$ where fj is a function applied j times, the remainder $R(a_j)$ is a's dependence but is independent of x, then $V_\pm(x,a_0)$ is said to be shaped invariant. The energy eigenvalue of the Hamiltonian H_- is given by [18]

$$E_n^{(-)}=\sum\nolimits_{k=1}^{n}R(a_k) \tag{10}$$

and by using equations (8) and (10) we get the energy spectra of the system given as,

$$E_n=E_n^{(-)}+E_0 \tag{11}$$

Based on the characteristics of lowering operator, the ground state wave function of H-, whose ground state energy is zero, is obtained from condition that,

$$H_-\psi_0^{(-)}=0\rightarrow A\psi_0^{(-)}=0 \tag{12}$$

Subsequently, the excited wave function, $\psi_1^{(-)}(x,a_0),..\psi_n^{(-)}(x,a_0)$ of H_- are obtained by using raising operator operated to the lower wave function [19], given as

$$\psi_n^{(-)}(x;a_0) \approx A^+(x;a_0)A^+(x.a_1)....A^+(x,a_{n-1})\psi_0^{(-)}(x;a_n) \tag{13}$$

The SUSY QM and the idea of shape invariance potential are suitable to be used in solving one dimensional Schrödinger equation problems. By obtaining the super-potential, the potential partners, $V_+(x,a_0)$ and $V_-(x,a_0)$, and the SUSY operators, A^+ and A are obtained and so the energy spectra and the wave functions.

3. Separation Variables Method for The Case

Schrödinger equation is defined by

$$\frac{-\hbar^2}{2m}\nabla^2\psi(\bar{r})+V(\bar{r})\psi(\bar{r})=E\psi(\bar{r}) \tag{14}$$

Doubled ring-shaped coulomb oscillator potential is changed by applying cylindrical coordinates so it becomes

$$V(r,\theta,z)=\frac{1}{2}M\omega^2r^2+\frac{\hbar^2}{2m}\left(\frac{b}{r^2\cos^2\theta}+\frac{c}{r^2\sin^2\theta}\right)+\frac{\alpha}{z^2}+\frac{\beta}{z} \tag{15}$$

By substituting this potential into Schrodinger equation, given as

$$\frac{-\hbar^2}{2m}\nabla^2\psi(\bar{r})+\left[\frac{1}{2}M\omega^2r^2+\frac{\hbar^2}{2m}\left(\frac{b}{r^2\cos^2\theta}+\frac{c}{r^2\sin^2\theta}\right)+\frac{\alpha}{z^2}+\frac{\beta}{z}\right]\psi(\bar{r})=E\psi(\bar{r}) \tag{16}$$

Then apply ∇^2 for cylindrical coordinate given by

$$\nabla^2=\frac{1}{r}\frac{\partial}{\partial r}\left(r\frac{\partial}{\partial r}\right)+\frac{1}{r^2}\frac{\partial^2}{\partial\theta^2}+\frac{\partial^2}{\partial z^2} \tag{17}$$

By substituting equation (18) to equation (17) and the wave function is $\psi(\bar{r})=R(r)P(\theta)\Phi(z)$ so the axial part could be separated

$$-\frac{\hbar^2}{2m}\left(\frac{1}{Rr}\frac{\partial}{\partial r}\left(r\frac{\partial R}{\partial r}\right)+\frac{1}{Pr^2}\frac{\partial^2P}{\partial\theta^2}\right)+\frac{1}{2}M\omega^2r^2+\frac{\hbar^2}{2m}\left(\frac{b}{r^2\cos^2\theta}+\frac{c}{r^2\sin^2\theta}\right)-E$$
$$=\frac{\hbar^2}{2m}\frac{1}{Z}\frac{\partial^2Z}{\partial z^2}-\left(\frac{\alpha}{z^2}+\frac{\beta}{z}\right)=\lambda_1 \tag{18}$$

The left side of equation (18) can be separated if it is multiplied with r^2, so we get

$$-\frac{\hbar^2}{2m}\frac{r}{R}\frac{\partial}{\partial r}\left(r\frac{\partial R}{\partial r}\right)+\frac{1}{2}M\omega^2r^4-(E-\lambda_1)r^2=\frac{\hbar^2}{2m}\frac{1}{P}\frac{\partial^2P}{\partial\theta^2}-\frac{\hbar^2}{2m}\left(\frac{b}{\cos^2\theta}+\frac{c}{\sin^2\theta}\right)=\lambda_2 \tag{19}$$

So the Schrodinger equation in Eq.(17) that has been separated may be written by

- Radial part

$$-\frac{\hbar^2}{2m}\frac{r}{R}\frac{\partial}{\partial r}\left(r\frac{\partial R}{\partial r}\right)+\frac{1}{2}M\omega^2r^4-(E-\lambda_1)r^2=\lambda_2 \tag{20}$$

- Angular part

$$-\frac{\hbar^2}{2m}\frac{\partial^2P}{\partial\theta^2}+\frac{\hbar^2}{2m}\left(\frac{b}{\cos^2\theta}+\frac{c}{\sin^2\theta}\right)P=\lambda_2P \tag{21}$$

- Axial part

$$-\frac{\hbar^2}{2m}\frac{\partial^2 Z}{\partial z^2}+\left(\frac{\alpha}{z^2}+\frac{\beta}{z}\right)Z=\lambda_1 Z \tag{22}$$

4. Solutions to The Each Part of The Three Dimensional Schrodinger Equation

Supersymmetric quantum mechanics is used to solve one dimensional Schrodinger equation with any shape invariance potentials. After has been separated into three parts, each part is solved by using SUSY QM method.

4.1. Solution of Axial Part

Assume that $\alpha=\frac{\hbar^2}{2m}l(l+1)$, $\beta=-e^2$ and $\lambda_1=E_z{}'$. So equation (22) can be written as follows

$$-\frac{\hbar^2}{2m}\frac{\partial^2 Z}{\partial z^2}+\left(\frac{\hbar^2}{2m}\frac{l(l+1)}{z^2}-\frac{e^2}{z}\right)Z=E_z{}'Z \tag{23}$$

Equation (23) is Schrodinger equation for axial part which have the effective potential is coloumb potential. We hypothesize that the superpotential is given by $\phi_o(z,a_0)=\frac{A}{z}+\frac{B}{A}$, by applying equation (8) and equation (23) we get the superpotential, super-partner potentials and ground state energy are

$$\phi_o(z,a_0)=-\frac{\hbar}{\sqrt{2m}}\frac{(l+1)}{z}-\frac{\sqrt{2m}e^2}{2\hbar(l+1)} \tag{24}$$

$$V_-(z,a_0)=\frac{\hbar^2}{2m}\frac{l(l+1)}{z^2}-\frac{e^2}{z}+\frac{me^4}{2\hbar(l+1)^2} \tag{25}$$

$$V_+(z,a_0)=\frac{\hbar^2}{2m}\frac{(l+1)(l+2)}{z^2}-\frac{e^2}{z}+\frac{me^4}{2\hbar(l+1)^2} \tag{26}$$

$$E_0{}'=-\frac{me^4}{2\hbar(l+1)^2} \tag{27}$$

The raising and lowering operator of axial part are given by

$$\mathrm{A}^+=-\frac{\hbar}{\sqrt{2m}}\frac{d}{dz}-\frac{\hbar}{\sqrt{2m}}\frac{(l+1)}{z}-\frac{\sqrt{2m}e^2}{2\hbar(l+1)} \tag{28}$$

$$\mathrm{A}=\frac{\hbar}{\sqrt{2m}}\frac{d}{dz}-\frac{\hbar}{\sqrt{2m}}\frac{(l+1)}{z}-\frac{\sqrt{2m}e^2}{2\hbar(l+1)} \tag{29}$$

Using the lowering operator equation (29) and applying equation (12) we get the ground state of wave function is

$$Z_0=C'z^{(l+1)}e^{\frac{\sqrt{2m}e^2}{2\hbar(l+1)}z} \tag{30}$$

And the first excited wave function is given by

$$Z_1(z,a_0)=C(l+1)z^l e^{\frac{\sqrt{2m}e^2}{2\hbar(l+1)}z} \tag{31}$$

With $a_0=l, a_1=l+1,...,a_n=l+n$ so we can get the next excited wave functions of axial part by using raising equation (28) and lowering operator equation (29) and applying equation (14).

The eigen value of Schrödinger equation for axial part can be obtained by using equations (9) and (10) is given by

$$E_{n_z} = \frac{me^4}{2\hbar(l+n_z+1)^2} \tag{32}$$

So we get the first separation variable constant λ_1 is like equation (33).

4.2. Solution of Angular Part

Assume that $b = v(v-1)$, $c = \kappa(\kappa-1)$ and $E_\theta" = \lambda_2$. So equation (22) can be written as follows

$$-\frac{\hbar^2}{2m}\frac{\partial^2 P}{\partial\theta^2} + \frac{\hbar^2}{2m}\left(\frac{v(v-1)}{\cos^2\theta} + \frac{\kappa(\kappa-1)}{\sin^2\theta}\right)P = E"P \tag{33}$$

Like in the subsection 4.1. We have to define the superpotential, by using equation (21) we hypothesizes that the superpotential is given by $\phi_0(\theta) = A\cot\theta + B\tan\theta$, so

$$\phi_0(\theta) = \frac{\hbar}{\sqrt{2m}}(v-1)\cot\theta - \frac{\hbar}{\sqrt{2m}}(\kappa-1)\tan\theta \tag{34}$$

And we get the ground state energy for angular part is given by

$$E_0" = \frac{\hbar^2}{2m}(v-\kappa)^2 \tag{35}$$

The raising and lowering operator for angular part are given by

$$A^+(\theta, a_0) = -\frac{\hbar}{\sqrt{2m}}\frac{d}{d\theta} + \frac{\hbar}{\sqrt{2m}}(v-1)\cot\theta - \frac{\hbar}{\sqrt{2m}}(\kappa-1)\tan\theta \tag{36}$$

$$A(\theta, a_0) = \frac{\hbar}{\sqrt{2m}}\frac{d}{d\theta} + \frac{\hbar}{\sqrt{2m}}(v-1)\cot\theta - \frac{\hbar}{\sqrt{2m}}(\kappa-1)\tan\theta \tag{37}$$

Using the lowering operator equation (38) and applying equation (12) we get the ground state of wave function is

$$P_0 = D'\left(\cos^{(v-1)}\theta\right)\left(\sin^{(\kappa-1)}\theta\right) \tag{38}$$

And the first excited wave function for angular part is

$$P_1 = D\left[(v-\kappa)\sin^{(v+\kappa-3)}\theta + (v-1)\cos^{\kappa}\theta\sin^{(v-2)}\theta - (\kappa-1)\cos^{(\kappa-2)}\theta\sin^{v}\theta\right] \tag{39}$$

With $a_0 = v;\kappa, a_1 = (v-1);(\kappa-1),\ldots,a_n = (v-n);(\kappa-n)$ so we can get the next excited wave functions by using raising and lowering operator equations (37) and (38) and applying equation (14).
The super-partner potentials for angular part are given as

$$V_-(\theta, a_0) = \frac{\hbar^2}{2m}\left(\frac{v(v-1)}{\cos^2\theta} + \frac{\kappa(\kappa-1)}{\sin^2\theta}\right) - \frac{\hbar^2}{2m}(v-\kappa)^2 \tag{40}$$

$$V_+(\theta, a_0) = \frac{\hbar^2}{2m}\left(\frac{(v-1)(v-2)}{\cos^2\theta} + \frac{(\kappa-1)(\kappa-2)}{\sin^2\theta}\right) - \frac{\hbar^2}{2m}(v-\kappa)^2 \tag{41}$$

The eigen value of Schrödinger equation for angular part can be obtained by using Eq.(9) and Eq.(10) is given by

$$E_{n_\theta}" = -\frac{\hbar^2}{2m}(v-\kappa)^2 \tag{42}$$

So the second variable constant is as like equation (43).

4.3. Solution of Radial Part

Radial part in Eq.(18) have to be reduced into Schrodinger equation by assumed that $R=\frac{U}{\sqrt{r}}$, $M=m$ and $E'''=E+\lambda_1$. Then the radial part has been reduced into Schrödinger equation, given as

$$-\frac{\hbar^2}{2m}\frac{\partial^2 U}{\partial r^2}+\left(\frac{1}{2}m\omega^2 r^2-\frac{\hbar^2}{2m}\frac{\lambda_2+\frac{1}{4}}{r^2}\right)U=E'''U \tag{43}$$

We hypothesize that the superpotential for the radial part is $\phi_0(r)=Ar+\frac{B}{r}$, by using equation (21) we get the superpotential is

$$\phi_0(r)=\omega\sqrt{\frac{m}{2}}r-\frac{\hbar}{\sqrt{2m}}\frac{\sqrt{-\lambda_2}+\frac{1}{2}}{r} \tag{44}$$

And we get the ground state energy is given by

$$E_0'''=\hbar\omega\left(\sqrt{-\lambda_2}+\frac{1}{2}\right) \tag{45}$$

The raising and lowering operator for radial part are given by

$$A^+(r,a_0)=-\frac{\hbar}{\sqrt{2m}}\frac{d}{dr}+\omega\sqrt{\frac{m}{2}}r-\frac{\hbar}{\sqrt{2m}}\frac{\sqrt{-\lambda_2}+\frac{1}{2}}{r} \tag{46}$$

$$A(r,a_0)=\frac{\hbar}{\sqrt{2m}}\frac{d}{dr}+\omega\sqrt{\frac{m}{2}}r-\frac{\hbar}{\sqrt{2m}}\frac{\sqrt{-\lambda_2}+\frac{1}{2}}{r} \tag{47}$$

Using the lowering operator equation (48) and applying equation (12) we get the ground state of wave function is

$$U_0=F'r^{\sqrt{-\lambda_2}+\frac{1}{2}}e^{-\frac{\omega m}{\hbar}r} \tag{48}$$

And the first excited wave function for radial part is given by

$$U_1=F\left[\frac{\omega\sqrt{m}}{\hbar}r^{\sqrt{-\lambda_2}+\frac{3}{2}}-\frac{2\hbar}{\sqrt{2m}}r^{\sqrt{-\lambda_2}-\frac{1}{2}}-\frac{\omega m}{\hbar}r^{\sqrt{-\lambda_2}+\frac{1}{2}}\right]e^{-\frac{\omega m}{\hbar}r} \tag{49}$$

With $a_0=\sqrt{-\lambda_2}, a_1=\sqrt{-\lambda_2}+1,\ldots,a_n=\sqrt{-\lambda_2}+n$ so we can get the next excited wave functions by using raising and lowering operator equations (47) and (48) and applying equation (14).
The super-partner potentials for radial part are given as

$$V_-(r,a_0)=\frac{1}{2}m\omega^2r^2-\frac{\hbar^2}{2m}\frac{\lambda_2+\frac{1}{4}}{r^2}-\hbar\omega\left(\sqrt{-\lambda_2}+\frac{1}{2}\right) \tag{50}$$

$$V_+(r,a_0)=\frac{1}{2}m\omega^2r^2-\frac{\hbar^2}{2m}\frac{\left(\sqrt{-\lambda_2}+\frac{1}{2}\right)\left(\sqrt{-\lambda_2}+\frac{3}{2}\right)}{r^2}-\hbar\omega\left(\sqrt{-\lambda_2}+\frac{1}{2}\right) \tag{51}$$

The eigen value of Schrodinger equation for radial part can be obtained by using equations (9) and (10) is given by

$$E_{n_r}''' = -\hbar\omega\left(\sqrt{-\lambda_2} + n_r + \frac{1}{2}\right) \tag{52}$$

So we get the energy spectra of the system is

$$E = \frac{me^4}{2\hbar(l+n_z+1)^2} - \hbar\omega\left(\frac{\hbar}{\sqrt{2m}}(\nu-\kappa) + n_r + \frac{1}{2}\right) \tag{53}$$

The energy spectra of Schrodinger equation for DRSCO potential that expressed in equation (2) is given in equation (54). There are three parts of the energy spectra, the first part is associated with the axial part of the potential, and the second part is associated with radial and polar parts of potential. Without the presence of axial part of potential, the energy spectra reduces to the energy spectra of harmonic oscillator potential.

The total wave functions, the un-normalized ground state and first excited state ones, are obtained from equations (31), (32), (38), (39), (48), and (49) given as

$$\psi_0(r,\theta,z) = G' r^{\sqrt{-\lambda_2}} e^{-\frac{\omega m}{\hbar}r}\left(\cos^{(\nu-1)}\theta\right)\left(\sin^{(\kappa-1)}\theta\right) z^{(l+1)} e^{\frac{\sqrt{2}me^2}{2\hbar(l+1)}z} \tag{54}$$

$$\psi_1(r,\theta,z) \cong \left[\frac{\omega\sqrt{m}}{\hbar} r^{\sqrt{-\lambda_2}+\frac{3}{2}} - \frac{2\hbar}{\sqrt{2m}} r^{\sqrt{-\lambda_2}-\frac{1}{2}} - \frac{\omega m}{\hbar} r^{\sqrt{-\lambda_2}+\frac{1}{2}}\right] e^{-\frac{\omega m}{\hbar}r}$$
$$\times\left[(\nu-\kappa)\sin^{(\nu+\kappa-3)}\theta + (\nu-1)\cos^{\kappa}\theta\sin^{(\nu-2)}\theta - (\kappa-1)\cos^{(\kappa-2)}\theta\sin^{\nu}\theta\right] \tag{55}$$
$$\times(l+1) z^l e^{\frac{\sqrt{2}me^2}{2\hbar(l+1)}z}$$

with λ_2 expressed in equation (43) respectively.

5. Conclusion

By applying a cylindrical coordinate system, the Schrodinger equation for DRSCO potential reduces to a perfectly variable separable Schrodinger equation. Three dimensional Schrodinger equation reduces to one radial Schrodinger equation, one angular Schrodinger equation, and one axial Schrodinger equation. These three one dimensional Schrodinger equation are exactly solvable since each of Schrodinger equation with shape invariant potential. By using SUSY Quantum Mechanics method and the idea of shape invariance the energy spectra are calculated and the total wave function is obtained which are combination of axial, angular and radial parts.

References

[1] Tao, Terence, Schrodinger Equation, *Preprints UCLA* (2016)
[2] Y. Aharonov and D. Bohm, *Phys. Rev.* 115 (1959) 485
[3] L. V. Hoang, L.X. Hai, L.I. Komarov and T. S. Romanova, *J. Phys. a* 25 (1992) 6461
[4] C.Y. Chen, Y.W. Liu, *HEP & NP* 23 (1999) 865
[5] C. Quesne, *J. Phys. A* 21 (1988) 3093
[6] A.S. Zhedanov, *J.Phys, A* 26 (1993) 4633
[7] C.Y. Chen, D.S. Sun, *Acta Photon. Sin.* 30 (2001) 104
[8] D.S. Sun, C.Y. Chen, *Acta Photon. Sin.* 30 (2001) 539
[9] T.L Chen, C.Y. Chen, *Acta Sin. Quantum-Opt.* 7 (2001) 67
[10] S.H. Dong, G.H. Sun, M. Lozada-Cassou, *Phys. Lett. A* 328 (2004) 299
[11] J.Y. Guo, J.C. Han, R.D. Wang, *Phys. Lett. A* 353 (2006) 378
[12] G.H. Liu, K.X. Guo, Superlattices Microstruct. 52 (2012) 997

[13] M.V Carpio-Bernido, C.C. Bernido, *Phy. Lett. A* 134 (1989) 395
[14] M.V Carpio-Bernido, *J. Math. Phys.* 32 (1991) 1799
[15] M.V Carpio-Bernido, C.C. Bernido, *Phys. Lett. A* 137 (1989) 1
[16] B.W. Huang, *HEP & NP* 27 (2003) 770
[17] F.L. Lu, C.Y. Chen, D.S. Sun, *Chin. Phys.* 14 (2005) 463
[18] F. Yasuk, A. Durmus, *Phys. Scr.* 77 (2008) 015005
[19] E. Witten, Dynamical Breaking of Supersymmetry, *Nucl. Phys. B* Vol.188 (1981) 513
[20] L.E.Gendenshtein, Derevation of exact spectra of Schrodinger Equation by means Supersymmetry, *JETP Lett.* Vol. 38 (1983) 356
[21] R. Dutt, A. Gangopadhyaya, and U.P. Sukhatme, Non-central potentials and Spherical Harmonics Using Supersymmetry and Shape Invariance, *Am. J. Phys*. 65 5 (1997)
[22] A. Gangopadhyaya, J.V.Mallow, C.Rasinariu, and U.P. Sukhatme, Exact Solution of the Schrodinger Equation : Connection between Supersymmetric Quantum Mechanics and spectrum Generating Algebras, *Chin. J. Phys.* Vol. 39 no 2 (2001)

IOP Conf. Series: Journal of Physics: Conf. Series **1127** (2019) 012006 doi:10.1088/1742-6596/1127/1/012006

Diffusion and Interaction between ion Ca^{2+} and ion Gd^{3+} in a Model Synapse: A Monte Carlo Study

Adita Sutresno[1,2,*], Freddy Haryanto[1], Sparisoma Viridi[1], Idam Arif[1]
[1]Nuclear Physics and Biophysics Research Group, Physics Department, Institut Teknologi Bandung, Jalan Ganesha 10 Bandung 40132, Indonesia,
[2]Physics Department, Faculty of Science and Mathematics, Satya Wacana Christian University, Jalan Diponegoro 52-60, Salatiga 50711, Indonesia

*Corresponding Author: adita1204@gmail.com

Abstract. Gadolinium (Gd^{3+}) is a commonly used in MRI. The role of Gd^{3+} as contrast agent greatly helps to improve the quality of an image and Gadolinium entrance in a body through a diffusion process after administration. Diffusion process takes place from outside pre-synapse to inside pre-synapse through Ca^{2+} channels. This process causes a delay of Ca^{2+} in passing Ca^{2+} channels to pre-synapse. This study aims to determine equilibrium concentration of Ca^{2+} and Gd^{3+} so that Ca^{2+} can pass Ca^{2+} channels. To address the effect of Gd^{3+} concentration around Ca^{2+} channels, we developed a Monte Carlo simulation on Ca diffusion and interaction with Gd. We find Gd^{3+} concentration in a certain threshold can be potential as a blocker Ca^{2+} diffuse entrance into pre-synapse. In another hand, an amount of Ca^{2+} channels and pre-synapse terminal to continuing diffusion of vesicles to clef was important. And the trapping Gd^{3+} in pre-synapse effects in diffusion delays also. These simulations identify characteristics of the interaction between Ca^{2+} and Gd^{3+} in Ca channels which entrance into pre-synapse through diffusion process depend on concentration with each other.

Key words: Monte Carlo Cell, Synapse, Diffusion, Contras Agents

1. Introduction

Gadolinium is chemical substances which it popularly used in magnetic resonance imaging as contrast media. The first gadolinium-based contrast agent (GBCA) available for clinical was in 1988 [1]. When Gadolinium injected into the body of the patient will be enhanced between pathological and normal anatomical structures facilitated the detection and assessment of the disease and improves the quality of the MRI images [1,2]. Behind the advantages of Gadolinium as a contrast agent, it has minor disadvantages especially for the patient with kidney disorders and has a potent blocking Calcium channel.

In 1982, G. W Bourne and J.M. Trifaro [3] found that Gadolinium ion a potent inhibitor of catecholamine and the possibility of an inhibition by Ga^{3+} of the Ca^{2+} mediated effects. Specifically, Gd^{3+} had a parallel shift to the right of the Ca^{2+} concentration [^{3}H]noradrenaline output relationship suggesting a competitive antagonism between Ca^{2+} and Gd^{3+}. The presence of Gd^{3+} a potent block both N-type and non-N-type current of Ca Channel and in a complex fashion also [4]. A subsequent study [5], the block by Gd^{3+} follows one prediction of this model, the reciprocal of the open time was a linear function of the blocking ion concentration which gadolinium ions produce some distinct kinds of block ion channels in Xenopus oocytes i.e. a concentration-dependent reduction in channel open, in open current, and reversible inhibition of channel opening.

Diffusion is one important mechanism in daily life especial in our body. Many substances in our body transported from one part to another part by the diffusion mechanism. Likewise, many substances injected in our body distributed by diffusion including Gadolinium. This mechanism is important to explore the functions of gadolinium with respect to the effect of Gd^{3+} in the body. One of a disadvantage Gd^{3+} is block Ca channel in a synapse. Consequently, transport substance in a body disturbed. Using this information, this study wanted to know the mechanism of Gadolinium in our body especially interaction between Gadolinium (Gd^{3+}) and Calcium (Ca^{2+}) and interaction each other in a synapse. This question is addressed using Monte Carlo simulation of the synapse which consist of a part which one are presynaptic and postsynaptic. The simulation is carrried out using a variety number of Gd^{3+} and Ca^{2+}. The result indicated the ratio between Gd^{3+} and Ca^{2+} can be used to determine the speed blocking of Ca channel.

2. Methods

Monte Carlo Cell simulation was carried out using MCell 3.3 running on desktop computer 3.40 GHz Intel Core i7 (Windows 7). Visualization of the synapse (pre-synapse and post-synapse) and movies was made with the output of MCell using blender 2.76 which add-on with Cellblender 1.0. (www.mcell.org).

The model illustrated in Fig. 1.a represents a sketch of a synapse which consisting of two part a presynaptic and postsynaptic where two kinds ions Gd^{3+} and Ca^{2+} around presynaptic. Figure 1.b was a model synapse design with blender 2.76 consist of the presynaptic is on top with volume 0.02 μm^3 and surface area 1.85 μm^2 and the postsynaptic is below with volume 0.21 μm^{3} $^{+}$ and surface area 2.07 μm^2. Ca^{2+} competes with Gd^{3+} to diffuse into presynaptic. Ca^{2+} diffusion goes into synapses through Ca channel that on a surface of presynaptic. The presynaptic model has 23 Ca channel with 500 ions as buffer distributed randomly. Gd^{3+} diffusion outside presynaptic and attach at ions buffer as blocking the channel. Depend on time amount of Gd^{3+} totally blocking Ca channel which affect to a speed of Ca^{2+} diffusion. Physically, Gd^{3+} has the same crystallographic radius as Ca^{2+} [6] which Gd^{3+} has an ionic radius (0.938 Å) and Ca^{2+} has the ionic radius (0.99 Å) [5].

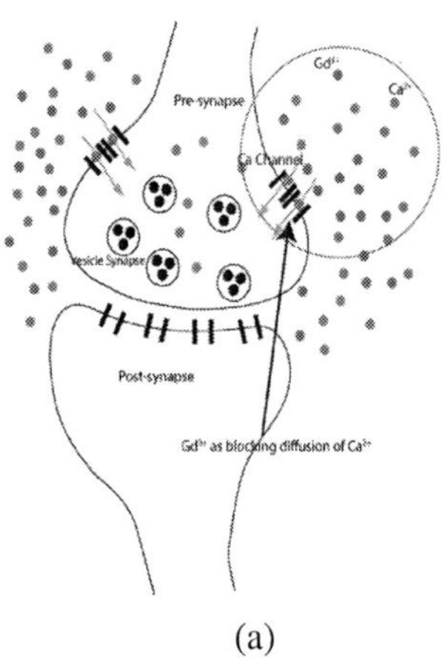

(a)

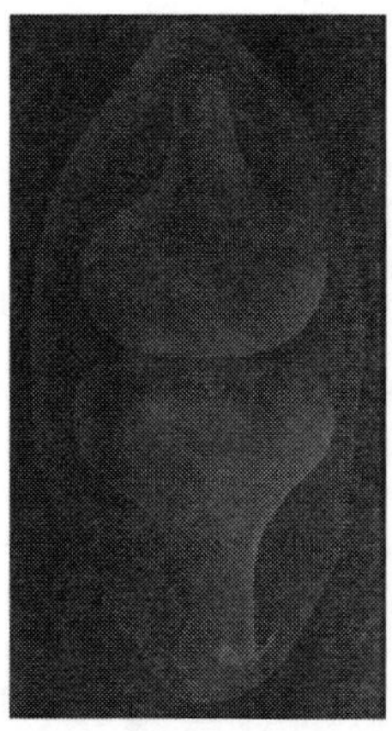

(b)

Figure 1. The model synapse consisting a presynaptic and a postsynaptic (a) synapse model with ion Gd^{3+} and Ca^{2+} which Gd^{3+} blocking Ca channel of Ca^{2+} base on Neuroscience [7], (b) synapse model used in the MCell simulator.

IOP Conf. Series: Journal of Physics: Conf. Series **1127** (2019) 012006 doi:10.1088/1742-6596/1127/1/012006

This study used a two kinds diffusion process. The first process describes diffusion Gd^{3+} to Ca channel and finally attach to Ca channel with forward rate constant *fw*. The second process in the same time with the previous process, Ca^{2+} diffusion into presynaptic through Ca channel with forward rate constant *fw* and backward constant *bw*. The simulation was carried out using a set of standard parameters values (Tabel 1) and a time step of $1x10^{-6}$ s was used for simulation with a standard parameter.

Table 1. Standard values of the parameters used in the simulations

Parameter	Standard value
Forward rate	$3x10^{8}$ $cm^{2}s^{-1}$
Backward rate	$2x10^{7}$ s^{-1}
Diffusion coefficient Ca^{2+}	$6x10^{-6}$ $cm^{2}s^{-1}$ [8] [9]
Diffusion coefficient Ga^{3+}	$1x10^{-6}$ $cm^{2}s^{-1}$
Diffusion coefficient buffer	0 $cm^{2}s^{-1}$
Ca Channel density	10^{-4} μm^{2} [10]

3. Results and Discussion

Blocking Ca Channel

Calcium and Gadolinium diffusion can be described mathematically using a combination of Fick's second law of diffusion. Diffusion through or attach on Ca channel of a presynaptic alike diffusion to N disk-like absorbers on the surface of a presynaptic which shape of a presynaptic base on a sphere. In this study, standard Ca channel density around a 10^{-4} μm^{2} distribute into 23 channel (Table 1) and compared five simulations with different random number seeds for ratio 1:19 for Gd^{3+} of Ca^{2+} and found $t_{50} = 11.9 \pm 0.2$ ms (mean ± SD).

Fig. 2 shows how the number of Ca^{2+} diffused from outside presynaptic (black line) into presynaptic (red line) through Ca channel with ratio 1:9 for Gd^{3+} of Ca^{2+}. Ca^{2+} declines quickly in the first $5.00x10^{-3}$ s, then decline more gradually and reaches a steady state from 0.20 s which half-saturation occurs at $t_{50} = 6.95x10^{-3}$ s. Moreover, the second part of the Fig. 2 was diffusion of Ga^{3+} attaches on Ca channel become a block of Ca^{2+}. Ga^{3+} decline gradually from an initial state and at $7.50x10^{-2}$ blocking 66.8 % of total area Ca channel and $t_{50} = 2.19x10^{-2}$ s. In short, Ca^{2+} diffused smooth into presynaptic although Gd^{3+} running well as blocking Ca^{2+}. It has been shown that, concentration ratio 1:9 for Gd^{3+} of Ca^{2+} was not enough to totally blocking Ca^{2+}. Ca^{2+} diffused perfectly until steady state condition before Gd^{3+} totally block the channel, indeed t_{50} of Gd^{3+} block Ca channel longer than t_{50} of Ca^{2+} diffused.

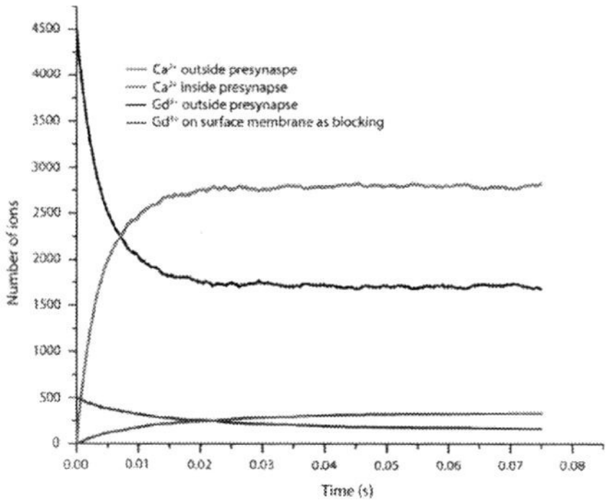

Figure 2. Result simulation of MCell ratio Gd^{3+} and Ca^{2+} 1:9. Ca^{2+} diffused from outside presynaptic into presynaptic through Ca channel and t_{50} Ca^{2+} faster than t_{50} Gd^{3+} blocking Ca Channel.

Effect of Gd^{3+}

Fig. 3 shows the relation between the logarithms of the Ca^{2+} number and output of Ca^{2+} number. In this study, a number of Ca^{2+} increase from 500 to 5000 in the absent as well as presence of Gd^{3+} and complementary to this, number Gd^{3+} was 500. The presence of Gd^{3+} makes diffusion of Ca^{2+} disturbed because wide of Ca channel be smaller. Thus, there was a shift to the right side of the Ca^{2+} output of absent Gd^{3+} in the presence of Gd^{3+}. The result of the simulation suggested the possibility of an inhibitor by Gd^{3+} of Ca^{2+} mediated effect and competitive antagonism between Ca^{2+} and Gd^{3+}.

The increased of Gd^{3+} in synapse had the effect at wide a shift between absent and a presence of Gd^{3+}. It has been shown that the presence of Gd^{3+} had a disadvantage in an excessive number because it can because delays transport Ca^{2+} diffuse into presynaptic.

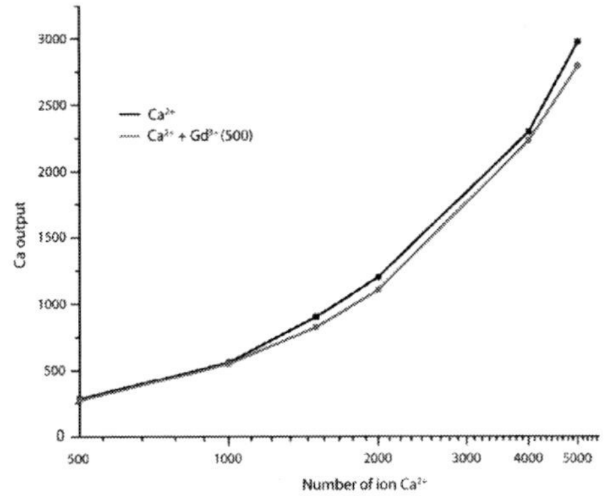

Figure 3. Effect of Gd^{3+} attached to Ca Channel and blocking diffusion of Ca^{2+} at 0.01 s. In the figure, black line was absence Gd^{3+} and red line was presence Gd^{3+} in Ca^{2+} which Ca^{2+} increase stepwise.

Speed of Blocking

Fig. 4 summarizes the result of simulation Gd^{3+} block of Ca^{2+} diffused into presynaptic. As synapse gets more, Ca^{2+} collide each other Ca^{2+} and Gd^{3+} as elastic collisions. Furthermore, Ca^{2+} diffused into presynaptic through Ca channel suddenly collide with ions buffer. In contrast, Gd^{3+} inelastic collision with buffer ions and produce a blocking buffer on Ca channel. It is apparent that increasing number of Gd^{3+} in synapse had an effect at speed blocking Ca^{2+} pass Ca channel.

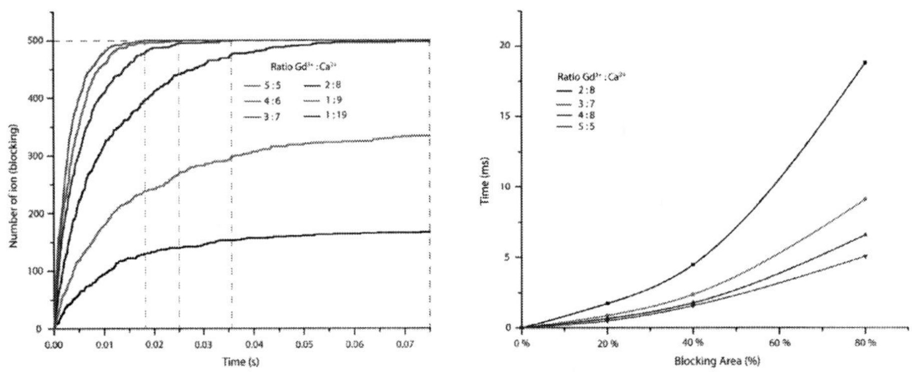

(a) (b)

Figure 4. Result of simulation dependent time of Gd^{3+} blocking Ca channel at variety ratio Gd^{3+} and Ca^{2+} (a) all range time (0-0.075s), (b) blocking Ca channel stepwise 20%

The variation ratio Gd^{3+} and Ca^{2+} from lower to the higher number of Gd^{3+} shows a result that higher number of Gd^{3+} has the higher speed for blocking. Fig. 4.a shows a relation between times and numbers of ion Gd^{3+} blocking Ca channel. In the lowest and a second lowest ratio of Gd^{3+}: Ca^{2+} (1:19) and (1:9), all Ca^{2+} successfully diffused into presynaptic and reach steady state before all Gd^{3+} blocked Ca channel. The ratio of Gd^{3+}: Ca^{2+} (2:8) until (5:5) more than enough was able to block all of Ca channel albeit with different duration. The increasing number of Gd^{3+} resulted in the duration of totally blocking of Ca channel shorter. On the other hand, the relation between blocking area in percent and time shows at Fig.4. b. The higher the percentage of Gd^{3+} to Ca^{2+} affects the speed closing Ca channel and at the low and high Gd^{3+} gave exponential function relation to every doubling area blocking come up linear function.

Thus, the results at Fig. 4 possible to calculation amount of Gd3+ to injection getting the best image in MRI treatment.

4. Conclusion

Monte Carlo successfully simulated interaction between Ca^{2+} and Gd^{3+} which antagonism of Ca^{2+} by Gd^{3+} and Gd^{3+} as blocker diffusion process in synapse. The result of this investigation provides a framework for determining how specific ratio between Gd^{3+} and Ca^{2+} so that diffusion still happen but Gd^{3+} can give another advantage. The speed of diffusion Ca^{2+} into presynaptic and speed of Gd^{3+} blocking of Ca channel depends on ratio between Ca^{2+} and Gd^{3+}. The higher number of Gd^{3+}, then an impact on the speed blocking Ca channel increase.

References

[1] J Lohrke, T Frenzel, J Endrikat 2016 *Advancesin therapy,* 33 pp. 1-28

[2] K M Hasebroock and N J Serkova 2009 *Expert Opinion on drug Metabolism & Toxicology,* vol. 5, no. 4, pp. 403-16,.

[3] G W Bourne and J M Trifaro 1982 *Neuroscience,* vol. 7, pp. 1615-22

[4] L M Boland, T A Brown and R Dingledine 1991 *Brain Research,* vol. 563 pp. 142-150

[5] X C Yang and F Saches 2004 *Science,* vol. 243 pp. 1068-71

[6] D Purves 2004 Neuroscience (Sunderland, Sinauer Associates Inc)

[7] F Sala and A Hernandez-Cruz 1990 *Biophysics Journal,* vol. 57 pp. 313-24

[8] B S Donahue and R F Abercrombie 1987 *Cell Calcium,* vol. 8 pp. 437-47

[9] J P Dilger 2010 *Biophysical,* pp. 959-67

[10] Y A Ermakov, K Kamaraju, K Sengupta and S Sukharev 2010 *Biophysical Journal,* vol. 98pp. 1018-27

Acknowledgement

This work is partially supported by Grant of PDD 2017 (001/K6/KM/SP2H/PENELITIAN/2017) and Grant of PUPT 2017 (009/SP2H/LT/DRPM/IV/2017) from Ministry of Research, Technology and Higher Education

IOP Conf. Series: Journal of Physics: Conf. Series **1127** (2019) 012007 doi:10.1088/1742-6596/1127/1/012007

Extracting new inflation model from the most general scalar-tensor theory

Getbogi Hikmawan[1], **Agus Suroso**[1,2], **and Freddy P. Zen**[1,2]

[1] Theoretical Physics Laboratory, THEPI Division, Faculty of Mathematics and Natural Sciences, Institut Teknologi Bandung, Jl. Ganehsa 10 Bandung 40132, Indonesia
[2] Indonesia Center of Theoretical and Mathematical Physics (ICTMP), Faculty of Mathematics and Natural Sciences, Institut Teknologi Bandung, Jl. Ganesha 10 Bandung 40132, Indonesia

E-mail: getbogihikmawan@s.itb.ac.id

Abstract. Inflation is now the strongest scenario to be the candidate of early universe. Until recently, there are many models tried to give explanation about this phenomenon, one of them is called Horndeski theory, the most general scalar-tensor theory. In this paper, we investigate the no-ghost and Laplacian stability conditions for scalar and tensor perturbation for Horndeski theory, and obtains exactly the same conditions as the actual stability conditions from some particular gravity coupled scalar field cosmological model. Therefore, there are big possibility to using the conditions obtained to become a constraint for extracting the new cosmological model for inflation scenario.

1. Introduction

The accepted viable phenomenological scenario describing the exponential expansion in the early time is the inflationary universe [1, 2] that in first place developed to become a possible solution to the horizon and flatness problems in cosmology. There was several model established to explain this phenomena [3, 4, 5, 6, 7], but until this time, inflation is still become a speculative scenario, although the observation of temperature fluctuation of the Cosmological Microwave Background (CMB) by WMAP [8] and COBE [9] bolster the slow-roll inflationary universe scenario driven by single scalar degree of freedom, because of the lack of knowledge about the origin of the scalar field.

Cosmological model where scalar field interact with curvature tensor via various interaction model have given favorable solutions for some problem in cosmology, such as for late-time acceleration [10], dark Matter and dark Energy [11], and inflation [12]. In 1974, Horndeski [13] derived the most general scalar-tensor theories, with lagrangian,

$$L = \sum_{i=2}^{5} L_i, \tag{1}$$

where, with $X = \partial_\mu \phi \partial^\mu \phi$,

$$L_2 = \mathcal{K}(\phi, X), \tag{2}$$
$$L_3 = -G_3(\phi, X)\Box\phi, \tag{3}$$
$$L_4 = G_4(\phi, X)R - 2G_{4X}(\phi, X)[(\Box\phi)^2 - \phi^{;\mu\nu}\phi_{;\mu\nu}], \tag{4}$$
$$L_5 = G_5(\phi, X)G_{\mu\nu}\phi^{;\mu\nu} + \frac{1}{3}G_{5X}(\phi, X)[(\Box\phi)^3 - 3(\Box\phi)(\phi_{;\mu\nu}\phi^{;\mu\nu}) + 2(\phi_{;\mu\nu}\phi^{;\mu\sigma}\phi^{;\nu}_{;\sigma})], \tag{5}$$

with d'Alembertian or $\Box$ (square) operator is defined as $\Box = \partial^\mu\partial_\mu = \frac{\partial^2}{\partial t^2} - \frac{\partial^2}{\partial x^2} - \frac{\partial^2}{\partial y^2} - \frac{\partial^2}{\partial z^2}$, where we consider natural unit, $c = 1$. This theory comprises all the models with gravity-coupled scalar field that previously studied by choosing particular coefficient functions $(\mathcal{K}(\phi, X), G_i(\phi, X))$, such as minimally coupled scalar field [14], Brans-Dicke theory [15], Dilaton gravity [16], $f(R)$ gravity [17], derivative couplings [18], Gauss-Bonnet couplings [6] and many others. Although the Horndeski theory is not fully satisfactory, this theory may indeed offer alluring solution for cosmology features with the possibility to obtain a new theory of inflation by adjusting the coefficient functions.

The organization of the paper is as follows, In section 2, we spell out the no-ghost and Laplacian stability condition of General Gravitational Theories from [19], then use it for Horndeski Lagrangian in Section 3. In Section 4, we try to find the stability conditions for some model (in this works for Brans-Dicke theory and Gauss-Bonnet couplings) and compare the result with actual stability conditions obtained before. The final chapter is for conclusion.

2. No-Ghost and Laplacian Stability of General Gravitational Theories

The ADM metric is given by,

$$ds^2 = g_{\mu\nu} = -N^2dt^2 + h_{ij}(dx^i + N^i dt)(dx^j + N^j dt), \tag{6}$$

with N is lapse function, N^i is shift vector and h_{ij} is the three-dimensional metric. Therefore we can get $g_{\mu\nu} = -N^2 + N^i N_i, N_i, h_{ij}$. An orthogonal vector at constant t of hypersurface Σ_t is given by $n_\mu = -Nt_{;\mu} = (-N, 0, 0, 0)$, so that $n^\mu = (\frac{1}{N}, \frac{-N^i}{N})$, so this orthogonal vector is timelike, $n^\mu n_\mu = -1$. The three-dimensional metric on the hypersurface, $h_{\mu\nu}$, has relation with four-dimensional metric $g_{\mu\nu}$, $h_{\mu\nu} = g_{\mu\nu} + n^\mu n^\nu$, satisfies orthogonal relation $n^\mu h_{\mu\nu} = 0$.

Extrinsic curvature is defined by,

$$K_{\mu\nu} = h^\lambda_\mu n_{\nu;\lambda} = n_{\nu;\mu} + n_\mu \dot{n}_\nu, \tag{7}$$

where $n^\mu K_{\mu\nu} = 0$, that means $K_{\mu\nu}$ is a quantity on hypersurface. The Geometry of the hypersurface can be expressed by $\mathcal{R}_{\mu\nu} =^{(3)} R_{\mu\nu}$, as,

$$R = \mathcal{R} + K_{\mu\nu}K^{\mu\nu} - K^2 + 2(Kn^\mu - \dot{n}^\mu)_{;\mu}, \tag{8}$$

with some scalar quantity defined as,

$$K \equiv K^\mu_\mu;\ \mathcal{S} \equiv K_{\mu\nu}K^{\mu\nu};\ \mathcal{R} \equiv \mathcal{R}^\mu_\mu;\ \mathcal{Z} \equiv \mathcal{R}_{\mu\nu}\mathcal{R}^{\mu\nu};\ \mathcal{U} \equiv \mathcal{R}_{\mu\nu}K^{\mu\nu}. \tag{9}$$

Therefore lagrangian from general gravitational theory depends of this scalar can be written as,

$$S = \int d^4x\sqrt{-g}L(N, K, \mathcal{S}, \mathcal{R}, \mathcal{Z}, \mathcal{U}; t). \tag{10}$$

If the Lagrangian above is expanded to second order, we can get,

$$L = \bar{L} + L_N\delta N + L_K\delta K + L_\mathcal{S}\delta\mathcal{S} + L_\mathcal{R}\delta\mathcal{R} + L_\mathcal{Z}\delta\mathcal{Z} + L_\mathcal{U}\delta\mathcal{U} \\ + \frac{1}{2}(\delta N\frac{\partial}{\partial N} + \delta K\frac{\partial}{\partial K} + \delta\mathcal{S}\frac{\partial}{\partial\mathcal{S}} + \delta\mathcal{R}\frac{\partial}{\partial\mathcal{R}} + \delta\mathcal{Z}\frac{\partial}{\partial\mathcal{Z}} + \delta\mathcal{U}\frac{\partial}{\partial\mathcal{U}})^2 L. \quad (11)$$

where bar denotes the background quantity. To get the no-ghost and Laplacian stability, we consider the second-order of Lagrangian (11), then if we consider flat homogenous and isotropic (FLRW) metric

$$ds^2 = -dt^2 + a^2\delta_{ij}dx^i dx^j, \quad (12)$$

as background metric, using relation,

$$\sqrt{-g} = \sqrt{\bar{h}} + \sqrt{\bar{h}}\delta N + \delta\sqrt{h} + \delta N\delta\sqrt{h} = a^3 + a^3\delta N + \delta\sqrt{h} + \delta N\delta\sqrt{h}, \quad (13)$$

the second order Lagrangian density can be obtained as,

$$\mathcal{L}_2 = a^3 L_2 + a^3\delta N L_1 + \delta N\delta\sqrt{h}L_0 + L_1\delta\sqrt{h} \\ = \delta\sqrt{h}[(\dot{\mathcal{F}} + L_N)\delta N + \mathcal{E}\delta_1\mathcal{R}] + a^3[(L_N + \frac{1}{2}L_{NN})\delta N^2 + \mathcal{E}\delta_2\mathcal{R} + \frac{1}{2}\mathcal{A}\delta K^2 \\ + \mathcal{B}\delta K\delta N + \mathcal{C}\delta K\delta_1\mathcal{R} + (\mathcal{D} + \mathcal{E})\delta N\delta_1\mathcal{R} + \mathcal{E}\delta_2\mathcal{R} + \frac{1}{2}\mathcal{G}\delta_1\mathcal{R}^2 + L_\mathcal{S}\delta K^\mu_\nu\delta K^\nu_\mu + L_\mathcal{Z}\delta\mathcal{R}^\mu_\nu\delta\mathcal{R}^\nu_\mu]. \quad (14)$$

with,

$$\mathcal{A} = L_{KK} + 4H^2 L_{\mathcal{SS}} + 4HL_{\mathcal{S}K},\ \mathcal{B} = L_{KN} + 2HL_{\mathcal{S}N}, \\ \mathcal{C} = L_{K\mathcal{R}} + 2HL_{\mathcal{SR}} + \frac{1}{2}L_\mathcal{U} + HL_{K\mathcal{U}} + 2H^2L_{\mathcal{SU}},\ \mathcal{D} = L_{N\mathcal{R}} - \frac{1}{2}\dot{L}_\mathcal{U} + HL_{N\mathcal{U}}, \\ \mathcal{E} = L_\mathcal{R} + \frac{1}{2}\dot{L}_\mathcal{U} + \frac{3}{2}HL_\mathcal{U},\ \mathcal{F} = L_K + 2HL_\mathcal{S},\ \mathcal{G} = L_{\mathcal{RR}} + 2HL_{\mathcal{RU}} + H^2L_{\mathcal{UU}}. \quad (15)$$

and $H \equiv \frac{\dot{a}}{a}$ is the Hubble parameter and dot denotes a derivative with respect to time.

2.1. Scalar Perturbation

For scalar pertubation, we take the gauge $h_{ij} = a^2 e^{2\zeta}\delta_{ij}$, then the second-order Lagrangian density (14) for scalar perturbation can be obtained as,

$$\mathcal{L}_2^{(s)} = a^3\{\frac{1}{2}(2L_N + L_{NN} + 9H^2\mathcal{A} - 6H\mathcal{B} + 6H^2L_\mathcal{S})\delta N^2 \\ + [\mathcal{W}(3\dot{\zeta} - \frac{\partial^2\psi}{a^2}) + 4(3H\mathcal{C} - (\mathcal{D} + \mathcal{E}))\frac{\partial^2\zeta}{a^2}]\delta N - (3\mathcal{A} + 2L_\mathcal{S})\dot{\zeta}\frac{\partial^2\psi}{a^2} - 12\mathcal{C}\dot{\zeta}\frac{\partial^2\zeta}{a^2} \\ + (\frac{9}{2}\mathcal{A} + 3L_\mathcal{S})\dot{\zeta}^2 + 2\mathcal{E}\frac{(\partial^2\zeta)^2}{a^4} + \frac{1}{2}(\mathcal{A} + 2L_\mathcal{S})\frac{(\partial^2\psi)^2}{a^4} + 4\mathcal{C}\frac{(\partial^2\psi)(\partial^2\zeta)}{a^4} + 2(4\mathcal{G} + 3L_\mathcal{Z})\frac{(\partial^2\zeta)^2}{a^4}\}. \quad (16)$$

Variations of the second-order action $S_2^{(s)} = \int d^4x\mathcal{L}_2^{(s)}$ with respect to δN and $\partial^2\psi$ lead to,

$$(2L_N + L_{NN} + 9H^2\mathcal{A} - 6H\mathcal{B} + 6H^2L_\mathcal{S})\delta N + \mathcal{W}(3\dot{\zeta} - \frac{\partial^2\psi}{a^2}) + 4(3H\mathcal{C} - (\mathcal{D} + \mathcal{E}))\frac{\partial^2\zeta}{a^2} = 0, \quad (17)$$

$$\mathcal{W}\delta N + (3\mathcal{A} + 2L_\mathcal{S})\dot{\zeta} - (\mathcal{A} + 2L_\mathcal{S})\frac{(\partial^2\psi)}{a^2} - 4\mathcal{C}\frac{(\partial^2\zeta)}{a^2} = 0. \quad (18)$$

There exist high-order spatial derivatives, so impose conditions,

$$\mathcal{A} + 2L_{\mathcal{S}} = 0;\ \mathcal{C} = 0;\ 4\mathcal{G} + 3L_{\mathcal{Z}} = 0, \tag{19}$$

to get rid of this high-order spatial derivatives. Using this conditions, The second-order Lagrangian density for scalar perturbation (16) can be written as,

$$\mathcal{L}_2^{(s)} = a^3 \mathcal{Q}_s[\dot{\zeta}^2 - \frac{c_s^2}{a^2}(\partial\zeta)^2]; \tag{20}$$

$$\mathcal{Q}_s \equiv \frac{2L_{\mathcal{S}}[3\mathcal{B}^2 + 4L_{\mathcal{S}}(2L_N + L_{NN})]}{\mathcal{W}^2};\ c_s^2 \equiv \frac{2}{\mathcal{Q}_s}(\dot{\mathcal{M}} + H\mathcal{M} - \mathcal{E}). \tag{21}$$

Therefore, to avoid the existence of ghost and Laplacian instability, these conditions must be satisfied,

$$\mathcal{Q}_s > 0 \ \rightarrow \frac{2L_{\mathcal{S}}[3\mathcal{B}^2 + 4L_{\mathcal{S}}(2L_N + L_{NN})]}{\mathcal{W}^2} > 0 \rightarrow 9\mathcal{W}^2 + 8L_{\mathcal{S}}\omega > 0 \tag{22}$$

$$c_s^2 > 0 \ \rightarrow \dot{\mathcal{M}} + H\mathcal{M} - \mathcal{E} > 0, \tag{23}$$

with,

$$\mathcal{W} = L_{KN} + 2HL_{\mathcal{S}N} + 4HL_{\mathcal{S}};\ \omega = 3L_N + \frac{3}{2}L_{NN} - 9H(L_{KN} + 2HL_{\mathcal{S}N} - 18H^2L_{\mathcal{S}}). \tag{24}$$

$$\mathcal{M} \equiv \frac{4L_{\mathcal{S}}(\mathcal{D} + \mathcal{E})}{\mathcal{W}}. \tag{25}$$

2.2. Tensor Perturbation

For tensor perturbation, we take the three-dimensional metric including tensor perturbation γ_{ij},

$$h_{ij} = a^2(\delta_{ij} + \gamma_{ij} + \frac{1}{2}\gamma_{il}\gamma^{lj});\ \gamma_{ii} = \partial_i\gamma_{ij} = 0. \tag{26}$$

From equation (14) the second-order Lagrangian density for tensor perturbation can be obtained as,

$$\mathcal{L}_2^{(h)} = \frac{a^3}{4}\mathcal{Q}_t[\dot{\gamma_{ij}}^2 - \frac{c_t^2}{a^2}(\partial_k\gamma_{ij})^2];\ \mathcal{Q}_t \equiv L_{\mathcal{S}};\ c_t^2 \equiv \frac{\mathcal{E}}{L_{\mathcal{S}}}. \tag{27}$$

Therefore, to avoid the existence of ghost and Laplacian instability, these conditions must be satisfied,

$$\mathcal{Q}_t > 0 \ \rightarrow L_{\mathcal{S}} > 0;\ c_t^2 > 0 \ \rightarrow \mathcal{E} > 0. \tag{28}$$

3. Horndeski Theory in ADM Variable

To get the condition to avoid the existence of ghost and Laplacian instability for Horndeski Lagrangian, firstly we write the Horndeski Lagrangian in ADM variable. Using the way we through in section 2, we can obtain the total Lagrangian for Horndeski theory in ADM variable as,

$$\begin{aligned} L = {} & \mathcal{K} + 2(-X)^{\frac{3}{2}}F_{3X}K - F_{3\phi}X + G_4\mathcal{R} + (2XG_{4X} - G_4)(K^2 - \mathcal{S}) - 2\sqrt{-X}G_{4\phi}K \\ & + \sqrt{-X}[F_5(\frac{1}{2}K\mathcal{R} - \mathcal{U}) - G_{5X}(2H^2 - 2KH + K^2 - \mathcal{S})] + \frac{1}{2}[(G_{5\phi} - F_{5\phi})\mathcal{R} + G_{5\phi}(K^2 - \mathcal{S})], \end{aligned} \tag{29}$$

where we take,

$$G_3 = F_3 + 2XF_{3X};\ G_{5X} = F_5/2X + F_{5X}, \tag{30}$$

with $X = \partial_\mu\phi\partial^\mu\phi = -\dot{\phi}^2$, $G_{iX} \equiv \partial G_i/\partial X$ and $G_{i\phi} \equiv \partial G_i/\partial\phi$.

Therefore we can obtain the conditions to avoid the existence of ghost and Laplacian Instability for Horndeski theory as, for tensor perturbation,

$$L_S = G_4 - 2XG_{4X} - H\dot{\phi}XG_{5X} - \frac{1}{2}XG_{5\phi} > 0, \tag{31}$$

$$\mathcal{E} = G_4 + \frac{1}{2}XG_{5\phi} - XG_{5X}\ddot{\phi} > 0 \tag{32}$$

and for scalar perturbation, is same as equation (22)-(23).

4. The Stability Condition for Some Particular Gravity-Coupled Scalar Field Cosmological Model

Now we try to using this condition to get the stability conditions for an example model that can be obtained from choosing the K, G_3, G_4, G_5 functions. In this works we try to find the stability condition for two cosmological model, The Brans-Dicke theory and Gauss-Bonnet coupling.

4.1. The Brans-Dicke (BD) Theory

The Lagrangian of Brans-Dicke theory is given by,

$$K = -\frac{M_{pl}\omega_{BD}X}{2\phi} - V(\phi);\ G_3 = 0;\ G_4 = \frac{1}{2}M_{pl};\ G_5 = 0. \tag{33}$$

There are several model that equivalent to BD Theory such as $f(R)$ gravity that equivalent to BD theory with $\omega_{BD} = 0$ based on the choice of the coefficient function and dilaton gravity that correspond to $\omega = -1$. For this model, the conditions (31) and (32) are satisfied for

$$\phi > 0. \tag{34}$$

Other than that, the value of (22) for this model can be written as,

$$\dot{\phi}^2(9 + 6\omega_{BD}) > 0. \tag{35}$$

Because $\dot{\phi}^2$ always have positive quantity (real scalar field), so for BD theory, we can get the no-scalar-ghost condition,

$$\omega_{BD} > -3/2. \tag{36}$$

Therefore, we can say that some model which equivalent with BD theory mentioned before, satisfied the no-scalar-ghost condition.

4.2. Gauss-Bonnet Coupling

There are several works with Gauss-Bonnet term [6, 7] and Hikmawan, et. al. [20] found that this model is stable for tensor perturbation if the squared effective speed of sound is positive,

$$c^2_{sound} \equiv \frac{1 + \ddot{f}}{1 + H\dot{f}} \backsimeq \frac{\ddot{f}}{H\dot{f}} = \frac{f^{(1)}\ddot{\phi} + f^{(2)}\dot{\phi}^2}{H\dot{\phi}f^{(1)}} > 0;\ f^{(n)} \equiv \partial^n f(\phi)/\partial\phi^n, \tag{37}$$

where we get the second relation when inflation occurs, the Ricci(gravity) term is dominated by the Gauss-Bonnet term, so the Ricci (gravity) term can be neglected [6, 7]. The cosmological model with Gauss-Bonnet term can be obtained from Horndeski theory by the choice [21],

$$\mathcal{K} = -2f^{(4)}X^2[3 - \ln(-X/2)];\ G_3 = 2f^{(3)}X[7 - 3\ln(-X/2)],$$
$$G_4 = 2f^{(2)}X[2 - \ln(-X/2)];\ G_5 = 4f^{(1)}\ln(-X/2), \tag{38}$$

then using equation (31) and (32) we can get the exactly same stability condition for tensor perturbation,

$$c_t^2 \equiv \frac{\mathcal{E}}{L_\mathcal{S}} = \frac{\ddot{\phi}f^{(1)} - Xf^{(2)}}{H\dot{\phi}f^{(1)}} = \frac{\ddot{\phi}f^{(1)} + \dot{\phi}^2 f^{(2)}}{H\dot{\phi}f^{(1)}} > 0. \tag{39}$$

5. Conclusion

We have analyzed an obtained the no-ghost and Laplacian stability conditions of general gravitational theories and then implementing the general form to obtain the no-ghost and Laplacian stability conditions for Horndeski Theory. Using the conditions obtained, we try to find the stability conditions for some particular gravity-coupled scalar field cosmological model (BD theory and Gauss-Bonnet coupling) and found exactly the same conditions as the actual stability conditions from the models. As indicated above, one can find the new and more realistic cosmological model by adjusting the coefficient function $(K(\phi, X), G_i(\phi, X))$ in Horndeski Lagrangian. Therefore, there are big possibility to using the no-go and Laplacian stability conditions obtained above to become a constraint for extracting the new cosmological model, for inflation solution, or for the other big problems in cosmology. We leave this excitement to become motivation of future works.

Acknowledgments

This work was supported by "Riset Inovasi KK ITB 2016", "Riset Desentralisasi ITB 2016", and "Riset PMDSU 2016" from Ministry of Research, Technology, and Higher Education of the Republic of indonesia.

References

[1] Starobinsky A A 1980 *Phys. Lett.* B **91** 99
[2] Guth A H 1981 *Phys. Rev.* D **23** 347
[3] Alishahiha M, Silverstein E and Tong D 2004 *Phys. Rev.* D **70** 123505 (*preprint* hep-th/0404084)
[4] Arkani-Hamed N, Creminelli P, Mukohyama S and Zaldarriaga M 2004 *JCAP* **0404** 001 (*preprint* hep-th/0312100)
[5] Cheung C, Creminelli P, Fitzpatrick A L,Kaplan J and Senatore L 2008 *JHEP* **0803** 014 (*preprint* hep-ph/0709.0293)
[6] Kanti P, Gannouji R and Dadhich N 2015 *Phys. Rev.* D **92** 041302 (*preprint* hep-th/1503.01579)
[7] Kanti P, Gannouji R and Dadhich N 2015 *Phys. Rev.* D **92** 083524 (*preprint* hep-th/1506.04667)
[8] Komatsu E et al 2009 *Astrophys. J. Suppl.* **180** (*preprint* astro-ph/0803.0547)
[9] Smoot G F et al 1992 *Astrophys. J.* **396**
[10] Caldwell R R, Dave R and Steinhardt P J 1998 *Phys. Rev. Lett.* **80** 1582 (*preprint* astro-ph/9708069)
Sushkov S V 2009 *Phys. Rev.* D**80** 103505 (*preprint* gr-qc/0910.0980)
[11] Zen F P, Arianto, Gunara B E, Triyanta and Purwanto A 2009 *Eur. Phys. J.* C **63** 477 (*preprint* hep-th/0809.3847v3)
[12] Capozziello S, Lambiase G and Schmidt H J 2000 *Annalen. Phys.* **9** 39 (*preprint* gr-qc/9906051)
Tsujikawa S 2012 *Phys. Rev.* D **85** 083518 (*preprint* astro-ph/1201.5926v2)
[13] Horndeski G W 1974 *Int. J. Theor. Phys.* **10**363
[14] Armendariz-Picon C, Damour T and Mukhanov V F 1999 *Phys. Lett.* B **458** 219 (*preprint* hep-th/9904075)
[15] Brans C and Dicke R H 1961 *Phys. Rev.* **124** 925
[16] Gasperini M and Veneziano G *Phys. Rept.* **373** 1 (*preprint* hep-th/0207130)
[17] Sotiriou TP and Faraoni V 2010 *Rev. Mod. Phys.* **82** 451 (*preprint* gr-qc/0805.1726)

[18] Amendola L 1993 *Phys. Lett.* B **301** 175 (*preprint* gr-qc/9302010v1)
[19] Gleyzes J, Langlois D, Piazza F and Vernizzi F *JCAP* **1308** 025 (*preprint*hep-th/1304.4840)
[20] Hikmawan G, Soda J, Suroso A and Zen F P 2016 *Phys. Rev.* D **93** 068301 (*preprint* hep-th/arXiv:1512.00222)
[21] Kobayashi T, Yamaguchi M and Yokoyama J 'i *Prog. Theor. Phys.* **126** 511 (*preprint* hep-th/1105.5723)

IOP Conf. Series: Journal of Physics: Conf. Series **1127** (2019) 012008 doi:10.1088/1742-6596/1127/1/012008

Simulation of Neural Network-Multicriterion Optimization Image Reconstruction Technique (NN-MOIRT) for imaging using a 32-channel Brain ECVT sensor

N Handayani[1,2], K F Hidayanti[2], M R Baidillah[3], I Arif[1], S N Khotimah[1], F Haryanto[1], W P Taruno[3]
[1]Department of Physics, Institut Teknologi Bandung, Jl. Ganesa No. 10, Bandung, Indonesia
[2]Department of Physics, UIN Sunan Kalijaga, Jl. M. Adisucipto No. 1, Yogyakarta, Indonesia
[3]CTech Labs PT Edwar Technology, Jl. Jalur Sutera Blok 23 BC, Tangerang, Indonesia

E-mail: nitahandayani@yahoo.com

Abstract. NN-MOIRT has been proposed earlier as an alternative for existing image reconstruction algorithms that can accurately show a volumetric image in a cylindrical sensor vessel. The Brain ECVT sensor has a different shape and dimensions compared with a common ECVT sensor, which has a cylindrical shape. Using a different sensor changes the image reconstruction algorithm parameters. Thus, the image reconstruction algorithm should be modified to be able to properly make a reconstruction. In this study, a simulation of the reconstruction of an image from a 32-channel Brain ECVT sensor using NN-MOIRT was conducted. The simulation was performed by varying the position and number of objects in the helmet-shaped Brain ECVT sensor. The alpha parameter (penalty factor) was varied from 10 to 150 with the number of iterations from 1 to 200. The RMSE (root mean square error) was calculated based on the difference between the permittivity distribution of the objects and the reconstructed image. It was found that the NN-MOIRT algorithm is more convergent and more stable for image reconstruction than the ILBP algorithm.

1. Introduction

Electrical capacitance volume tomography (ECVT) was developed based on the electrical capacitance tomography (ECT) technique to obtain a volumetric image. The volumetric effect is not obtained by stacking 2-D images but the system was designed to produce a 3-D image directly and in real time [1]. Known as volume tomography, this technique produces simultaneous information from the permittivity distribution of the area enclosed by the volumetric sensor. The ECVT system consists of three main components, namely a capacitance sensor system, a data acquisition system and a computer for image reconstruction, interpretation and display of the image.

Successfull implementation of tomography depends on the selection of the appropriate sensor design in relation to the application and image reconstruction algorithm. Selection of the image reconstruction technique in accordance with the sensor model and its application is very important because it determines the quality of the resulting image. A reliable technique for the reconstruction of volumetric and dynamic objects is the neural network-multicriterion optimization image reconstruction technique (NN-MOIRT). NN-MOIRT has been used successfully for imaging two- and three-phase flow systems using ECT-shaped column with a 12-electrode sensor [2]. NN-MOIRT shows robustness to noise and its accuracy and consistency are higher compared with the iterative linear back projection (ILBP) technique and simultaneous image reconstruction technique (SIRT). For example, NN-MOIRT can distinguish small bubbles or particles in oil.

Furthermore, NN-MOIRT can be applied to linear process tomography (ultrasonic computed tomography) and non-linear process tomography (ECT) as reported in [3]; the results of the simulation and experiment showed improvement in accuracy, consistency and robustness to noise compared with the LBP and ILBP techniques. Another advantage of this technique is that it reduces the computational time and parallelism inherent in the computation. In 2007, NN-MOIRT was used experimentally to reconstruct the image of a moving object for the first time. Hereafter, volumetric imaging in real time is called ECVT [1].

In this study, NN-MOIRT was applied to reconstruct the image of a helmet-shaped Brain ECVT sensor with 32 electrodes. Brain ECVT is one of the design developments of the ECVT sensor. It is used for imaging the electrical activity of the human brain based on the permittivity distribution in the cortical area. The results of image reconstruction using NN-MOIRT were compared with those from the ILBP technique.

2. Methods

An important point in ECVT is the process of image reconstruction from the measured capacitance values between electrode pairs that is based on Poisson's equation [4, 5]:

$$\varepsilon(x,y)\,\nabla^2\phi(x,y) + \nabla\varepsilon(x,y)\nabla\phi(x,y) = 0 \tag{1}$$

where $\varepsilon(x,y)$ is the dielectric constant distribution and $\phi(x,y)$ is the potential distribution. The measured capacitance between pairs of electrodes can be calculated by the equation:

$$C_i = -\frac{1}{\Delta V_i}\oint_{\Gamma} \varepsilon(x,y)\nabla\phi(x,y)\cdot\hat{n}\,dl \tag{2}$$

where ΔV_i is the voltage difference between the source and the detector electrodes of the i-th pair of electrodes. ECVT techniques generally consist of a forward problem and an inverse problem.

2.1 *Forward Problem*

The forward problem includes data collection of the capacitance from the electrodes that are placed around the outside of the vessel [1]. In this study, the forward problem was solved by a linearization technique using sensitivity models [4, 6]. This technique is based on the superposition theorem of electricity networks. The sensitivity model can be written as [7]:

$$S_{ij} \cong V_{0j}\frac{\vec{\boldsymbol{E}}_{si}(x,y,z)\cdot\vec{\boldsymbol{E}}_{di}(x,y,z)}{V_{si}V_{di}} \tag{3}$$

where V_{0j} is the volume of the j-th voxel. $\boldsymbol{E}_{si}$ $(= -\nabla\phi)$ is the vector of the electric field distribution when the source electrode pair i-th is activated by voltage V_{si} and $\boldsymbol{E}_{di}$ is the vector of the electric field distribution when the detector electrode pair is activated by voltage V_{di}. Equation (3) is in linear form and can be written in matrix form as follows:

$$\boldsymbol{C} = \boldsymbol{SG} \tag{4}$$

where $\boldsymbol{C}$ is an M-dimension capacitance data vector, $\boldsymbol{S}$ is a sensitiviy matrix, $\boldsymbol{G}$ is an N-dimension image vector, N is the number of voxels in the domain (32768 for $32 \times 32 \times 32$) and M is the number of electrode pairs ($[n \times (n-1)]/2$ with n is the number of electrodes).

2.2 *Inverse Problem*

The process of image reconstruction is the inversion that involves estimating the distribution of the permittivity of the measured capacitance data. The simplest way to estimate the image vector is by using a back projection technique [4,6]:

$$\boldsymbol{G} = \boldsymbol{S}^T\,\boldsymbol{C} \tag{5}$$

where $\boldsymbol{S}^T$ is the matrix transpose of $\boldsymbol{S}$.

The process of iterative image reconstruction involves a method for estimating image vector $\boldsymbol{G}$ from capacitance data vector $\boldsymbol{C}$ and minimizing the error between the measured capacitance and the estimated value, such that:

$$\boldsymbol{SG} \leq \boldsymbol{C} \tag{6}$$

An iterative method that is widely used to solve problems in 2-D ECT is the Landweber technique, called the linear iterative back projection (ILBP) technique. ILBP is commonly used in optimization theory [8]. This technique aims to determine the image vector by minimizing the least square error function:

$$f(\boldsymbol{G}) = \frac{1}{2} \|\boldsymbol{SG} - \boldsymbol{C}\|^2 = \frac{1}{2} (\boldsymbol{SG} - \boldsymbol{C})^T (\boldsymbol{SG} - \boldsymbol{C}) \tag{7}$$

The iteration procedure, which is based on a steepest gradient descent technique, can be written as:

$$\boldsymbol{G}^{k+1} = \boldsymbol{G}^k - \alpha^k \nabla f(\boldsymbol{G}^k) = \boldsymbol{G}^k - \alpha^k \boldsymbol{S}^T(\boldsymbol{SG}^k - \boldsymbol{C}) \tag{8}$$

where α^k is the penalty factor of the k-th iteration, which is usually a constant. A problem of the use of the Landweber technique is that the reconstructed image depends on the number of iterations and convergence is not always guaranteed. To produce an accurate 3-D reconstructed image it takes more than one objective function. Therefore, a multi-criteria optimization technique has been developed using more than one objective function, called NN-MOIRT.

2.3 *Multicriterion Optimization Image Reconstruction Technique (MOIRT)*

This image reconstruction technique based on multi-criterion optimization was developed by Warsito and Fan to solve the inverse problem of 2-D ECT into 3-D ECT [2, 3]. Optimization of the image vector by simultaneously minimizing four objective functions includes [1]:

a. The negative entropy function

$$f_1(\boldsymbol{G}) = \gamma_1 \delta_1 \boldsymbol{G} \ln \boldsymbol{G} \;, \delta_1 = \begin{cases} 1, & if\ G_j > 0 \\ 0, & if\ G_j = 0 \end{cases} \tag{9}$$

b. The least square error function

$$f_2(\boldsymbol{G}) = \frac{1}{2} \gamma_2 \|\boldsymbol{SG} - \boldsymbol{C}\|^2 \tag{10}$$

c. The smoothness and small peakedness function

$$f_3(\boldsymbol{G}) = \frac{1}{2} \gamma_3 (\boldsymbol{G}^T \boldsymbol{XG} + \boldsymbol{G}^T \boldsymbol{G}) \tag{11}$$

where $\boldsymbol{X}$ is an N by N nonuniformity matrix.

d. The 3-D to 2-D matching function

$$f_4(\boldsymbol{G}) = \frac{1}{2} \gamma_4 \|\boldsymbol{H}_{2D} \boldsymbol{G} - \boldsymbol{G}_{2D}\|^2 \tag{12}$$

H^{2D} is a projection matrix from 3-D onto 2-D with an $N \times N_{2D}$ dimension. N_{2D} is the number of voxels in a layer of the volumetric image vector $\boldsymbol{G}$. $\gamma_1, \gamma_2, \gamma_3$ and γ_4 are normalized constants between 0 and 1. Multi-criteria optimization for image reconstruction problems is to choose an image vector that has a multi-objective function value (global objective function) $\boldsymbol{F}(\boldsymbol{G}) = [f_1(\boldsymbol{G}), f_2(\boldsymbol{G}), f_3(\boldsymbol{G}), f_4(\boldsymbol{G})]^T$ are minimized simultaneously.

2.4 *Hopfield Neural Network*

Hopfield neural network models (Hopfield nets) have been successfully used to solve a number of difficult optimization problems, including image reconstruction for tomography [9, 10]. To solve the problem of image reconstruction, the value of an image voxel G_j to be reconstructed is mapped to the variable output neurons v_j in Hopfield nets.

Hopfield neural network behavior is characterized by the time evolution of the state of the neuron u_j according to the following differential equation [11]:

$$C_{0j} \frac{du_j}{dt} = -\frac{\partial E(\boldsymbol{G})}{\partial G_j} \tag{13}$$

where C_{0j} is the capacitance of j-th neuron, and $\boldsymbol{E}(\boldsymbol{G})$ is the total energy of the Hopfield nets. The evaluation time constant is defined as:

$$\tau = R_{0j} C_{0j} \tag{14}$$

where R_{0j} is the resistance of j-th neuron. The energy functions for the network optimization problem is stated as [1]:

$$E(\boldsymbol{G}) = \gamma_1\delta_1\, \boldsymbol{G}\ln \boldsymbol{G} + \frac{1}{2}\gamma_2\|\boldsymbol{z}_1\|^2 + \frac{1}{2}\gamma_3(\boldsymbol{G}^T\boldsymbol{X}\boldsymbol{G} + \boldsymbol{G}^T\boldsymbol{G}) + \frac{1}{2}\gamma_4\|\boldsymbol{z}_2\|^2 + \Psi(\alpha_1\boldsymbol{z}_1) + \Psi(\alpha_2\boldsymbol{z}_2) + \sum_{j=1}^{N}\frac{1}{R_j}\int_0^{G_j} f_{\Sigma}^{-1}(\boldsymbol{G})d\boldsymbol{G} \quad (15)$$

The equation for selection of the penalty parameter α_k is as follows:

$$\alpha_k(t) = \alpha_{0,k} + \varsigma_k exp(-\eta_k t) \quad (16)$$

α_{0k}, ς_k *and* η_k are positive constants. The penalty parameter provides a mechanism for removing local minima by varying the direction of motion of the neurons in such a way that the ascent step is taken largely by the penalty function in the first step.

2.5 Simulation of Brain ECVT sensor

A simulation of the reconstruction of an image from a Brain ECVT sensor using NN-MOIRT was conducted by varying the position of the objects and the number of objects as shown in Figure 1. The value of relative permittivity (ε_r) of each given object was 604.55, which corresponds to the permittivity of the gray matter in the brain at a frequency of 2.5 MHz.

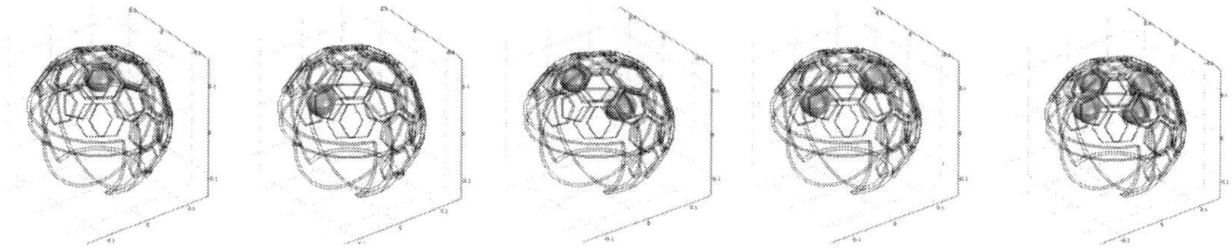

Figure 1. Variations in the number of objects in the Brain ECVT sensor.

The stages of the process of image reconstruction using the neural-network approach based on multi-criteria optimization include [2]:

a. Initialization step

Set the initial state of the neuron $u_j(0) = 0$; $v_j(0) = f\Sigma(u_j(0))$. Set steepness gain factor $\beta = 2$. Initialization of penalty parameter α and gain parameter ζ, which affect the performance of convergence. Use $\alpha_0 = \zeta$ for all reconstructions. Adding Gaussian noise in the simulation to test the robustness of the NN-MOIRT reconstruction technique: $C_i = [1 + N(\mu, \sigma^2)]C$
where $N(\mu, \sigma^2)$ is a Gaussian distribution function with mean μ and variance σ^2. Set $\mu = 0$ and $\sigma =$ 0,05. The initial weights are selected to be $\omega_1^{(0)} = \omega_2^{(0)} = \omega_3^{(0)} = \omega_4^{(0)} = 1/4$.

b. Updating step

The weights $\omega_1, \omega_2, \omega_3$ and ω_4 for every iteration are updated by the following equation:

$$\omega_i^{(t+\Delta t)} = \frac{\Delta\omega_1^{(t)}/\Delta\omega_i^{(t)}}{\sum_{i=1}^{4}\Delta\omega_1^{(t)}/\Delta\omega_i^{(t)}} \qquad \Delta\omega_i^{(t)} = f_i(\mathbf{G}(t+\Delta t)) - f_i(\mathbf{G}(t)) \quad (17)$$

with $i = 1,2,3,4$.

c. Stopping step

The rule for stopping is determined by $|\boldsymbol{G}_j(t+\Delta t) - \boldsymbol{G}_j(t)|^2 \ll 1$ for all voxels (neurons).

In this study, reconstruction of an image with NN-MOIRT was done by iterating 200 times and variation of alpha from 10 to 150. The number of iterations and penalty factor α were varied to get a stable value in the process of image reconstruction. The squared error value was calculated by the following equation [8]:

$$RMSE = \frac{\sum|G^{rec} - G^{ref}|^2}{\sum|G^{ref}|^2} \quad (18)$$

$\boldsymbol{G}^{rec}$ is the reconstructed image vector and $\boldsymbol{G}^{ref}$ is the reference image vector.

3. Results

The distribution of normalized sensitivity for all 496 electrode pairs of the Brain ECVT sensor is shown in Figure 2. The sensitivity is influenced by the electric field intensity, which is spread in the domain of the ECVT sensor. The area that has many slope variations along the axial direction is an area that has a high sensitivity.

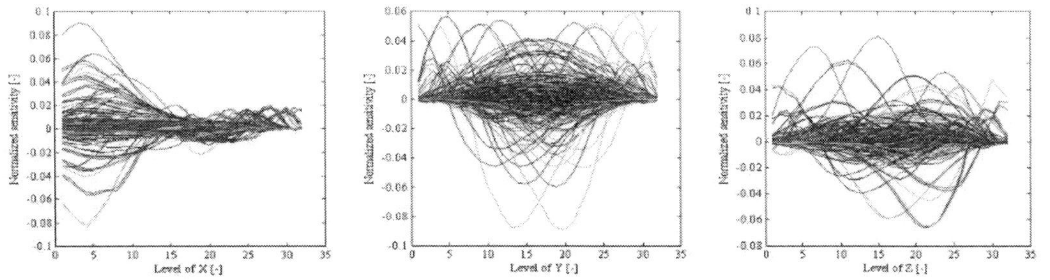

Figure 2. Distribution of normalized sensitivity for all electrode pairs of Brain ECVT sensor.

Results of image reconstruction with NN-MOIRT by varying the number of objects and object positions are shown in Figure 3.

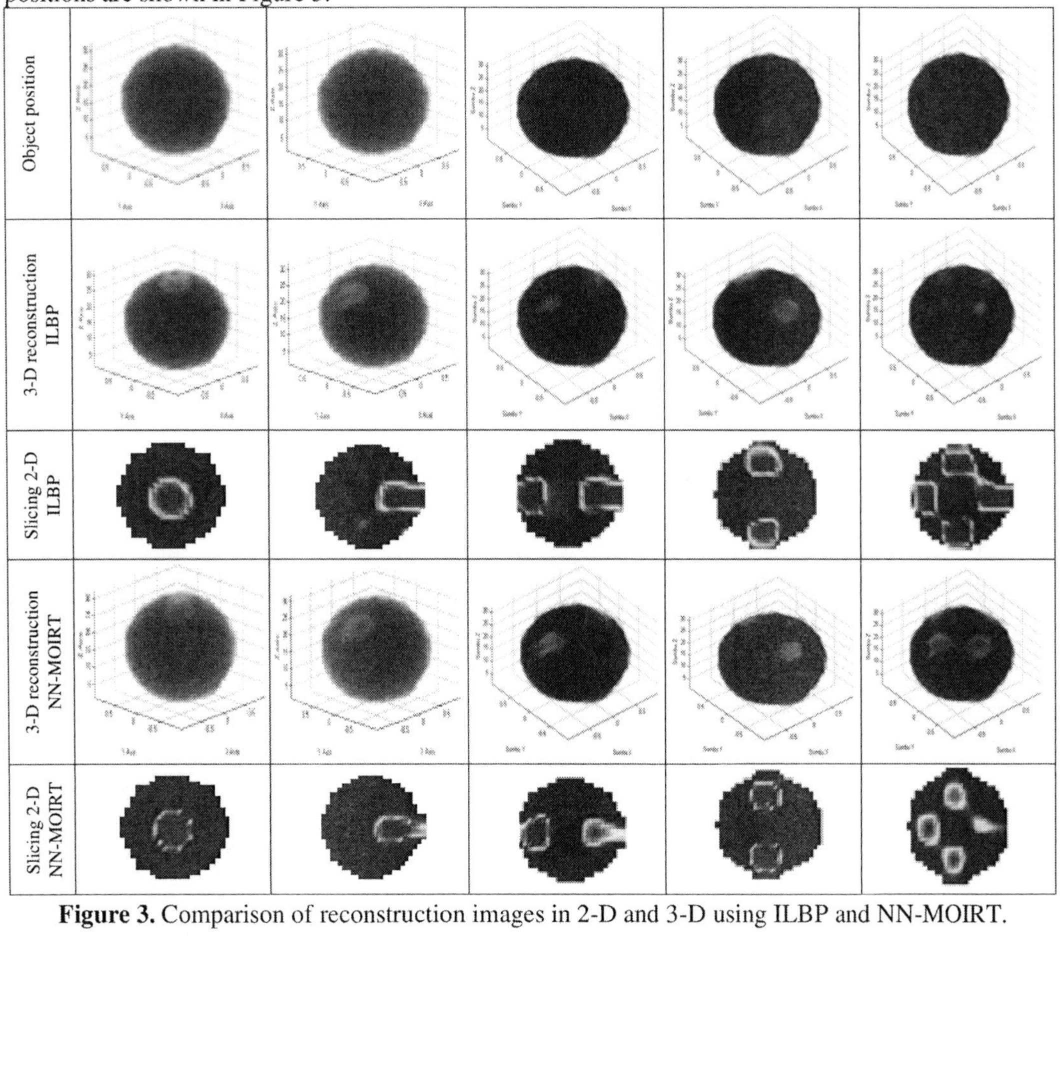

Figure 3. Comparison of reconstruction images in 2-D and 3-D using ILBP and NN-MOIRT.

4. Discussion

The sensitivity matrix representing the sensitivity of the change in capacitance is measured on a pair of electrodes to change the value of the permittivity of the area covered by the sensor [12]. The sensitivity distribution is strongly influenced by the design of the sensors, such as sensor geometry, shape and size of the electrodes and the permittivity of the object in the sensing domain. In addition, the sensitivity distribution map is also affected by the number of voxels and sensor electrodes. The convergence of the iterative reconstruction technique is determined by the smoothness and the variation of the slope of the sensitivity distribution curve along the axial direction. The smoother the curve slope and the more variation in it, the more easily the reconstruction process will achieve convergence [1].

Based on the simulation results (Figure 3), it appears that image reconstruction using NN-MOIRT is more accurate than using ILBP technique. The object's position in the sensing domain greatly affects the quality of the resulting image, because the sensitivity in the sensing domain varies according to Figure 2. For areas with high sensitivity the image produced tends to clear, while if the object is in a dead zone the image becomes blurred. If the object is located near the sensor, an elongation effect appears. This effect is more visible if the object is located within the sensing area with higher sensitivity rather than in the surrounding area. The parameters that affect the quality of the reconstructed image are the selection of the penalty factor and the number of iterations. In reconstruction with NN-MOIRT the smallest error was 0.4684 occurs during alpha 130 and number of iteration 116. NN-MOIRT is capable of maintaining a stable condition when the number of iterations and alpha parameter are increased.

5. Conclusions

The volumetric imaging technique NN-MOIRT was successfully applied to imaging of an object using a helmet-shaped Brain ECVT sensor with 32 electrodes. The use of multicriterion optimization produces an image that is more accurate than with other iterative reconstruction techniques, such as ILBP. NN-MOIRT is feasible for real-time volumetric imaging.

References

[1] Warsito W, Marashdeh Q and Fan L S 2007 *IEEE Sensor Journal* **7(4)** pp 525-35
[2] Warsito W and Fan L S 2001 *Meas. Sci. Technol.* **12** 2198-210
[3] Warsito W and Fan L S 2003 *Chem. Eng. Proc.* **48** 663-74
[4] Xie C G, Huang S M, Hoyle B S, Thorn R, Lean C, Snowden D and Beck M S 1992 *IEE Proc. G* **139** 89-98
[5] Xie C G, Reinecke N, Beck M S, Mewes D, Williams R A 1995 *Chem. Eng. J.* **56** 127-33
[6] Huang S M, Plaskowski A, Xie C G and Beck M S 1989 *J. Phys. E: Sci. Instrum.* **22** 173-7
[7] Lionheart W R B 2001 *Proc. 2nd World Congr. Industrial Process Tomography* 4-11
[8] Yang W Q and Peng L 2003 *Meas. Sci. Technol.* **14** R1-13
[9] Wang Y 1997 *Opt. Commun.* **143** 279-86
[10] Wang Y 1998 *Int. J. Imaging Sys. Technol.* **9** 381-7
[11] Hopfield J and Tank D 1985 *Biol. Cybern.* **52** 141-52
[12] Marashdeh Q and Teixeira F L 2004 *IEEE Trans. Magn.* **40** 1204-7

Acknowledgments

The authors would like to thank Almushfi Saputra, researcher at CTech Labs Edwar Technology Tangerang, for his assistance in the simulation work. This work was partially supported by PUPT Ristekdikti 2016 with contract number 584o/I1.C01/PL/2016.

Investigating Starobinsky's Inflation Model Using Arnowitt-Deser-Misner (ADM) Formalism for Scalar Perturbation in The Cosmic Inflation Epoch

J Pebralia[1,a), A Suroso[2,b) and F P Zen[2,c)

[1]Theoretical Physics Laboratory, THEPI Division, Faculty of Mathematics and Natural Sciences, Institut Teknologi Bandung, Jl. Ganesha 10 Bandung 40132, Indonesia.
[2]Indonesia Center for Theoretical and Mathematical Physics(ICTMP), Faculty of Mathematics and Natural Sciences, Institut Teknologi Bandung, Jl. Ganesha 10 Bandung 40132, Indonesia.

a) jesipebralia@students.itb.ac.id
b) agussuroso@fi.itb.ac.id
c) fpzen@fi.itb.ac.id

Abstract. In this paper, we investigate Starobinky's inflation model for scalar perturbation during cosmic inflation epoch by using Arnowitt-Deser-Misner (ADM) formalism. We implemented this formalism to the action of curvature-scalar of Starobinsky's model, which is one of the models of cosmic inflation. As the result of this investigation, we found the power spectrum of scalar perturbation in the form of Starobinsky's inflation model which has a linear correlation to the comoving Hubble radius which is shrinking during inflation epoch.
Keyword. Inflation, Starobinsky's inflation model, Scalar Perturbation, ADM Formalism, Cosmology

1. Introduction

The correction of the photons's temperature in the cosmic microwave background (CMB) and the formation of large-scale structures observed today is believed come from a quantum fluctuation that growth very rapidly during inflation epoch. These perturbations include perturbation in metric and also perturbation in the matter. Perturbation of the metric can be a scalar perturbation, a vector perturbation or a tensor perturbation. The equations of motion of the perturbation can be derived by Einstein's equation or through action approach.

The application of Arnowitt-Deser-Misner (ADM) metric has been widely used to describe the dynamics of quantum fluctuation. ADM formalism is a way to describe space-time into the slices of hypersurface for a constant time. Inflation requires a negative pressure to produce the accelerated expansion. Therefore, constituent that generate inflation is not the type of ordinary matter that exists today [1-4]. There are a lot of models of inflation which has been filed with the different theoretical motivation and different predictions of observations.
One of the best way to describe the interaction of a gravitational field that is using the principle of action. The action of each perturbation can be expanded to any order. Since the perturbation is a quantum fluctuation, a quantization of classical observable to the quantum observable is needed. The equation of motion which was expanded up to second order has a form that similar to the form of a harmonic oscillator's equation of motion. Therefore, the quantization process of scalar perturbation can be done in the same way with harmonic oscillator by using the creation and annihilation operator [1, 5].

Statistical properties of scalar perturbation can be determined by deriving the two-point correlation function or the power spectrum where it is the Fourier transform of the correlation function of the perturbation. Two-point correlation function can be derived by calculating the probability amplitude of each perturbation [1, 2, 6, 7].

In this work, we modify the parts of scalar curvature by implementing the Starobinsky's inflation model [8, 9]. We interested to investigate the cosmological perturbation in term of the scalar perturbation by using modify scalar curvature action for Starobinsky's model and ADM formalism during cosmic inflation epoch.

2. Arnowitt-Deser-Misner (ADM) Formalism

The Arnowitt-Deser-Misner (ADM) formalism describes space-time by dividing the space-time into the slices of space with a constant time.

$$ds^2 = -\left(N^2 - N^i N_i\right)dt^2 + 2N_i dx^i dt + h_{ij}dx^i dx^j. \tag{1}$$

The ADM formalism was implemented by replacing the variables in the action with ADM variables. The most common action to describe the dynamics of space-time is the Einstein-Hilbert action as follow

$$S = \frac{1}{2}\int d^4x\sqrt{-g}R. \tag{2}$$

After some integral by part was done to the equation (2), the Einstein-Hilbert action in term of ADM formalism would be obtained,

$$S = \frac{1}{2}\int d^4x\sqrt{h}N\left[R^{(3)} - K_i^i K_j^j + K_{ij}K^{ij}\right], \tag{3}$$

where $K_{ij} = \frac{1}{2N}\left(\nabla_i^{(3)}N_j + \nabla_j^{(3)}N_i - \dot{h}_{ij}\right)$ is the extrinsic curvature.

3. ADM Formalism for Scalar Field Inflation

The action of scalar field form,

$$S = \frac{1}{2}\int d^4x\sqrt{-g}[R - (\nabla\Phi)^2 - 2V(\Phi)]. \tag{4}$$

IOP Conf. Series: Journal of Physics: Conf. Series **1127** (2019) 012009 doi:10.1088/1742-6596/1127/1/012009

The action of scalar field in the ADM formalism is:

$$S = \frac{1}{2}\int d^4x \sqrt{h}N\left[R^{(3)} + \frac{1}{N^2}\left(E_{ij}E^{ij} - E^2\right) + \frac{1}{N^2}\left(\dot{\Phi} - N^i\partial_i\Phi\right)^2 - h^{ij}\partial_i\Phi\partial_j\Phi - 2V(\Phi)\right], \quad (5)$$

where $E_{ij} \equiv -NK_{ij}$. A hamiltonian density of a system defined by

$$\mathcal{H} = \pi^{ij}\dot{h}_{ij} - \mathcal{L}, \quad (6)$$

where π^{ij} is conjugate momentum. Therefore the Hamiltonian density of the action of scalar field is

$$\begin{aligned}\mathcal{H} = {} & 2\pi^{ij}E_{ij} - 2N_j\nabla_i\left(h^{-\frac{1}{2}}\pi^{ij}\right) \\ & - \sqrt{h}N\left[R^{(3)} + \frac{1}{N^2}\left(E_{ij}E^{ij} - E^2\right) + \frac{1}{N^2}\left(\dot{\Phi} - N^i\partial_i\Phi\right)^2 - h^{ij}\partial_i\Phi\partial_j\Phi \right. \\ & \left. - 2V(\Phi)\right]. \end{aligned} \quad (7)$$

Then we have have two constraint equation: the Hamiltonian constraint (which is given by derivation of Hamiltonian with respect to the lapse function),

$$R^{(3)} - \frac{1}{N^2}\left(E_{ij}E^{ij} - E^2\right) - \frac{1}{N^2}\left(\dot{\Phi}\right)^2 - 2V(\Phi) = 0. \quad (8)$$

and the Gauss constraint (which is given by derivation of Hamiltonian with respect to the shift vector),

$$\nabla_i\left[\frac{1}{N}\left(h_{il}E^{ij} - \delta_l^j E\right)\right] = 0. \quad (9)$$

To obtain solution of constrains equation, we use perturbation of lapse function and shift vector and also we use the Maldacena gauge, respectively

$$N \equiv 1 + \alpha_1 + \cdots, \text{and } N_i \equiv \partial_i\psi + \tilde{N}_i^{(1)} + \cdots \quad (10)$$

$$h_{ij} = a^2\left(e^{2\varsigma}\delta_{ij}\right) \quad (11)$$

The solution of Hamiltonian constraint at zeroth order is

$$H^2 = \frac{1}{6}\dot{\phi}^2 + \frac{1}{3}V(\phi), \quad (12)$$

which is the Friedmann equation. For Gauss constraint, the zeroth order solution is zero. Meanwhile, the lapse function and shift vector at the first order respectively are

$$\alpha_1 = \frac{\dot{\varsigma}}{H}, \text{ and } \psi^{(1)} = \frac{1}{2}a^2\frac{\dot{\Phi}^2}{H^2}\partial^{-2}\dot{\varsigma} - \frac{1}{H}\varsigma. \quad (13)$$

Then, the second order of scalar-tensor action in the ADM formalism is

$$S = \frac{1}{2}\int d^4x\, a^3\frac{\dot{\Phi}^2}{H^2}\left[\dot{\varsigma}^2 - a^{-2}(\partial\varsigma)^2\right]. \quad (14)$$

Power spectrum of scalar perturbation is

$$P_\varsigma(k) = \frac{H^2}{\dot{\Phi}^2}\frac{H^2}{2k^3}. \quad (15)$$

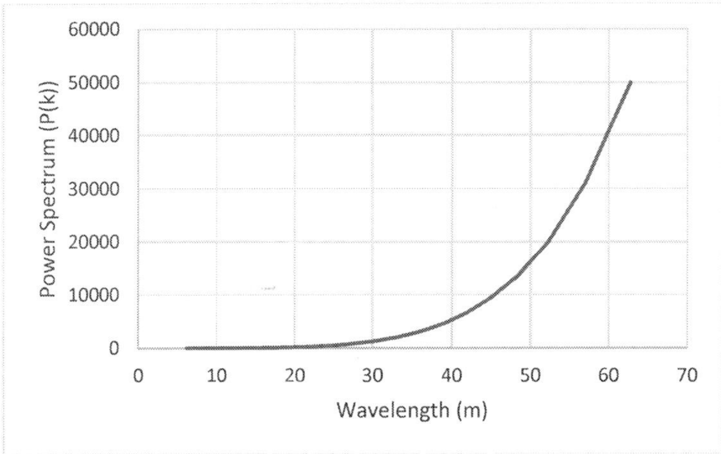

Figure 1. Plot Power Spectrum of Scalar Perturbation of Scalar Field Inflation Model to the Wavelength Perturbation

The power spectrum of scalar perturbation of scalar field inflation's model grows exponentially with respect to the wavelength of perturbation.

4. ADM Formalism for Starobinsky's Inflation

One of the simplest theory of modified gravity is the $f(R)$ theory. Starobinsky's gravity model is a modified model of gravity by adding a factor of R^2 in the action of the general scalar curvature action. The addition of R^2 factor is to provide a higher correction in the Einstein-Hilbert action [8-11]. The action of scalar curvature in the Starobinsky's model is

$$S = \frac{1}{2}\int d^4x\sqrt{-g}\left[R + \frac{R^2}{6M^2}\right]. \tag{16}$$

The action of Starobinsky's model in the ADM formalism is:

$$S = \frac{1}{2}\int d^4x\sqrt{h}N\left[R^{(3)} + \frac{{R^{(3)}}^2}{6M^2} + \frac{1}{N^2}\left(E_{ij}E^{ij} - E^2\right) + \frac{1}{3M^2N^2}R^{(3)}\left(E_{ij}E^{ij} - E^2\right) + \frac{1}{6M^2N^4}\left(E_{ij}E^{ij} - E^2\right)^2\right]. \tag{17}$$

Using the same procedures by scalar field action, we get the Hamiltonian density of the action of Starobinsky's model is

$$\mathcal{H} = 2\pi^{ij}E_{ij} - 2N_j\nabla_i\left(h^{-\frac{1}{2}}\pi^{ij}\right) - \frac{1}{2}\sqrt{h}N\left[R^{(3)} + \frac{{R^{(3)}}^2}{6M^2} + \frac{1}{N^2}\left(E_{ij}E^{ij} - E^2\right) + \frac{1}{3M^2N^2}R^{(3)}\left(E_{ij}E^{ij} - E^2\right) + \frac{1}{6M^2N^4}\left(E_{ij}E^{ij} - E^2\right)^2\right]. \tag{18}$$

Then the Hamiltonian constraint is

$$R^{(3)} + \frac{{R^{(3)}}^2}{6M^2} - \frac{1}{N^2}\left(E_{ij}E^{ij} - E^2\right) - \frac{1}{3M^2N^2}R^{(3)}\left(E_{ij}E^{ij} - E^2\right) - \frac{1}{2M^2N^4}\left(E_{ij}E^{ij} - E^2\right)^2 = 0, \tag{19}$$

and the Gauss constraint is

$$\nabla_i \left\{ \frac{1}{N}\left(h_{il}E^{ij} - \delta_l^j E\right) + \frac{1}{3M^2N} R^{(3)}\left(h_{il}E^{ij} - \delta_l^j E\right) + \frac{1}{3M^2N^3}\left(E_{ij}E^{ij} - E^2\right)\left(h_{il}E^{ij} - \delta_l^j E\right) \right\} = 0. \quad (20)$$

The solution of Hamiltonian constraint at zeroth order is

$$H^2 = \frac{M^2}{3}. \quad (21)$$

The solution in the equation (21) describe that the magnitude of H is constant. Based on the magnitude of H, it is indicated that $\frac{dH}{dt} = 0$. This condition agree with the inflation condition that $\frac{dH}{dt} = 0$. For Gauss constraint, the zeroth order solution is zero. Based on the solution of Gauss constraint and Hamiltonian constraint at zeroth order, we generated the first order perturbation of lapse function and shift vector, respectively

$$\alpha_1 = \frac{\dot{\varsigma}}{H} + \frac{8}{9H^2a^2}\partial_i\partial_i\varsigma, \quad \text{and} \quad \psi^{(1)} = -\frac{1}{H}\varsigma. \quad (22)$$

Finally, the action of scalar-curvature of order, is Starobinsky's model in the ADM formalism is expanded to second is

$$S = \frac{1}{2}\int d^4x \left[-\frac{4a^2}{3}(\partial_i\varsigma)^2 + 18H^2a^4\varsigma^2 \right]. \quad (23)$$

The equation of motion of scalar perturbation was derived directly by using Euler-Lagrange equation,

$$\frac{27}{2}H^2a^2\varsigma + \partial_i\partial_i\varsigma = 0. \quad (24)$$

One can also get the power spectrum of scalar perturbation,

$$P_\varsigma(k) = \sqrt{\frac{1}{54}\frac{\hbar}{aH}\frac{(2\pi)^3}{k}}. \quad (25)$$

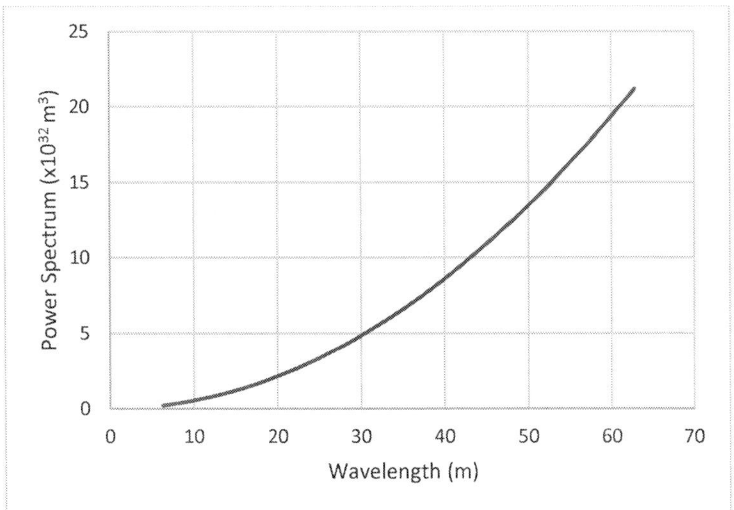

Figure 2. Plot Power Spectrum of Scalar Perturbation of Starobinsky's Inflation Model to The Wavelength Perturbation.

The power spectrum of scalar perturbation of Starobinsky inflation's model grow exponentially respect to the wavelength of perturbation.

5. Conclusion

Using the same procedures of scalar field inflation, ADM formalism is implement to the action of Starobinsky's model expanded to the second order action. The equation of motion of inflationary Starobinsky's model differs from equation of motion of scalar field because there were different approach for each model. Scalar field inflation model implemented modified matter approach, while for the Starobinsky's inflation model implemented modified gravity approach. In the Starobinsky's inflation model, the equation of motion of scalar perturbation had the same form of harmonic oscillator equation of motion. Meanwhile, in the scalar field inflation, the equation of motion of scalar perturbation had the same form of spherical Bessel function at first order.

Furthermore, the power spectrum of scalar perturbation of inflationary Starobinsky's model is analyzed. It had linier correlation to the commoving Hubble radius $\left(\frac{1}{aH}\right)$ with $\sqrt{\frac{1}{54}}$ as an additional factor. During cosmic inflation epoch, comoving Hubble radius shrink until at the end of inflation. After the inflationary period is over, comoving Hubble radius grow up along with the growth of universe. Based on the power spectrum obtained from the Starobinsky's model in the equation (25), at the cosmic inflation period, the rate of shrinking comoving Hubble radius is more rapidly than shrinking amplitude of scalar perturbation. Consequently, at the cosmic inflation epoch, there is possibility that amplitude of scalar perturbation on outside of comoving Hubble radius.

References

[1] Dodelson, S. (2003). Modern cosmology. Academic press.
[2] Baumann, D. (2009). Tasi lectures on inflation. arXiv preprint arXiv:0907.5424.
[3] Zen, F.P., Gunara, B.E. and Purwanto, A., (2009). Cosmological evolution of interacting dark energy in Lorentz violation. *The European Physical Journal C, 63*(3), pp.477-490.
[4] Hikmawan, G., Soda, J., Suroso, A. and Zen, F.P., (2016). Comment on "Gauss-Bonnet inflation". *Physical Review D, 93*(6), p.068301.
[5] Jacobson, T. (2005). Introduction to quantum fields in curved spacetime a hawking effect. In Lectures on Quantum Gravity, pages 39–89. Springer.
[6] Schneider, P. (2014). Extragalactic astronomy and cosmology: an introduction. Springer.
[7] Amendola, L. dan Tsujikawa, S. (2010). Dark energy: theory and observations. Cambridge University Press.
[8] Starobinsky, A. A. (1980). A new type of isotropic cosmological models without singularity. Physics Letters B, 91(1):99–102.
[9] Kehagias, A., Dizgah, A. M., dan Riotto, A. (2014). Remarks on the starobinsky model of inflation and its descendants. Physical Review D, 89(4):043527.
[10] Starobinskii, A. (1983). The perturbation spectrum evolving from a nonsingular initially de-sitter cosmology and the microwave background anisotropy. Soviet Astronomy Letters, 9:302–304.
[11] Starobinsky, A. A. (2007). Disappearing cosmological constant in f (r) gravity. JETP Letters, 86(3):157–163.

Analytical Solution of Dirac Equation for Hyperbolic Manning-Rosen-like Potential Using Hypergeometric Method for Exact Spin Symmetry

Oky Rio Pamungkas[1], Miftahul Ma'arif[1], Suparmi[2], Cari[2]

[1]Physics Departement, Undergraduate Study Program, Faculty of Mathematics and Natural Sciences, Sebelas Maret University, Surakarta, Indonesia
[2]Physics Departement, Post Graduate Study Program, Faculty of Mathematics and Natural Sciences, Sebelas Maret University, Surakarta, Indonesia
E-mail : okyriopamungkas@student.uns.ac.id

Abstract. Analytical solution of Dirac equation for hyperbolic Manning-Rosen-like potential is obtained using hypergeometric method. The analytical solution is aimed to determine energy spectrum and radial wave function for this potential with exact spin symmetry. Behavior of atomic particles can be clearly understood if the energy spectrum and wave function of particle are known. Dirac equation for exact spin symmetry is reduced into a second order differential equation like Schrödinger equation. Hypergeometric method for Dirac equation with hyperbolic Manning-Rosen-like potential is the same as for Dirac equation with general hyperbolic Manning-Rosen potential. Energy spectrum is exactly obtained in the closed form and the radial wave functions are expressed in the form of hypergeometric method.
Keywords: Dirac equation, hyperbolic Manning-Rosen-like potential, Hypergeometric Method.

1. Introduction

In quantum mechanics, a particle that moving around in a strong potential field must pay attention to relativistic effect, one example is the particles in an accelerator[1]. When a particle notice relativistic effects, the behavior of a particle can be described using Dirac or Klein-Gordon equation[2,3].
A charge of particle will be influenced by some potentials energy. Potentials that affect are Coulomb, Manning-Rosen[4,5], Morse, Symmetrical Top, Rosen-Morse[6], Pöschl-Teller, etc. In this paper one kind of potentials will be analyzed, that is hyperbolic Manning-Rosen-like potential which suffered by an electron in an atom. In addition to motion around the nucleus, electrons also do rotate, has a multi-electron structure of the atom, molecules interactions, core deflected, and correlation of quantum states that are affected by such potential like Manning-Rosen potential.
Dirac equation is a difficult equation to be solved in exponential form, need specific functions to simplify the solution of Dirac equation[7]. One method that can be used is the hypergeometric method. Dirac equation is reduced to Schrödinger equation that for nonrelativistic problems like Dirac equation. With separation of radial variable method, from the reduced Dirac equation then obtained a second-order differential equation hypergeometric type. If hypergeometric equation has been obtained, spectrum energy levels and wave function can be acquired easily[8].
Dirac equation with vector potential $V(\bar{r})$ and scalar potential $S(\bar{r})$ shown in equation (1)

$$[-\hbar^2\nabla^2 + 2V(r)(E+M)]S(r) = (E+M)(E-M)S(r) \tag{1}$$

Assuming $\hbar = 1, E = M, V(r) = \frac{1}{2}V$, and $E - M = E_{NR}$ so that equation can be expressed in equation (2)

$$[-\nabla^2 + 2V(r)M]S(r) = 2ME_{NR}S(r) \tag{2}$$

Equation (2) is the Dirac equation that reduced to be similar to the Schrödinger equation for non-relativistic with exact spin symmetry and equation (1) will be used to analysis waves function and spectrum energy levels for particle in hyperbolic Manning-Rosen-like potential with hypergeometric method[9].

2. Dirac Equation for Hyperbolic Manning-Rosen-like Potential

Manning-Rosen potential is a one dimensional potential order of Dirac equation that could be solved by a specific function. Energy and wave function that affected by hyperbolic Manning-Rosen potential. The hyperbolic Manning-Rosen potential shown in equation (3)

$$V(r) = C^2\left(\frac{Q(Q-1)}{sinh^2Cr} - 2qcothCr\right) \tag{3}$$

A constant C is the centrifugal factor that used to obtain potential length, Q and q are non-dimensional parameter in hyperbolic Manning-Rosen potential[9].

In this paper, we solve hyperbolic Manning-Rosen-like potential with exact spin symmetry with hypergeometric method. For hyperbolic Manning-Rosen-like potential is expressed in equation (4)

$$V(r) = \frac{V_0\,\alpha^2\sinh^4\alpha r + V_1\,\alpha^2\sinh^3\alpha r\,\alpha^2\cosh\alpha r + V_2\,\alpha^2\sinh^2\alpha r\cosh^2\alpha r}{\sinh^4\alpha r} \tag{4}$$

Equation (4) can be simplify into equation (5)

$$V(r) = \alpha^2(V_0 + V_1\coth\alpha r + V_2\coth^2\alpha r) \tag{5}$$

Equation (5) is hyperbolic Manning-Rosen-like potential.

In the relativistic case for exact spin symmetry that the Dirac equation shown in equation (6)

$$[-\nabla^2 + V(r)(E+M)]\psi(r) = (E^2 - M^2)\psi(r) \tag{6}$$

The Laplacian operator can be write like equation (7) for spherical coordinate.

$$\nabla^2 = \frac{1}{r^2}\frac{\partial}{\partial r}\left(r^2\frac{\partial}{\partial r}\right) + \frac{1}{r^2\sin\theta}\frac{\partial}{\partial\theta}\left(\sin\theta\frac{\partial}{\partial\theta}\right) + \frac{1}{r^2\sin^2\theta}\frac{\partial^2}{\partial^2\varphi} \tag{7}$$

For this case, we assuming $\hbar = 1, V(r) = \frac{1}{2}V$. Then we solve in radial part only. So that, the radial part shown in equation (8)

$$\left[\frac{1}{r^2}\frac{\partial}{\partial r}\left(r^2\frac{\partial}{\partial r}\right) - (E+M)V(r)\right]\psi(r) - \frac{l(l+1)}{r^2}\psi(r) = (E^2 - M^2)\psi(r)$$

$$\left[\frac{1}{r^2}\frac{\partial}{\partial r}\left(r^2\frac{\partial}{\partial r}\right) - \alpha^2(E+M)(V_0 + V_1\coth\alpha r + V_2\coth^2\alpha r)\right]\psi(r) - \frac{l(l+1)}{r^2}\psi(r) = (E^2 - M^2)\psi(r) \tag{8}$$

If we assuming that wave function in radial part is $\psi(r) = R$ so we have equation (9)

$$\left[\frac{1}{r^2}\frac{\partial}{\partial r}\left(r^2\frac{\partial}{\partial r}\right) - (E+M)\left(\alpha^2(V_0 + V_1\coth\alpha r + V_2\coth^2\alpha r)\right)\right]R - \frac{l(l+1)}{r^2}R = (E^2 - M^2)R \tag{9}$$

Where $R = \frac{U}{r}$ and $(E^2 - M^2) = \alpha^2E'$so the equation (9) becomes to equation (10)

$$\frac{d^2U}{dr^2} - \alpha^2(E+M)(V_0 + V_1\coth\alpha r + V_2\coth^2\alpha r)U - \frac{l(l+1)U}{r^2} = \alpha^2E'U \tag{10}$$

If $\frac{1}{r^2} = \frac{\alpha^2}{\sinh^2\alpha r}$, so equation (10) becomes to equation (11)

$$\frac{d^2U}{dr^2} - \alpha^2\left[(E+M)(V_0 + V_2 + V_1 \coth\alpha r) + \frac{(E+M)V_2 - l(l+1)}{\sinh^2\alpha r}\right]U = \alpha^2 E'U \tag{11}$$

Equation (11) is Dirac equation for Hyperbolic Manning-Rosen like potential. This potential will be solve with hypergeometric method to find energi and wave function.

3. Hypergeometric Method for Dirac Equation with Hyperbolic Manning-Rosen-like Potential

In the principle, the hypergeometric method finds a variable form that will be substituted to the Dirac equation so that it becomes a second-orde differential equation hypergeometric function. Equation (12) is the second-order differential equation hypergeometric function

$$z(1-z)\frac{d^2\varphi}{dz^2} + (c - (a+b+1)z)\frac{d\varphi}{dz} - ab\varphi = 0 \tag{12}$$

Equation (12) has two regular singular points

$$z(1-z) = 0 \tag{13}$$

Solution for equation (13) are $z = 0$ and $= 1$. In this paper we use $z = 0$ term because at $z = 0$ the equation (12) is more simple to be solved. Solution of equation (12) can be expressed in series form like equation (14)

$$\varphi = z^s \sum a_n z^n = a_0 z^s + a_1 z^{s+1} + a_2 z^{s+2} + \cdots \tag{14}$$

For solve this problem, we substituded equation (14) to equation (12) and we have equation (15)

$$z(1-z)(a_0 s(s-1)z^{s-2} + a_1 s(s+1)z^{s-1} + a_2(s+2)(s+1)z^s + \cdots) + (c - (a+b+1)z)(a_0 s z^{s-1} + a_1(s+1)z^s + a_2(s+2)z^{s+1} + \cdots) - ab(a_0 z^s + a_1 z^{s+1} + a_2 z^{s+2} + \cdots) = 0 \tag{15}$$

So we find simple form and equation (15) is the identity equation. Coefficient from each part of z should be equal to zero, so acquiring equation (16)

$$a_n = \frac{a(a+1)\ldots(a+n-1).b(b+1)\ldots(b+n-1)}{c(c+1)\ldots(c+n-1)n!} \tag{16}$$

So the solution of the hypergeometric differential equation is shown equation (17)

$$F_1^2(a;b;c;z) = \varphi(z) = \sum_{n=0} \frac{(a)_n(b)_n}{n!(c)_n} \tag{17}$$

In this paper, we will solve hyperbolic Manning-Rosen like potential with hypergeometric method. In the previous part, we find Dirac equation for hyperbolic Manning-Rosen like potential in equation (11). By hyperbolic identity, we can calculated equation (11) in order to equation (12). The hyperbolic identity expressed in equations (18), (19), (20).

$$\coth\alpha r = 1 - 2z, \qquad \operatorname{csch}^2\alpha r = 4z(1-z) \tag{18}$$

$$\frac{\partial z}{\partial r} = 2\alpha z(z-1), \qquad \frac{d}{dr} = \frac{\partial}{\partial z}\frac{\partial z}{\partial r} = 2\alpha z(z-1)\frac{\partial}{\partial z} \tag{19}$$

$$\frac{d^2}{dr^2} = 4\alpha^2 z^2(z-1)^2\frac{\partial^2}{\partial z^2} + 4\alpha^2 z(z-1)(2z-1)\frac{\partial}{\partial z} \tag{20}$$

We substituted equations (18), (19), (20) in to (12) therefore becomes equation (21)

$$\left[4\alpha^2 z^2(z-1)^2\frac{\partial^2}{\partial z^2} + 4\alpha^2 z(z-1)(2z-1)\frac{\partial}{\partial z}\right]U - \alpha^2\big[(E+M)\big(V_0 + V_2 + V_1(1-2z)\big) + ((E+M)V_2 - l(l+1))(4z(1-z))\big]U = \alpha^2 E'U \tag{21}$$

In this case, we assuming the parameter

$V_3 = (E+M)(V_0+V_2) V_4 = (E+M)V_1$ $\qquad V_5 = (E+M)V_2 - l(l+1)$

So that equation (21) becomes to equation (22)

$$\left[4\alpha^2 z^2 (z-1)^2 \frac{\partial^2}{\partial z^2} + 4\alpha^2 z(z-1)(2z-1)\frac{\partial}{\partial z}\right] U - \alpha^2 [V_3 + V_4(1-2z) + V_5(4z(1-z))] U = E'U \tag{22}$$

Equation (24) divided by $4\alpha^2 z(z-1)$ thus obtained equation (23)

$$z(z-1)\frac{\partial^2 U}{\partial z^2} + (2z-1)\frac{\partial U}{\partial z} - \left[\frac{(V_4 - V_3 - E\prime)}{4z(1-z)} - \frac{2V_4}{4(1-z)} - V_5\right] U = 0 \tag{23}$$

The equation (23) like the equation (12) is the second-order differential equation that has two singular points in $z = 0$ and $z = 1$. This equation (23) have the solution was approached by equation (24)

$$U_z = z^\alpha (1-z)^\beta f_z \tag{24}$$

$$\frac{\partial U_z}{\partial z} = z^\alpha (1-z)^\beta \left[\frac{\alpha}{z} f_z - \frac{\beta}{1-z} f_z + f_z{}'\right] \tag{25}$$

$$\frac{\partial^2 U_z}{\partial z^2} = z^\alpha (1-z)^\beta \left[\frac{\alpha(\alpha-1)}{z^2} f_z + \frac{\beta(\beta-1)}{(1-z)^2} f_z - \frac{2\alpha\beta}{z(1-z)} f_z + \frac{2\alpha}{z} f_z' - \frac{2\beta}{(1-z)} f_z' + f_z{}''\right] \tag{26}$$

Substituted equations (24), (25), (26) in to equation (23) so we obtained equation (27)

$$-\frac{\alpha(\alpha-1)(1-z)}{z} f_z + \frac{\alpha(2z-1)}{z} f_z - \frac{\beta z(\beta-1)}{(1-z)} f_z - \frac{\beta(2z-1)}{(1-z)} f_z - \frac{((V_4 - V_3 - E\prime))}{4z(1-z)} f_z + \frac{2V_4}{4(1-z)} f_z + V_5 f_z + 2\alpha\beta f_z - 2\alpha(1-z) f_z' + 2\beta z f_z' + (2z-1) f_z' - z(1-z) f_z'' = 0 \tag{27}$$

Taken every coefficients of f_z to eliminate the parameter z in the numerator that expressed in equation (28) and (29)

$$-\frac{\alpha(\alpha-1)(1-z)}{z} = \alpha(\alpha-1) - \frac{\alpha(\alpha-1)}{z}, \qquad \frac{\alpha(2z-1)}{z} = 2\alpha - \frac{\alpha}{z} \tag{28}$$

$$-\frac{\beta z(\beta-1)}{(1-z)} = \beta(\beta-1) - \frac{\beta(\beta-1)}{(1-z)}, \qquad -\frac{\beta(2z-1)}{(1-z)} = 2\beta - \frac{\beta}{(1-z)} \tag{29}$$

Subtituted equations (28), (29) to equation (27) so we have a simple equation that expressed in equation (30)

$$z(1-z) f_z'' + [(2\alpha+1) - (2\alpha+2\beta+2)z] f_z' - [V_5 + (\alpha+\beta)(\alpha+\beta+1)] f_z + \frac{4\alpha^2 - (-V_4 + V_3 + E')}{4z} f_z + \frac{4\beta^2 - (V_3 + E' + V_4)}{4(1-z)} f_z = 0 \tag{30}$$

The equation (30) is identically same as the general form of the hypergeometric equation

$$z(1-z)\frac{d^2\varphi}{dz^2} + (c - (a+b+1)z)\frac{d\varphi}{dz} - ab\varphi = 0$$

With

$$\alpha^2 = \frac{1}{4}(V_3 + E' - V_4), \qquad \alpha = \sqrt{\frac{1}{4}(V_3 + E' - V_4)} \tag{31}$$

$$\beta^2 = \frac{1}{4}(V_3 + E' + V_4), \qquad \beta = \sqrt{\frac{1}{4}(V_3 + E' + V_4)} \tag{32}$$

$$(\alpha+\beta)^2 = \alpha^2 + \beta^2 + 2\alpha\beta = \frac{1}{2}(V_3 + E') + \frac{1}{2}\sqrt{V_3 + E')^2 - V_4{}^2} \tag{33}$$

$$\alpha + \beta = \pm\sqrt{\frac{1}{2}\sqrt{(V_3 + E')^2 - V_4{}^2} - \frac{1}{2}(V_3 + E')} \tag{34}$$

Subtituted equations (31), (32), (33), (34) back to equation (30) so that we obtained equation (35)

$$z(1-z) f_z'' + [(2\alpha+1) - (2\alpha+2\beta+2)z] f_z' - [V_5 + (\alpha+\beta)(\alpha+\beta+1)] f_z = 0 \tag{35}$$

IOP Conf. Series: Journal of Physics: Conf. Series **1127** (2019) 012010 doi:10.1088/1742-6596/1127/1/012010

With

$$a' = \left(\alpha + \beta + \tfrac{1}{2}\right) - \sqrt{\tfrac{1}{4} - V_5} = -n \qquad b' = -n2\sqrt{\tfrac{1}{4} - V_2(E+M) + l(l-1)}$$

$c' = 2\alpha + 1$

So that the spectrum energy as one of the solution for Analytical solution of Dirac equation for Hyperbolic Manning-Rosen-like potential using Hypergeometric method for exact spin symmetry is obtained

$$E' = \frac{{V_4}^2}{4\left[\left(\sqrt{\frac{1}{4} - {V_5}^2}\right) + \frac{1}{2} + n\right]^2} + \left[\left(\sqrt{\frac{1}{4} - {V_5}^2}\right) + \frac{1}{2} + n\right]^2 - V_3 \tag{36}$$

$$E' = \frac{[(E+M)V_1]^2}{4\left[\left(\sqrt{\frac{1}{4} - \{(E+M)V_2 - l(l+1)\}^2}\right) + \frac{1}{2} + n\right]^2} + \left[\left(\sqrt{\frac{1}{4} - \{(E+M)V_2 - l(l+1)\}^2}\right) + \frac{1}{2} + n\right]^2 - [(E+M)(V_0 + V_2)] \tag{37}$$

Thus obtained the wave function for radial part is shown in equation (38)

$$U_r = \left(\frac{1 - \coth \alpha r}{2}\right)^{\alpha} \left(\frac{1 + \coth \alpha r}{2}\right)^{\beta} f(a', b', c', z) \tag{38}$$

For function $f(a', b', c', z)$ it can be expressed as

$$f_n = 1 + \frac{a'b'}{c'} z + \frac{a'(a'+1)b'(b'+1)}{c'(c'+1)} \frac{z^2}{2!} + \cdots$$

Furthermore, wave function for radial part also obtained that expressed in equation (39). The energy bound state $n_r = 0$ than $f_0 = 1$. Therefore the bound state wave function is shown in equation (39)

$$U_r = \left(\frac{1 - \coth \alpha r}{2}\right)^{\frac{-V_4}{4\left[-\left(\sqrt{\frac{1}{4} - {V_5}^2}\right) - \frac{1}{2} - n\right]} + \frac{1}{2}\left[-\left(\sqrt{\frac{1}{4} - {V_5}^2}\right) - \frac{1}{2} - n\right]} \left(\frac{1 + \coth \alpha r}{2}\right)^{\frac{V_4}{4\left[-\left(\sqrt{\frac{1}{4} - {V_5}^2}\right) - \frac{1}{2} - n\right]} + \frac{1}{2}\left[-\left(\sqrt{\frac{1}{4} - {V_5}^2}\right) - \frac{1}{2} - n\right]} \tag{39}$$

4. Conclucion

The Dirac equation can be reduce to Shrodinger equation. The Schrodinger equation with Manning-Rosen potential can be solve with hypergeometric method.

In this work, we can find energy and wave function to describe condition of particles. Energy spectrum is exactly obtained that shown in equation (37) and wave function is expressed in equation (39). For spherical coordinate, the Dirac equation for hyperbolic Manning-Rosen-like potential in the scheme of centrifugal term approximation is exactly solved using hypergeometric method with spin symmetric cases.

References

[1] Eckmann, V.D Waerden. *Practical Quantum Mechanics I*. New York : Springer.(1971)

[2] C.S. Jia. *Physics Letters A* 377 (2013) 682-686.

[3] Suparmi, Cari. *IOP Publishing, Journal of Physics: Conference Series* 539 (2014) 012004.

[4] G.F. Wei. *Physics Letters A* 373 (2008) 49-53.

[5] T Jayanti. Surakarta : Ilmu Fisika Pasca Sarjana UNS.(2014)

[6] Suparmi, Cari, Angraini. *AIP Conference Proceedings* 1615, 111 (2014)

[7] S. M.Ikhdair. *Applied Mathematics And Computation* 218 (2012) 10082-10093.

[8] A Kurniawan. Surakarta : Fisika FMIPA UNS.(2014)

[9] Suparmi. *Mekanika Kuantum II*. Surakarta : Jurusan Fisika FMIPA UNS. (2011)

Stationary Gravitation Field Equation With Constant Ricci Scalar in Four Dimension

Rosikhuna Fikri Assafari[1], Abednego Wiliardy[1], and Bobby Eka Gunara[1,2]

[1] Theoretical Physics Laboratory, Theoretical High Energy Physics and Instrumentation Research Group, Faculty of Mathematics and Natural Sciences, Institut Teknologi Bandung, Jl. Ganeca no. 10 Bandung,Indonesia, 40132
[2] Indonesia Center for Theoretical and Mathematical Physics (ICTMP)

rosikhuna@s.itb.ac.id, abednego.wiliardy@s.itb.ac.id, bobby@fi.itb.ac.id

Abstract. The solution for vacuum Einstein field equation for stationary spherical black hole was first published by Kerr. In this paper, the modified Kerr metric that satisfies Einstein's condition, is considered. Furthermore, the metric can be modified to obtain stationary black hole metric with arbitrary constant Ricci scalar as generalization from Einstein condition. This condition implies a non-zero Einstein field equation and could be interpreted generally as a stationary black hole placed on a non-vacuum curved space-time background. Characterization of the geometry is to analyze singularity of the manifold, including the physical singularity and the formed horizon. The critical condition of the black hole is also considered. Moreover, surface gravity and surface area of the black hole are also calculated for further research.

1. Introduction

Another solutions of Einstein's field equation was published 1963 by Roy Kerr. The solution describing the curvature around a rotating black hole. But these solution was done without cosmological constant involved. The cosmological term was added by Einstein to compensate the matters' gravitational pull In static universe model. After the discovery of expanding universe, the interpretation of cosmological constant changed to explain the repelling effect of matters in the universe.

Considering the cosmological constant, Hassanudin had solved the corresponding Einstein equation for stationary black hole in his thesis [1]. This solution is called Kerr-de Sitter or Kerr-Anti-de Sitter depending on the sign of cosmological constant.

In this paper the solution stationary axisymmetric spacetime on a constant Ricci scalar condition will be calculated by modifying the solution from Einstein condition. The calculation and analysis of curvature singularity by considering some possibilities of the formed horizon, and energy density of the geometry are included. Besides, in order to consider the black hole thermodynamics, the surface gravity and surface area of the black hole are also calculated.

1.1. Axisymmetric stationary spacetime

In a general space, we can consider a point A in a coordinate system x^μ and another point B that has coordinate interval dx^μ with A. so from these two points it can be known the infinitesimal distance

$$ds^2 = g_{\mu\nu} dx^\mu dx^\nu , \qquad (1)$$

with $g_{\mu\nu}$ is called metric tensor. Metric tensor is determined by the geometry of the considered spacetime and the coordinate system, The metric related to stationary axisymmetric spacetime can be represented as follow [2]

$$ds^2 = -e^{2\nu}(dt)^2 + e^{2\psi}(d\varphi - \omega dt)^2 + e^{2\mu_2}(dx^2)^2 + e^{2\mu_3}(dx^3)^2. \tag{2}$$

1.2. Einstein's field equation and considered conditions

Based on general covariance principle and tensor algebra, Eintein formulated his field equation as follows [3],

$$R_{\mu\nu} - \frac{1}{2}g_{\mu\nu}R = \frac{8\pi G}{c^4}T_{\mu\nu}, \tag{3}$$

where $T_{\mu\nu}$ is energy-momentum distribution.

Einstein added the cosmological term to his equation. Einstein's field equation with an added cosmological constant is written as follows [3],

$$R_{\mu\nu} - \frac{1}{2}g_{\mu\nu}R + \Lambda g_{\mu\nu} = \frac{8\pi G}{c^4}T_{\mu\nu}. \tag{4}$$

We consider calculation in the vacuum region, so the energy momentum momentum, $T_{\mu\nu}$, is zero. The last equation can be written to

$$R_{\mu\nu} = \Lambda g_{\mu\nu}. \tag{5}$$

The term Λ was interpreted as repulsive effect that compensates the gravitational pull. Thus preventing the spacetime to fall. Although Einstein later stated that it as no longer needed to use this term, it was still used to obtain the cosmological model more generally. Equation (5) is called Einstein's condition. The solutions obtained by Hassanudin [2] are

$$g_{00} = -\frac{\Delta_r - a^2\Delta_\theta}{\rho^2}, \tag{6}$$

$$g_{01} = \frac{a}{\rho^2}\left(\sin^2(\theta)\Delta_r - \left(a^2 + r^2\right)\Delta_\theta\right), \tag{7}$$

$$g_{11} = \frac{\left(a^2 + r^2\right)^2\Delta_\theta - a^2\sin^4(\theta)\Delta_r}{\rho^2}, \tag{8}$$

$$g_{22} = \frac{\rho^2}{\Delta_r}, \tag{9}$$

$$g_{33} = \frac{\rho^2\sin^2(\theta)}{\Delta_\theta}. \tag{10}$$

with coordinate (t, φ, r, θ), where

$$\rho^2 = r^2 + a^2\cos^2\theta, \tag{11}$$

$$\Delta_r = \frac{1}{3}\Lambda r^2\left(r^2 + a^2\right) + r^2 - 2Mr + a^2, \tag{12}$$

$$\Delta_\theta = \left(1 - \frac{1}{3}\Lambda a^2\cos^2\theta\right)\sin^2\theta. \tag{13}$$

The corresponding inverse metric components are

$$g^{00}=\frac{a^2\sin^4(\theta)}{\rho^2\Delta_\theta}-\frac{\left(a^2+r^2\right)^2}{\rho^2\Delta_r}, \tag{14}$$

$$g^{01}=\frac{a}{\rho^2}\left(\frac{\sin^2(\theta)}{\Delta_\theta}-\frac{a^2+r^2}{\Delta_r}\right), \tag{15}$$

$$g^{11}=\frac{1}{\rho^2\Delta_\theta}-\frac{a^2}{\rho^2\Delta_r}, \tag{16}$$

$$g^{22}=\frac{\Delta_r}{\rho^2}, \tag{17}$$

$$g^{33}=\frac{\Delta_\theta\csc^2(\theta)}{\rho^2}. \tag{18}$$

The Kerr solution is assymptotically Minkowski, but the solution obtained by Hassanudin is assymptotically de Sitter. This implies a rotating black hole on de Sitter background.

Beside setting condition for Ricci tensor, we can also consider the some other conditions. One of them is to set the condition for Ricci scalar to be a constant

$$R=g^{\mu\nu}R_{\mu\nu}=k, \tag{19}$$

where k is a constant. This condition is more general than Einstein's. This can be interpreted as a stationary black hole placed on a non-vacuum curved background. The non-vacuity could be caused by the presence of any kind of matter, energy, or field.

2. Calculation on Constant Ricci Scalar Condition

To find the solution for this condition, The solution for Einstein's condition will be used and modified. The modification occurs in Δ_r and Δ_θ shown in equation (12) and (13) in the following form

$$\begin{aligned}\Delta_r&=\frac{1}{3}\Lambda r^2\left(r^2+a^2\right)+r^2-2Mr+a^2+A(r),\\ \Delta_\theta&=\left(1-\frac{1}{3}\Lambda a^2\cos^2\theta\right)\sin^2\theta+B(\theta).\end{aligned} \tag{20}$$

Using the solution for Einstein's condition to calculate Ricci scalar yields

$$k\rho^2=2\left[\Delta_r^{\frac{1}{2}}\left(\Delta_r^{\frac{1}{2}}\right)_{,r}\right]_{,r}+\frac{2}{\sin\theta}\left[\frac{\Delta_\theta^{\frac{1}{2}}}{\sin\theta}\left(\Delta_\theta^{\frac{1}{2}}\right)_{,\theta}\right]_{,\theta}. \tag{21}$$

The notation $A_{,\mu}$ is used to represent partial diferentiation,

$$A_{,\mu}=\partial_\mu A=\frac{\partial A}{\partial x^\mu}. \tag{22}$$

Substituting the modified solution in equation (20), equation (21) becomes

$$\left(k-4\Lambda\right)\rho^2=2\left[A^{\frac{1}{2}}\left(A^{\frac{1}{2}}\right)_{,r}\right]_{,r}+\frac{2}{\sin\theta}\left[\frac{B^{\frac{1}{2}}}{\sin\theta}\left(B^{\frac{1}{2}}\right)_{,\theta}\right]_{,\theta}. \tag{23}$$

It can be separated into 2 equations:

$$\left(k-4\Lambda\right)r^2=2\left[A^{\frac{1}{2}}\left(A^{\frac{1}{2}}\right)_{,r}\right]_{,r}-C_1, \tag{24}$$

$$\left(k-4\Lambda\right)a^2\cos^2\theta=\frac{2}{\sin\theta}\left[\frac{B^{\frac{1}{2}}}{\sin\theta}\left(B^{\frac{1}{2}}\right)_{,\theta}\right]_{,\theta}+C_1, \tag{25}$$

which have solutions

$$\begin{aligned}A(r)&=\frac{1}{12}\left(k-4\Lambda\right)r^4+\frac{1}{2}C_1r^2+C_2r+C_3,\\ B(\theta)&=\frac{1}{12}\left(k-4\Lambda\right)a^2\cos^4\theta-\frac{1}{2}C_1\cos^2\theta-C_4\cos\theta+C_5.\end{aligned} \tag{26}$$

Thus the value of Δ_r and Δ_θ for constant Ricci scalar condition is written as follows.

$$\begin{aligned}\Delta_r&=\frac{1}{3}\Lambda r^2a^2+r^2-2mr+a^2+\frac{1}{12}kr^4+\frac{1}{2}C_1r^2+C_2r+C_3,\\ \Delta_\theta&=1-\frac{\Lambda}{3}a^2\cos^2\theta-\cos^2\theta+\frac{1}{12}ka^2\cos^4\theta-\frac{1}{2}C_1\cos^2\theta-C_4\cos\theta+C_5.\end{aligned} \tag{27}$$

where C_1, C_2, C_3, C_4, C_5, are constants.

3. The Singularity

There are two kinds of black hole singularity including coordinate singularity and the real one. In a certain coordinate, the singularity can be related to the event horizon, but the real singularity is still the same whatever the coordinate is and can be observed through its curvature.

3.1. Event Horizon

This singularity is observed in Boyer-Lindquist coordinates when the g_{rr} is singular, that is

$$\Delta_r=r^2\left(\frac{C_1}{2}+1-\frac{a^2\Lambda}{3}\right)+a^2+r(C_2-2M)+C_3-\frac{kr^4}{12}=0, \tag{28}$$

or in other form

$$-\frac{12}{k}\Delta_r=r^4+Ar^2+Br+C=0, \tag{29}$$

where

$$A=-\frac{12}{k}\left(\frac{C_1}{2}+1-\frac{a^2\Lambda}{3}\right), \tag{30}$$

$$B=-\frac{12}{k}\left(C_2-2M\right),\quad\text{and} \tag{31}$$

$$C=-\frac{12}{k}\left(a^2+C_3\right). \tag{32}$$

The solutions of this polynomial are

$$r_{1,2,3,4}=\pm_{(1)}\frac{\sqrt{Q}}{2}\pm_{(2)}\frac{1}{2}\sqrt{-\left(2A+Q\pm_{(1)}\frac{2B}{\sqrt{Q}}\right)}, \tag{33}$$

where

$$P=2A^3-72AC+27B^2+\sqrt{\left(2A^3-72AC+27B^2\right)^2-4\left(A^2+12C\right)^3},\quad\text{and} \tag{34}$$

$$Q=\frac{\sqrt[3]{2}\left(A^2+12C\right)}{3\sqrt[3]{P}}-\frac{2A}{3}+\frac{\sqrt[3]{P}}{3\sqrt[3]{2}}. \tag{35}$$

These solutions represent r_+ and r_- which is associated with black hole event horizon, and r_c and r_{--} which is associated with cosmological event horizon [5]. If k is very small and positive, then r_c, r_+, and r_- are positive, but the r_{--} is negative, e.g. Kerr-de Sitter black hole.

For more information about the critical value of these solutions, we consider discriminant of the polynomial

$$D = 16A^4C - 4A^3B^2 - 128A^2C^2 + 144AB^2C - 27B^4 + 256C^3. \quad (36)$$

The equation $D = 0$ gives three conditions, those are polynomial with quadruple root, triple root, and double root(s). These conditions are related to critical values of the black hole.

1. Quadruple root : $A = B = C = 0$
2. Triple root : $A = 2i\sqrt{3}\sqrt{C}$, $B = \pm\frac{8\sqrt[4]{-1}C^{3/4}}{3^{3/4}}$, $C < 0$
3. Two double roots : $A = -2\sqrt{C}$, $B = 0$, $C > 0$

   A double and two real roots : $A < -2\sqrt{C}$, $-\infty < C < \infty$

   $$B = \pm\frac{1}{3}\sqrt{\frac{2}{3}}\sqrt{-A^3 \pm \sqrt{\left(A^2 + 12C\right)^3} + 36AC}$$

   A double and two complex roots : $A > -2\sqrt{C}$, $-\infty < C < \infty$

   $$B = \pm\frac{1}{3}\sqrt{\frac{2}{3}}\sqrt{-A^3 + \sqrt{\left(A^2 + 12C\right)^3} + 36AC}$$

If the value of A, B, and C are returned, then we will get explicite form of the critical value.

For Kerr-Newman-de Sitter case, Dehghani and KhajehAzad [4] suggest that there are exist two possible critical conditions. First condition, r_- coincides with r_+ which is related to black hole rotation and charges, and the other one, r_+ coincides with r_c which is related to the curvature of the background space time.

4. Black Hole Thermodynamics

Hawking showed that the dynamics of a black hole are similar to the thermodynamics processes [6]. For a rotating black hole, the first law can be written as

$$dM = \frac{1}{8\pi}\kappa dA + \Omega_H dJ. \quad (37)$$

The change of energy is correlated with the change of mass and the change of entropy is correlated with the change of surface area. The work term is correlated with the rotation of the black hole and the temperature is correlated with the surface gravity. The more general form of the first law is given by Gibbons [7] as

$$dM = \frac{1}{8\pi}\kappa dA + \Omega_H dJ + \psi^\Lambda dq_\Lambda + \chi_\Lambda dp^\Lambda + \left(\frac{\partial M}{\partial \phi^a}\right) d\phi^a. \quad (38)$$

This equation also includes the possibility of any other works due to the presence of any charges or any fields, e.g. electric or magnetic charge in electric or magnetic field.

4.1. Surface gravity

Killing vector for a axisymmetric black hole is given by

$$\chi = \partial_t + \Omega_H \partial_\phi \quad (39)$$

with Ω_H is the black hole's rotation speed at surface,

$$\Omega_H = -\frac{g_{t\phi}}{g_{\phi\phi}} = \frac{a\left(\left(a^2 + r^2\right)\Delta_\theta - \sin^2(\theta)\Delta_r\right)}{\left(a^2 + r^2\right)^2 \Delta_\theta - a^2 \sin^4(\theta)\Delta_r}. \quad (40)$$

Norm of the Killing vector is given by

$$V = \sqrt{-\chi_\mu \chi^\mu} = \sqrt{\frac{\Delta_\theta \Delta_r \left(a^2\left(-\sin^2(\theta)\right) + a^2 + r^2\right)^2}{\left(a^2\cos^2(\theta) + r^2\right)\left(\left(a^2 + r^2\right)^2 \Delta_\theta - a^2 \sin^4(\theta)\Delta_r\right)}}, \tag{41}$$

Finally, we can calculate the surface gravity as

$$\begin{aligned}\kappa &= \sqrt{\nabla_\mu V \nabla^\mu V} = \sqrt{g^{\mu\nu}\partial_\mu V \partial_\nu V} \\ &= \frac{1}{6}\sqrt{\frac{\left(r\left(2a^2\Lambda - 3C1 + kr^2 - 6\right) - 3C2 + 6M\right)^2}{\left(a^2 + r^2\right)^2}}\end{aligned} \tag{42}$$

Surface gravity is evaluated at the surface, therefore we take $r = r_+$ as the result.

4.2. *Surface area*

A black hole's surface area is formed at the outer event horizon. It can be calculated as follows.

$$A = \int_{\phi=0}^{2\pi}\int_{\theta=0}^{\pi} \sqrt{\gamma}\, d\theta\, d\phi, \tag{43}$$

where

$$\gamma = \left(a^2 + r_+^2\right)^2 \sin^2(\theta). \tag{44}$$

If the integral in equation (43) is evaluated, we obtain

$$A = \int_{\phi=0}^{2\pi}\int_{\theta=0}^{\pi} \left(a^2 + r_+^2\right)\sin(\theta)\, d\theta\, d\phi = 4\pi\left(a^2 + r_+^2\right) \tag{45}$$

5. Conclusion

The solution for stationary black hole under constant Ricci scalar condition has been calculated. It yields more general solution than Kerr-dS or Kerr-AdS with several constants that have not yet been interpreted physically. The horizon is calculated to be the roots of a polynomial. The surface gravity, and its surface area have also been calculated.

Accknowledgments

This work is funded by Riset Inovasi ITB 2016 and Riset Desentralisasi DIKTI-ITB 2016.

References

[1] Hassanudin 2012 *Ruang waktu empat dimensi stasioner dan axis-simetrik dengan metrik Einstein* (Tesis Program Magister, Institut Teknologi Bandung)

[2] S Chandrasekhar 1978 *The Kerr Metric and Stationary Axis-Symmetric Gravitational Field* (Proc. Roy. Soc., A **358**, 405-420).

[3] R M Wald 1984 *General Relativity* (The University of Chicago Press)

[4] M H Dehghani and H KhajehAzad 2003 *Thermodynamics of a Kerr Newman de Sitter black* hole (Canadian Journal of Physics, **81(12)**, 1363-1375 (arXiv:hep-th/0209203v3))

[5] L Huaifan, Z Shengli, W Yueqin, Z Lichun, dan Z Ren 2009 *Hawking radiation of Kerr-Newman-de Sitter black hole* (The European Physical Journal C, **63**, 133-138 (arXiv:0906.3680v1 [gr-qc]))

[6] J M Bardeen, B Carter, dan S W Hawking 1973 *The four laws of black hole mechanics* (Commun. Math. Phys., **31**, 189-196)

[7] G Gibbons, R Kallosh, dan B Kol 1996 Moduli, scalar charges, and the first law of black hole thermodynamics (Physical Review Letters, **77**, 25, 4992-4995)

Consistency Test Calculation Porosity of Logging Resistivity Through the Measurement of Common Resistant in Artificial Core Sample

Yuliora E[1], **Hendrajaya L**[2]

[1] *Mechanical Engineering Research Group, Polytechnic of Bengkalis, Jl. Bathin Alam, Sungai Alam, Bengkalis,Riau, Indonesia, 28711.*
[2] *Physics of the Earth and Complex Systems Research Group, Faculty of Mathematics and Natural Science, Bandung Institute of Technology, Jl. Ganesha no. 10 Bandung, Indonesia, 40132.*

Abstract. Research has been done about petrophysics of reservoir as an accumulated hydrocarbon. Every rocks have unequal of resistivity, from this resistivity the porosity (ϕ) quantifiable with using Archie equation. Core sample are made for consistency test calculation porosity of logging resistivity data through the measurement of common resistant in artificial core samples with variation of composite. The result measurement resistivity of shaly sands core samples is 1,19 Ωm consist of clean formation to sandstone 1,29 Ωm, shale 2,58 Ωm and fresh water 0,734 Ωm with measurable porosity is 39,38 % (shaly sands), 19,09 % (sandstone), 42,51 % (shale) and formation factor (F) of shaly sands core samples is 1,62 (shaly sands), 1,75 (sandstone), 3,51 (shale) with this results indicated that decrease in formation factor of shaly sands core samples comparable with increase of porosity of shaly sands core samples. So the data can be used to analyze oil deposit on a reservoir having specific characteristics, so as to increase productivity oil ans natural gas.

1. Introduction

Reservoir rocks is Igneous rocks as container filled and saturated of gas and oil. The main requirement rocks reservoir is pore in rocks to accumulated gas and oil is porosity.To calculate porosity and resistivity of layers we used characteristics of petrophysic rocks. characteristics of petrophysic rocks can be analyzed with result of logging process. Well logging is record process result measurement of the rocks based on the characteristics physical [2].

Evaluation rock formations is a process of analysis features and the nature of the rocks reservoir by using the measurement result the borehole [4] .The purpose of did evaluation formation is to the knowledge of the rocks is in reservoir and then the knowledge of the hydrocarbon (oil , gas or kondensat) , how many hydrocarbon can be produced , how the properties of , the state of rocks and fluid contained in it , is the amount of the womb hydrocarbon there is enough economical to be produced .To answer all the questions above then done evaluation formation covering cutting evaluation , coring , and well logging that .

Resistivity is capability of materials to hinder the course of the tide electricity in through those rocks [3]. Logs resistivity used to mendeterminasi zone hydrocarbons and zone water, identify zone permeable with mendeterminasi porosity resistivity , because rocks and matrik not conductive so the rocks capability to conduct an electric current depends on fluid and pore those rock.

Coring is method to take core from borehole [2]. Coring important to calibration modelling of petrophysics and to get information not be able in log. After drilling, core is saved because Core representing such the condition of rocks its place upon its was and are not was suffering from a so many information that can be acquired.

To ensure consistency well logging data , we need an evaluation directly .A used to evaluate the state of a formations directly can be done by coring or core of sampling .In this study author trying to make sample artificial core porous rock as a media to know consistency log entry resistivity. Knowing porosity and resistivity measurable of a sample of artificial core we can know consistency of data by compares the results plot porosity by a factor of the formation of porous rock obtained from the data well logging. So the data can be used to analyze oil deposit on a reservoir having specific characteristics, so as to increase productivity oil ans natural gas.

2. Theory

The comparison between volume an empty space / pore total not filled by a solid object (pores , cracks) that is between elements a mineral of rocks , to the volume of total rocks:

$$\phi_{\text{tot}} = \frac{V_b - V_s}{V_b} = \frac{V_p}{V_b} x100\% \tag{1}$$

Resistivity is the ability of an ingredient for can send an electric current that flows him .If an ingredient easy flow an electric current so resistivitynya low and if the material having resistivity high so material is not easy flow an electric current .Equation common used to calculate resistivity observable between 2 side of any surface at a given temperature expressed as:

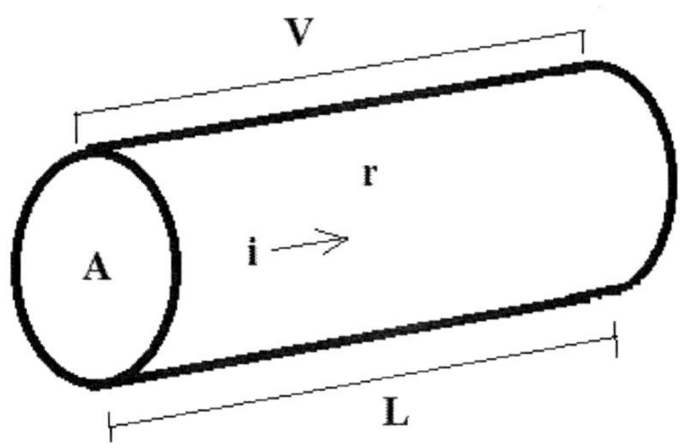

Figure 1. Resistivity of cylinder conductor.

$$R = \frac{r\,A}{L} \tag{2}$$

$$r = \frac{V}{I} \tag{3}$$

IOP Conf. Series: Journal of Physics: Conf. Series **1127** (2019) 012012 doi:10.1088/1742-6596/1127/1/012012

$$R = \frac{VA}{IL} \tag{4}$$

Gustave e .Archie , that voters importance in log analysis is the relations between porosity , resistivity, and saturation hydrocarbon of a rock a reservoir .The equation worn by archie or archie's equation [1] is:

$$F = \frac{R_0}{R_w} \tag{5}$$

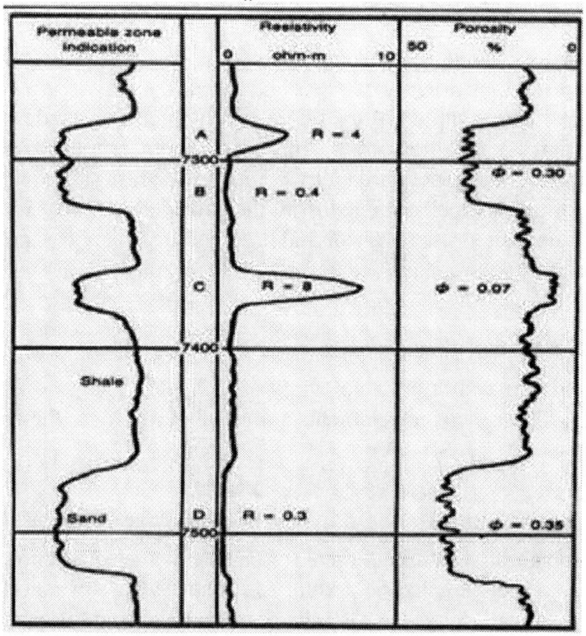

Figure 2. Data of logging resistivity [5]

Figure 2 is an example of a data logging that resistivity that is indicative of the price porosity and resistivity rock formations reservoir .Of the equation model pipe kapilaritas archie developed into law:

$$\frac{R_o}{R_w} = a\ \phi^{-m} \tag{6}$$

$$F = a\ \phi^{-m} \tag{7}$$

The value of F this is an impediment in electrical conductivity in rocks , can be seen that the total amount of ro will be more bigger than before with increase the value of F. In his postulate Archie said that the value of m would not be equal and vary depending on rocks of cementation [1] .

3. Experiment Method

To making core sample Csd (core) sandstone, Cs (core cement), Ccy (core clay), Cr-1 (shaly sands), Cr-2 (limy sands), and Cr-3 (clayey sands) used molding cylindrical from a pipe PVC in diameter 5,8 centimeters and 15 inches tall.For a sample of core csh (core shale), clm (core limestone) not using mold pipe pvc but those rocks cut to shaped beams with size 3x0,9x5 centimeters and 3,7x3,2x3,9 cm.

To make core formations sample combined polluter 65 % sandstone, 5 % adhesive of cement and 30 % defilers (shale, limestone and clay). Making core clean sample formation done by dissolving a starter (sandstone, clay, cement) with water and then inserted into the mold pipe with the same pressure and dried using an oven electricity temperatures 150 °C for 2 hours. After core dry weigh dry perfect mass then wading into a container contains water for 10 hours , it is called with the penjenuhan .Then core saturated by water again weigh mass the core to calculate porosity the core. figure 3 to describe experiment method.

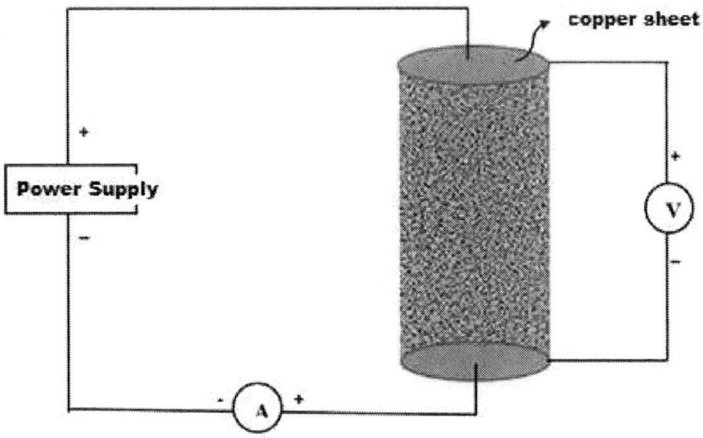

Figure 3. The shceme measurement of the core sample resistance

4. Experiment Result

Measurement porosity in samples clean core formations performed on 5 kind of rock were sample 10 core . So that the porosity and resistivity of rocks dirty formation is presented in Figure 4 and Figure 5

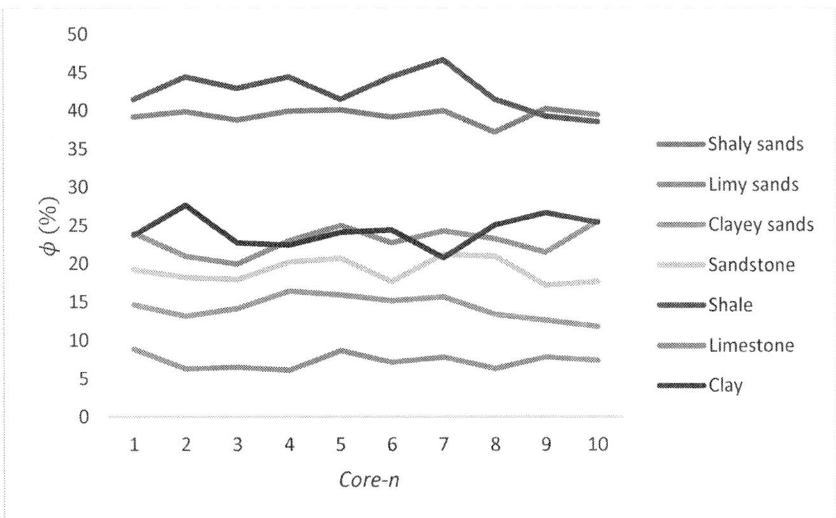

Figure 4. Porosity of dirty formation sample core

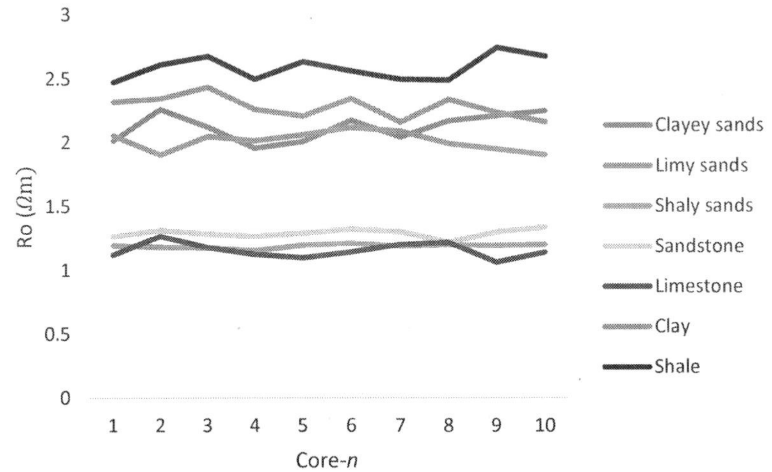

Figure 5. Resistivity of clean formation sample core

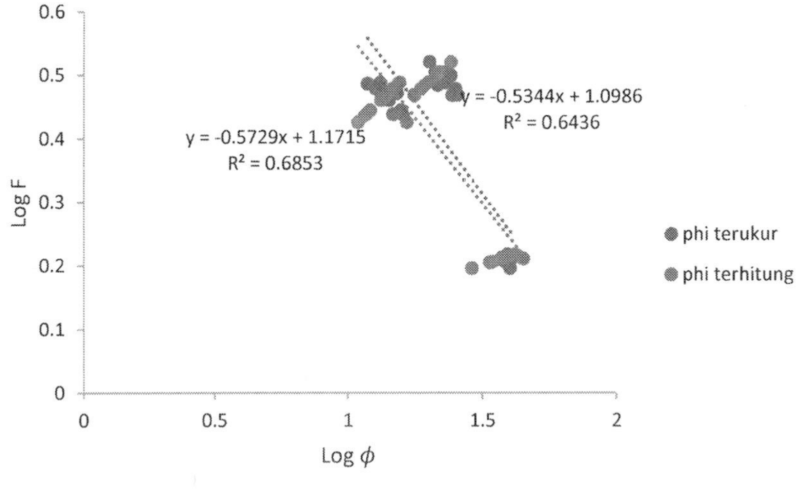

Figure 6. Plot of Log F to Log ϕ from experiment

In Figure 6 to Describing consistency results eksperiment with the data logging resistivity.

from

1. Sampel *core* Cr_{sd} $F=\frac{1{,}03}{\phi^{0{,}19}}$
2. Sampel *core* Cr_{sh} $F=\frac{1{,}01}{\phi^{0{,}33}}$
3. Sampel *core* Cr_{lm} $F=\frac{0{,}9}{\phi^{0{,}21}}$
4. Sampel *core* Cr_{cy} $F=\frac{1{,}01}{\phi^{0{,}31}}$
5. Sampel *core* Cr-1 $F=\frac{1{,}18}{\phi^{0{,}13}}$
6. Sampel *core* Cr-2 $F=\frac{1{,}14}{\phi^{0{,}37}}$
7. Sampel *core* Cr-3 $F=\frac{1{,}16}{\phi^{0{,}41}}$

Based on data porosity and resistivitas above , using the equation archie and obtained factors masing-masing core formation , with information factors this formation then we can analyzed core or reservoir which can accommodate oil reserves.

5. Conclusion

The log ϕ to log F in dirty formation, shaly sands showed that sample core it has porosity higher than elementary substance sandstone but smaller than material shale, while for limy sands having its porosity is greater than porosity elementary limestone , and to clayey sands having its porosity less than the prositas elementary substance (sandstone) and materials clay.The log ϕ to log F in shaly sands showed that sample core it has factors the formation of rock formations (F) smaller than Felementary substance (sandstone) and materials shale, while for limy sands having F greater than the for basic sandstone and limestone, and to clayey sands having its porosity is higher than F elementary substance (sandstone) and materials clay. So the data can be used to analyze oil deposit on a reservoir having specific characteristics, so as to increase productivity oil ans natural gas.

6. References

[1] Archie G E 1942 *Electrical Resistivity Log as an Aid in Determining Some Reservoir Characteristics*. Trans.AIME 146,54-62. [2]
[2] Batemen, R 1985 *Open-Hole Analysis and Formation Evaluation*. Boston : IHRDC.
[3] Darling, T 2005 *Well Logging and Formation Evaluation*. USA : GPP.
[4] Harsono, A 1997 *Evaluasi Formasi dan Aplikasi Log*. Jakarta : Schlumberger Oilfield Services
[5] Schlumberger 1989 *Log Interpretation Principles/Applications*. Texas : Sclumberger

Effect of Solvent on Stretching and Twisting of DNA

Rizki Khairani[1], Donny Dwiputra[1], Wahyu Hidayat[1,2], Freddy P. Zen[1,2]

[1] Theoretical Physics Laboratory, Faculty of Mathematics and Natural Sciences, Institut Teknologi Bandung Jl. Ganesha no. 10 Bandung, 40132, Indonesia

[2] Indonesian Center of Theoretical and Mathematical Physics (ICTMP), Indonesia

E-mail: rizkikhairani028@gmail.com

Abstract. We extend the Peyrard-Bishop model into helicoidal DNA case in presence of solvent. Our model focuses on twisting and stretching DNA base pair motions. The dynamics of this system are drawn from the solitary wave solution by using multiple scale expansion to the case of the vectorial lattice. The breather soliton is the solution for the stretching motion and the kink soliton is the solution for the twisting motion. A positive value of the solvent strength factor stretches the hydrogen bond and the strand arc length which makes the DNA more winded.

Keywords: DNA-Solvent interaction, breather-soliton, kink-soliton.

1. Introduction

The conformational changes of DNA due to the environment is not fully understood yet. The complex dynamics of DNA have been investigated. Peyrard-Bishop (PB) presented simple model focusing on the denaturation which is initialized by the bubble formation, without considering the DNA helicity in a vacuum [1]. In this model, PB interpret the hydrogen bond as a Morse potential. Next, Dauxois extended the PB model by considering the DNA helicity, assumed the bases in a different strand do not only interact in transversal direction but also in longitudinal direction [2]. Still, in helicoidal DNA case, Barbi et al [3] extended the PB model by generalizing the curvature of the DNA double helix, assuming the helicity can be presented by observing the interaction of three body diks (nucleotida). Because of that, Barbi uses two degrees of freedom, radial and angular (polar coordinates) and modifies the Hamiltonian PB model by adding the helicoidal term [3].

In this paper, we include the effect of the environment, in our case is the solvent, to realistically model the dynamics of DNA. PB, Dauxois, and Barbi have modeled the DNA dynamics but still cannot explain the real DNA condition. In fact, DNA is immersed in solvent, one of the environment factors affecting the dynamics of DNA. According to Zoli, solvent will stabilize the open state or the base pairs opening [4]. Solvation causes weakening and lengthening of the hydrogen bond from base pairs [5] while experimentally it is the transition of DNA from B-DNA to A-DNA due to increased solvent concentration [6]. Some experimental facts of DNA-solvent interaction cannot be fully explained. Because of that, we present a model of DNA in solvent for helicoidal DNA by adopting the helicoidal model of Barbi by adding the solvent potential term in Lagrangian.

IOP Conf. Series: Journal of Physics: Conf. Series **1127** (2019) 012013 doi:10.1088/1742-6596/1127/1/012013

2. Lagrangian Model

The helicoidal model of DNA is described in polar coordinate system with two degrees of freedom; r_n dan φ_n , radial dan angular variable, both of them will produce the stable helical (variable coupled) structure. The distance of a base from the hydrogen bond center in equilibrium condition (B-DNA) is denoted by R_0. The distance between the neighbouring base pairs denoted by h. In our model, the adjacent site height h is assumed constant. The length of backbone between disks in one strand when in equilibrium state is denoted by $L = \sqrt{h^2 + 4R_0^2 \sin^2\left(\frac{\Theta_0}{2}\right)}$ which is larger than h. The twist angle between base pair in equilibrium state is $\varphi_n - \varphi_{n-1} = \pm\Theta_0$. The values of h, R_0, and Θ_0 adopted from B-DNA parameters [3]. When the opening base pair processing, the backbone interpreted by a spring will stretch, L_d. The strands are hydrophobic, making the water molecules around base pairs eliminated from the core of DNA, by forming helicoidal backbone. The competition between the rigidity of backbone and the stacking interaction of the neighboring base pairs stabilizes the double helix structure of DNA. To get the effect of helicity, we should take three body curvature, and adding to the Lagrangian in third terms (1), where G_0 is a curvature constant.

The DNA-solvent interaction described by the solvent potential, in the last terms of the Lagrangian (1) with f_s is the solvent strength, and l_s is the width of solvent potential [4] that expresses the effective range. Thus the Lagrangian of our model is

$$\begin{aligned}\mathcal{L} &= \sum_n m\left(\dot{r}_n^2 + r_n^2\dot{\varphi}_n^2\right) - \sum_n D\left(e^{-\alpha(r_n - R_0)} - 1\right)^2 - \sum_n G_0\left(\varphi_{n+1} + \varphi_{n-1} - 2\varphi_n\right)^2 \\ &\quad - \sum_n K\left(\sqrt{h^2 + r_{n-1}^2 + r_n^2 - 2r_{n-1}r_n\cos\left(\varphi_n - \varphi_{n-1}\right)} - L^2\right) \\ &\quad + \sum_n Df_s\left(\tanh\left(\frac{r_n - R_0}{l_s}\right) - 1\right). \qquad (1)\end{aligned}$$

The mass of a disk, m, the depth of the Morse potential, D and the width of the Morse potential, α, are adopted from the PB model [3], m = 300 u.m.a., D = 0.04 eV, α =4.45Å^{-1}, while K = 1.0 eVÅ^{-2} and $G_0 = KR_0^2/2$ are adopted from the helicoidal DNA model [3]. The value of parameter geometry for B-DNA are $R_0 \approx$ 10.0Å, $\Theta_0 = 36°$ and h = 3.4Å. In this model, the assumed constant h means that before and after solvation h is constant.

To simplify the Lagrangian, we present some dimensionless variables, $r'_n = \alpha r_n$, $t' = t(\sqrt{\frac{D\alpha^2}{m}})$, $K' = \frac{K}{D\alpha^2}$, $\mathcal{G}' = \frac{G_0}{D}$, $R'_0 = \alpha R_0$, $h' = \alpha h$, $L' = \alpha L$, $l'_s = \alpha l_s$. For simplicity hereafter, the new variables are written without prime. The shift of base, y_n, or the arc length $y_n = r_n - R_0$ and the length of the disk position in angular, $\phi_n = R_0(\varphi_n - n\Theta_0)$.

The stacking interaction (fourth term) is expanded to second order around the equilibrium position, yielding $\frac{R_0^2}{L^2}\left\{(y_n + y_{n-1})(1 - \cos\Theta_0) + (\phi_n - \phi_{n-1})\sin\Theta_0\right\}^2$. The Morse potential and the solvent potential are expanded to forth order, so we get the equation of motion:

$$\begin{aligned}\ddot{y}_n &= \left(\frac{y_n}{R_0} + 1\right)\frac{\dot{\phi}_n^2}{R_0} - \left(y_n - \frac{3}{2!}y_n^2 + \frac{7}{3!}y_n^3\right) - \frac{K_{y\phi}}{2}\left(\phi_{n+1} - \phi_{n-1}\right) \\ &\quad - K_{yy}\left(y_{n+1} + 2y_n + y_{n-1}\right) + \frac{f_s}{2l_s}\left(1 - \frac{y_n^2}{l_s^2}\right), \qquad (2)\end{aligned}$$

and

$$\ddot{\phi}_n = K_{\phi\phi}\left(\phi_{n+1} + \phi_{n-1} - 2\phi_n\right) + \frac{K_{y\phi}}{2}\left(y_{n+1} - y_{n-1}\right)$$

$$-G\left(6\phi_n - 4\phi_{n+1} - 4\phi_{n-1} + \phi_{n+2} + \phi_{n-2}\right)$$
$$-\frac{2}{R_0}\dot{y}_n\dot{\phi}_n - \frac{2}{R_0}y_n\dot{\phi}_n - \frac{2}{R_0^2}y_n\dot{y}_n\dot{\phi}_n - \frac{1}{R_0^2}y_n^2\ddot{\phi}_n, \quad (3)$$

where $K_{\phi\phi} = \frac{KR_0^2}{L^2}\sin^2\Theta_0; K_{y\phi} = 2\frac{KR_0^2}{L^2}\sin\Theta_0\left(1-\cos\Theta_0\right); K_{yy} = \frac{KR_0^2}{L^2}\left(1-\cos\Theta_0\right)^2; G = \frac{\mathcal{G}}{R_0^2}$. K_{yy} and $K_{\phi\phi}$ are the effective elastic constant from the shifting or stretching of disk and twist rotation, $K_{y\phi}$ is the coupling constant between the stretching and twisting motion.

3. The envelope soliton solution for small amplitude

The soliton solution is obtained by using the analitycal method developed by Cocco in [7]. We use the well-known technic to get the solution of envelope soliton for small amplitude use the Multiple Scale Expansion (MSE), based on the perturbation theory. The model of helicoidal DNA in solvent has two component wave vectors $\vec{E}$; y_n and ϕ_n. First, we solve the linear order. Here the wave is packed with a weak dispersion. From (2) and (3) the equation of motion is written:

$$\frac{\partial^2}{\partial t^2}\begin{pmatrix} y_n \\ \phi_n \end{pmatrix} = -\sum_{n',\alpha'} J\left(n-n',\alpha,\alpha'\right)\begin{pmatrix} y_n \\ \phi_n \end{pmatrix} \quad (4)$$

with

$$\hat{J}(q) = \begin{pmatrix} 1+2K_{yy}(1+\cos q) & iK_{y\phi}\sin q \\ -iK_{y\phi}\sin q & \begin{matrix}2K_{\phi\phi}(1-\cos q) \\ +G\left(6-8\cos q+2\cos 2q\right)\end{matrix} \end{pmatrix} \equiv \begin{pmatrix} a & c \\ c^\star & b \end{pmatrix}.$$

Now, we take care of the nonlinearity of the system to get the small amplitude envelope soliton. To increase the accuracy of the nonlinear term gradually, we present an expansion parameter ϵ. We introduce the coordinates $x_1 = \epsilon x$, $t_1 = \epsilon t$, $t_2 = \epsilon^2 t$. The nonlinearty solution form is

$$\vec{E}(n) = \epsilon e^{i(q_0 n_0 - \omega_+ t_0)}\left(\vec{V}^+ - i\epsilon\vec{V}^{(1)+}\frac{\partial}{\partial x_1}\right)A\left(x_1,t_1,t_2\right) + c.c + \epsilon\vec{\sigma}\left(x_1,t_1,t_2\right)$$
$$+\epsilon^2 e^{2i(q_0 n_0 - \omega_+ t_0)}\vec{\gamma}\left(x_1,t_1,t_2\right) + c.c + \epsilon^2\vec{\mu}\left(x_1,t_1,t_2\right)$$
$$+\vec{\eta}\left(x_0,x_1,t_0,t_1,t_2\right). \quad (5)$$

with $A\left(x_1,t_1,t_2\right)$, $\vec{\sigma}\left(x_1,t_1,t_2\right)$, $\vec{\gamma}\left(x_1,t_1,t_2\right)$, $\vec{\mu}\left(x_1,t_1,t_2\right)$, $\vec{\eta}\left(x_0,x_1,t_0,t_1,t_2\right)$ are the slowly varying amplitudes that envelope the wave packet. The velocity of the envelope is smaller than the velocity of the carrier wave. The carrier wave is $e^{i(q_0 n_0 - \omega_+ t_0)}$ in first order and $e^{2i(q_0 n_0 - \omega_+ t_0)}$ in second order.

The wave packet is interpreted by first order $O(\epsilon)$, the second term arise because the linear system produce the nonzero solution on the force constant matrics $\hat{J}(0)$ at $q_0 = 0$, $\hat{J}(0) = \begin{pmatrix} 1+4K_{yy} & 0 \\ 0 & 0 \end{pmatrix}$ where $1+4K_{yy}$ is the non zero solution of $\hat{J}(0)$. Hence there must be another new basis for non oscilating term in the solution, denoted by $\vec{\sigma}\left(x_1,t_1,t_2\right)$. The second order $O(\epsilon^2)$ consists of the second harmonic term and the constant term, these terms are due to the quadratic term of the equations of motion. The last term appears because in the equation of motion present $O(\epsilon^0)$ term (2).

From the $O(\epsilon^0)$ and (e^0) term, we obtain the solutions for η_1 and η_2 are $\eta_1 = \frac{f_s}{2l_s(1+4K_{yy})}$, $\eta_2 =$ const. We choose $\eta_2 = 0$. From the second order, O (ϵ^2), we get the solution for γ, is $\vec{\gamma} = \gamma_c A^2$. The solution of the constant term (e^0), $\vec{\mu}$ and $\vec{\sigma}$, are obtain from the second order and the third order. Checking the second order, we get $-2K_{y\phi}\frac{\partial}{\partial x_1}\sigma_1 = 0$ *and* $\sigma_1 = \text{const.}$. From the non oscillating term of the third order we get the solution for μ_2 can be written

$$\mu_2 = \frac{4\omega_+\omega_+^{(1)}}{K_{y\phi}R_0}\left|V_2^+\right|^2 q_1 \int |A|^2\, dx_1 = \mu_{2c}\int |A|^2\, dx_1, \tag{6}$$

we get the solution μ_1 and σ_2, $\mu_1 = \mu_{1c}\left|A\right|^2$, $\sigma_2 = \sigma_{2c}\int |A|^2\, dx_1$.

Now we are done with the envelopes, η, $\vec{\sigma}$, $\vec{\gamma}$, $\vec{\mu}$, unless those are dependent on A. The nonlinear term $\mathrm{O}(\epsilon^3)$ in $\exp i\,(q_0 n_0 - \omega_+ t_0)$ gives the equation for A in form of Nonlinear Schrödinger (NLS). We found the NLS equation for envelope expression A in moving frame with velocity $\omega_+^{(1)}$ $(S_1 = x_1 - \omega_+^{(1)} t_1,\ t_2)$,

$$E\frac{\partial^2 A}{\partial S_1^2} + Fi\frac{\partial A}{\partial t_2} + Hi\frac{\partial A}{\partial S_1} + IA + Q\left|A\right|^2 A = 0, \tag{7}$$

This simplifies to $PA_{s_1 s_1} + iA_{t_2} + \mathcal{H}iA_{s_1} + \mathcal{I}A + \mathcal{Q}\left|A\right|^2 A = 0$, with $P = \frac{E}{F}$, $\mathcal{H} = \frac{H}{F}$, $\mathcal{I} = \frac{I}{F}$, $\mathcal{Q} = \frac{Q}{F}$. When $P\mathcal{Q} > 0$, the bright soliton case, the envelope solution will be [8]

$$A\left(s_1, t_2\right) = \mathcal{A}\sec h\left[\frac{1}{\mathcal{L}_e}\left(s_1 - u_e t_2\right)\right]\exp\left[i\left(\frac{u_e - \mathcal{H}}{2P}\right)\left(s_1 - u_c t_2\right)\right], \tag{8}$$

where u_e is the velocity of the envelope and u_c is the velocity of carrier, while the amplitude is $\mathcal{A} \equiv \sqrt{\frac{u_e^2 - 2u_c u_e + 2u_c\mathcal{H} - \mathcal{H}^2}{2P\mathcal{Q}} + \frac{4\mathcal{I}}{\mathcal{Q}}}$, and $\frac{1}{\mathcal{L}_e} \equiv \sqrt{\frac{u_e^2 - 2u_c u_e + 2u_c\mathcal{H} - \mathcal{H}^2}{4P^2} + \frac{\mathcal{I}}{P}}$ where $\mathcal{L}_e$ is the inverse width of the envelope.

The complete solution is

$$\begin{aligned} y \;=\;& 2\epsilon V_1^+\mathcal{A}\sec h\left[\frac{1}{L_e}\epsilon\left(x - V_e t\right)\right]\cos\left(\mathcal{K}x - \Omega t\right) + \epsilon\sigma_{1c} + \eta_1 \\ & + 2\epsilon^2\gamma_{1c}\mathcal{A}^2\sec h^2\left[\frac{1}{L_e}\epsilon\left(x - V_e t\right)\right]\cos\left(2\mathcal{K}x - 2\Omega t\right) \\ & + \epsilon^2\mu_{1c}\mathcal{A}^2\sec h^2\left[\frac{1}{L_e}\epsilon\left(x - V_e t\right)\right] + \epsilon^2 V_1^{(1)}\mathcal{A}\sec h\left[\frac{1}{L_e}\epsilon\left(x - V_e t\right)\right] \\ & \left\{-2\frac{1}{L_e}\tanh\left[\frac{1}{L_e}\epsilon\left(x - V_e t\right)\right]\sin\left(\mathcal{K}x - \Omega t\right) + \left(\frac{u_e - \mathcal{H}}{P}\right)\cos\left(\mathcal{K}x - \Omega t\right)\right\} + \mathcal{O}\left(\epsilon^3\right), \end{aligned} \tag{9}$$

and

$$\begin{aligned} \phi \;=\;& \epsilon\left[\sigma_{2c} + \epsilon\mu_{2c}\right]\frac{1}{\mathcal{L}_e}\mathcal{A}^2\tanh\left[\frac{1}{L_e}\epsilon\left(x - V_e t\right)\right] + \eta_2 \\ & + 2i\epsilon\mathcal{A}V_2^+\sec h\left[\frac{1}{L_e}\epsilon\left(x - V_e t\right)\right]\sin\left(\mathcal{K}x - \Omega t\right) \\ & + 2i\epsilon^2\gamma_{2c}\mathcal{A}^2\sec h^2\left[\frac{1}{L_e}\epsilon\left(x - V_e t\right)\right]\sin\left(2\mathcal{K}x - 2\Omega t\right) + i\epsilon^2 V_2^{(1)}\mathcal{A}\sec h\left[\frac{1}{L_e}\epsilon\left(x - V_e t\right)\right] \\ & \left\{2\frac{1}{L_e}\tanh\left[\frac{1}{L_e}\epsilon\left(x - V_e t\right)\right]\cos\left(\mathcal{K}x - \Omega t\right) + \left(\frac{u_e - \mathcal{H}}{P}\right)\sin\left(\mathcal{K}x - \Omega t\right)\right\} + \mathcal{O}\left(\epsilon^3\right) \end{aligned} \tag{10}$$

where $V_e = \omega_+^{(1)} + \epsilon u_e$; $V_c = \omega_+^{(1)} + \epsilon u_c$; $\mathcal{K} = q_0 + \epsilon\frac{u_e - \mathcal{H}}{2P}$; $\Omega = \omega_+ + \epsilon\frac{u_e - \mathcal{H}}{2P}V_c$. The variables are still dimensionless. We take the wave vector, $q_0 = 0.1$ and we choose $\epsilon = 0.1$, $u_e = -0.1$, $u_c = 0.1$.

IOP Conf. Series: Journal of Physics: Conf. Series **1127** (2019) 012013 doi:10.1088/1742-6596/1127/1/012013

4. Effect of solvent on the stretching motion y_n

The analytical method produces the soliton solution like a breather. The value of the non oscillating factor σ_{1c} is taken small to prioritize the solvent effect, $\sigma_{1c} = 0.001$. To understand the base pairs opening, the analytical solution is plotted in (x, t) plot under the varying solvent strength factor f_s. The stretch-inverse stretch plot y_n vs. $-y_n$ is plotted to give a clear view of the base opening process.

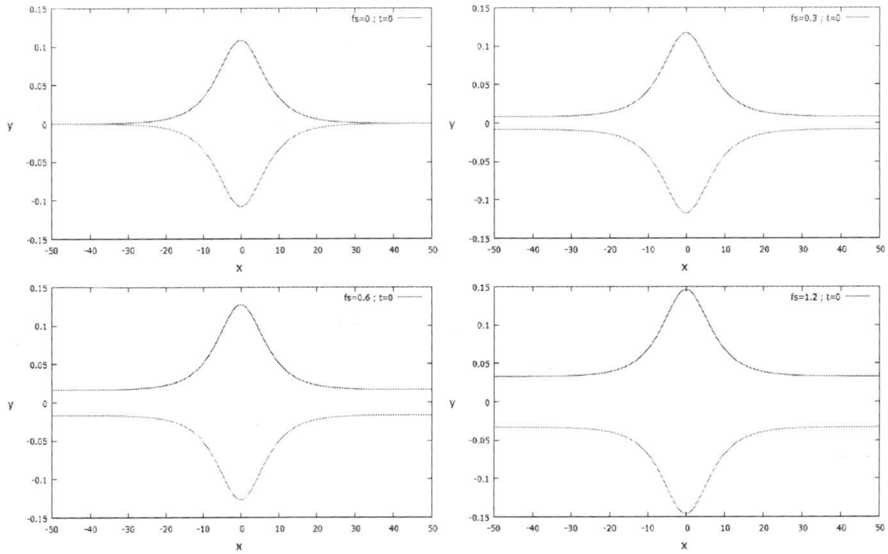

Figure 1. y and its inverse $(-y)$ vs. x for $t = 0$ with the variation of f_s. We obtain the positive value of a solvent strength factor make the hydrogen bond stretch.

Figure 1 shows the base pair opening after solvation with positive value of f_s. The base pair opening (Y_n) for a solvent having strength f_s at $t = 0$ is $Y_n = y_n - (-y_n) = 2y_n$. The more positive value of the f_s the more the hydrogen bonds are stretched. The value of the base pair opening in Figure 1 are still dimensionless. The real value of the base pair with $f_s = 0.1$ is $Y_n = 2 \times \left(0.0006\,\text{Å} < y_n < 0.025\,\text{Å}\right)$. If we compare to the diameter of B-DNA, 20 Å, the solvent effect in the stretching motion is small.

For each $x = -50$ and $x = 50$ without a solvent $y = -y = 0$ that means the base pairs opening is zero $Y = 0$. After solvation $Y(x = 50) > 0$ indicated that the minimum point of the bound state (the bound state is the DNA breathing condition) no more at $Y = 0$ (the base pair in the stable condition). The shifting of the stable condition is caused by the Constant Force of the solvent, $S_F = f_s/2\alpha l_s$, the value of the constant force depends on the type of the solvent (f_s). The constant force is obtained from $O(\epsilon^0)$ of the equation of motion. It forces the base outward the DNA axis.

Effect of solvent on the twisting motion

The solvation affects DNA gets more winded (Figure 2) and increasingly steep of the kink-soliton solution. The arc length of the base pair for $x = 0$ is $\phi_0 = 0$, then the arc length of the neighbours have the same value but have an opposite direction.

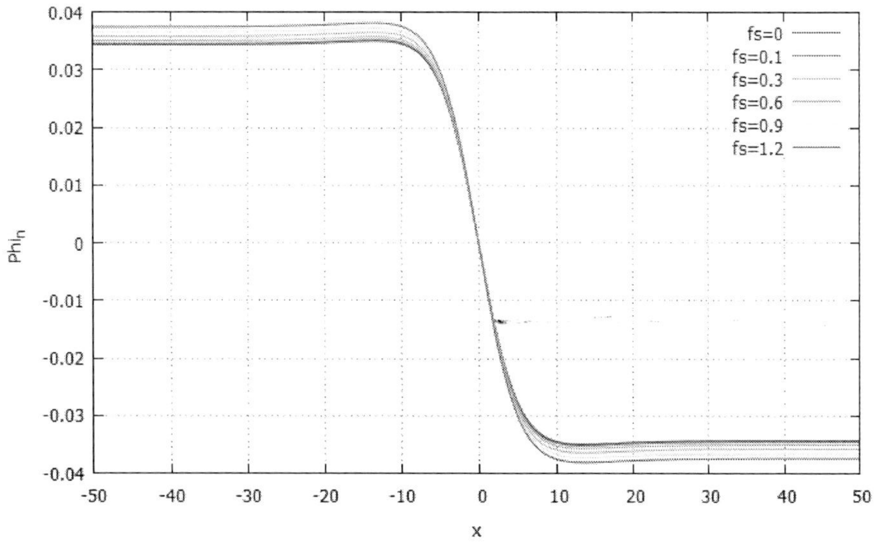

Figure 2. The analytical solution for arc length ϕ_n with the variation of f_s

5. Conclusion

The solvent effect on the helicoidal DNA dynamics especially for stretching motion changes the stability of the open state, it is showed in the shifting of the minimum point to a positive value, $y_n > 0$. The shifting is caused by the constant force due to the existence of solvent. For twisting motion, the solvation caused the arc length of the disk to be more positive in value and more negative in value with the initial condition at $x = 0$. This means that DNA after solvation is more winded for the more positive value of the solvent barrier factor.

The hydrophobic effect of DNA strand forces the structure to be double helix. If the environment of DNA consist of water effect the backbone winding. Because of the behaviour of the base pair, it can be concluded that DNA will be more winded if there is some water around the DNA. Related to the analytical solution of this model, that the positive value of f_s lets the DNA more winded, we get can suggest that a positive f_s correspond to the situation around DNA with high water content.

References

[1] Peyrard M 2004 *Nonlinearity* **17** R1
[2] Dauxois T 1991 *Physics Letters A* **159** 390–395
[3] Barbi M, Cocco S and Peyrard M 1999 *Physics Letters A* **253** 358–369
[4] Zoli M 2011 *The Journal of chemical physics* **135** 115101
[5] Poater J, Swart M, Guerra C F and Bickelhaupt F M 2012 *Computational and Theoretical Chemistry* **998** 57–63
[6] Fang Y, Hoh J H and Spisz T S 1999 *Nucleic acids research* **27** 1943–1949
[7] Cocco S, Barbi M and Peyrard M 1999 *Physics Letters A* **253** 161–167
[8] Remoissenet M 2013 *Waves called solitons: concepts and experiments* (Springer Science & Business Media)

Comparison between EGSnrc and MCNPX for X-ray target in 6 MV photon beam

S Yani[1,2], R Tursinah[3], M F Rhani[4], F Haryanto[1] and I Arif[1]

[1] Department of Physics, Faculty of Mathematics and Natural Science, Institut Teknologi Bandung, Jalan Ganesa 10 Bandung Indonesia
[2] Akademi Kebidanan Pelita Ibu Kendari, Jalan Syech Yusuf No.2, Kendari Indonesia
[3] National Nuclear Energy Agency, Jalan Tamansari 71 Bandung, Indonesia
[4] Department of Radiation Oncology, National University Cancer Institute Singapore, Singapore/Singapore

E-mail: sitti.yani@s.itb.ac.id

Abstract. The Monte Carlo codes, EGSnrc and MCNPX, have been specifically developed to study the particles transport in linear accelerator used in radiotherapy. A simple X-ray target 6 MV photon beam geometry was chosen to highlight similarities and differences of particles transport algorithms implemented in these Monte Carlo codes. The purpose of this study was to compare the particles transport in X-ray target 6 MV photon beam using EGSnrc and MCNPX codes. This target has 0.2484 cm thickness consisting of 0.0899 cm Tungsten and 0.1575 Copper. The global cut-off energies used in the simulations for electron and photon were 500 KeV and 10 KeV, respectively. Monte Carlo simulations were performed for 6.4 MeV monoenergetic beam. The parameters used in this simulation were using the DEFAULT setting. The particles fluence, spectral distribution after target and simulation time were analysed in this simulation. The results showed that the spectrum from EGSnrc and MCNPX was not well-matched. There was under 5% difference, especially in the build-up region. Meanwhile, the EGSnrc was 2 to 3 times more efficient than MCNPX code to achieve certain statistical precision. This happened because the differences of transport algorithm and approximation in these MC codes. EGSnrc just follow photons and electrons during simulation, while MCNPX can simulate transport 34 types of particles, including neutrons. It also causes the simulation time difference between these two codes. MCNPX simulation took longer than the EGSnrc. However, in some cases, differences in results as well as simulation time between 2 codes can be ignored.

1. Introduction

The Monte Carlo codes have been specifically developed to study the particles transport in linear accelerator used in radiotherapy [1, 2, 3, 4, 5]. The particle transport is a complicated process depending on the energy and material interact with it. A simple X-ray target 6 MV photon beam geometry was chosen to highlight similarities and differences of particles transport algorithms implemented in these Monte Carlo codes. The X-ray target is one of the important component in photon beam linear accelerator. In order to simulate the particles transport in X-Ray target of 6 MV photon beam, the Monte Carlo codes used in medical physics were adopted; EGSnrc and MCNPX. The methods for tracking particles in the volume of interest used in both of these codes were different from each other. Many researchers had conducted some study in benchmarking some Monte Carlo code system package (EGSnrc, PENELOPE, Geant4 and MCNPX) for many kinds of cases [6, 7, 8, 9]. Steinfelds (2007) found that EGSnrc is much faster than MCNP at modelling primary electrons and MCNP is more versatile than EGSnrc [6].

The objective of this study was to compare the output of X-Ray target of Varian Clinac iX 6 MV photon beam using EGSnrc and MCNPX code system (spectral distribution, statistical uncertainty and simulation time).

2. Method

The X-Ray target 6 MV photon beam was modeled in EGSnrc and MCNP package according to the geometry specifications provided by the Varian Monte Carlo Data package. This package describes the geometry and materials of the target. This target has 0.2474 cm thickness consisting of 0.0899 cm Tungsten and 0.1575 Copper [10].

2.1. EGSnrc

The EGS (Electron Gamma Shower) code system is a general purpose package for radiation transport simulation. EGSnrc was one of the code developed by the Ionizing Radiation Standards Group, Institute for National Measurement Standards, National Research Council of Canada. This code is able to simulate electron and photon for particle energies ranging from 1 keV to 10 GeV in homogeneous and inhomogeneous materials [11].

Figure 1 show the sketch of X-ray target modeled using EGSnrc code. The scoring plane sets just below the Copper material. The global cut-off energies used in the simulations for electron and photon were 521 keV and 10 keV, respectively. Monte Carlo simulations were performed for 6.4 MeV monoenergetic beam with FWHM 1.0 cm and field size 5×5 cm^2. The parameters used in this simulation were using the DEFAULT setting. The scoring plane in EGSnrc was stored in phase space (phsp) files containing the information and identity of particles stored in it (charge, energy, position, direction, etc). BEAMDP data processor was used to analyze this output to obtain particles fluence, angular and spectral distribution.

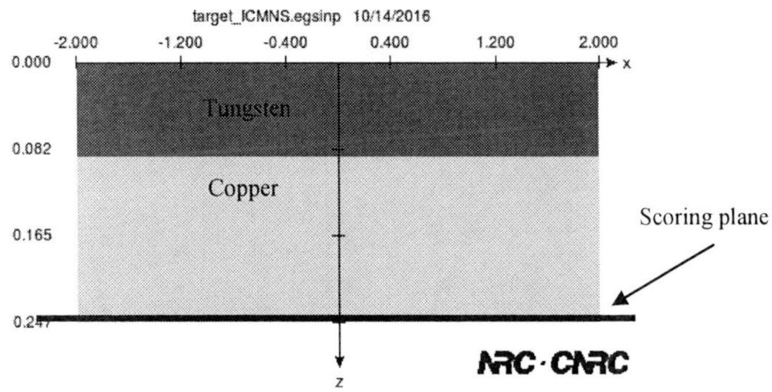

Figure 1 Sketch of X-ray target was simulated by EGSnrc code.

2.2. MCNPX

The Monte Carlo N-Particle X-tended (MCNPX) code system was developed by the Los Alamos National Laboratory over the past 30 years [12]. MCNP is a general purpose code that can be used to simulate neutron, photon and electron transport. MCNP has a wide range of capabilities which make it useful for medical physics applications. MCNPX code uses a three-dimensional inhomogeneous geometry and transports photons, electrons and neutrons in the energy range from 1 keV to 100 MeV.

The MCNP input file consist of cell cards, surface cards and surface cards. Specifies of output called F4 tally at the bottom of the target to obtain the spectral distribution. As in the previous simulation using EGSnrc, the source were performed for 6.4 MeV monoenergetic beam. The sketch of X-ray target modelled by MCNPX was shown in Figure 2.

Figure 2 Sketch of X-ray target was simulated by EGSnrc code.

The spectral distribution in EGSnrc and MCNPX was compared to analyze how much were the difference between these results, the statistical uncertainty and simulation time. The output particles analyzed in this simulation was photon.

3. Results and discussion

Production of X-rays in target material of linear accelerator (linac) was extremely inefficient. X-ray production in high energy linac photon beam has a higher conversion of electrons into photons. The total number of electrons converted to heat is 99% and only 1% of the electrons were converted to bremsstrahlung photons. Tungsten material has high melting point so it would not melt, despite exposing to very large temperature changes. And the electron interact with the Copper material to produce X-ray bremsstrahlung. X-ray in target was produced by electromagnetic radiation emitted when an electron losses energy as a consequence of coulomb interaction with the nucleus of an atom.

The statistical uncertainty both of the EGSnrc and MCNPX simulations were below 1 %. Figure 3 below shows the angular distribution and fluence (number of photons or electrons per MeV per incident electron on the target) at the scoring plane obtained from EGSnrc code. Only the particles which were inside the defined field were counted.

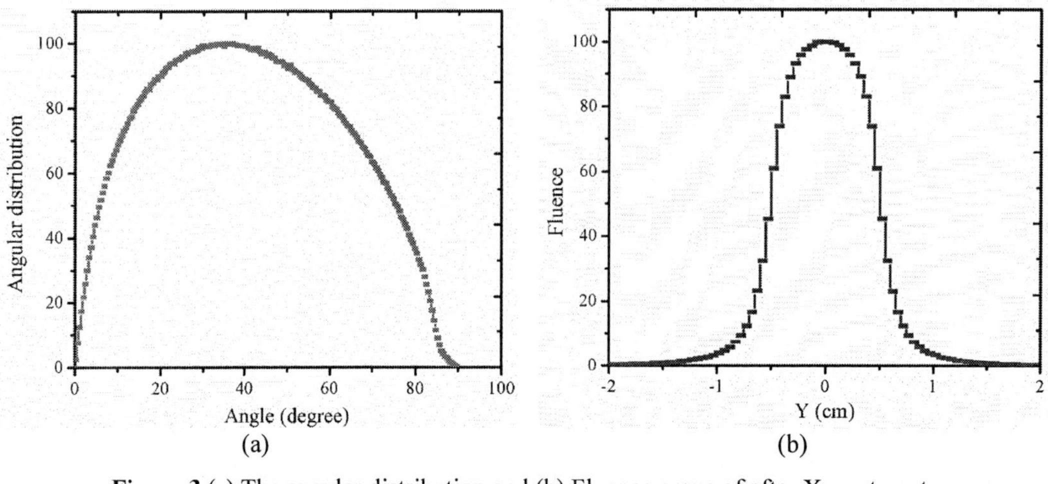

Figure 3 (a) The angular distribution and (b) Fluence curve of after X-ray target.

The angular distribution and fluence only scanned in the Y direction. The X and Y direction is similar and identical to each other, so the graphs just scored in one direction. The particles interaction in X-ray target and the products that were scattered have the relative distribution to the beam axis. The angular distribution is distribution relative to the scattered particles and product of its interaction. Most of photon bremsstrahlung scattered in the wide angle (20 – 60 degree).

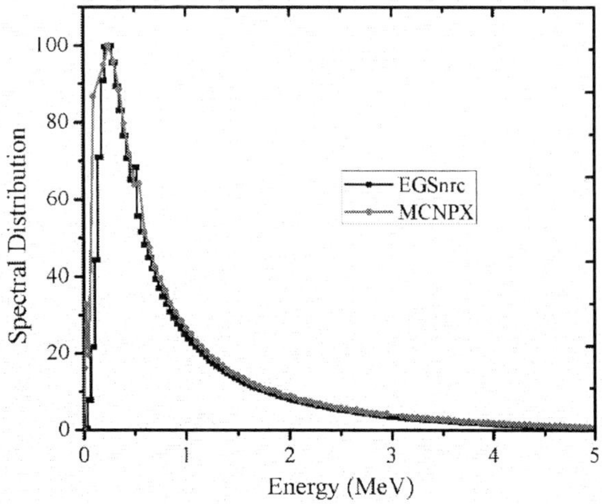

Figure 4 Spectral distribution comparison of X-ray target was simulated by EGSnrc and MCNPX code.

The results show that the spectrum from EGSnrc and MCNPX was not well-matched. There was under 5% difference, especially in the build-up region. The spectral distribution chart is a visual representation of the X-ray spectrum produced by a 6 MV photon beam target. This graph shows the relative intensities of X-ray at MeV energy. The 100% peak of EGSnrc and MCNPX spectral distribution was 0.245 and 0.25 MeV, respectively. Meanwhile, the spectrum has the same shape and value in the tail region.

In the simulation process, EGSnrc was more efficient than MCNPX code to achieve certain statistical precision (Table 1). This happened because the differences of transport algorithm and approximation in these MC codes. EGSnrc just follow the photons and electrons during the simulation, while MCNPX can simulate transport 34 types of particles, including neutrons. It also causes the simulation time difference between these two codes. MCNPX simulation took longer than the EGSnrc.

Table 1. Simulation parameters of EGSnrc and MCNPX.

Parameter \ Software	EGSnrc	MCNPX
Number of histories	1.000.000.000 electrons	30.000.000 electrons
Simulation time	21.66 hours	21.22 hours
Statistical uncertainty	0.0001%	0.13%

4. Conclusions

The X-ray spectrum in the target material of 6 MV linear accelerator (linac) simulated by EGSnrc and MCNPX was identical in tail region but it was not well-match in build-up region. And the EGSnrc was 2 to 3 times more efficient than MCNPX to achieve certain statistical precision in this simulation.

Acknowledgements

The authors thank Varian Medical Systems for providing technical data of linear accelerator to be modelled with Monte Carlo. This study was fully supported by Riset Kerjasama Luar Negeri Dikti 2016 (077/SP2H/PL/LT/II/2016) and Hibah Disertasi Doktor Dikti 2016.

References

[1] Ahnesjo A and Aspradakis M M 1999 Dose Calculations for External Photon Beams in Radiotherapy *Phys. Med. Biol.* **44**

[2] Yani S, Dirgayussa I G E, Rhani M F, Soh R C X, Haryanto F and Arif I 2016 Monte Carlo study on electron contamination and output factors of small field dosimetry in 6 MV photon beam *Smart Science* **4(2)** pp 1–8 (http://dx.doi.org/10.1080/23080477.2016.1195609)

[3] Yani S, Tursinah R, Rhani M F, Soh R C X, Haryanto F and Arif I 2016 Neutron contamination of Varian Clinac iX 10 MV photon beam using Monte Carlo simulation *Journal of Physics: Conference Series* **694** 012020 (doi:10.1088/1742-6596/694/1/012020)

[4] Medina A L, Teijeiro A, Garcia J, Esperon J, Terron J A, Ruiz D P and Carrion M C 2005 Characterization of electron contamination in megavoltage photon beams *Med. Phys.* **32** pp 1281–92

[5] Fippel M, Haryanto F, Dohm O, Nusslin F and Kriesen S 2003 A virtual photon energy fluence model for Monte Carlo dose calculation *Med. Phys.* **30(3)** pp 301–311

[6] Steinfelds E 2007 Comparison of predictions by MCNP and EGSnrc of radiation dose imparted to various material targets by beams and small volumetric sources *Computational Medical Physics Working Group Workshop II* Florida USA Sep 30 – Oct 3, 2007

[7] Pacilio M, Lanconelli N, Meo S L, Betti M, Montani L, Aroche L A T and Pérez M A C 2009 Differences among Monte Carlo codes in the calculations of voxel S values for radionuclide targeted therapy and analysis of their impact on absorbed dose evaluations *Med. Phys.* **36 (5)** 1543–52

[8] Nedaie H A, Mosleh-Shirazi M A, Gharaati H, Allahverdi M 2006 Assessment of different MCNP Monte Carlo codes in electron absorbed dose *Rep Pract Oncol Radiother* **11(6)** pp 293–298

[9] Archambault J P and Mainegra-Hing E 2015 Comparison between EGSnrc, Geant4, MCNP5 and Penelope for mono-energetic electron beams *Phys. Med. Biol.* **60** pp 4951–62 (doi:10.1088/0031-9155/60/13/4951)

[10] Varian Medical Systems Confidential, Monte Carlo Data Package High Energy Accelerator

[11] Kawrakow I and Rogers D W O 2011 The EGSnrc Code System: Monte Carlo Simulation of Electron and Photon Transport. Ottawa: National Research Council Canada

[12] Pelowitz D B 2008 MCNPX Users Manual. New Mexico: Los Alamos National Laboratory

Analysis Of Three-Component Seismograms To Obtain The Result Of Moment Tensor Of Earthquake In Manokwari

I Setyowidodo[1,a), B J Santosa[2], A D Handayani[1], S Widodo[1]

[1]Universitas Nusantara PGRI Kediri, Jl. K.H Ahmad Dahlan No.76, Kediri 64112, Indonesia
[2]Institut Teknologi Sepuluh Nopember, Keputih, Sukolilo, Surabaya, 60111, Indonesia

[a)]E-mail: irwansetyo@unpkediri.ac.id

Abstract. This study analyzed three-component waveform inversion of the earthqake that occurred in Manokwari, Papua on January 13, 2010 at 17:18:11 GMT with a magnitude of 5.7 Mw which the epicenter was at -0.72 lattitude, longitude 133.33 and depth of 42 km. The study used the local seismic data downloaded from the data of earthquake IA. Then, inversion of the seismic data were done by using iterative deconvolution methods. This method was implemented in ISOLA software which was developed to obtain the parameters of the epicentre. Furthermore, the results of parameters were used to determine the actual form of the fault by using the HC method. Synthetic seismograms were computed with ISOLA that the input was a model of the earth and the seismogram recorded by seismological stations MWP, SWI and LBM. Interpretation of the analysis of three-component seismogram waveform showed that the orientation of the field of earthquake fault in Manokwari, Papua on January 13, 2010 had a dip angle of 77° to the horizontal plane which caused the fault zone in the area which was easy to occur a shift and an earthquake. By this analysis, it was known that the cause of this earthquake faults were oblique strike-slip faults moving from the northwest to the southeast.

1. Introduction

Geologically, Indonesian archipelago is located at the confluence of two major seismic zones, namely the circum-Pacific belt and The Alpine belt. Because of its location, the Indonesian archipelago is in an area that has relatively high earthquake activity. Tectonic earthquake is shaking movement of the Earth which is caused by collisions among tectonic plates. Convection currents of the Earth's core flow into the upper mantle. Those currents also control the movements of the plates. The continuance of continental plates' impulse driven by convection currents in the upper mantle causes earthquakes and produces advanced deformations on the surface of continents [1]

Many studies have been developed to understand the structure and dynamics of the Earth as well as the mechanism of the earthquake through the parameter estimation of the epicentre Centroid Moment Tensor (CMT). CMT determination of the earthquake has been using single-component Green's functions on the Z-axis only. Seismic waves move from the source to the observation station in three-dimensional space. Therefore, it is necessary to develop three-component Green's function to estimate the earthquake source parameter carefully. This study compares station and synthetic waveform counted by using the Green's function based on three-component waveform.

IOP Conf. Series: Journal of Physics: Conf. Series **1127** (2019) 012015 doi:10.1088/1742-6596/1127/1/012015

To understand the characteristics of this earthquake, it can be done by modelling the moment tensor of earthquakes [2]. The moment tensor model can be done by using inversion method with waveform [3]. Data analysis used in this study is local seismic data taken from seismic data IA (Internet Accelerograph). The results of this analysis are in the form of earthquake parameters which include scales, depth and energy of earthquakes as well as fault models causing earthquakes.

2. Theoretical Review

2.1. Moment Tensor

Seismic waves transmit elastic strain energy which comes out of earthquake's seismic source. The velocity of seismic vibration is determined by the elastic modulus and the density of the material passed. Researchers who focus on terrestrial field conclude that the main cause of earthquakes originates from the force of movement in the Earth's interior to push the crust, so that when the crust is not tough in responding to the force, it will create a fault and produce an earthquake. Model of fault motion and fault character of the earthquake cause can be recognized by the moment tensor of earthquakes. This moment tensor is used to describe the direction of the force which causes earthquakes [2]. Component of seismic recordings of a point source can be stated as follows:

$$U_k(x,t) = \sum_{i=1}^{6} G_{ki}(x, x_s, t) * f_i(t) \tag{1}$$

The concept of moment tensor can give a complete description of the force of seismic point source. Couples of the moment tensor component is described in the following figure 1.

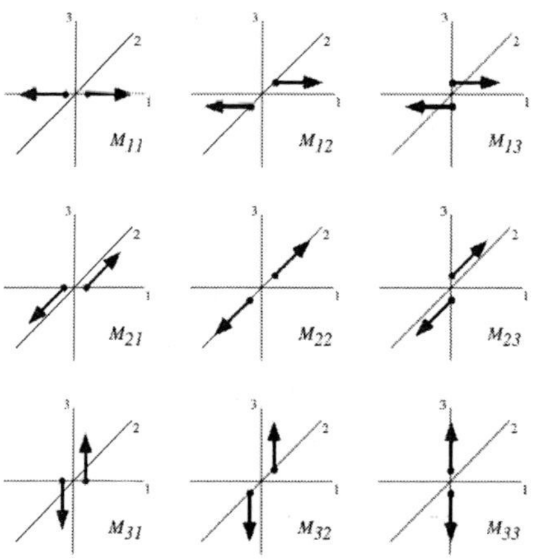

Figure 1. The force couples that make up the components of the moment tensor

The nature of the moment tensor is symmetrical because M_{ij} equals M_{ji}. M_{ij} component value can be used to determine the parameters of the strike (φ), dip (δ) and rake (λ) causing earthquakes. Because $M_{ji} = M_{ij}$, so there are 6 independent elements of the moment tensor left among 9 moment tensor [2]. The relationship among the moment tensor, a strike (φ), dip (δ) and rake (λ) is expressed in the following equation:

$$M_{11} = M_{xx} = -M_0(\sin\delta \cos\lambda \sin 2\phi + \sin 2\delta \cos\lambda \sin 2\phi)$$
$$M_{12} = M_{xy} = M_0(\sin\delta \cos\lambda \cos 2\phi + 0.5 \sin 2\delta \cos\lambda \sin 2\phi) = M_{21} = M_{yx}$$
$$M_{13} = M_{xz} = -M_0(\cos\delta \cos\lambda \cos\phi + \cos 2\delta \sin\lambda \sin\phi) = M_{31} = M_{zx}$$
$$M_{22} = M_{yy} = M_0(\sin\delta \cos\lambda \sin 2\phi - \sin 2\delta \sin\lambda \cos 2\phi)$$
$$M_{23} = M_{yz} = -M_0(\cos\delta \cos\lambda \sin\phi - \cos 2\delta \sin\lambda \cos\phi) = M_{32} = M_{xy}$$
$$M_{33} = M_{zz} = -(M_{11} + M_{22}) = M_0(\sin 2\delta \sin\lambda)$$

The moment tensor can be used to measure earthquake strength by using seismic moment parameter (M_0). Furthermore, variance reduction (V_r) and correlation (C) are calculated by equation:

$$V_r = 1 - \left[\frac{|d - s|^2}{|d|^2} \right] \qquad C = \sqrt{V_r} \tag{2}$$

$$M_0 = \frac{1}{\sqrt{2}} \left[\sum_{ij} M_{ij}^2 \right]^{1/2} \tag{3}$$

in which: d = observation data; s = synthetic data

Weak zones are around fault ones. Hence, these zones are prone areas to get geological disasters, especially if the formed faults are still active or the areas are in the active tectonic zones, such as faults in Papua (Hannekam Fault Zone, Zaagkam Fault, Wanagon Fault Zone, Fault Meren Valley) that can cause catastrophic earthquakes [4].

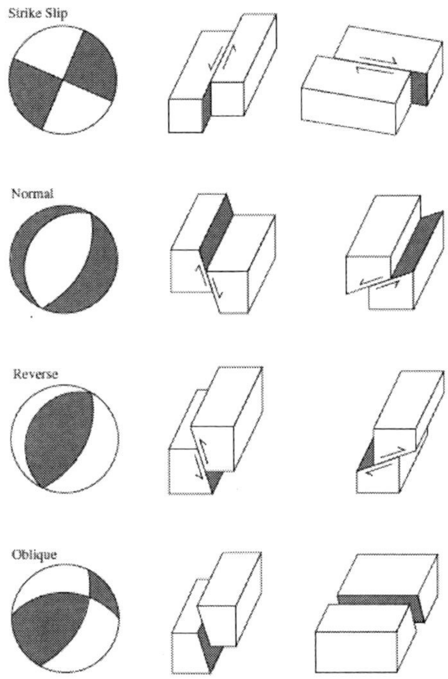

Figure 2. Focal spheres and their corresponding fault geometries [2]

2.2. Geology of Papua

Papua is formed of the interaction of two plates, namely the Australian plate and the Pacific one. These elements are resulted from the compression force south-westwards to north-eastwards, in the direction of the collision. Papua is also formed because of a collision with an oblique angle between the Pacific Oceanic plate and Caroline plate moving southwards at velocity of around 110 mm - 125 mm/year.

The main crust involved in Papua is the Australian continental crust and Pacific oceanic one. The Australian continental crust is steady and becomes the foundation of the south while the Pacific oceanic crust is the base of northern coast, including Cendrawasih Bay. At that time, the tectonic processes in this area began to be encouraged so that it produced the position of collisions heading towards the southwest is more intensive. Very prominent parts of this tectonic are transform faults which go to the left, that is Sorong-Yapen fault zone [5].

3. Research Methods

The analysis method used in this study was three-component local waveform. The parameters of the earthquake source were estimated by using inversion model to achieve three-component fitting waveform properly. Good inversion was based on the results of matching of the observation data and synthetic data. Good results occurred if the observation data and synthetic data overlapped. Seismogram was recorded by network IA (Internet Accelerograph), then it was inverted by using Green's functions which were calculated by the discretization method of number wave performed iteratively [6].

The data inversion process of three-component waveform used iterative deconvolution methods [7]. This method was implemented in ISOLA software developed to obtain earthquake source parameters [3]. The parameters of this earthquake were illustrated by using Centroid Moment Tensor (CMT). After that, the results of the parameters were used to know fault-plane by using H-C (hipocenter-centroid).

4. Data Analysis

The data used in this study are local seismic data taken from seismic data IA (Internet Accelerograph), namely earthquakes in Manokwari of Papua on January 13, 2010 at 17:18:11 GMT and with the hypocentre at -0.72 latitude, 133.33 longitudes at a depth of 42 km.

There are three selected stations because their locations are close to the source of the earthquake. Those stations are MWP, SWI, and LBM. The data analysis uses three-component waveform (BHN, BHE and BHZ). The parameters of the earthquake source are estimated by using inversion model to achieve proper fitting. This is based on the results of matching of the observation data and synthetic data in the inversion result which overlap each other. Comparison of the results of observation data and synthetic data as shown in the following figure 3.

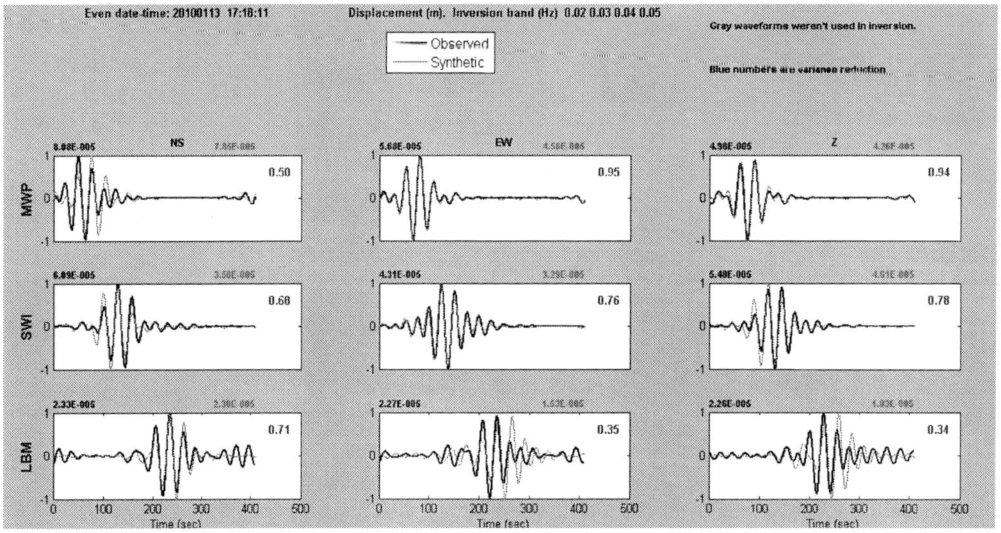

Figure 3. Waveform of Observation and Synthetic Data

Next, the interpretation is done on the Centroid Moment Tensor (CMT). This is a description of the fault model which causes earthquakes. CMT is described like beach ball which has a physical meaning that is, the bright part is the origin of the force pressing towards the dark region [2]. The CMT results of earthquake data processing in Manokwari Papua on January 13, 2010 showed that the kind of the earthquake was oblique strike-slip fault. The inversion results show that the earthquake magnitude is 5.7 Mw and the reduction variance is 0.70 with DC 54.50%, and 45.50% CLVD.

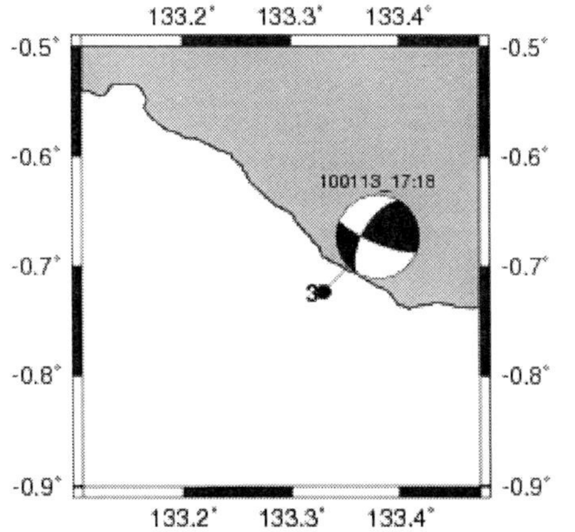

Figure 4. Earthquake Moment Tensor

If the result of earthquake source parameter is compared with BMG data and USGS obtained from Emergency Response Team of the Center for Volcanology and Geological Hazard Mitigation, Geological Agency, and DESDM, it ill be seen in the following table 1.

Table 1. Comparison of Earthquake Source Parameters of BMG, USGS, and Writer

Agency	Lattitude	Longitude	Event	Dept (km)	Magnitude
BMG	- 0.83	133.36	00:18:12 WIB	26	6.2 SR
USGS	- 0.69	133.31	17:18:08 GMT	10	5.7 Mw
Writer	- 0.72	133.33	17:18:11 GMT	42	5.7 Mw

The depiction of fault-plane which causes earthquakes is performed by using hc-plot software based on the method of H-C [3]. Input entered on the hc-plot software appears as follows:

Hipocenter	: lattitude = -0.69 ;	longitude = 133.33;	depth = 32
Centoid	: lattitude = -0.72 ;	longitude = 132.84;	depth = 42
Plane 1	: strike = 210	; dip = 60	
Plane 2	: strike = 112	; dip = 77	

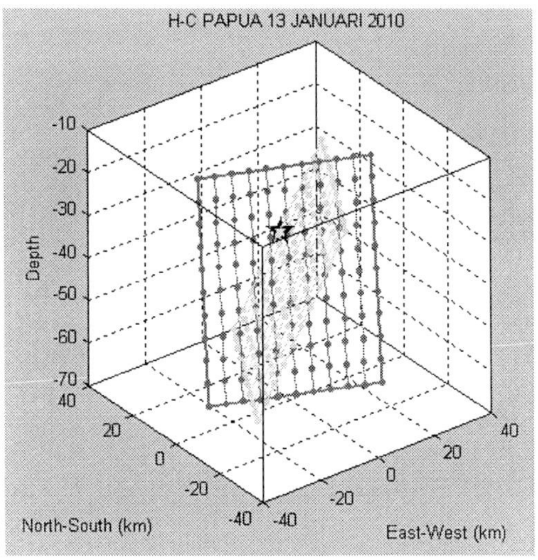

Figure 5. Fault-Plane of CMT

Later, the result presents output centroid hypocenter = 10.64, the minimum distance from the field 1 = 7.19, and the minimum distance of the field 2 = 1.80, so that the depiction of the earthquake fault-plane in Manokwari Papua on January 13th, 2010 was earthquake cause faults are red. It means that these faults head towards the north-west – the south-east.

The earthquake on January 13th, 2010 occurred with magnitude 5.7 Mw whose epicenter was at lattitude -0.72, longitude 133.33 and located at a depth of 42 km in the northwest of Manokwari. The area has high seismic intensity, especially with shallow depth of active fault systems. This is because it is close to subduction lane of Pacific oceanic plate and Australian continental crust in the northernward. As a result of the tectonic evolution, there are several active faults, they are Sorong Fault, Fault Ransiki, Aiduna Tarera Fault, and Jayawijaya Mountains Fault. In addition, the condition of quake-affected areas is generally composed of pre-tertiary metamorphic rocks, tertiary sedimentary rocks, and quarter sedimentary rocks. The pre-tertiary and tertiary rocks have experienced weathering and quarter rocks are generally loose, soft, unconsolidated and reinforcing effects of earthquake vibration so they are vulnerable to earthquake shaking [8]. Furthermore, the cause of the earthquake January 13th, 2010 is Sorong Active Fault system activity located in the northern part of Manokwari.

5. Conclusion

Based on the results of the analysis, it can be concluded that the earthquake occurred in Papua on January 13th, 2010 has a magnitude of 5.7 Mw. Epicenter is at -0.72 lattitude, longitude 133.33 and a depth of 42 km in the northwest of Manokwari. The analysis also shows that the faults which cause earthquakes belongs to the oblique strike-slip faults that move north-westwards to south-eastwards.

References

[1] Bagus Jaya S 2006 *Seismogram Analyzing of Earthquake b122696c Papua at Surrounding Observasional Stasions* (Surabaya: Departemen of Physics ITS)

[2] Peter M. *Shearer* 2009 *Introduction To Seismology Second Edition (*Cambridge: Cambridge University Press)

[3] Zahradník J and Sokos 2008 *ISOLA a Fortran Code and a Matlab GUI to Perform Multiple-Point Source Inversion of Seismic Data (*Computers & Geosciences) p 967–977

[4] Carmen Scholz 1990 *The Mechanics of Earthquakes and Faulting* (Cambridge: Cambridge University Press) p 439

[5] Wahyuni L I 2008 *Geology Irian Jaya (*Malang: Departement of Geeography UM)

[6] Bouchon M 1981 *A Simple Method to Calculate Green's Functions for Elastic Layered Media* vol 71 (America: Bulletin of the Seismological Society) **p** 959-971

[7] Kikuchi M and H Kanamori 1991 *Inversion Of Complex Body Waves – III* vol 81 (America: Bulletin of the Seismological Society) **p** 2335-2350

[8] Center for Volcanology and Geological Hazard Mitigation, Geological Agency 2009 Manokwari Earthquake Reports

IOP Conf. Series: Journal of Physics: Conf. Series **1127** (2019) 012016 doi:10.1088/1742-6596/1127/1/012016

An investigation of spatial resolution and noise in reconstructed CT images using iterative reconstruction (IR) and filtered back-projection (FBP)

C Anam[1,2], F Haryanto[2], R Widita[2], I Arif[2] and G Dougherty[3]

[1] Department of Physics, Faculty of Mathematics and Natural Sciences, Diponegoro University, Indonesia.
[2] Department of Physics, Faculty of Mathematics and Natural Sciences, Bandung Institute of Technology, Indonesia.
[3] Applied Physics and Medical Imaging, California State University Channel Islands, California, USA.

E-mail: anam@fisika.undip.ac.id

Abstract. This study compared the spatial resolution and the amount of noise in CT images reconstructed using the iterative reconstruction (IR) and filtered back-projection (FBP) techniques. We used a circular homogeneous phantom (pixel value of 0.5) containing a small circle at its center, with a diameter of 0.53-cm and a pixel value of 1.0. In order to reconstruct the image, we used the simultaneous iterative reconstruction technique (SIRT) with 5, 15, 25, 35 and 45 iterations respectively, and FBP with a Ram-Lak filter. The spatial resolution of the reconstructed image was characterized by the modulation transfer function (MTF) and image noise was characterized by the standard deviation within the homogeneous area. We found that the spatial resolution using SIRT with 35 iterations is similar to using FBP. The spatial resolution using SIRT tends to be worse using less than 35 iterations, and it tends to be better using more than 35 iterations. Using SIRT, image noise decreased with increasing iterations. The standard deviation for the SIRT technique using 45 iterations was about 0.02 similar to the FBP technique. Thus, SIRT could potentially produce better spatial resolution than the FBP technique with comparable image noise.

1. Introduction

In tomography, there are two main techniques used to reconstruct the images, namely filtered back-projection (FBP) and iterative reconstruction (IR) [1]. Iterative reconstruction (IR) was used in the early CT scans. However, the main drawback of the IR technique was that it needs very long computation times [2]. This led to it being replaced by the FBP technique in clinical CT applications. The FBP technique produces high quality images with a fast computation time, and it is considered the standard technique for CT image reconstruction. There are many filters that can be used in FBP, including the Ram-Lak, Shepp-Logan, Butterworth, cosine, Hamming and Hann filters [3]. In clinical settings, the filters are given straightforward names, such as "bone filter", "soft tissue filter" and so on.

Nowadays, there is considerable concern about CT dose [4-6]. Compared to other radiological modalities, CT delivers considerably more dose to the patient [7]. Consequently, efforts have been made to reduce the dose in a CT examination. One way is by using modified IR [8]. However, the modified IR technique is considered proprietary, with very little information on the details and algorithmic principles [2]. There are at least two methods of modified IR: statistical iterative reconstruction [9] and model-based iterative reconstruction [10].

The algorithms of standard IR that have been widely known are the algebraic reconstruction technique (ART) [11], simultaneous ART (SART) [12], and simultaneous iterative reconstruction technique (SIRT) [13]. In this study, we compare the efficacy of SIRT with standard FBP by analyzing the resulting spatial resolution and image noise.

2. Method

2.1. Overview
The steps in this study are shown in Figure 1. The first step is to design a point source phantom. The second step is to perform projections with a parallel beam, using angular increments of 1°. The third step is to reconstruct the image using both FBP and SIRT with iterations of 5, 15, 25, 35 and 45. The spatial resolution of the resulting images were then analyzed using modulation transfer function (MTF) curves, and the noise in the images was computed using the standard deviations within homogeneous areas. All the steps were implemented using MatLab software.

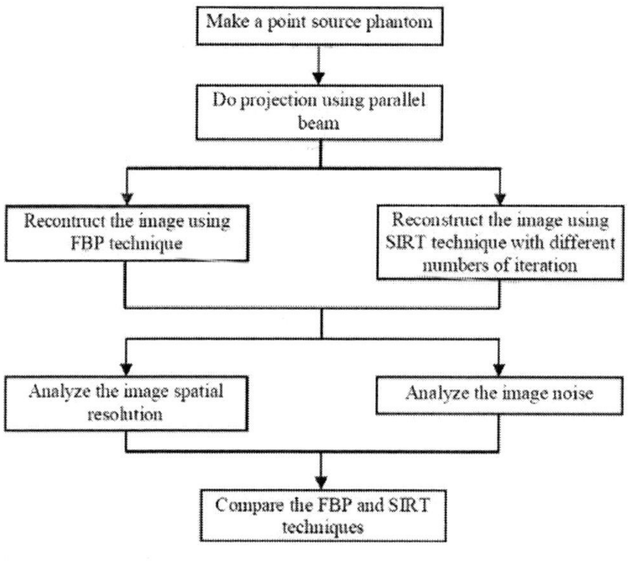

Figure 1. Flow chart diagram of this study

2.2. Design of Phantom
A point source phantom was used for the comparison of spatial resolution and noise image. The phantom consists of three circles with different diameters and different pixel values. The diameter of largest circle was 32.00 cm, the diameter of the second circle was 31.07 cm, and the diameter of the smallest circle was 0.53 cm. Pixel values in the smallest circle was 1.0, the pixel values outside the smallest circle and inside the second circle was 0.5, and the pixel values outside the second circle and inside the largest circle was 1.0. The total number of pixels was 512 x 512, representing a field of view of 32 cm.

2.3. The FBP Technique
Currently the standard algorithm for image reconstruction is filtered back projection (FBP). Back-projection can be expressed as the sum of the x-rays passing through a point (x, y) at any angle θ. Back projection aims to fill the voxels of an image with the values of the projection data. The main

drawback of this technique is that the resulting image will appear blurry due to star-like artifacts. This blur can be eliminated by performing mathematical filtering of the projection data (i.e., the sinograms), before doing back-projection. The mathematical filter used in this study is the cropped ramp filter or Ram-Lak filter.

2.4. The SIRT Technique

Another technique for image reconstruction with a low blur is the iterative reconstruction (IR) technique. The first CT scanners used this reconstruction, but because it was computationally very slow it was later abandoned. Actually, the IR technique can produce a relatively good image, even with a low current-time product (mAs). As a result, the CT radiation dose can be reduced significantly. More recently, the IR technique has been modified using statistical methods (statistical iterative reconstruction) and using model-based methods (model-based iterative reconstruction). There are several standard algorithms in the IR. In this study, we used simultaneous iterative reconstruction technique (SIRT). In the SIRT, image enhancement is performed iteratively after a complete projection for all angles.

2.5. MTF and Noise Calculation

In this study, the spatial resolution image was analyzed using the PSF and MTF curves [14], and image noise calculated as the standard deviation in homogeneous areas. Region of interests (ROIs) for the calculation of MTF and noise are shown in Figure 2. The ROI for the calculation of PSF and MTF was at the center of the image (256, 256), with a pixel size of 128 x 128.

The flow chart for the PSF and MTF calculations is shown in Figure 3. The image reconstructions, from both the FBP and SIRT techniques are read, and the images cropped at their center to give a pixel size of 128 x 128. The pixels are then summed along the y- direction to obtain the profile (summation of pixels) for the x-direction. After that is the zeroing process, so that the values of PSF at both edges of the profile are zero. The resulting curve is known as the point spread function (PSF). The PSF is then normalized so that the total of all data is equal to one, and Fourier transformation used to transform the data to the frequency domain to give a plot of the modulation transfer function (MTF).

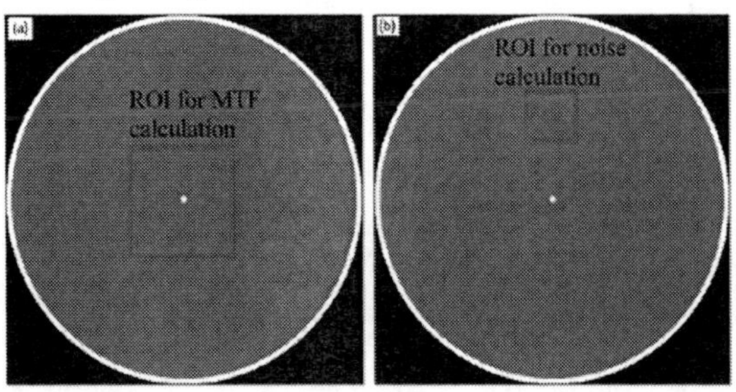

Figure 2. ROI for (a) PSF and MTF calculations and (b) ROI for noise calculation

Noise is calculated as the standard deviation of a homogeneous area. In this study, we used an area of 50 x 50 pixels, centered at position (256, 128). The reconstructed image was cropped to this region, and the standard deviation calculated.

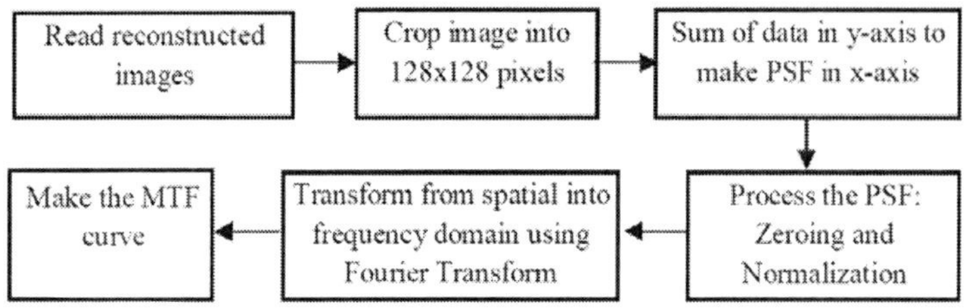

Figure 3. Flow chart of PSF and MTF calculations

3. Results

3.1. Reconstructed Images

Figure 4 is the result of the reconstruction of the object of a point source phantom using FBP and SIRT for various iterations. Figure 4a is a result of FBP reconstruction, and Figs. 4b-f are the results of SIRT for the iterations 5, 15, 25, 35 and 45 respectively. The FBP method produces a clear image, without any appreciable blur or noise. The reconstructed images using the SIRT technique are strongly dependent on the number of iterations. At the 5th iteration, blurring is still apparent, but it decreases with increasing iterations until at the 25th iteration, image quality using the SIRT technique is comparable to the FBP technique.

3.2. PSF and MTF Curve

PSF curves of the reconstructed image from FBP and SIRT for various numbers of iterations are shown in Figure 5 (left). The PSF peak value increases with increasing iterations. At the 35th iteration of the SIRT technique the PSF curve is similar to the FBP technique, and at iterations above 35 the peak of the PSF curve is much higher than with the FBP technique. However, at the tails, the PSF using SIRT do not directly equal zero, but are negative. To return to the zero position requires a shift of approximately 30 mm.

The corresponding MTF curves are shown in Figure 5 (right). The MTF curve shows clearly that the spatial resolution of the reconstructed image by SIRT increases with the number of iterations. At the 35th iteration, the MTF curves using SIRT and FBP are similar, while above 35, the spatial resolution from SIRT is higher than from FBP. The cut-off spatial frequency for all curves remains at approximately 2 cycles / mm.

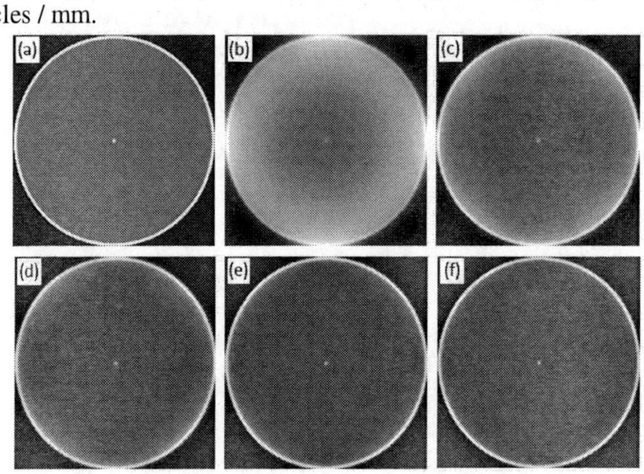

Figure 4. Reconstructed images of a point souce phantom using FBP and SIRT for various iterations.a. FBP, b. SIRT 5, c. SIRT 15, d. SIRT 25, e. SIRT 35, f. SIRT 45

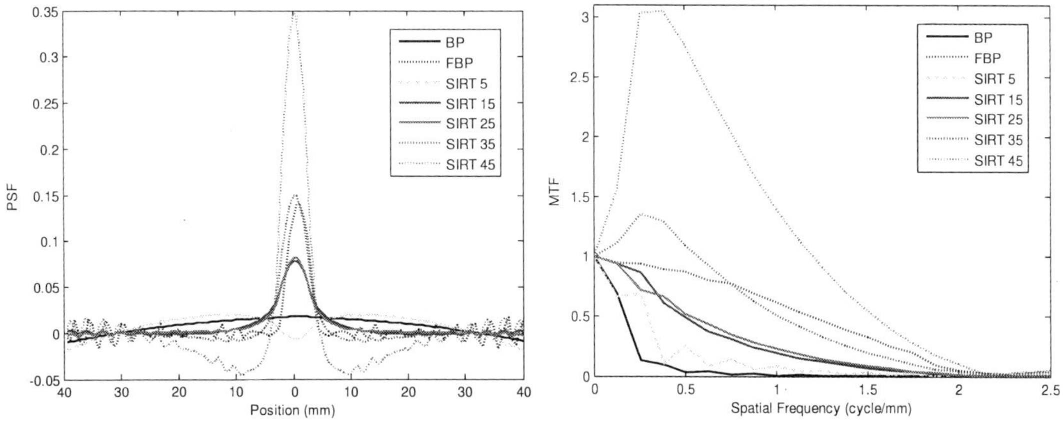

Figure 5. PSF curves (left) and MTF curves (right) of reconstructed images using FBP and SIRT for various numbers of iterations

3.3. Noise of Image

The average pixel values and standard deviations (noise) for a number of reconstructed images are shown by Table 1. The value of the original pixel in our point source phantom is 0.5. The FBP technique generates pixel values (0.49 ± 0.01) close to the original value of phantom. For the SIRT technique the values are also influenced by the number of iterations. The greater number of iterations, the closer the average pixel value is to 0.5, and the smaller is the standard deviation. For example, at iteration 45, the pixel value is 0.51, and the standard deviation (noise) is 0.02.

Table1. The average pixel value and its standard deviation for reconstructed images by FBP and SIRT for various iterations

Pixel Value	BP	FBP	SIRT 5	SIRT 15	SIRT 25	SIRT 35	SIRT 45
Average	386.23	0.49	1.51	-0.10	0.32	0.46	0.51
Standard deviation	5.10	0.01	0.69	0.06	0.04	0.03	0.02

4. Discussion

This study aims to compare the spatial resolution and noise of reconstructed images using the SIRT and FBP techniques. The results indicate that in the SIRT technique, the spatial resolution and noise are strongly influenced by the number of iterations. As the number of iterations increase, the spatial resolution increases and the noise decreases. Our results show that using the SIRT technique with 35 iterations, the spatial resolution is similar to that obtained with FBP. If the number of iterations is greater than 35, the SIRT produces a higher spatial resolution. The noise produced by the SIRT technique using 45 iterations is similar to the noise produced by FBP. If the number of iterations is greater than 45, SIRT produces less noise. Thus the SIRT method can potentially produce images with better spatial resolution and noise, if a relatively large number of iterations (>45) is used. Clinically, the number of iterations can be set as required.

However, although SIRT can produce higher spatial resolution and lower noise, the characteristics of the reconstructed images are very different from those reconstructed by FBP. This is evident from the very different shapes of the underlying PSF curves. For instance, the tail area of the PSF curves for FBP are always positive, but the PSF curves for the SIRT technique have negative values, and this area becomes larger with the increase in the number of iterations. This difference direct impacts the shape of the MTF curves. It should be noted that the SIRT method produces artifactual values outside the phantom, which is cause for concern.

The major advantage of iterative reconstruction is that noise visible in the sinogram can be reduced, so that small mAs can be used, with a consequent reduction in radiation dose to the patient. In subsequent studies, we will add synthetic noise to the projection data to evaluate the effect of noise in images reconstructed using SIRT.

5. Conclusions

The spatial resolution and noise in the image reconstructed using the SIRT method is influenced by the number of iterations. Larger numbers of iterations result in greater spatial resolution and less noise. At the 35th iteration the spatial resolution is similar to using FBP reconstruction, and at the 45th iteration the noise is similar to using FBP.

Acknowledgements

This work was funded by the Research and Innovation Program, Institut Teknologi Bandung (ITB) No. 006n/I1.C01/PL/2016.

References

[1] Koch W, Hamann C, Welsch J, Po¨pperl G, Radau P E, and Tatsch K 2005 Is iterative reconstruction an alternative to filtered backprojection in routine processing of dopamine transporter SPECT Studies? *J. Nucl. Med.* **46** 1804–1811

[2] Beister M, Kolditz D, Kalender W A 2012 Iterative reconstruction methods in X-ray CT *Phys. Medica.* **28** 94-108

[3] Jun-Yan R, Guo-Tao F, Cun-Feng W, Cui-Li S, and Long W 2010 Measurement of spatial resolution of the micro-CT system *Chinese Physics C.* **34** 412-416

[4] Brenner D J and Hall E J 2007 Computed tomography-An increasing source of radiation exposure *N. Engl. J. Med.* **357** 2277-2284

[5] Anam C, Haryanto F, Widita R, Arif I, Dougherty G, and McLean D 2017 The impact of patient table on size-specific dose estimate (SSDE) *Australas Phys. Eng. Sci. Med.* **40** 153–158

[6] Anam C, Haryanto F, Widita R, Arif I, and Dougherty G 2016 A fully automated calculation of size-specific dose estimates (SSDE) in thoracic and head CT examinations *J. Phys.: Conf. Ser.* **694** 012030.

[7] Bauhs J A, Vrieze T J, Primak A N, Bruesewitz M R, McCollough C H 2008 CT dosimetry: comparison of measurement techniques and devices *RadioGraphics.* **28** 245-253

[8] Kalender W A 2014 Dose in x-ray computed tomography *Phys. Med. Biol.* **59** R129-R150

[9] Hussain F A, Mail N, Shamy AM, Alghamdi S, and Saoudi A 2016 A qualitative and quantitative analysis of radiation dose and image quality of computed tomography images using adaptive statistical iterative reconstruction *J. Appl. Clin. Med. Phys.* **17** 419-432

[10] Pickhardt P J, Lubner M G, Kim D H, Tang J, Ruma J A, del Rio A M, and Chen G 2012 Abdominal CT with model-based iterative reconstruction (MBIR): Initial results of a prospective trial comparing ultralow-dose with standard-dose imaging *Am. J. Roentgenol.* **199** 1–9

[11] Gordon R, Bender R, and Herman G T 1970 Algebraic reconstruction techniques (ART) for three-dimensional electron microscopy and x-ray photography *J. Theor. Biol.* **29** 471-482

[12] Andersen A H and Kak A C 1984 Simultaneous algebraic reconstruction technique (SART): a superior implementation of the ART algorithm *Ultrason Imaging* **6** 81-94

[13] Gilbert P 1972 Iterative methods for the three-dimensional reconstruction of an object from projections *J. Theor. Biol.* **36** 105-117

[14] Kayugawa A, Ohkubo M, and Wada S 2013 Accurate determination of CT point-spread-function with high precision *J. Appl. Clin. Med. Phys.* **14** 216-226

IOP Conf. Series: Journal of Physics: Conf. Series **1127** (2019) 012017 doi:10.1088/1742-6596/1127/1/012017

The formulation of $f(R)$-gravity on singular semi-Riemannian manifolds

I Robiyana and M F Rosyid

Kelompok Penelitian Kosmologi, Astrofisika, Fisika Matematik dan Fisika Partikel (KAMP)
Departemen Fisika, Universitas Gadjah Mada, Yogyakarta, 55281, Indonesia

E-mail: iqbal.robiyana@mail.ugm.ac.id

Abstract. In singular semi-Riemannian manifolds, the metric is allowed to be degenerate, i.e. it is not necessarily non-degenerate. The formulation of $f(R)$-gravity on this kind of manifold has been studied. The field equation is obtained via variational principle approach. The quantities which is involved in the formulation is a smooth quantity even when the metric becomes degenerate. These quantities are the densitized energy-momentum tensor, the Ricci curvature tensor, and the arbitrary function of scalar curvature. The resulting filed equation is a more general field equation. It reduces to the standard field equation when the metric is non-degenerate. The extension of the modified Friedmann equations has also been obtained. The generalized mass (energy) density and the generalized pressure density can be constructed based on the definition of the densitized energy-momentum tensor. These quantities remain smooth even when the metric is degenerate. The modified-$f(R)$ Schwarzschild black holes has also been considered. By coordinate transformation, the metric which is semi-regular has been obtained. The consequences of this formulation to the spacetime singularity has also been discussed.

1. Introduction

General Relativity by its cosmological model known as Λ-CDM (Lambda-Cold Dark Matter) still cannot gives an adequate explanation concerning accelerated expansion of the universe. In the framework of Λ-CDM model, this acceleration is caused by an unknown form of energy that exist abundantly in the universe. The term "dark energy" is assigned to describe this kind of energy. Until now, the origin and the nature of dark energy is still become one of mysteries in our universe.

An alternative way to understand that phenomena is by considering the modification of geometric term of the field equation. It is done by replacing Langrangian density of Einstein-Hilbert action formerly Ricci scalar, R becomes an arbitrary function that depends on R. This idea is known as $f(R)$-gravity. Based on this theory, the accelerated expansion of the universe is understood as a nature of spacetime, i.e it is purely geometrical property of spacetime.

It is well known that the mathematical framework of general relativity is semi-Riemannian manifold. That is a differentiable manifold endowed with a symmetric, non-degenerate bilinear form on its tangent bundle. In 1996, Kupeli introduced singular semi-Riemannian manifold as a generalization of semi-Riemannian manifold[1]. That is a differentiable manifold having on its tangent bundle a symmetric bilinear form which is not necessarily non-degenerate, i.e. it is allowed degenerate.

The application of singular semi-Riemannian geometry to general relativity is studied extensively by Stoica [2]. Stoica generalized the work of Kupeli by dropping the assumption of constant signature of the metric and then formulated general relativity on it [3]. More over, Stoica considered the spacetime singularity [4],[5]. It is defined as a point in spacetime where the metric is degenerate.

In this paper, we are intereseted in the formulation of $f(R)$-gravity on singular semi-Riemannian manifolds. By extending field equation that includes degenerate point, we can consider spacetime singularity in the theory of $f(R)$-gravity.

2. A brief review on $f(R)$-gravity

Action for $f(R)$-gravity in vacuum is given as follows

$$S = \int d^4x\sqrt{-g}f(R). \tag{1}$$

By applying the variational principle, $\delta S = 0$, we can finally obtain modified field equations in vacuum,

$$f'(R)R_{\mu\nu} - \frac{1}{2}f(R)g_{\mu\nu} + g_{\mu\nu}\nabla^\alpha\nabla_\alpha f'(R) - \nabla_\mu\nabla_\nu f'(R) = 0. \tag{2}$$

When we consider the present of matter, the full field equations is written as

$$f'(R)R_{\mu\nu} - \frac{1}{2}f(R)g_{\mu\nu} + g_{\mu\nu}\Box f'(R) - \nabla_\mu\nabla_\nu f'(R) = \kappa T_{\mu\nu}, \tag{3}$$

where $\Box := g^{\alpha\beta}\nabla_\beta\nabla_\alpha$ is Laplace-Beltrami operator and $T_{\mu\nu}$ is energy-momentum tensor. The first modified Friedmann equation is

$$3f'(R)H^2 = \frac{f'(R)R - f(R)}{2} - 3H\dot{f}'(R) + \kappa\rho, \tag{4}$$

and the second modified Friedmann equation is

$$-2\dot{H}f'(R) = \ddot{f}'(R) - H\dot{f}'(R) + \kappa(\rho + P). \tag{5}$$

In the framework of standard model of cosmology (Λ-CDM), the accelerated expansion of the universe is caused by an unknown form of energy whose parameter of state is $\omega < -\frac{1}{3}$. The term "dark energy" is assigned to describe this kind of energy. On the other hand, in $f(R)$-gravity, that parameter of state can be obtained naturally from field equations.

$f(R)$-gravity theory is not only succeed to give a satisfied explanation on the accelerated expansion of the universe, but also can gives an explanation about the other phenomena that has been predicted by Einstein general relativity. One of them is collapsing star that become black holes. The metric of modified Schwarzschild black holes for the case $R = R_0$ is written as follows.

$$ds^2 = -\left(1 - \frac{2m}{r} - \frac{\lambda}{3}r^2\right)dt^2 + \left(1 - \frac{2m}{r} - \frac{\lambda}{3}r^2\right)^{-1}dr^2 + r^2d\Omega^2 \tag{6}$$

where m is mass of the object, $\lambda = \frac{f_0}{2f'_0}$ is the ratio between value of $f(R)$ and its derivative when evaluated at $R = R_0$ and $d\Omega^2 = d\theta^2 + \sin^2\theta d\phi^2$, is the metric of the sphere S^2.

IOP Conf. Series: Journal of Physics: Conf. Series **1127** (2019) 012017 doi:10.1088/1742-6596/1127/1/012017

3. Singular semi-Riemannian manifolds

In this section, we will give a bird's eye view on singular semi-Riemannian manifold. We shall be closely following Stoica [7]. More detailed and rigorous exposition can be found in [2].

A *singular semi-Riemannian manifold* (M, g) is a differentiable manifold M endowed with a symmetric bilinear form g named metric. Note that the metric is not necessarily non-degenerate. This condition, in fact, bring some obstacles. We cannot define an isomorphism between the tangent space T_pM and the dual T_p^*M. It is induce the absence of the inverse metric. Thus, the geometric objects which is defined with the help of inverse metric cannot be constructed.

The concept of radical in the vector space can be extended to manifold by considering the tangent space which possess the property of vector space. This concept is needed to construct the covariant contraction which is normally defined with the help of inverse metric.

The *radical* of T_pM is denoted by $(T_pM)_o = (T_pM)^\perp$. It is the set of tangent vectors in T_pM which its inner product to any tangent vector in T_pM is zero. An inner product g_p on T_pM is non-degenerate if and only if $(T_pM)_o = \{\theta\}$. *The radical* of TM is defined as $T_oM = \bigcup_{p\in M}(T_pM)_o$. *The radical-annihilator* of TM is then defined as $T^\bullet M = \bigcup_{p\in M}(T_pM)^\bullet$, where $(T_pM)^\bullet \subseteq T_p^*M$ is the space of covectors at p which can be expressed as $\omega_p(X_p) = \langle Y_p, X_p\rangle$ for some $Y_p \in T_pM$ and any $X_p \in T_pM$. The sections of $T^\bullet M$ can be defined, in the general case, by

$$\mathcal{A}^\bullet(M) := \{\omega \in \mathcal{A}^1(M) | \omega_p \in (T_pM)^\bullet \text{ forany } p \in M\} \tag{7}$$

A unique non-degenerate inner product g_p on $T^\bullet M$ is defined by $g_\bullet(\omega, \tau) := \langle X, Y\rangle$ where $X^\bullet = \omega, Y^\bullet = \tau, X, Y \in \mathfrak{X}(M)$

Let T be a tensor type (r, s). Tensor T is called radical in the k-th contravarian slot if $T \in \mathcal{T}_0^{k-1}M \otimes_M T_oM \otimes_M \mathcal{T}_s^{r-k}M$. It is called radical-annihilator in the l-th covariant slot if $T \in \mathcal{T}_{l-1}^{r}M \otimes_M T^\bullet M \otimes_M \mathcal{T}_{s-l}^{0}M$. The covariant contraction between two covariant indices in which a tensor is radical-annihilator can be defined by the inner product $g_\bullet$. The contraction $C_{kl}T$ of a tensor field T is defined by $T(\omega_1, ..., \omega_r, v_1, ..., \bullet, ..., \bullet, ...v_s)$.

The Koszul form is defined as $\mathcal{K} : \mathfrak{X}(M)^3 \to \mathcal{C}^\infty(M)$,

$$\mathcal{K}(X, Y, Z) := \frac{1}{2}\{X\langle Y, Z\rangle + Y\langle Z, X\rangle - Z\langle X, Y\rangle - \langle X, [Y, Z]\rangle + \langle Y, [Z, X]\rangle + \langle Z, [X, Y]\rangle\} \tag{8}$$

The Koszul formula becomes $\langle \nabla_X Y, Z\rangle = \mathcal{K}(X, Y, Z)$. For non-degenerate metric, the unique Levi-Civita connection is obtained by raising the 1-form, $\mathcal{K}(X, Y, \_)$. It is $\nabla_X Y = \mathcal{K}(X, Y, \_)^\sharp$. In a coordinate basis, the Koszul form reduce to,

$$\mathcal{K}_{\mu\nu\alpha} = \mathcal{K}(\partial_\mu, \partial_\nu, \partial_\alpha) = \frac{1}{2}(\partial_\mu g_{\nu\alpha} + \partial_\nu g_{\alpha\mu} - \partial_\alpha g_{\mu\nu}) = \Gamma_{\mu\nu\alpha}, \tag{9}$$

which are Christoffels symbols of the first kind.

The properties of the Koszul form is correspond to the Levi-Civita connection which is only defined when the metric is non-degenerate. Thus, the construction of the Koszul form is then used to define covariant derivative in singular semi-Riemannian manifolds.

The lower covariant derivative of a vector field Y in the direction of a vector field X is the differential 1-form $\nabla^\flat_X Y \in \mathcal{A}^1(M)$ defined as $(\nabla^\flat_X Y)(Z) := \mathcal{K}(X, Y, Z)$ for any $Z \in \mathfrak{X}(M)$. The lower covariant derivative operator is the operator $\nabla^\flat : \mathfrak{X}(M) \times \mathfrak{X}(M) \to \mathcal{A}^1(M)$ which associates to each $X, Y \in \mathfrak{X}(M)$ the differential 1-form $\nabla^\flat_X Y$.

Singular semi-Riemannian manifold (M, g) is *radical-stasionary* if it satisfies the condition $\mathcal{K}(X, Y, \_) \in \mathcal{A}^\bullet(M)$ for any vector fields $X, Y \in \mathfrak{X}(M)$. The covariant derivative of a radical-annihilator 1-form $\omega \in \mathcal{A}^\bullet(M)$ in the direction of a vector field $X \in \mathfrak{X}(M)$ defined as $\nabla : \mathfrak{X}(M) \times \mathcal{A}^\bullet(M) \to \mathcal{A}^1_d(M)$,

$$(\nabla_X \omega)(Y) := X(\omega(Y)) - g_\bullet(\nabla^\flat_X Y, \omega), \tag{10}$$

IOP Conf. Series: Journal of Physics: Conf. Series **1127** (2019) 012017 doi:10.1088/1742-6596/1127/1/012017

where $\mathcal{A}^1_d(M)$ denotes the set of 1-forms which are smooth on the regions of constant signature.

For a radical-stationary semi-Riemannian manifold (M, g), then can be defined the vector spaces of differential forms having smooth covariant derivatives:

$$\mathscr{A}^{\bullet 1}(M) = \{\omega \in \mathcal{A}^{\bullet}(M) | (\forall X \in \mathfrak{X}(M)) \nabla_X \omega \in \mathcal{A}^{\bullet}(M)\} \tag{11}$$

A semi-regular semi-Riemannian manifold is a singular semi-Riemannian manifold (M, g) which satisfies $\nabla^{\flat}_X Y \in \mathscr{A}^{\bullet 1}(M)$ for any vector fields $X, Y \in \mathfrak{X}(M)$. A radical-stationary semi-Riemannian manifold (M, g) is semi-regular if and only if for any $X, Y, Z, T \in \mathfrak{X}(M)$ satisfies $\mathcal{K}(X, Y, \bullet)\mathcal{K}(Z, T, \bullet) \in \mathscr{F}(M)$.

The Riemann curvature tensor of a singular semi-Riemannian manifold (M, g) is defined as $R : \mathfrak{X}(M) \times \mathfrak{X}(M) \times \mathfrak{X}(M) \times \mathfrak{X}(M) \to \mathcal{C}^{\infty}(\mathcal{M})$,

$$R(X, Y, Z, T) := \nabla_X \nabla^{\flat}_Y Z - \nabla_Y \nabla^{\flat}_X Z - \nabla^{\flat}_{[X,Y]} Z, \tag{12}$$

for any $X, Y, Z, T \in \mathfrak{X}(M)$

The Riemann curvature of a semi-regular semi-Riemannian manifold (M, g) is a smooth tensor field $R \in \mathcal{T}^0_4 M$. In a coordinate basis, the components of the Riemann curvature tensor are given by

$$R_{\beta\mu\nu\alpha} = \partial_\beta \mathcal{K}_{\mu\nu\alpha} - \partial_\mu \mathcal{K}_{\beta\nu\alpha} + g^{\rho\lambda}_{\bullet} (\mathcal{K}_{\beta\nu\rho}\mathcal{K}_{\mu\alpha\lambda} - \mathcal{K}_{\mu\nu\rho}\mathcal{K}_{\beta\alpha\lambda}). \tag{13}$$

The Ricci curvature tensor is defined as the covariant contraction of the Riemann curvature tensor, $Ric(X, Y) := R(X, \bullet, Y, \bullet)$ and *the Ricci scalar* is defined as the covariant contraction of the Ricci curvature tensor, $R := Ric(\bullet, \bullet)$.

4. The formulation and its relation to spacetime singularity

4.1. Extended field equation

Let us reconsider action for $f(R)$-gravity in vacuum as follows

$$S = \int d^4x \sqrt{-g} f(R). \tag{14}$$

By considering the construction of Ricci scalar on singular semi-Riemannian manifold, the variation of action becomes

$$\delta S = \int d^4x \sqrt{-g} [f'(R) R_{\mu\nu} - \frac{1}{2} f(R) g_{\mu\nu}] \delta g^{\mu\nu}_{\bullet} + \int d^4x \sqrt{-g} f'(R) g^{\mu\nu}_{\bullet} \delta R_{\mu\nu}. \tag{15}$$

The last term in the equation 15 above must be formed as variation of $\delta g^{\mu\nu}_{\bullet}$. By calculating the expression of $g^{\mu\nu}_{\bullet} \delta R_{\mu\nu}$, the last term of 15 is written as follows

$$\int d^4x \sqrt{-g} f'(R) g^{\mu\nu}_{\bullet} \delta R_{\mu\nu} = \int d^4x \sqrt{-g} g_{\mu\nu} g^{\alpha\beta}_{\bullet} \partial_\beta \partial_\alpha f'(R) \delta g^{\mu\nu}_{\bullet} - \int d^4x \sqrt{-g} \partial_\mu \partial_\nu f'(R) \delta g^{\mu\nu}_{\bullet}. \tag{16}$$

By joining the above term to the total action and then applying variational principle, $\delta S = 0$, we arrive to the extended field equation as follows

$$f'(R) R_{\mu\nu} \sqrt{-g} - \frac{1}{2} f(R) g_{\mu\nu} \sqrt{-g} + g_{\mu\nu} \Box f'(R) \sqrt{-g} - \nabla_\mu \nabla_\nu f'(R) \sqrt{-g} = 0, \tag{17}$$

where $\Box := g^{\alpha\beta}_{\bullet} \nabla_\beta \nabla_\alpha$ is Laplace-Beltrami operator in singular semi-Riemannian manifolds.

When the present of matter is considered, the Lagrangean density is also contain the component of volume form, $\sqrt{-g}$. Therefore, the full field equations is written as

$$f'(R) R_{\mu\nu} \sqrt{-g} - \frac{1}{2} f(R) g_{\mu\nu} \sqrt{-g} + g_{\mu\nu} \Box f'(R) \sqrt{-g} - \nabla_\mu \nabla_\nu f'(R) \sqrt{-g} = \kappa T_{\mu\nu} \sqrt{-g} \tag{18}$$

In obtaining the field equations above, it is not just refrain to divide by $\sqrt{-g}$. But, we have to ensure that at every step in the formulation we have only smooth quantities.

4.2. Extension of Modified Friedmann equation

The construction below is obtained by following the method developed by Stoica [5]. We extend the result to the $f(R)$-gravity.

Consider the first modified Friedmann equation which is expressed in term mass (energy) density and pressure density as follows.

$$\rho = \frac{3}{\kappa}\left[\left(\frac{\dot{a}}{a}\right)^2 f'(R) - \frac{f'(R)R - f(R)}{6} + \frac{\dot{a}}{a}\dot{f}'(R)\right]. \tag{19}$$

and the second modified Friedmann equation

$$\rho + 3P = -\frac{3}{\kappa}\left[2\frac{\ddot{a}}{a}f'(R) + \frac{\dot{a}}{a}\dot{f}'(R) - \frac{2}{6}\left(f'(R)R - f(R)\right) - \ddot{f}'(R)\right]. \tag{20}$$

Two equations above shows that when $a \to 0$, mass (energy) density and pressure density appears to tend to infinity. In other words, a finite amount of matter occupies a very small volume. At that situation, mass (energy) density becomes infinite and then "bang!" spanned spacetime. That event is called by bigbang.

The construction of field equations in the previous section shows that the more fundamental object is in fact, energy-momentum tensor weight +1, that is $T_{\mu\nu}\sqrt{-g}$. This implies to introduce the definition of generalized mass (energy) density $\rho\sqrt{-g}$ and generalized pressure density $P\sqrt{-g}$ as a natural quantity in fluid equation of the model of the universe. Both quantities is expressed as ρ and P only in an orthonormal frame, where the determinant of the metric equals -1 so that component $\sqrt{-g}$ is negligible. But, when $a = 0$, an orthonormal frame would become singular, because $g = 0$. Therefore, we avoid the usage of ρ and P because they cannot be define there.

This idea is in accordance with the study of differential form. Both quantities ρ and P are scalar fields as seen by an observer moving in an orthonormal frame. The invariant quantities is in fact must involve the volume form which its component in a coordinate system are $\rho\sqrt{-g}$ and $P\sqrt{-g}$.

In a non-singular coordinate system, g has to be variable, as it is in the comoving coordinate system of the flat FLRW model which is expressed by $\sqrt{-g} = a^3$. Based on those idea, we can define *densitized* energy-momentum tensor as

$$T_{\mu\nu}\sqrt{-g} = (\tilde{\rho} + \tilde{P})u_\mu u_\nu + \tilde{P}g_{\mu\nu}, \tag{21}$$

where $\tilde{\rho} = \rho\sqrt{-g} = \rho a^3$ and $\tilde{P} = P\sqrt{-g} = Pa^3$. By substituting the expression and to the modified Friedmann equation, we finally obtain the modified Friedmann equation which is smooth even at $a = 0$ as follows

$$\tilde{\rho} = \frac{3}{\kappa}\left[\dot{a}^2 a f'(R) - \frac{a^3}{6}\Big(f'(R)R - f(R)\Big) + \dot{a}a^2\dot{f}'(R)\right] \tag{22}$$

and

$$\tilde{\rho} + 3\tilde{P} = -\frac{3}{\kappa}\left[2\ddot{a}a^2 f'(R) + \dot{a}a^2\dot{f}'(R) - \frac{2a^3}{6}\Big(f'(R)R - f(R)\Big) - a^3\ddot{f}'(R)\right]. \tag{23}$$

4.3. Semi-regularity of modified Schwarzschild black holes

Following Stoica [4], we develop here the coordinate transformation for modified Schwarzschild black holes. By coordinate change $r = \tau^2$ and $t = \xi\tau^T$, we can obtain a new metric which is written as follows

$$ds^2 = -\frac{12\tau^4}{6m - 3\tau^2 + \lambda\tau^6}d\tau^2 + \left(\frac{6m - 3\tau^2 + \lambda\tau^6}{3}\right)\tau^{2T-4}(T\xi d\tau + \tau d\xi)^2 + \tau^4 d\Omega^2. \tag{24}$$

By finding the value of T which satisfy the semi-regularity condition, we obtain $T = 4$ as a unique solution. Finally, we obtain a metric which is an analytic extension of the modified Schwarzschild metric. That is a metric which is semi-regular and analytic including at the point $r = 0$ by a coordinate transformation $r = \tau^2$ and $t = \xi\tau^4$ as follows

$$ds^2 = -\left(\frac{12\tau^4}{6m - 3\tau^2 + \lambda\tau^6}\right)d\tau^2 + \left(\frac{6m - 3\tau^2 + \lambda\tau^6}{3}\right)\tau^4(4\xi d\tau + \tau d\xi)^2 + \tau^4 d\Omega^2. \qquad (25)$$

5. Conclusion and discussion

In the framework of the metric, there are two types of singularities[6]:

(i) *malign singularity*: some of the components of the metric are divergent, $g_{\mu\nu} \to 0$

(ii) *benign singularity*: the components of the metric $g_{\mu\nu}$, are smooth and finite, but its determinant, $g \to 0$

An obvious definition of singularity would seem to be that it is a point where the metric is singular i.e fails to be suitably differentiable[8]. Based on this idea, such spacetime singularity refers to *malign singularity*. That is because the usual spacetime is modeled as a differentiable manifold which is equipped by the metric that must be non-degenerate. On the other hand, the other type, *benign singularity* is not used to define spacetime singularity.

In the previous section, we have shown the work of Stoica [2] for the mathematical construction of manifold which is equipped by the metric that can be degenerate. The concept of *benign singularity* is then used to consider spacetime singularity. Spacetime singularity is defined as a point where the metric becomes degenerate. By this point of view, the field equation that has been obtained above is more general because it includes singularities. The extension of the modified Friedmann equation that includes degenerate point can be viewed as the extension to the bigbang singilarity. We also obtain a new netric for modified Schwarzschild black holes. It can be viewed as a transformation from malign singilarity to benign singularity. In the new metric, the singularity at $r = 0$ is benign. The consideration of singularity problem in $f(R)$-gravity as in [9] is left for future work.

Acknowledgments

The present work has been partially supported by hibah Penelitian Unggulan Perguruan Tinggi (PUPT) No. 015/SP2H/LT/DRPM/II/2016; I.R. is supported by scholarship from Indonesia Endowment Fund for Education (LPDP), Ministry of Finances, Republic of Indonesia.

References

[1] Kupeli D 1996 *Singular Semi-Riemannian Geometry* (Amsterdam: Kluwer Academic Publishers Group)
[2] Stoica O C 2014 *Int. J. Geom. Methods Mod. Phys.* **11**(5) 1450041
[3] Stoica O C 2014 Singular General Relativity *Preprint* gr-qc/13012231v4
[4] Stoica O C 2012 *Eur. Phys. J. Plus* **127**(7) 18
[5] Stoica O C 2016 *Int. J. Theor. Phys.* **55**(1) 71-80
[6] Stoica O C 2012 *An. Şt. Univ. Ovidius Constanţa* **20** 213-238
[7] Stoica O C 2015 *J. Phys.: Conf. Series* **626** 1-8
[8] Hawking S W 2014 *Eur. Phys. J. H* **39**(4) 413-503
[9] Frolov A V 2008 *Phys. Rev. Let* **101**(6)

IOP Conf. Series: Journal of Physics: Conf. Series **1127** (2019) 012018 doi:10.1088/1742-6596/1127/1/012018

Inflation due to non-minimal coupling of $f(R)$ gravity to a scalar field

Romy H.S. Budhi

Physics Department, Universitas Gadjah Mada, Indonesia

E-mail: romyhanang@ugm.ac.id

Abstract. In this work we investigate a inflationary scenario generated by a large scalar field ϕ that non-minimally couples to a $f(R)$ modified gravity model. For a Starobinsky's like model, it is found that along a particular flat direction, the scalar potential takes a simple form $V = (M_p^4/4)[V(\phi)/\alpha(\phi)^2]$ where $\alpha(\phi)$ is a non-minimal coupling to Ricci scalar R in the model. The inflation, therefore, is effectively represented as a single field inflaton scenario. For a specific example, such as a scalar potential $V(\phi) = \mu_1\phi^2 + \mu_2\phi^4$, we found that the predictions match nicely in the 1σ confidence level of *Plank* TT, TE, EE+lowP combination data of Planck 2015 CMB data for $0 < \mu_3 \leq 100$, where $\mu_3 := |\mu_1|/(\mu_2 M_p^2)$. For example, taking $\mu_3 = 0.01$ the scalar-to-tensor ratio $r = 0.0004$ and and spectral index $n_s = 0.96985$ for $N_* = 50$ while taking $\mu_3 = 100.0$ produces $r = 0.03$ and $n_s = 0.96359$ for $N_* = 60$.

1. Introduction

Inflation, a rapid accelerated expansion in the early universe, was originally introduced to solve the flatness and horizon problems in the big bang cosmology. The inflation provides a mechanism to generate density fluctuation which later evolve into large scale structures in the universe. Its prediction to the almost scale-invariant spectrum of cosmological perturbations is in remarkable agreement with the high-precision CMBR observations such as the COBE, WMAP, PLANCK,and BICEP2. The recent Planck-2015 mission measures the scalar spectral index as $n_s = 0.968 \pm 0.006$ and combined with the BICEP2/Keck Array CMB polarization experiments have put a bound on the tensor-to-scalar ratio as $r < 0.12$ (95% CL). These constraints imply that a single field inflation models with quartic $\lambda\phi^4$ and quadratic $m^2\phi^2$ potentials are ruled out from the observations as they produce large tensor-to-scalar ratio $r \simeq 0.26$ and $r \simeq 0.13$, respectively, compared to the inflationary scenario motivated from modified gravitational sector view point, such as R^2 correction known as Starobinsky model that predicts smaller scalar-to-tensor ratio [1]. A novel scenario from particle physics view point that is mathematically equivalent to the R^2 inflation is now known as the Higgs inflation scenario [2], where the inflaton field ϕ is non-minimally coupled to the curvature scalar R. However, due to a very large non-minimal coupling to the curvature scalar to obtain the observed CMB amplitude, such a model however encounters the problem of unitarity violation in Higgs-Higgs scattering via graviton exchange at Planck energy scales [3, 4]. On the other hand, a generalization of scalar-curvature coupling $\xi\phi^2 R$ of Higgs inflation to $\xi\phi^a R^b$ eliminates this large coupling problem [5]. However, this kind of generalized non-minimally coupled model with an additional quantum corrected quartic potential $\lambda\phi^{4(1+\gamma)}$, which is equivalent to the inflation model $R + R^\beta$ [6] produces large

$r \simeq 0.2$, and thus is disfavored by present CMB observation [1]. Inspired from this trend, in this paper, we would like elaborate non-minimal coupling of a scalar field to the curvature scalar in the context of general modified gravity $f(R)$ theory. To have a comparison with the previous studies, a specific $f(R)$ and scalar potential will be considered in the detail.

2. Action of $f(R)$ gravity with non-minimal coupling to a scalar

The action for a scalar field ϕ in a $f(R)$ gravity theory can be written as

$$\mathcal{S} = \int d^4x\sqrt{-g}\left[f(R) - g^{\mu\nu}\partial_\mu\phi\partial_\nu\phi - V(\phi)\right] \tag{1}$$

in the *Jordan frame*, where here $g^{\mu\nu}$ denotes metric tensor of the space. If there are non-minimal couplings between $f(R)$ gravity with the scalar ϕ, to accommodate such couplings, the function $f(R)$ should include any term containing Ricci scalar R and ϕ. Therefore, $f(R)$ in the action can be modified to be $F(R,\phi)$ [5]. The gravitational part of the action may be assumed to be generated by an auxiliary scalar field χ [7, 8, 9], so that in this context $F(R,\phi) = F(\chi,\phi)$. Without losing generality, if we work in arbitrary D dimension space-time, the associated gravitational action term satisfies

$$\int d^Dx\sqrt{-g}\left[F(R,\phi)\right] = \int d^Dx\sqrt{-g}\left[F_\chi(R-\chi) + F\right], \tag{2}$$

where $F_\chi := \partial F(\chi,\psi)/\partial\chi$. Varying the action (2) with respect to χ yields conditions $F_{\chi\chi} = 0$ atau $\chi = R$, showing that it is classically equivalent to the original model if we take $\chi = R$. Thus, rearrangement of equation (1) leads into this following form

$$\mathcal{S} = \int d^Dx\sqrt{-g}\left[F_\chi R - g^{\mu\nu}\partial_\mu\phi\partial_\nu\phi - (V(\phi) - (F - \chi F_\chi))\right]. \tag{3}$$

The action (3) can be rewritten to the Einstein frame by conformal transformation $\hat{g}_{\mu\nu} = \Omega^2 g_{\mu\nu}$ [10], to gives rise to the standard Einstein-Hilbert action containing $(M_p^2R/2)$ term where M_p denotes (reduced) Planck mass in D dimension which is $M_p = (8\pi GN)^{-1/2} = 2.4\times10^{18}$ GeV in $D = 4$. Here, we use a caret to indicate quantities in the Einstein frame. One finds some transformed quantities that are useful in the calculation:

$$\hat{g}^{\mu\nu} = \Omega^{-2}g^{\mu\nu}, \tag{4}$$

$$\sqrt{-\hat{g}} = \Omega^D\sqrt{-g}, \tag{5}$$

$$\hat{R} = \Omega^{-2}\left[R - \frac{2(D-1)}{\Omega}\Box\Omega - \frac{(D-1)(D-4)}{\Omega^2}g^{\mu\nu}\nabla_\mu\Omega\nabla_\nu\Omega\right] \tag{6}$$

where $\Box\Omega := g^{\mu\nu}\nabla_\mu\nabla_\nu\Omega = \frac{1}{\sqrt{-g}}\partial_\mu\left[\sqrt{-g}g^{\mu\nu}\partial_\nu\Omega\right]$. It is possible to get rid of the non-minimal coupling remaining in action (3) if we identify

$$\Omega^{D-2} = \frac{2}{M_p^{D-2}}F_\chi. \tag{7}$$

As a result, the action in the Einstein frame is

$$\mathcal{S}_E = \int d^Dx\sqrt{-\hat{g}}\left[\frac{M_p^{D-2}\hat{R}}{2} - \frac{M_p^{D-2}}{2}\frac{(D-1)}{(D-2)}\frac{\hat{g}^{\mu\nu}}{F_\chi^2}\hat{\nabla}_\mu F_\chi\hat{\nabla}_\nu F_\chi - \frac{M_p^{D-2}}{2}\frac{\hat{g}^{\mu\nu}}{F_\chi}\hat{\nabla}_\mu\phi\hat{\nabla}_\nu\phi - V_E\right], \tag{8}$$

where we have introduced the transformed potential from the original one

$$V_E := \frac{M_p^D}{2^{D/(D-2)}} \frac{[V(\phi) - (F - \chi F_\chi)]}{F_\chi^{D/(D-2)}}. \tag{9}$$

The transformed action (8) still contains a non-canonical kinetic term in the second term of the equation and a mixing term to the scalar field ϕ's kinetic term. The former can be further simplified to the canonical kinetic term by doing the field redefinition

$$\psi := \sqrt{\frac{(D-1)}{(D-2)}} \ln(F_\chi), \quad \tilde{\psi} := \sqrt{\frac{D-2}{D-1}}\psi. \tag{10}$$

Hence, after some treatments, we obtain the final action:

$$\mathcal{S}_E = \int d^D x\sqrt{-\hat{g}} \left[\frac{M_p^{D-2}}{2}\hat{R} - \frac{M_p^{D-2}}{2}\hat{g}^{\mu\nu}\hat{\nabla}_\mu\psi\hat{\nabla}_\nu\psi - \frac{M_p^{D-2}}{2}e^{-\tilde{\psi}}\hat{g}^{\mu\nu}\hat{\nabla}_\mu\phi\hat{\nabla}_\nu\phi - V_E\right]. \tag{11}$$

3. Starobinsky model with non-minimal coupling to a scalar field

In this section, we will investigate a particular example of $f(R)$ gravity theory that contains R^2 term called Starobinsky model [11]. If we consider a nonminimally coupling to a scalar field, the gravitational Lagrangian density can be written as $f(R) = \alpha(\phi)R + \beta(\phi)R^2$. Thus, we can find the associated Lagrangian density in term of an auxiliary field χ as $F(\chi, \phi) = \alpha(\phi)\chi + \beta(\phi)\chi^2$ and we have $\psi = \sqrt{3/2}\ln[\alpha + 2\beta\chi]$. The potential in the Einstein frame from equation (9) take this following form

$$V_E := \left(\frac{M_p^2}{2}\right)^2 \left[e^{-2\tilde{\psi}}V(\phi) - \frac{\left(1 - \alpha e^{-\tilde{\psi}}\right)^2}{2\beta}\right]. \tag{12}$$

Therefore, there is a flat direction along $\tilde{\psi} = \ln(\alpha)$ that the inflaton proceeds it on. Along this direction, the potential is given as

$$V = \frac{M_p^4}{4}\frac{V(\phi)}{\alpha(\phi)^2} \tag{13}$$

showing that the inflation is effectively generated by a single field ϕ thus formulation corresponding to single field inflation can be used.

For a specific chase, we consider a scalar field with a general potential form $V(\phi) = \mu_1\phi^2 + \mu_2\phi^4$ where $\mu_1 < 0$ to guarantee the field having a vacuum expectation value (VeV) at $v := \sqrt{-\mu_1/2\mu_2}$. In this model, it is also assumed a condition for non-minimal coupling to gravity to be $\alpha(\phi) = a\phi^2$. The potential (12), therefore, can be written to be

$$V = V_0\left[1 - \mu_3\tilde{\phi}^2\right] \tag{14}$$

where here we have defined $V_0 := \frac{\mu_2 M_p^4}{4a^2}$, $\mu_3 := \frac{|\mu_1|}{\mu_2 M_p^2}$ and $\tilde{\phi} := \frac{M_p}{\phi}$. This is a kind of inverse power law inflation that is promising to explain Planck experimental data for CMB and having a similar features as the mutated hybrid inflationary models [12]. Slow-roll parameters for this single field inflation are:

$$\varepsilon = \frac{M_p^2}{2}\left[\frac{V'}{V}\right]^2 = \frac{2\mu_3^2\tilde{\phi}^6}{(1 - \mu_3\tilde{\phi}^2)^2}, \quad \eta = M_p^2\left[\frac{V''}{V}\right] = -\frac{6\mu_3\tilde{\phi}^4}{(1 - \mu_3\tilde{\phi}^2)}, \tag{15}$$

Table 1: Numerical results for several variations of μ_3 and N_*.

μ_3	N_*	n_{s*}	r_*	VeV (v) $\times M_p$	a $\times 10^4$
0.01	50.0	0.96985	0.00040	0.07070	13.1755
0.01	60.0	0.97489	0.00030	0.07070	15.1061
1.00	50.0	0.96850	0.00400	0.70711	4.16645
1.00	60.0	0.97386	0.00304	0.70711	4.77697
10.0	50.0	0.96526	0.01265	2.23607	2.34297
10.0	60.0	0.97139	0.00962	2.23607	2.68629
30.0	50.0	0.96178	0.02191	3.87298	1.78027
30.0	60.0	0.96875	0.01667	3.87298	2.04114
50.0	50.0	0.95939	0.02828	5.00000	1.56684
50.0	60.0	0.96693	0.02152	5.00000	1.79643
75.0	50.0	0.95701	0.03464	6.12372	1.41580
75.0	60.0	0.96512	0.02635	6.12372	1.62326
100.0	50.0	0.95500	0.04000	7.07107	1.31755
100.0	60.0	0.96359	0.03043	7.07107	1.51061

where $V' := dV/d\phi$ and $V'' := d^2V/d\phi^2$. Those slow-roll parameters show that the inflation starts, as indicated by condition $\varepsilon \ll 1$, at scale $(\mu_3\tilde{\phi}^2) \ll 1$ and terminates while $\varepsilon \simeq 1$ or at the scale $(\mu_3\tilde{\phi}_e^2) \simeq 1 + \sqrt{2}$. Observational parameters, the scalar spectrum index n_s and the ratio of the tensor perturbation to the scalar perturbation r, can be represented by using those two slow-roll parameters as follows [13]

$$n_s = 1 - 6\varepsilon + 2\eta, \quad r = 16\varepsilon \tag{16}$$

During that stage, the inflaton evolves from a particular value ϕ into its value at the end of inflation ϕ_e as given by e-folding number

$$N = -\frac{1}{M_p^2}\int_{\phi}^{\phi_e} \frac{V}{V'} d\phi = \frac{1}{2\mu_3}\left[\frac{\tilde{\phi}^{-4}}{4} - \frac{\mu_3\tilde{\phi}^{-2}}{2}\right]_{\tilde{\phi}_e}^{\tilde{\phi}} := N(\tilde{\phi}) - N(\tilde{\phi}_e) \tag{17}$$

where here, we have defined

$$N(\tilde{\phi}) := \frac{1}{2\mu_3}\left[\frac{\tilde{\phi}^{-4}}{4} - \frac{\mu_3\tilde{\phi}^{-2}}{2}\right]. \tag{18}$$

The spectrum of scalar perturbation predicted by the inflation is expressed as [13]

$$\mathcal{P}_{\mathcal{R}}(k) = \Delta_{\mathcal{R}}^2 \left(\frac{k}{k^*}\right)^{n_s - 1}, \qquad \Delta_{\mathcal{R}}^2 = \frac{V}{24\pi^2 M_{\rm pl}^4 \varepsilon}\bigg|_{k^*}. \tag{19}$$

The CMB observations give the normalization such that $\Delta^2_{\mathcal{R}} \simeq 2.43 \times 10^{-9}$ at $k_* = 0.002\ \mathrm{Mpc}^{-1}$. This constrains the value of V/ε at the time when the scale characterized by the wave number k_* exits the horizon [15]. On the other hand, the remaining e-foldings N_* of the inflation after the scale k_* exits the horizon is dependent on the reheating phenomena and others as [13, 14]

$$N_* \simeq 61.4 - \ln\frac{k_*}{a_0 H_0} - \ln\frac{10^{16}\ \mathrm{GeV}}{V_{k_*}^{1/4}} + \ln\frac{V_{k_*}^{1/4}}{V_{\mathrm{end}}^{1/4}} - \frac{1}{3}\ln\frac{V_{\mathrm{end}}^{1/4}}{\rho_{\mathrm{reh}}^{1/4}}. \tag{20}$$

Taking account of this uncertainty, N_* is usually considered to take a value in the range 50 - 60. Here we also use the values in this range and we represent a value of ϕ which gives the e-foldings N_* as ϕ_*. The observational observables, tensor to scalar ratio r and the spectral index n_s, need to be expressed at this value to be able have comparison with the observation. These values will be denoted as r_* and n_{s*}.

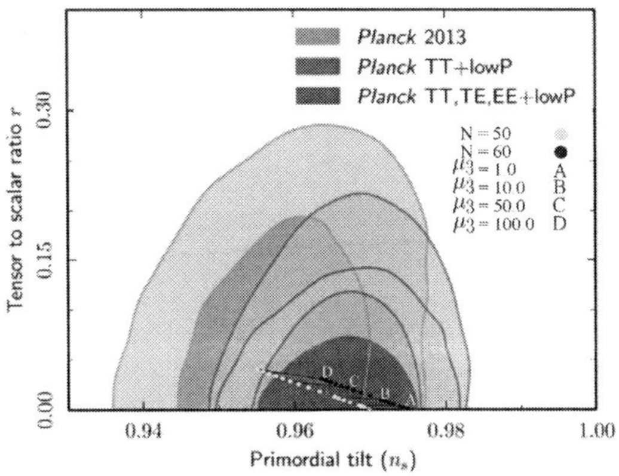

Figure 1: Predicted values of (n_s, r) for various μ_3 values in the interval $(0, 100)$ as given in Table 1. Contours given as Fig. 6 in Planck Collaboration [1] are used here.

4. Predictions

To discuss the features of the model, numerical calculations are needed particularly to consider all possible terms in the formula. However, in this model, the inflation ends at scale $\tilde{\phi}_e \simeq \sqrt{(1+\sqrt{2})/\mu_3}$. Therefore, for $\mu_3 < 10^2$, $N_e = N(\tilde{\phi}_e)$ is negligible thus e-folding number can be approximately given as it value when the comoving scale crosses the Hubble radius at the first time $N \simeq N_*$. Considering that the inflation starts at the scale $(\mu_3\tilde{\phi}^2) \ll 1$, the leading contributions of the e-folding number and slow-roll parameters are

$$N_* \simeq \frac{\tilde{\phi}^{-4}}{8\mu_3}, \quad \varepsilon_* \simeq 2\mu_3^2\tilde{\phi}^6 \simeq \sqrt{\frac{\mu_3}{2^7}}\frac{1}{N_*^{3/2}}, \quad \eta_* \simeq -6\mu_3\tilde{\phi}^4 \simeq -\frac{3}{4}\frac{1}{N_*}. \tag{21}$$

The scalar-to-tensor ratio r_* and the spectral index n_{s*} therefore will depend to N_* and μ_3 as follows

$$n_{s*} = 1 - 6\sqrt{\frac{\mu_3}{2^7}}\frac{1}{N_*^{3/2}} - \frac{3}{2}\frac{1}{N_*}, \quad r_* = 16\sqrt{\frac{\mu_3}{2^7}}\frac{1}{N_*^{3/2}}. \tag{22}$$

CMB maps provided by Planck 2015 constrains the tensor-to-scalar ratio and the spectral index given as $n_s = 0.968 \pm 0.006$ and $r < 0.11$ (95% CL) respectfully [1]. Due to these constraints, several numerical results are shown in the table 1 for some value of μ_3 with a fixed $N_* = 50, 60$, as well as in the Figure 1 for the predicted points (n_s, r) in the contour provided from Planck 2015.

The examples show that the predicted points match nicely in the 1σ confidence level of *Plank* TT, TE, EE+lowP combination data for $0 < \mu_3 \leq 100$, for which the vacuum expectation value v of the scalar field ϕ and non-minimal coupling constant to the R gravity term a can be estimated. The latter one can be approximated from the CMB normalization given in equation (19). The scalar generating the inflation, therefore, should evolved in the transPlanckian value and having vacuum expectation value $v = \sqrt{\mu_3/2}M_p$. The non-minimal coupling is typically in 10^4 order which is in the same order of non-minimal coupling of the Higgs inflation scenario [2]. This large value had been shown causing several difficulties such as unitarity problem [3, 4, 16], however some treatments can be used to cure the problem [17, 18]. The reason might be from the absence of the non-minimal coupling to R^2 gravity term when the inflaton takes the flat direction in the potential (14), therefore using other scenario in this context may treat the problem. Instead of the difficulty that still appears, the main purpose to save scalar power law potentials which are already disfavored by recent observations can be achieved.

References

[1] Planck Collaboration (P.A.R. Ade (Cardiff U.) et al.), *Planck 2015 results. XX. Constraints on inflation*, Astron.Astrophys. 594 (2016) A20, arXiv:1502.02114 [astro-ph.CO].
[2] Fedor L. Bezrukov and Mikhail Shaposhnikov, *The Standard Model Higgs boson as the inflaton*, Phys.Lett. B659 (2008) 703-706, arXiv:0710.3755 [hep-th].
[3] Burgess, C.P. et al. *Power-counting and the Validity of the Classical Approximation During Inflation*, JHEP 0909 (2009) 103, arXiv:0902.4465 [hep-ph]
[4] Barbon, J.L.F. et al, *On the Naturalness of Higgs Inflation*, Phys.Rev. D79 (2009) 081302, arXiv:0903.0355 [hep-ph].
[5] Girish Chakravarty, Subhendra Mohanty, Naveen K. Singh, *Higgs Inflation in $f(\phi, R)$ Theory*, Int. J. Mod. Phys. D 23, 1450029 (2014), arXiv:1303.3870 [astro-ph.CO].
[6] G. Chakravarty, S. Mohanty and N. K. Singh, Int. J. Mod. Phys. D 23, no. 4, 1450029 (2014); J. Joergensen, F. Sannino and O. Svendsen, arXiv:1403.3289 [hep-ph]; A. Codello, J. Joergensen, F. Sannino and O. Svendsen, arXiv:1404.3558 [hep-ph]; G. K. Chakravarty and S. Mohanty, Phys. Lett. B 746, 242 (2015).
[7] de la Cruz-Dombriz, Alvaro et al., *Spotting deviations from R^2 inflation*, JCAP 1605 (2016) no.05, 060 arXiv:1603.05537 [gr-qc].
[8] Torabian, Mahdi, *When Higgs Meets Starobinsky in the Early Universe*, IPM-PA-381, arXiv:1410.1744 [hep-ph].
[9] Sho Kaneda and Sergei V. Ketov, *Starobinsky-like two-field inflation*, Eur. Phys. J. C (2016) 76:26.
[10] Kaiser, David I, *Conformal Transformations with Multiple Scalar Fields*, Phys.Rev. D81 (2010) 084044 arXiv:1003.1159 [gr-qc] MIT-CTP-4125
[11] A. A. Starobinsky, Phys. Lett. B 91 (1980) 99.
[12] Zhun Lu, *Inflation in the Generalized Inverse Power Law Scenario*, JCAP 1311 (2013) 038, arXiv:1311.0348 [hep-th].
[13] D. H. Lyth and A. Riotto, Phys. Rep. **314** (1999) 1; A. R. Liddle and D. H. Lyth, *Cosmological Inflation and Large-Scale Structure* (Cambridge, 2000).
[14] D. S. Gorbunov, V. A. Rubakov, *Introduction to the theory of the early universe, Cosmological perturbations and inflationary theory*, World Scientific, 2011.
[15] WMAP Collaboration, D. N. Spergel, *et al.*, *First year Wilkinson Microwave Anisotropy Probe (WMAP) observations: Determination of cosmological parameters.*
[16] F. Bezrukov, A. Magnin, M. Shaposhnikov (ITPP, Lausanne), and S. Sibiryakov, *Higgs inflation: consistency and generalisations*, JHEP 1101 (2011) 016, arXiv:1008.5157 [hep-ph].
[17] Lerner, Rose N. et al., *A Unitarity-Conserving Higgs Inflation Model,* Phys.Rev. D82 (2010) 103525, arXiv:1005.2978 [hep-ph].
[18] Hyun Min Lee, *Running inflation with unitary Higgs*, Phys.Lett. B722 (2013) 198-206, arXiv:1301.1787 [hep-ph].

Preliminary study of bamboo-like tree structure based on granular particle-spring model: Relaxation and tortuosity

S Viridi[1,]*, S N Khotimah[2], R Kurniadi[3], Novitrian[4], K Basar[5], A Purqon[6]

[1-6]Physics Department, Faculty of Mathematics and Natural Sciences, Institut Teknologi Bandung, Jalan Ganesha 10, Bandung 40132, Indonesia
[1]National Center for Nanotechnology, Jalan Ganesha 10, Bandung 40132, Indonesia
[1-5]Nuclear Physics and Biophysics Research Division, Institut Teknologi Bandung, Jalan Ganesha 10, Bandung 40132, Indonesia
[6]Earth Physics and Complex System Research Division, Institut Teknologi Bandung, Jalan Ganesha 10, Bandung 40132, Indonesia

[1]dudung@fi.itb.ac.id, [2]nurul@fi.itb.ac.id, [3]rijalk@fi.itb.ac.id, [4]novit@fi.itb.ac.id, [5]khbasar@fi.itb.ac.id, [6]acep@fi.itb.ac.id

Abstract. Bamboo-like tree structure is modelled using granular particles connected with linear and bending springs. The first type of spring is responsible for contraction and relaxation in axial direction, while the second one is in angular direction. For simplicity the system is first considered only two-dimension in this work. Previous work showed that for two branches tree, wind influenced only one branch which has resonance with the wind velocity. Only relaxation stage is reported in this work, that can not yet achieved stable state.
Keywords: granular particle, spring, bamboo tree, relaxation stage

1. Introduction
It is interesting to study bamboo and its structure, since its die-back could influence tree regeneration [1] as important component of ecosystems [2], it can be used as functionally graded but affordable materials [3], its-like structure is observed in nanotubes growth [4], and also its interactions with environment that influences its morphogenesis [5]. Bamboo-like trees can be considered as a cylindrical beam or a wire, which has certain Youngs' modulus E and shear modulus G. For a wire a simple model using only G as parameter has been proposed [6], and also for two-branches-tree with using linear and bending springs [7]. In this work, bending spring is replaced by linear spring and to achieve similar effect the linear springs must have certain configuration explained in the model.

2. Theory

2.1. Model
In this part linear springs refers to push and pull spring, which is differed from bending spring. The first type of spring is related to Young's modulus E, while the second type to shear modulus G. A configuration of springs and masses proposed here uses only linear spring, but the system can hopefully can also accomodate shear modulus G in term of only liniear spring.

IOP Conf. Series: Journal of Physics: Conf. Series **1127** (2019) 012019 doi:10.1088/1742-6596/1127/1/012019

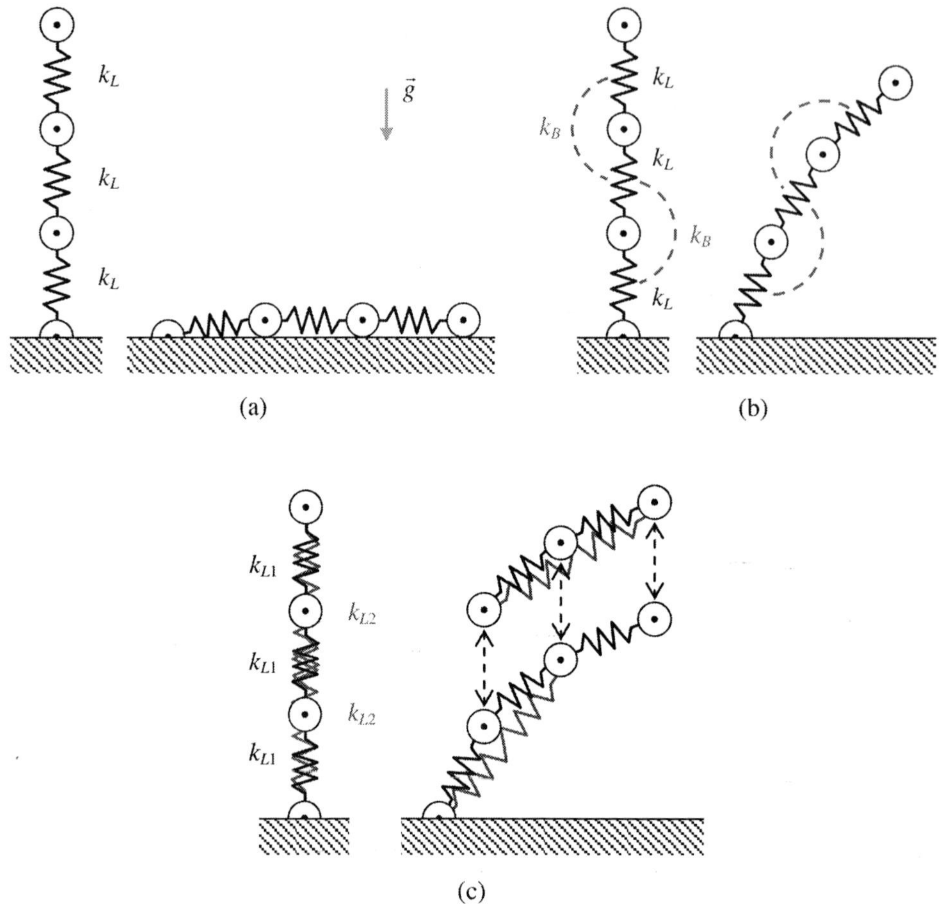

Figure 1. Several models for bending wire due to its on weight or bamboo-like tree system: (a) it fails when using only linear spring k_L between two adjacent particle masses, (b) using linear spring k_L and bending spring k_B [7], and (c) using two types of linear spring k_{L1} and k_{L2}, where connection between particles with odd and event indices are separated for clearer illustration.

Supposed that there are N particle masses with mass of each is

$$m = \frac{M}{N}, \tag{1}$$

with total mass of the system is M. Particle index is $i = 1, ..., N$, where first particle is fixed on the ground. In the model proposed in this work there are two types of spring, one with spring constant k_{L1} and the other with k_{L2}. Spring k_{L1} connects two adjacent particles with index i and i+1, while spring k_{L2} connects odd and even indices only, e.g. i and $i + 2$. If length of the system L then initial distance between particle mass is

$$\Delta l = \frac{L}{N-1}, \tag{2}$$

with initial position of particle i is

$$\vec{r}_i = (i-1)\Delta l\left(\hat{e}_y \cos\theta_0 + \hat{e}_x \sin\theta_0\right) \tag{3}$$

and θ_0 is initial angle measured from negative direction of gravity $\vec{g}$. This initial angle is the same as in model of wire curvature [6]. Gravitation force acted upon particle i is simple defined as

$$\vec{F}_i^G = m\vec{g} \tag{4}$$

and the spring force due to interaction other particle mass j is

$$\vec{F}_{ij}^L = -k_L\left(r_{ij} - l_{ij}\right)\hat{r}_{ij}, \tag{5}$$

where k_L could be k_{L1} or k_{L2}, depends on the interacting particles. Using Newton's second law of motion

$$\sum \vec{F} = m\frac{d^2\vec{r}}{dt^2}, \tag{6}$$

following differential equation can be obtained

$$\frac{d^2\vec{r}_i}{dt^2} + \frac{1}{m}\sum_{j=i-2}^{i+2}\sum_{k=1}^{2}\left[\delta_{k,|i-j|}k_{Lk}\left(r_{ij} - \Delta l\right)\hat{r}_{ij}\right] - \vec{g} = 0 \tag{7}$$

for particle i. Boundary conditions sourced from particles index i, which hold only from 1 to N, requires step function

$$u(x - x_0) = \begin{cases} 1, & x \geq x_0, \\ 0, & x < x_0. \end{cases} \tag{8}$$

Equation (7) must be modified using Equation (8) into

$$\frac{d^2\vec{r}_i}{dt^2} + \frac{1}{m}\sum_{j=i-2}^{i+2}\sum_{k=1}^{2}\left\{u(j-1)\left[1 - u(j-N-1)\right]\delta_{k,|i-j|}k_{Lk}\left(r_{ij} - \Delta l\right)\hat{r}_{ij}\right\} - \vec{g} = 0 \tag{9}$$

to assure the boundary conditions. In the condition where there is wind, another force must be added to Equation (6)

$$\vec{F}_i^D = -3\pi\eta D(\vec{v}_i - \vec{v}_w) \tag{10}$$

which is the drag force for spherical particle with diameter D. Wind velocity is represented by $\vec{v}_w$. Here, air viscosity must be also considered in the term of η.

2.2. Numerical method

Equation (9) will be solved using numerical method, e.g. Euler algorithm, which gives

$$\frac{d^2\vec{r}_i(t)}{dt^2} = \vec{g} - \frac{1}{m}\sum_{j=i-2}^{i+2}\sum_{k=1}^{2}\left\{u(j-1)\left[1 - u(j-N-1)\right]\delta_{k,|i-j|}k_{Lk}\left(r_{ij} - \Delta l\right)\hat{r}_{ij}\right\}, \tag{11}$$

and

$$\frac{d^n\vec{r}_i(t+\Delta t)}{dt^n} = \frac{d^n\vec{r}_i(t)}{dt^n} + \Delta t\frac{d^{n+1}\vec{r}_i(t)}{dt^{n+1}}, \quad n = 0,1. \tag{12}$$

Value of Δt must be chosen carefully, that is smaller than any physical time that could characterize the system, e.g 1/100 smaller than contact time as in granular simulation [8].

Final condition is achieved then position of every particle mass does not change, or the change

$$\varepsilon(t)=\sum_{i=1}^{N}\left|\vec{r}_i(t-\Delta t)-\vec{r}_i(t)\right|. \tag{13}$$

is smaller than a certain value. This value should be comparable any physical length that could characterize the system, e.g. 1/100 smaller than Δl.

2.3. Calculation steps

There are two steps in implementing the numerical. The first is relaxation step and the second is resonance step. Both step will require following equations

$$\frac{d^2\vec{r}_i(t)}{dt^2}=\vec{g}-\frac{3\pi\eta D}{m}(\vec{v}_i-\vec{v}_w)-\frac{1}{m}\sum_{j=i-2}^{i+2}\sum_{k=1}^{2}\left\{u(j-1)[1-u(j-N-1)]\delta_{k,|i-j|}k_{Lk}(r_{ij}-\Delta l)\hat{r}_{ij}\right\}, \tag{14.a}$$

$$\frac{d\vec{r}_i(t+\Delta t)}{dt^n}=\frac{d\vec{r}_i(t)}{dt^n}+\Delta t\frac{d^2\vec{r}_i(t)}{dt^2}, \tag{14.b}$$

$$\vec{r}_i(t+\Delta t)=\vec{r}_i(t)+\Delta t\frac{d\vec{r}_i(t)}{dt}. \tag{14.c}$$

i. Relaxation step. This step is required to set $\vec{v}_w=0$ and then iteratively performs Equations (14.a) – (14.c) until time t where condition of $\varepsilon(t)<\varepsilon_{\text{calc}}$ is reached.

ii. Resonance step. Using the relaxed particles configuration wind velocity can be varied e.g. $\vec{v}_w=\vec{v}_w(t)$. If there is a period or duration of the wind, value of time step Δt must be chosen to be smaller than that.

3. Results and Discussion

Following parameters are used in the simulation if not further specified.

Table 1. Simulation parameters.

Parameters	Value	Unit
ρ	500	kg/m^3
L	30	m
D	0.25	m
E	20	GPa
k	1.722×10^{18}	N/m
N	20	-
θ_0	10	°

Using parameters in Table 1 a typical relaxation process is given in Figure 2, which unfortunately does not reach a stable expected final state. As previously investigated [6], even not yet a stable final state, it is interesting to see the evolution of tortuosity T, which is given in Figure 3. It can be seen that gravitation potential energy $U\sim y$ is decreasing with time and also tortuosity. The first aspect indicates that it relaxes and but the second need further analysis for interpretation, such as that reduction of tortuosity indicates that the structure of bamboo-like trees becomes more curved than its initial structure, which could be addressed to value of k_L ($k_{L1}=k_{L2}=k_L$) that has not yet represent the right value of bending spring k_B or shear modulus G [6]. Not constant value of k_L could be proposed for next work and also some damping to vanish the oscillation.

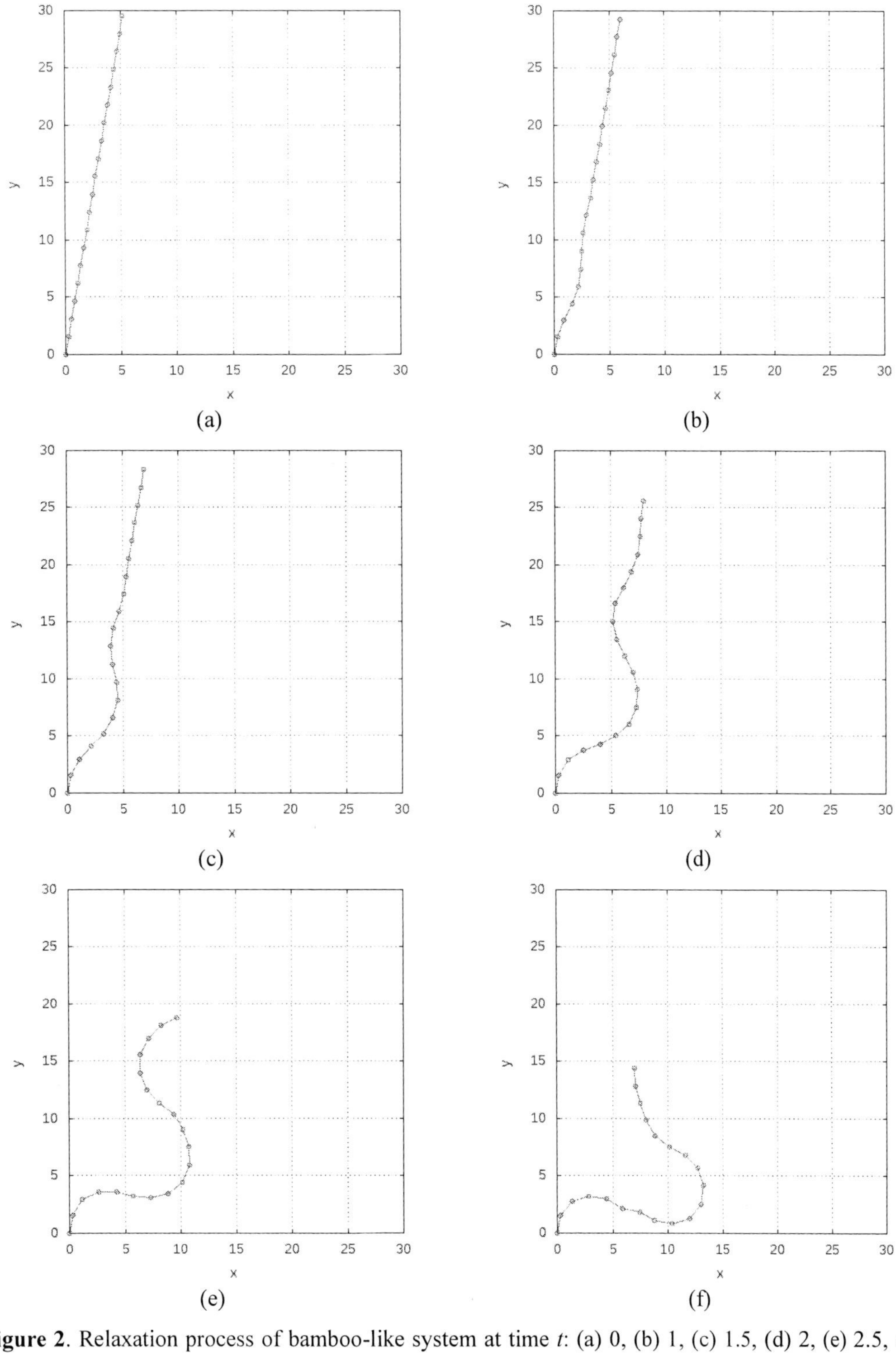

Figure 2. Relaxation process of bamboo-like system at time *t*: (a) 0, (b) 1, (c) 1.5, (d) 2, (e) 2.5, and (f) 2.8.

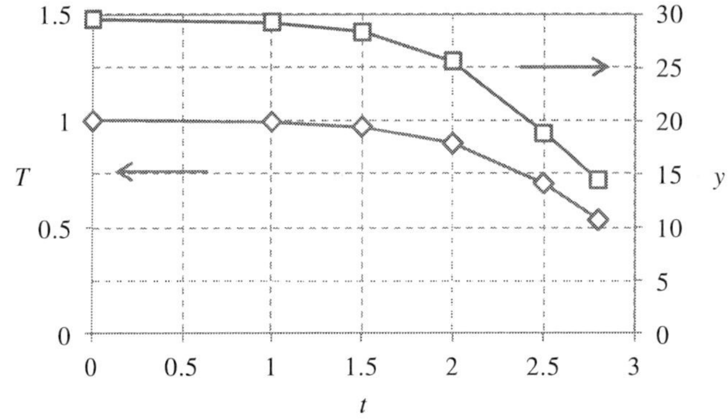

Figure 3. Change of tortuosity T (◊) and vertical position y (□) in relaxation process.

4. Conclusion

Structure of bamboo-like tree based on granular particle-spring model has been proposed, but unfortunately it can reach final stable state in the relaxation stage. It could be addressed to the same value of spring constant along the structure.

Acknowledgements

This work is supported by Institut Teknologi Bandung through Program Riset and Inovasi ITB with contract number 006q/I1.C01/PL/2016. Presentation of this work is supported by Committee of ICMNS 2016 and FMIPA ITB.

References

[1] Taylor A H, Zisheng Q 1992 *J. Veg. Sci.* **3** 253
[2] Campanello P I, Gatti M G, Ares A, Montti L, Goldstein G 2007 *For. Eco. Manage.* **252** 108
[3] Silva E C N, Walters M C, Paulino G H 2006 *J. Mater. Sci.* **41** 6991
[4] Chadderton L T, Chen Y 2002 *J. Cryst. Growth 240* 164
[5] Gielis J 2017 *Bamboo Leaves and Tree Rings* in The Geometrical Beauty of Plants, Atlantis Press, Paris, pp. 168-185
[6] Viridi S, Abdullah M and Amalia N 2016 *Res. Dev. Nanotechnol. Indonesia* **3** 30
[7] Viridi S, Patana P, Subrata S A, Hertiasa H and Abdullah M 2015 "In Searching of Trees Natural Vibration Frequency Based on Granular Particles Interactions and Vibration", *Proc. Int. Sem. Biological Sci.* (ISBS 2015), ed P. Paton *et al.*, Medan, Indonesia, 17 October 2015, pp. 113-122.
[8] Schäfer J, Dippel S and Wolf D E 1996 *J. Phys. France I* **6** 5

Fabrications of Tapered Optical Fibers by Laser Induced Photopolymerization Technique

Nikmatul Husna[1], Nina Siti Aminah[2], Rahmat Hidayat[1]

[1]Physics of Magnetism and Photonics Research Group, Faculty of Mathematics and Natural Sciences, Bandung Institute of Technology, Bandung, Indonesia

[2]Theoretical High Energy Physics and Instrumentation Research Group, Faculty of Mathematics and Natural Sciences, Bandung Institute of Technology, Bandung, Indonesia

Email: rahmat@fi.itb.ac.id

Abstract. Taper shaped optical fibers have been much investigated because it may enhance the evanescence wave formation on the surface of the fiber, which can be utilized for sensing. In this paper, we report the fabrication of tapered optical fiber by employing laser induced photo-polymerization (LIP) technique from hybrid organo-siloxane polymer precursor. This precursor was prepared by the sol-gel method for creating the inorganic cross-links. The photopolymerization process then made the organic cross-links and transformed it into the solid phase. The taper structure formation was strongly dependent on the laser power and time duration. A typical condition for taper structure formation was the laser power of 2 μW and the fiber drawing speed in the range of 20 μm/s - 40 μm/s. The resulted tapers lengths were around 1.5 mm–5.5 mm with the diameter around 100 μm at the beginning of the taper and less than 5 μm at the end of the taper. The optical propagation and evanescent wave profile in this kind of tapered optical fiber was also studied computationally by using Finite Difference Time Domain (FDTD) method.

Keywords hybrid polymer, tapered optical fiber sensor, optical waveguide, evanescent wave

1. Introduction

Generally, a sensor has a function to detect or measure a physical quantity such as temperature, field, heat, light etc., which is then commonly converted into an electrical quantity by a transducer. In biochemical sensors, the change in measurable physical quantities may be related to the presence or the change of a substance due to either physical process, such as adsorption or phase change, or chemical process, such as chemical reaction or redox reaction. A fiber optic sensor is functioning by measuring the change of light related physical quantities, such as light intensity or light polarization, which happen due to refractive index change of the surrounding medium. There are several fiber optic sensors that have been developed in more sophisticated configuration for improving the efficiency, such as in the form of interferometric configuration, Bragg grating structure or involving metal nanolayer or nano-particles for generating surface Plasmon resonance (SPR). However, there are still more simple modifications for improving the sensitivity, such as by using a bent fiber optic or tapered optical fiber structure [1]. In such structure, the evanescent wave may be extensively formed on the fiber surface into the surrounding medium. Figure 1 illustrates the formation of evanescence wave on the glass plate (n_1) that extent to the adjacent dielectric medium (n_2). Larger penetration depth of this evanescent wave may produce better sensing sensitivity.

The fabrication of taper structure in silica or glass based optical fiber requires high temperature. In addition, it is not easy to attach functional sensing or ligand molecules on the silica or glass surface. It is then interesting to develop the tapered optical fiber from polymer, which relatively easier to bind organic molecules. Such structure may be fabricated by laser induced polymer (LIP) or laser self writing (LSW) technique. There are many reports on the formation optical waveguide or fiber by this technique [2,3].

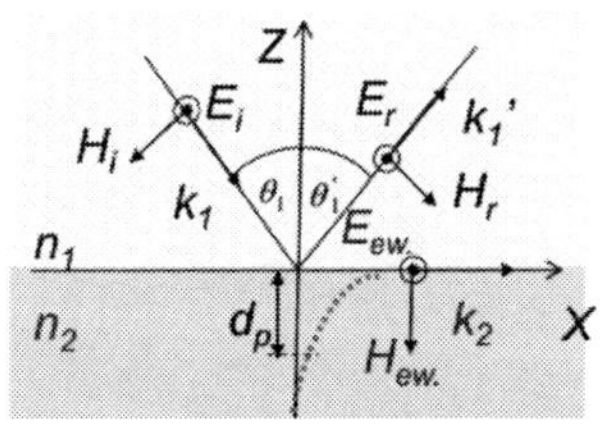

Figure 1. Evanescent wave profile in TE mode [3].

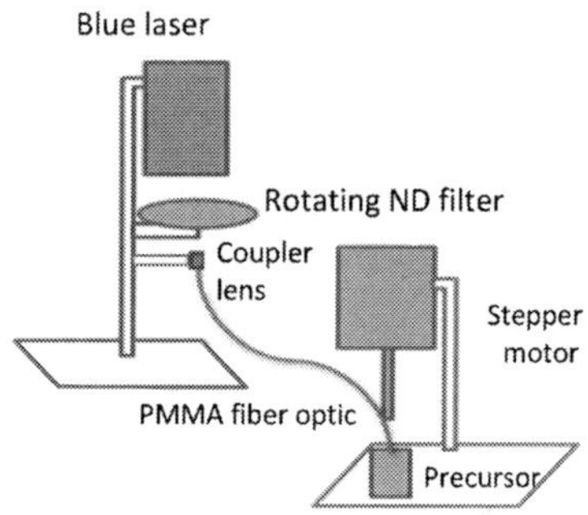

Figure 2. Experimental setup of laser-induced photo-polymerization

The electric field of evanescent wave E_{ew} is given by:

$$\tilde{E}_{ew} = \tilde{E}_{0_2} e^{-z/\delta} e^{-j(\omega t - k_2 x)} \quad (1)$$

where the depth penetration parameter is

$$\delta = d_p = \frac{\lambda_0}{2\pi\sqrt{n_1{}^2 \sin^2\theta_1 - n_2{}^2}} \quad (2)$$

2. Experiment

2.1. Hybrid polymer precursor preparations

Hybrid polymer precursor used in this work was made from a kind of organically modified siloxane monomer, namely 3-(Trimethoxysilyl) propyl methacrylate (TMSPMA, Aldrich), by sol-gel technique as reported in our previous work [4,5]. This hybrid polymer can be used to make various nano-structure with high transparency. Photo-initiator agent, namely Irgacure-369 (Ciba Speciality Chemical Inc), was dissolved into this hybrid polymer precursor solution in order to polymerize organic part of the monomer.

2.2. Laser-induced photopolymerization

Experimental setup for LIP process is presented in Figure 2. This experiment used a plastic optical fiber with 100 μm in diameter, where the taper structure was then formed at the one end of the taper. For that purpose, a blue laser producing light with 450 nm in wavelength was injected into the fiber to initiate photo-polymerization. The shape of resulted tapers was varied by varying the laser intensity by using a motorized variable neutral density (ND) filter and the drawing speed of the stepper motor.

3. Simulation

Finite Difference Time Domain (FDTD) is a method to calculate electromagnetic propagation by solving numerically the differential forms of Maxwell's equations. This method firstly introduced by

IOP Conf. Series: Journal of Physics: Conf. Series **1127** (2019) 012020 doi:10.1088/1742-6596/1127/1/012020

Yee, who proposed the numerical calculation algorithm for solving those Maxwell's equations [6]. In this algorithm, the electric field and magnetic field was calculated as followings,

$$E_x^{n+\frac{1}{2}}(k) = E_x^{n-\frac{1}{2}}(k) + \frac{}{\sqrt{\iota}} \quad (3)$$

$$H_y^{n+\frac{1}{2}}(k+1/2) = H_y^{n}(k+1/2) + \frac{\Delta t}{\sqrt{\mu_0 \varepsilon_0} \Delta z}(E_x^{,} \quad (4)$$

where the discretization is done in both spatial and time domains. In the present work, this FDTD calculation was carried out by using a software which is called as OptiFDTD. The calculations were done by providing parameter inputs such as the taper geometry and refractive indices as well as the light or laser wavelength. Based on Liu's simulation work [7], because the evanescence is effectively formed only on the tip of the taper structure, the simulation calculations here were then performed only for the tip region in order to reduce the memory use and time computation. The parameters used in the calculation of this tapered optical fiber were the taper length of 50 μm, the taper front face width of 10 μm, the taper end face of 1 μm, the light beam width of 5 μm with Gaussian beam profile, the light wavelength of 0.63 μm, the light power input of 10 mW, the core refractive index of 1.45, and the cladding refractive index of 1.3.

4. Results and Discussions

4.1. Experimental Results

The formation of cross-linked polymers as the result of photo-polymerization was firstly confirmed by FTIR spectroscopy. The FTIR spectrum after photo-polymerization (not shown here) shows the reduction of the absorption bands located at 880-995 cm^{-1} in comparison to that in the spectrum before photo-polymerization, which can be assigned to the transformation of C=C bond into C-C bond.

The fabrications parameters of the taper structure and their images resulted in this work are presented in Table 1 and Table 2. In this work, we used three steps of laser illuminations, namely:

1. The first illumination is to initiate radicals and generate polymerization seed on the end face of the optical fiber.
2. The second step is to provide photo-polymerization for taper growth, which is in combined with the fiber drawing. In this second step, we used relatively weaker light intensity in comparison to that in the first step, which was achieved by using a rotating neutral density filter to reduce the laser intensity.
3. The last illumination step is done to complete this fabrication by a termination step, which is required for just improving the polymer cross-linking without being followed further by taper growth. This step is done after removing the taper from the precursor solution.

Table 1 shows the taper fabrication parameters and the resulted tapers with drawing speed of 20 μm/s. Meanwhile, Table 2 shows for the drawing speed of 40 μm/s. The resulted tapers have around 100 μm in diameter at the beginning of taper. The diameter then decreases up to less than 5 μm at the end of the taper. The taper length varies depending on the laser power and drawing speed as well as laser illumination procedures. As shown by the images of the resulted taper structure in the tables, the taper shape may also vary, either in a linear, concave or convex shapes.

Table 1. Fabrication parameters and the images of the resulted taper structure with the drawing speed of 20 μm/s

Laser power (μW)	Fabrication parameters	Resulted taper length (mm)	Resulted taper image

1	0 s without filter, 5 minutes by filter, 5 minutes termination	± 2.0	
1.8	1 minute without filter, 5 minutes by filter, 1 minute termination	± 1.5	
3.8	15 s without filter, 5 minutes by filter, 5 minutes for termination	± 2.5	

Table 2. Similar table as Table 1 but with the drawing speed of 40 μm /s

Laser power (μW)	Fabrication parameters	Resulted taper length (mm)	Resulted taper image
2.2	3 minutes by filter, 5 minutes termination	± 2.0	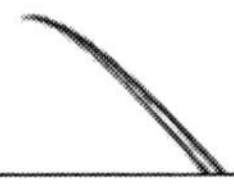
2.3	30 s without filter, 3 minutes by filter, 5 minutes termination	± 3.0	
3	5 minutes by filter, 5 minutes termination	± 5.5	

4.2. Simulation Results

We investigated three kinds of taper structures, namely linear, concave and convex shaped structure. As indicated in the previous section, such structures can be obtained by adjusting some fabrication parameters using this LIP technique. The taper structure dimensions used in the simulation did not match exactly with the actual structure of the fabricated tapers above. In the present stage, we just concern on the taper shapes. Figure 3 illustrates *xz* plane cross-section of the taper structure with concave shape used in the simulation calculation, where the vertical axis represents the refractive index pattern in the core and the cladding regions. The length of the taper is 50 μm with the right end face has a diameter of 10 μm and the left end face has a diameter of 1 μm. The core material is the hybrid polymer used in the experiment where the refractive index was supposed to be 1.45. The cladding material is water with the refractive index of 1.3, which is the medium to be sensed in the application of this sensor. The light wavelength was set to 0.63 μm, which is the wavelength of red diode laser. The simulation calculations were performed for TE mode, where the electric field was polarized along the *y*-axis.

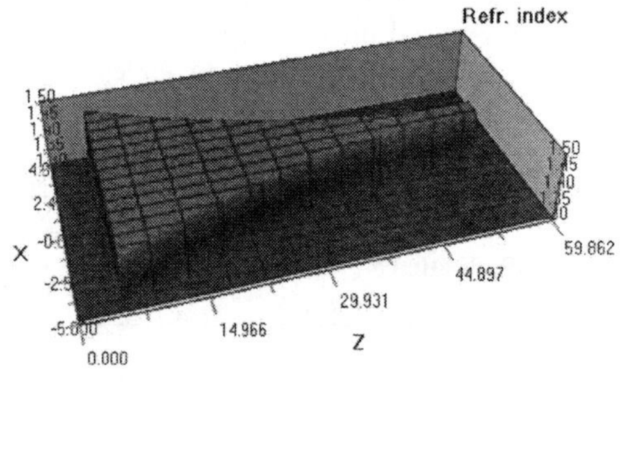

Figure 3. The *xz* plane cross-section of the taper structure with concave shape used in the simulation calculation, where the vertical axis represents the refractive index. (note that the scale is not proportional)

IOP Conf. Series: Journal of Physics: Conf. Series **1127** (2019) 012020 doi:10.1088/1742-6596/1127/1/012020

Figure 4 The E_y electric field distribution (in arbitrary scale) calculated for taper structures with (a) linear, (b) concave and (c) convex shapes. The taper was set from 5 μm up to 55 μm along the x-axis. (note that the geometrical scale is not proportional) shows the distribution of electric field in y direction (E_y), where the blue color represents zero E_y electric field while green and red colors represent medium and large electric field. In comparison to linear shaped taper, the concave shaped taper has more intense electric field distribution inside the structure as indicated by stronger green color. The distribution also may indicate more light leakage in comparison to other shapes. The convex shaped taper shows weaker electric field distribution inside the taper but strong electric field distribution at the surface near by the taper tip. However, the convex shaped taper seems showing the smallest light leakage in comparison to other shapes.

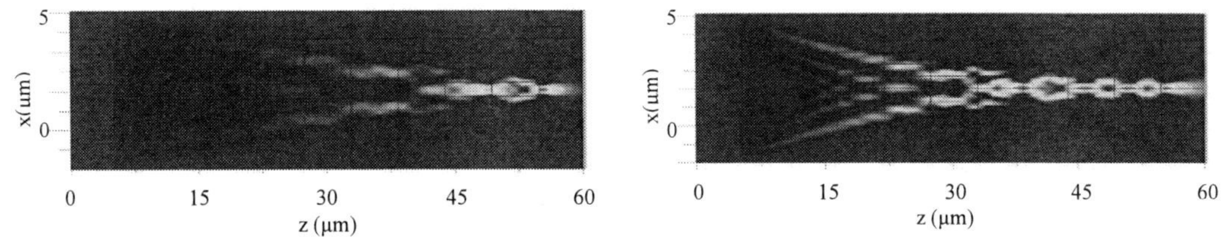

(b)

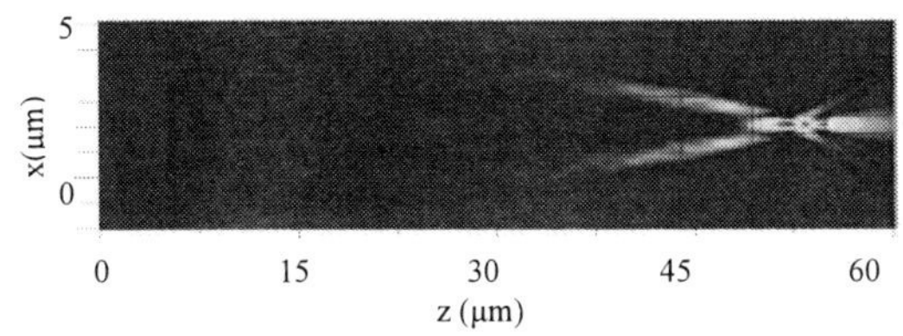

(c)

Figure 4 The E_y electric field distribution (in arbitrary scale) calculated for taper structures with (a) linear, (b) concave and (c) convex shapes. The taper was set from 5 μm up to 55 μm along the x-axis. (note that the geometrical scale is not proportional)

Figure 5 shows the distributions of electric field (in black color lines) showing evanescent wave and leakage wave at position z = 7 μm from the left end of the tip for (a) linear, (b) concave and (c) convex shaped taper structures. The blue lines indicate the profile of refractive index formed by the core and cladding regions at 7 μm from the left face of the taper. The widths of the cores are not same because of different shapes of those tapers. For all structures, we can see side peaks that may attributed to leakage waves, which are much larger in the concave shaped taper in comparison to other structures. However, this concave shape generates more extensive evanescence wave as indicated by penetration of electric field at the surface of the taper. It may be seen as the black line that crossing the vertical blue lines.

IOP Conf. Series: Journal of Physics: Conf. Series **1127** (2019) 012020 doi:10.1088/1742-6596/1127/1/012020

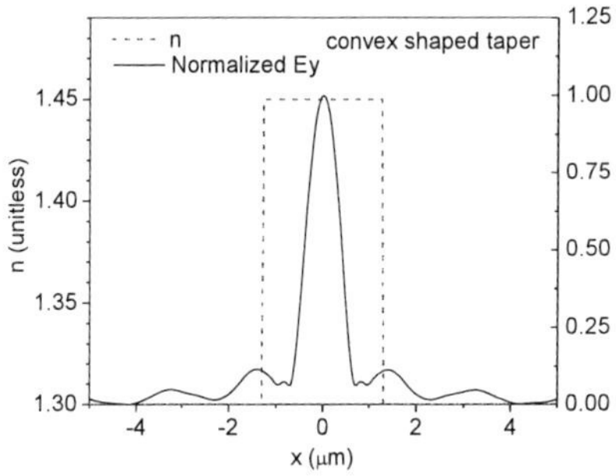

(a)

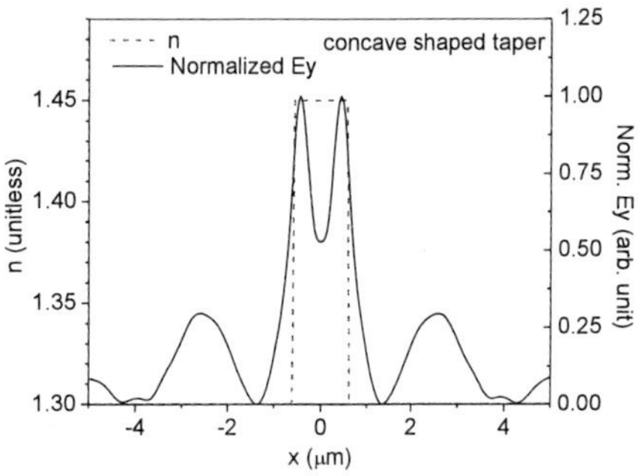

(b)

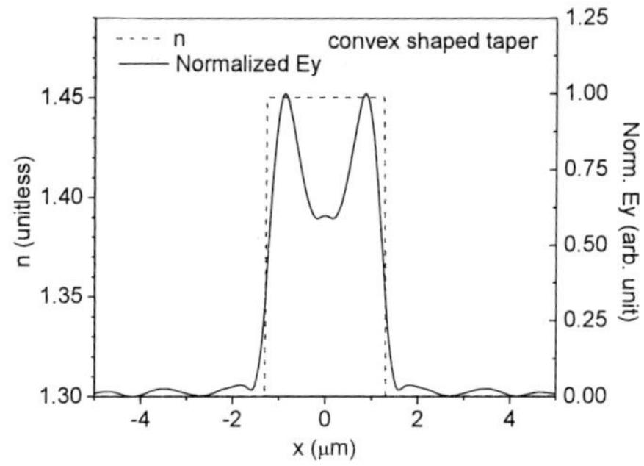

(c)

Figure 5. Distribution of electric field showing evanescent wave and leakage wave at position z=7 μm from the left end of the tip for (a) linear, (b) concave and (c) convex shaped taper structures.

5. Conclusions

We have demonstrated the formation of tapered optical fiber by LIP technique from hybrid a polymer precursor with the diameter less than 100 μm. Although further systematical works are required, at the present stage we can see that the length and shape of the taper can be varied by controlling some fabrication parameters, such as laser intensity and fiber drawing speed. The concave shape is interesting because the simulation results show more extensive evanescence formation in comparisons the other two shapes investigated here. Such structure may be useful for building more complex sensor structure such as microchannel based sensors or photonic noses.

Acknowledgement

The authors acknowledge the support from Program Riset dan Inovasi ITB 2012 and Program Beasiswa Tesis dan Disertasi (LPDP).

References

[1] Tian Y, Wang W, Wu N, Zou X and Wang X 2011 *Tapered Optical Fiber Sensor for Label-Free Detection of Biomolecules* Sensors **11** p 3780-3790; doi:10.3390/s110403780

[2] Sugihara O et al 2004 IEEE Photonics Technology Letters 16 no 3 p 804-6

[3] Yamashita T; Kawasaki A; Watanabe O; Kagami M; Tomiki M; and Sakata H 2012 *Light-induced self-written polymeric waveguides for low-cost integration of single-mode devices* IEEE Optical Interconnects Conference; DOI: 10.1109/OIC.2012.6224426

[4] Hidayat, R., Hidayat, S., Fitrilawati, F., Herman, Tjia, M. O., Fujii, A. and Ozaki, M. 2012

Distributed feedback grating fabricated from hybrid polymer precursor gel by employing short-pulse laser interference for photopumped polymer laser applications Polym. Adv. Technol. **23** p 1264–1270. doi:10.1002/pat.2039

[5] Hidayat R, Gomulya W, Pitriana P, Irmansyah R, Miranti R, Herman, Hidayat S, Fitrilawati, Fujii A and Ozaki M 2012 *Siloxane based Organic-Inorganic Hybrid Polymers and their Applications for Nanostructured Optical/Photonic Components* ITB Journal of Engineering Science **44** p 207-219. DOI: 10.5614/itbj.eng.sci.2012.44.3.1

[6] Sullivan DM 2013 *Electromagnetic simulation using the FDTD method* (John Wiley & Sons)

[7] Liu Z, Guo C, Yang J, and Yuan L 2006 *Tapered fiber optical tweezers for microscopic particle trapping* Opt. Expr. **14** p 12510

2D Interpretation Of Controlled Source Audio Magnetotelluric (CSAMT) Data Integrated With Borehole Data In Kamojang Geothermal Field West Java, Indonesia

Susilawati[1] and Enjang Jaenal Mustopa[2]

[1] Earth Physics and Complex System Research Divison, Faculty of Mathematics and Natural Sciences, Bandung Institute of Technology, Indonesia

E-mail: Susilawati.sobur@yahoo.com

Abstract. CSAMT is the one of electromagnetic method which uses unnatural source grounded dipole with the range of frequency between 0.1Hz-10 kHz. In order to qualify the response measured is plan wave, then the corrected source effect. The objective of this study was to obtain that have been corrected for source to have the same characteristic as MT data. CSAMT data that has been corrected, subsequently inverted to get the exact model and validated to well log data in the form of temperature data. After interpretation found that the some of the line measurement at geothermal field Kamojang has two layers. There are seven line, line-1 until line-7, line-2 until line-7 have the characteristic of geothermal, the first layer has a resistivity of 3-15 Ω.m and the average of thickness of 400-1500 meters. The second layer has a resistivity value of 20-300 Ω.m that is more resistive than the first layer, except for line 1, there is no low resistivity in the first layer. The first layer is a layer of clay, identified with resistivity values ranging from 2,5 - 15 Ω.m with 900-1400m thickness with temperature data on KMJ-12 range 180-200°C, and the next layer is more resistive layer than the second layer is 18 Ω.m -100 Ω.m with temperature data at KMJ-53 at a depth of 710 meters is 240°C.

Keywords: : CSAMT method, Resistivity, Temperature.

1. Introduction

A Controlled source audio frequency magnetotelluric (CSAMT) measurement, in the operating frequency range of 0.25 to 8192 Hz, the location transmitter was about 3-5 km in the southwest from the survey area, the sounding sites were distributed irregularly and the distance between each of them varied from 200-500 m [1]. The purpose of the measurements is to demonstrate the capability of the CSAMT method used for a geothermal exploration by locating an anomalous low resistivity zone associated with the existence of geothermal reservoir and validated with the temperature data. For information about the temperature and minerals obtained from the results of research [2].

The Kamojang geothermal field which is located in West Java. The CSAMT survey area is geographically between 07° 10'30'' – 07° 07'30''South latitude and 107° 45' 30" East longitude.

Kamojang is the geothermal that placed in Garut, West Java and operating since 1983, it has been producing a sufficient steam of the vapor dominated resources for generating electricity with the total installed capacity of about 200 MW [3].

Kamojang geothermal field is located in Garut approximately 42 km south-east of Bandung city. This area is vapor dominated system and high temperature around 235°-250°C. This heat interacts with fluids forming hydrothermal alteration rock that serves as a cap rock and reservoir in the Kamojang geothermal system. This area was first discovered by Dutch in 1920 and the early exploration in 1973 with the cooperation between the government of Indonesian and New Zealand [4]. After 10 years of exploration this field produces in 1983 to producing 140 MWe and in 1997 expanded to 220 MWe [5]. CSAMT method has successfully applied for ground water exploration in a suburban Beijing [11], also successfully mapped the results of uranium exploration [6]. CSAMT method has successfully mapped the distribution of resistivity in the geothermal field Takigami in Kyushu [7]. CSAMT method was used to map brine contamination in the vicinity of the Prue Sand Unit [8]. CSAMT method was successfully for uranium exploration, the data show a good correlation to geology [9]. Controlled source audio-frequency magnetotellurics (CSAMT) is a frequency-domain electromagnetic sounding technique which uses a fixed grounded dipole or horizontal loop as an artificial signal source.

The cooler upper zones are characterized by alteration of electrical conductive layer that is formed at temperatures above 70°C. At higher temperatures, illite of less conductive layer becomes interlayered with smectite. The proportion of illite increases with the temperature, forming about 70% of the mixed-layer clay at 180 °C. Above this temperature, the smectite content continued to decrease and pure illite commonly appears at greater than 220 °C [10].

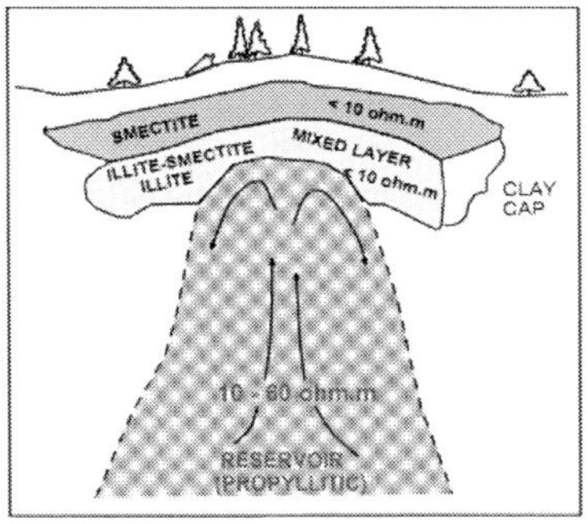

Figure 1. A generalized geothermal system

2. CSAMT Theory

The CSAMT is electromagnetic method which use unnatural source with the range of frequency is 0.1Hz-10kHz, unnatural source electromagnetic is the electric current which approach 10A that injection to the earth. Basic theoretical principle for CSAMT theory can be described in terms of Maxwell's equations.

$$\vec{\nabla}\times\vec{E} = -\frac{\partial B}{\partial t} \quad (1)$$

$$\vec{\nabla}\times\vec{H} = \vec{J} + \frac{\partial \vec{D}}{\partial t} \quad (2)$$

$$\vec{\nabla}\bullet\vec{D} = \rho_c \quad (3)$$

$$\vec{\nabla}\bullet\vec{B} = 0 \quad (4)$$

With the assumption of electromagnetic field behavior within the conducting half space in a cylindrical coordinat system and used modified of Bessel function:

$$E_\theta = \frac{Idl\sin\phi}{2\pi\sigma r^3}\left[2 - e^{-kr}\left(1 + ikr\right)\right] \quad (5)$$

$$H_r = \frac{Idl\sin\phi}{2\pi r^2}\left\{3I_1\left(\frac{ikr}{2}\right)k_1\left(\frac{ikr}{2}\right) + \frac{ikr}{2}\left[I_1\left(\frac{ikr}{2}\right)k_0\left(\frac{ikr}{2}\right) - I_0\left(\frac{ikr}{2}\right)k_1\left(\frac{ikr}{2}\right)\right]\right\} \quad (6)$$

Where *dl* is the dipole length, *r* is the source-sounding distance, I_m and K_m are modified Bessel function of the *m-th* order. Far from dipole source, equations (5) and (6) approach the following limits, with quasi static assumption:

$$E_\phi \approx \frac{Idl\sin\phi}{\pi \sigma r^3} \quad (7)$$

$$H_r \approx \frac{Idl\sin\phi}{\pi\sqrt{\mu\sigma\omega}\, r^3} e^{\frac{-i\pi}{4}} \quad (8)$$

As shown in equations (7) and (8) the horizontal fields in the far field zone all decay as $1/r^3$, the far field apparent resistivity can be determined by measuring the perpendicular E and H, that is [12]:

$$\rho_a = \frac{1}{\mu\omega}\left|\frac{E_\varphi}{H_r}\right|^2 \quad (9)$$

Because of the need to minimize the transmitter-receiver separation for maximum signal, the far-filed zone is restricted by this situation, therefore, most CSAMT data sets will contain some near-field and transition zone data. To overcome the source effect problem, a correction technique to CSAMT data should be applied to obtain apparent resistivity value equivalent to the far field situation. In this paper, a technique to correct near-field CSAMT apparent resistivity proposed by Yamashita et al [10]. The technique suggest that the plane wave apparent resistivity calculated by near field equation for the lower frequency and the apparent resistivity calculated by Cagniard equation for the high frequency.

The 2D inversion has been used for many years to interpret automatically CSAMT data. This technique makes the interpretation more objective and less time consuming then the trial and error approach but it does not yield a unique solution. Nevertheless, a common approach to fitting a 2D CSAMT data set is construct a cross-section of the area based on prior geological knowledge and the model parameterization to solve for the conductivities by least-square inversion. A 2D inversion scheme using the non-linier conjugate gradients (NLCG) algorithm proposed by Rodi and Mackie was applied to the corrected CSAMT data of transverse magnetic mode. The NLCG scheme minimizes an objective function that penalized data residuals and the second spatial derivatives of the resistivity.

Tikhonov's method defines a regularized solution of the inverse problem to the model that minimize to object function:

$$\psi(m) = (d-F(m))^T V^{-1}(d-F(m)) + \lambda \|L(m-m_o)\|^2 \quad (10)$$

Where d is observed data vector, m is an unknown model vector, mo is a priori model, F is a forward modeling operator, V is an error covariance matrix, L is a linier operator and λ is a regularization parameter. Each datum di is log amplitude or phase of transverse electric (TE) or transverse magnetic (TM) mode of complex apparent resistivity at a particular station and frequency. The model vector is also log resistivity as a function of position $(m(x) = \log \rho(x))$. The Laplacian operator can be written as follow:

$$\|L(m-mo)\|^2 = \int (\Delta(m(x)-m_o(x)))^2 dx \quad (11)$$

The model sequence is given by:

$$\psi(m_j + a_j h_j) = \min_a \psi(m_j + a h_j) \quad (12)$$

$$m_{j+1} = m_j + a_j h_j \quad (13)$$

$$h_j = -C_j g_j + \beta_j h_{j-1} \quad (14)$$

Where aj is a size step, hj is a search direction, Cj is a preconditioner, g_f is a gradient of objective function and βj is a scalar calculated as:

$$\beta_j = \frac{g_l^T c_l (g_l - g_{l-1})}{g_{l-1}^T c_{l-1} g_{l-1}} \quad (15)$$

3. CSAMT Data Interpretation

In this paper, a two-dimensional (2D) inversion technique based on a non linier conjugate gradient method using finite element discretization was applied to invert the corrected CSAMT data.

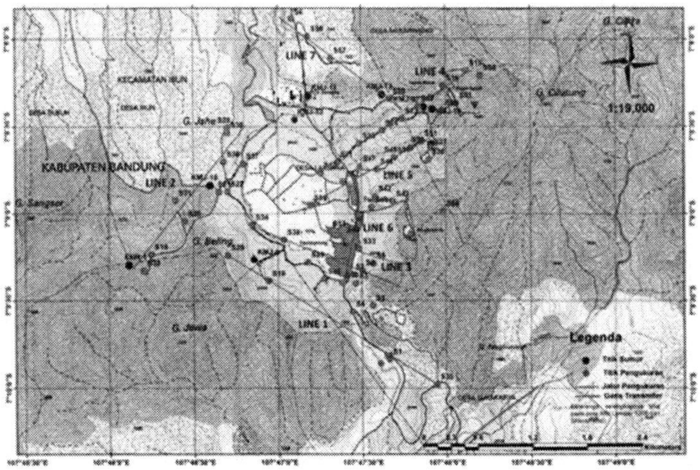

Figure 2. The location of the CSAMT survey in Kamojang area

The measuring of CSAMT method can described in the figure 2. There are seven lines (line 1-line 7) and there are 60 soundings. After the correction effect source is done to all sounding source, then we do the inversion process.

4. Correlation of 2-D Inversion Result with well Log Data

To confirm the 2-D CSAMT inversion results, the temperature data from wells KMJ-7, KMJ-10, KMJ-11, KMJ-59, KMJ -12, KMJ-52, KMJ-53 and KMJ CHR-1. That wells are spread in lines, KMJ-7 in line-4, KMJ-10 in line-2, KMJ-59 in line 4, KMJ-12 in line-7, KMJ-53 in line-6 and KMJ CHR-1 in line-2. The picture below is the resistivity section and the temperature values.

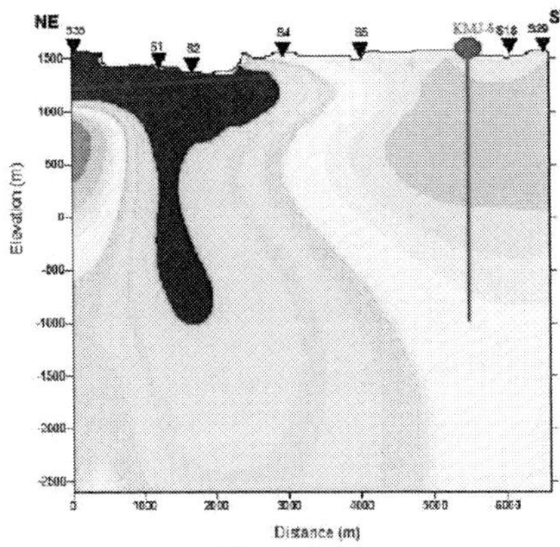

Figure 3. Line 1

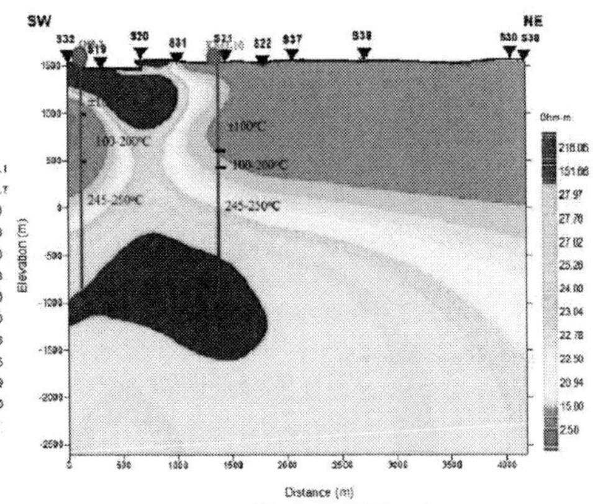

Figure 4. Line 2

According to the result obtained through the 2-D inversion result of CSAMT data in TM mode, the 2D resistivity mode section for each profile can be seen from the figure 3-9. The color in each area shows the resistivity value, the blue color indicates a high resistivity value, NE direction in Line 1 it has high resistivity and NE direction. In line 2 there are two wells, KMJ CHR-1 and KMJ-10, in KMJ

IOP Conf. Series: Journal of Physics: Conf. Series **1127** (2019) 012021 doi:10.1088/1742-6596/1127/1/012021

CHR-1 in the depth 1500-1000 m asl the temperature is ±100°C with the mineral is Crystobalite-Montmorillonite with resistivity's profile of 2D is 2.5-151 Ω.m.

In the depth 1000-500 m asl the temperature is 100-200°C with illite-montmorillonite the mineral is formed and the value of resistivity is around 2.5-22.5 Ω.m. On the sea level until 500m asl the temperature is 245-250 °C with the mineral is formed is Chlorite epidote with the resisitivity value above 15-151 Ω.m.

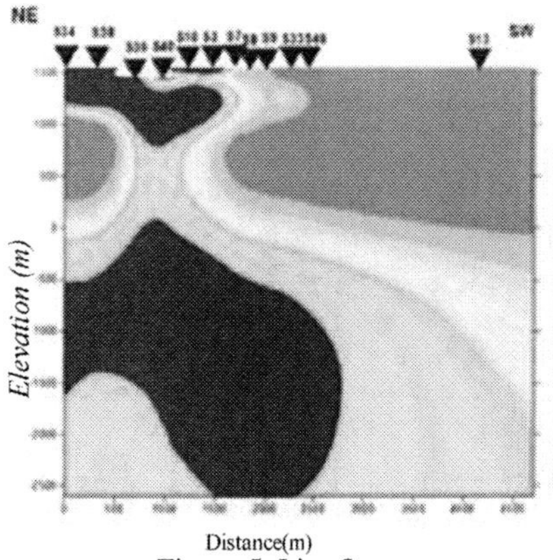

Figure 5. Line 3

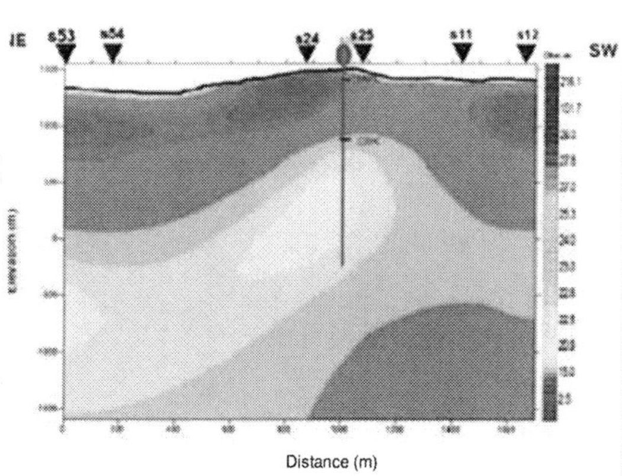

Figure 6. Line 4

Figure 5 shows the inversion result in line 3 and figure 6 is 2D inversion section for line 4, there is low resistivity value in SW direction of line 3 with a thickness 1500 m below the surface. In line 4 there is KMJ-7 with the temperature 220° in the depth 433 meter and the resistivity value is 10-15 Ω.m with the mineral's formed is laumontite. In the depth 96 meter the mineral's formed is wairakite with the temperature is 140°C and the value of resistivity is 2,5-10 Ω.m

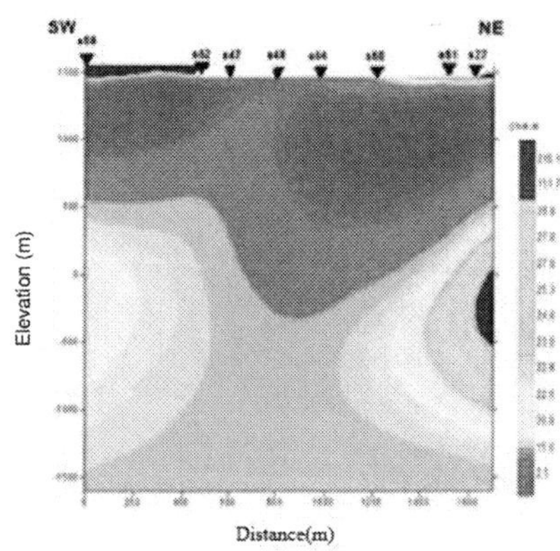

Figure 7. Line 5

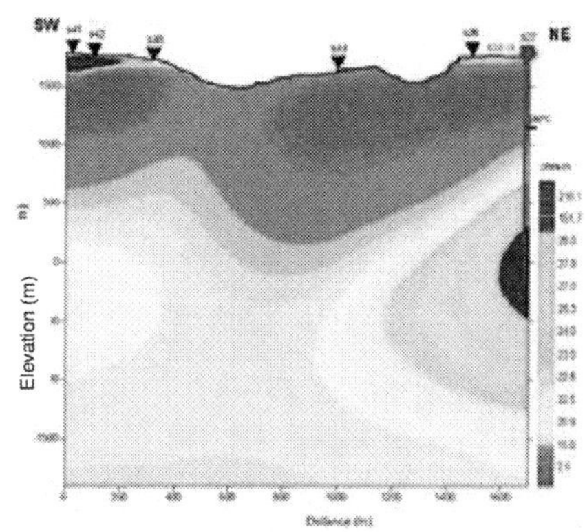

Figure 8. Line 6

The line 5 has a layer with a thickness of 1000 to 1500 m below the surface with resistivity value is 2.5-10 Ω.m and the line 6 there is KMJ-53 in the depth 710m below ground level with the temperature 240oC with the resistivity value between 5-15 Ω.m, the content of mineral is wairakite.

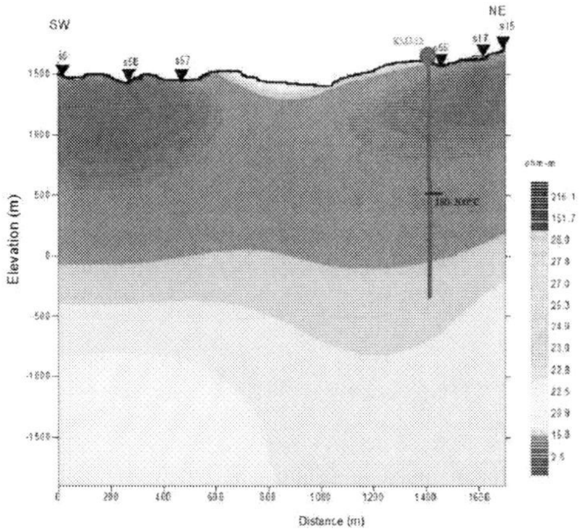

Figure 9. Line 7

In line 7 there is KMJ-12 in the depth 1000m below ground level with the temperature 180-200°C with the resistivity value is 3-15 Ω.m, the content of mineral is wairakite.

5. Conclusions

Based on the results of measurements of CSAMT in geothermal field, Kamojang. The CSAMT method applied was successful to locate an anomalously low resistivity zone that indicates a potential geothermal reservoir. Generally, the subsurface resistivity structures in Kamojang geothermal field are composed mainly by two types of resistivity feature obtained from 2D inversion results of CSAMT data. Generally, the Kamojang area is composed mainly of two layers of resistivity structure. Line 2 until line 7 has the characteristic of geothermal, the first layer has a resistivity of 3-15 Ω.m and the average of thickness of 400 -1500 meters. The second layer has a resistivity value of 20-300 Ω.m that is more resistive than the first layer, except for line 1, there is no low resistivity in the first layer. The electromagnetic exploration methods are more effective to detect a low resistivity zone on the vertical direction correlating with the existing alteration zone.

6. References

[1] Resistivity Structure in Kamojang Geothermal Field Derived from CSAMT Data. Mustopa, J.,M., at al. Institut Teknologi Bandung. Indonesian Journal of Physics. 2011.

[2] Utami, P., Characteristics of the Kamojang geothermal reservoir (West Java) as revealed by its hydrothermal alteration mineralogy.

[3] http://www.esdm.go.id/berita/45-panasbumi/3548-lapangan-kamojang-pltp-pertama-di indonesia.html, Wednesday, June 29, 2016.

[4] Suparka, E., at al. Current temperature studies in Kamojang Geothermal field West Java Indonesia. Proceedings, 2nd ITB Geothermal workshop 2013.

[5] Sudarman S, M. Boedihardi, Kris Pudyastuti, Bardan (1995), "Kamojang geothermal field 10 year operation experience", proc. WGC, 1773-1777.

[6] Zonge, K. L., and Hughes, L. J., *Controlled Sorce Audio-Frequency Magnetotellurics* (1991)

[7] Mustopa et al. *Resistivity Structure in Kamojang Geothermal Field Derived from CSAMT Data*: Indonesian Journal of Physics, Vol.11, No.1. (2011)

[8] Nabhigian, N., M., Electromagnetic method in applied Geophysics. Society of exploration Geophyschist. 2008.

[9] Yamashita, M., CSAMT (Controlled Source Audio Magnetotellurics). Phoenix . Geophysics Limited. 1984.

[10] E. Anderson, D. Crosby and G. Ussher, Bullseye! Simple resistivity imaging ti Reliably Locate the Geothermal Reservoir, Proc. WGC 2000, 909-914, 2000.

[11] Fu, C., at al. Application of the CSAMT method to groundwater exploration in a metropolitan environment. 2013.

[12] A.A. Kaufman and G. V. Keller, Frequency and Transient Sounding Method, Elsevier Sci. Publ. Co. Inc., 1983.

Dynamical System of Gauss-Bonnet Model with Vector Field

Azwar Sutiono[1], Bansawang B.J.[2], Agus Suroso[1,3], Tasrief Surungan[2], Freddy P. Zen[1,3]

[1] Department of Physics, Institut Teknologi Bandung, Bandung 40132, Indonesia
[2] Department of Physics, Universitas Hasanuddin, Makassar 90245, Indonesia
[3] Indonesia Center for Theoretical and Mathematical Physics (ICTMP) Faculty of Mathematics and Natural Sciences, Institut Teknologi Bandung, 40132, Indonesia

E-mail: azwarsu@gmail.com

Abstract. Modifying the Einsteins gravity is one of the interesting proposal to explain the dark energy that dominated our universe today. We investigated Gauss-bonnet model with vector field in the action function and we show the influence of the vector field to the density parameters of the model with a dynamical system approach. Parameter $\bar{\beta}$ that contributed by vector field give change to the evolution of density parameters. The model also can represent the condition of the present universe which dominated by dark energy.

Keywords: Modified gravity, Dynamical system, Gauss-Bonnet, Vector field.

1. Introduction

The cosmic acceleration of universe has been investigated by different observational data such as the supernovae type Ia [1-4], cosmic microwave background anisotropy [5], and the age of universe [6]. The interpretation of the observational data shows that this accelerated expansion is due to some kind of negative-pressure form of matter known as dark energy (DE). The analysis of cosmological observations also suggests that the universe is spatially flat, and consists of about 30% of dark matter, and 70% of homogeneously distributed dark energy with negative pressure. There are basically two approaches for construction of DE model. The first approach is called "modified matter models" where we modify the energy-momentum tensor $T_{\mu\nu}$ on the r.h.s. of the Einstein equations that contain matter source with a negative pressure. The second approach is called modified gravity models in which the Einstein tensor $G_{\mu\nu}$ on the l.h.s. of the Einstein equations is modified.

One of modification of gravity is add to the Einstein Lagrangian general functions of the Ricci and Riemann tensors, e.g., $f(R, R_{\mu\nu}R^{\mu\nu}, R_{\mu\nu\alpha\bar{\beta}}R^{\mu\nu\alpha\bar{\beta}}, ...)$[6]. There is a way to modify gravity with a combination of Ricci and Riemann tensors that keeps the equations at second-order in the metric and does not necessarily give rise to instabilities, namely a GaussBonnet (GB) term coupled to scalar field [7-9]. In addition, we combined vector field to the model and identified the influence of it to the evolution of dark energy density in the model.

IOP Conf. Series: Journal of Physics: Conf. Series **1127** (2019) 012022 doi:10.1088/1742-6596/1127/1/012022

The organization of the paper is as follows. In section II, we construct Gauss-Bonnet Model with the vector field. In section III, we investigate the evolution of density parameters of the model with a dynamical system approach. The final section is devoted to the conclusion.

2. Gauss-Bonnet Model with Vector Field

In this section, we reconstruct Gauss-Bonnet model with add the action of the vector field. We assumed that the Lorentz symmetry is spontaneously broken by getting the expectation values of a vector field u^μ as $\langle 0|u^\mu u_\mu|0\rangle = -1$ [10]. The action can be written as the sum of two distinct parts:

$$S = S_{GB} + S_u, \tag{1}$$

where the actions for the Gauss-Bonnet S_{GB}, is given by

$$S_{GB} = \int d^4x\sqrt{-g}\left[\frac{R}{2} - \frac{1}{2}(\nabla^2\phi) - V(\phi) - f(\phi)G\right] + S_m, \tag{2}$$

where $V(\phi)$ and $f(\phi)$ are functions of ϕ, and G is the GB term defined by

$$G = R^2 - 4R_{\mu\nu}R^{\mu\nu} + R_{\mu\nu\rho\sigma}R^{\mu\nu\rho\sigma}, \tag{3}$$

and the action of vector field S_u, is given by [11,12]

$$\begin{aligned} S_u &= \int d^4x\sqrt{-g}\left[\beta_1\nabla^\mu u^\nu\nabla_\mu u_\nu - \beta_2(\nabla_\mu u^\nu)^2 - \beta_3\nabla^\mu u^\nu\nabla_\nu u_\mu\right. \\ &\quad \left.+\lambda(u^\mu u_\nu + 1)\right]. \end{aligned} \tag{4}$$

In the above β_i $(i = 1, 2, 3, 4)$ are arbitrary parameters and λ is a Lagrange multiplier. For the time-like vector field, we impose a constraint

$$u^\mu u_\mu = -1. \tag{5}$$

Vector u^μ is a dimensionless vector. For the background solutions, we use the homogeneity and isotropy of the universe spacetime

$$ds^2 = -dt^2 + a^2(t)\left[\frac{dr^2}{1-Kr^2} + r^2(d\theta^2 + \sin^2 d\phi^2)\right], \tag{6}$$

The scale of universe determined by $a(t)$. We take the constraint [12],

$$u^\mu = (1, 0, 0, 0). \tag{7}$$

Varying the action (1) with respect to $g_{\mu\nu}$, we have field equation

$$\begin{aligned} T_{\mu\nu} &= 2R_{\mu\nu} - g_{\mu\nu}R + g_{\mu\nu}f(\phi)G - 4f(\phi)RR_{\mu\nu} + 8f(\phi)R^\rho_\mu R_{\nu\rho} \\ &\quad -4f(\phi)R_{\mu\nu\sigma\tau}R^{\rho\sigma\tau}_\nu + 8f(\phi)R_{\mu\rho\sigma\nu}R^{\rho\sigma} + 4f(\phi)(\nabla_\mu\nabla_\nu f(\phi))R \\ &\quad -4g_{\mu\nu}(\nabla^2 f(\phi))R - 8(\nabla^\rho\nabla_\mu f(\phi))R_{\nu\rho} - 8(\nabla^\rho\nabla_\nu f(\phi))R_{\mu\rho} \\ &\quad +8(\nabla^2 f(\phi))R_{\mu\nu} + 8g_{\mu\nu}(\nabla^\rho\nabla^\sigma f(\phi))R_{\rho\sigma} \\ &\quad -8(\nabla^\rho\nabla^\sigma f(\phi))R_{\mu\rho\nu\sigma}, \end{aligned} \tag{8}$$

where $T_{\mu\nu} = T^{(u)}_{\mu\nu} + T^{(\phi)}_{\mu\nu} + T^{(m)}_{\mu\nu}$ is the total energy-momentum tensor, $T^{(u)}_{\mu\nu}$, $T^{(\phi)}_{\mu\nu}$ dan $T^{(m)}_{\mu\nu}$ are the the energy-momentum tensors of vector, scalar fields and matter, respectively, defined by formula

$$\begin{aligned} T^{(u)}_{\mu\nu} &= 2\beta_1(\nabla_\mu u^\tau \nabla_\nu u_\tau - \nabla^\tau u_\mu \nabla_\tau u_\nu) - 2\nabla_\tau(u_{(\mu}J^\tau{}_{\nu)}) - 2\nabla_\tau(u^\tau J_{(\mu\nu)}) \\ &\quad + 2\nabla_\tau(u_{(\mu}J^\tau_{\nu)}) + 2u_\sigma \nabla_\tau J^{\tau\sigma} u_\mu u_\nu + g_{\mu\nu}\mathcal{L}_u, \end{aligned} \tag{9}$$

$$T^{(\phi)}_{\mu\nu} = \partial_\mu\phi\partial_\nu\phi - g_{\mu\nu}\left(\frac{1}{2}\partial^\rho\phi\partial_\rho\phi + V(\phi)\right), \tag{10}$$

$$T^{(m)}_{\mu\nu} = (\rho_m + p_m)\nu_\mu\nu_\nu + p_m g_{\mu\nu}, \tag{11}$$

where

$$J^\mu{}_\nu = -\beta_1\nabla^\mu u_\nu - \beta_2\delta^\mu_\nu\nabla_\tau u^\tau - \beta_3\nabla_\nu u^\mu. \tag{12}$$

We take the following exponential potential

$$V(\phi) = V_0 e^{-\lambda\phi}, \tag{13}$$

The GB coupling is generally given by the sum of the exponential terms

$$f(\phi) = (f_0/\mu)e^{\mu\phi}. \tag{14}$$

With assuming ϕ depends on time, the (tt) components of equation (8) give

$$3H^2 = \dot{\phi}^2/2 + V(\phi) + 24f_{,\phi}\dot{\phi}H^3 + \rho_m + \rho_r - 3\beta H^2, \tag{15}$$

where

$$\beta = \beta_1 + 3\beta_2 + \beta_3. \tag{16}$$

On the other hand, variation of equation (1) with ϕ give,

$$\ddot{\phi} + 3H\dot{\phi} - \lambda V + 24f_{,\phi}\dot{\phi}H^2(H^2 + \dot{H}) = 0, \tag{17}$$

3. Dynamical System of the Model

In order to study cosmological dynamics in the presence of a scalar field, Gauss-Bonnet term and a background fluid, from equation of motion (15), it is convenient to introduce the following dimensionless variables:

$$x_1 = \frac{\dot{\phi}}{\sqrt{6\bar{\beta}}H}, \quad x_2 = \frac{\sqrt{V(\phi)}}{\sqrt{3\bar{\beta}}H}, \quad x_3 = f_{,\phi}H^2, \quad x_4 = \frac{\sqrt{\rho_r}}{\sqrt{3\bar{\beta}}H}, \tag{18}$$

where $\bar{\beta} = \beta + 1$. Density parameters of scalar field, Gauss-Bonnet term, radiation and matter, we have, respectively

$$\Omega_\phi = x_1^2 + x_2^2, \quad \Omega_{GB} = 8\sqrt{6}x_1x_3, \quad \Omega_r = x_4^2, \quad \Omega_m = \frac{\rho_m}{3H^2}. \tag{19}$$

These satisfy the relation $\Omega_m + \Omega_r + \Omega_\phi + \Omega_{GB} = 1$ from equation (15). Differentiating the variables with respect to the number of e-foldings $N = ln\, a$ and take $\bar{\beta}$ as constant, we obtain the following autonomous equation

$$\frac{dx_1}{dN} = -3x_1 + \frac{\sqrt{6}}{2}\lambda\sqrt{\bar{\beta}}x_2^2 - 4\sqrt{6}x_3 - (x_1 + 4\sqrt{6}x_3)\frac{\dot{H}}{H^2}, \tag{20}$$

$$\frac{dx_2}{dN} = -x_2\left(\frac{\sqrt{6}}{2}\lambda x_1 + \frac{\dot{H}}{H^2}\right), \tag{21}$$

$$\frac{dx_3}{dN} = x_3\left(\sqrt{6}\mu\bar{\beta}x_1 + 2\frac{\dot{H}}{H^2}\right), \tag{22}$$

$$\frac{dx_4}{dN} = -x_4\left(2 + \frac{\dot{H}}{H^2}\right), \tag{23}$$

Taking the time derivative of Equation (15) and eliminating the term $\ddot{\phi}$ by using Equation (17), we find

$$\begin{aligned}&\left[6\bar{\beta} - 72\sqrt{6}\bar{\beta}x_1x_3 + \left(\sqrt{6\bar{\beta}}x_1 + 24\sqrt{\bar{\beta}}x_3\right)24\sqrt{\bar{\beta}}x_3\right]\frac{\dot{H}}{H^2} = \\ &\left(\sqrt{6\bar{\beta}}x_1 + 24\sqrt{\bar{\beta}}x_3\right)\left(-2\sqrt{6\bar{\beta}}x_1 + 3\lambda\bar{\beta}x_2^2 - 24\sqrt{\bar{\beta}}x_3\right) \\ &-3\bar{\beta}\sqrt{6\bar{\beta}}\lambda x_1x_2^2 + 144\mu\bar{\beta}\sqrt{\bar{\beta}}x_1^2x_3 - 9\left(1 - x_1^2 - x_2^2 - 8\sqrt{6}x_1x_3\right. \\ &\left.-x_4^2\right) - 12\bar{\beta}x_4^2.\end{aligned} \tag{24}$$

For the absence of radiation ($x_4 = 0$), with parameters $\beta = 0.1$, $\lambda = 2$, and $\mu = 7$, we got critical points like on the Table 1. The point (a) is matter dominated solution with density parameter $\Omega_m = 1$. The kinetic energy of scalar field is dominant for the points (b) and (c) because of the density parameter $\Omega_\phi = 1$ but it just was contributed by kinetic variable x_1 (variable $x_2 = 0$ as the variable of scalar field potential). Point (d) is the solution of scalar field dominated era with the most contributed variable is x_2.We identify the stable solution with see the trajectory of critical points and see which one become attractor point. Fig. 1.a is the plot of trajectory of phase space (x_2, x_3) by set $x_1 = 0$. We are able to see that point (d) is the attractor solution. So we can conclude that point (d) is stable.

In Fig. 1.b we plot the evolution of Ω_m, Ω_r, Ω_ϕ, and Ω_{GB} versus N. The evolution of each

Point	x_1	x_2	x_3	Ω_ϕ	Ω_m
a	0	0	0	0	1
b	-1	0	0	1	0
c	+1	0	0	1	0
d	0	0.965	0.074	$\simeq 1$	$\simeq 0$

Table 1. Critical points of Gauss-Bonnet model with vector field for $\bar{\beta} = 0.1$, $\lambda = 2$, and $\mu = 7$.

density parameter starting from the epoch of matter-radiation equality ($\Omega_r = \Omega_m = 0.5$). As we know before equality epoch, the universe was dominated by radiation. But after that era, the density parameter of radiation go down significantly till reach the minimum value $\Omega_r = 0.003$ at $N = 21$. Otherwise, after equality epoch, the density parameter of matter rise up and reach the maximum value $\Omega_m = 0.899$ at $N = 14$. We call this epoch is matter-dominated era. At last, the density parameter of matter go down to $\Omega_m = 0.31$ at $N = 21$. The density parameter of the scalar field started to rise up at $N = 9$ and dominated at $N = 21$ with $\Omega_\phi = 0.529$. The density parameter of Gauss-Bonnet term rise up at $N = 18$ and reach the maximum value $\Omega_{GB} = 0.157$ at $N = 21$. We assumed that the dark energy consist of scalar field and Gauss-Bonnet term, so we have shown the domination of dark energy in the present universe ($N = 21$).

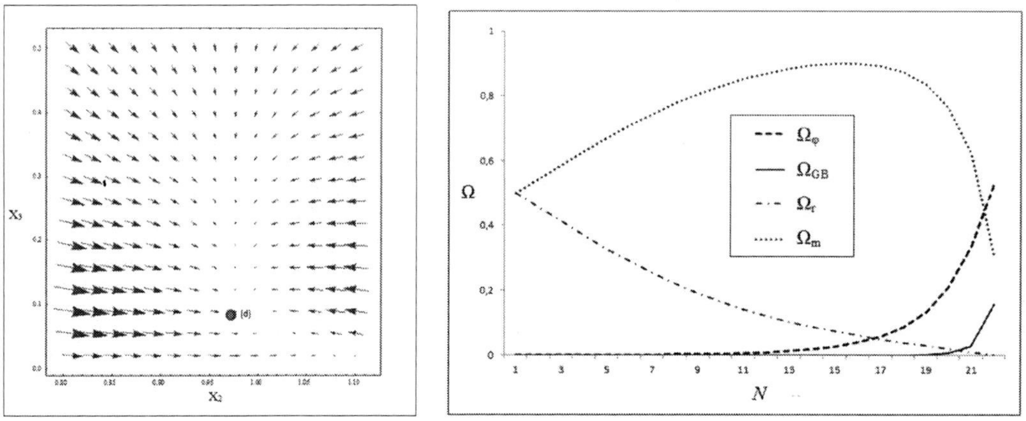

(a) (b)

Figure 1. (a) The trajectory of the solution of Gauss-Bonnet model with vector fieldfor $\bar{\beta} = 0.1$, $\mu = 7$, and $\lambda = 2$. (b) The evolution of variables Ω_ϕ, Ω_{GB}, Ω_r, and Ω_m for $\bar{\beta} = 11$, $\mu = 4$, and $\lambda = 23$. We choose initial condition $x_1 = 0.01$, $x_2 = 0.01$ $x_3 = 10^{-6}$, and $x_4 = 0.707$.

In Fig. 3 we show the evolution of each density parameters with variation of $\bar{\beta}$. At $N = 21$, the density parameter of each components for $\bar{\beta} = 12$ respectively $\Omega_r = 0.002$, $\Omega_m = 0.2273$, $\Omega_\phi = 0.447$, and $\Omega_{GB} = 0.276$. For $\bar{\beta} = 13$ at same N, the density parameter of each components are $\Omega_r = 0.001$, $\Omega_m = 0.196$, $\Omega_\phi = 0.346$, and $\Omega_{GB} = 0.456$. Evidently, the changed of $\bar{\beta}$ influence the value of each density parameters especially Ω_{GB} and Ω_m significantly (See Table 2). The result show that the increase of β will raise the density parameter value of Gauss-Bonnet term, but decrease the density parameter value of matter.

From the variations of $\bar{\beta}$ above, evolution of each density parameters for $\bar{\beta} = 12$ get near to the condition of present universe. The universe that showed at $N = 21$ is dominated by scalar field and Gauss-Bonnet term which represent the dark energy. The sum of density parameter both of them is $\Omega = 0.723$. The result is suitable with observations that show the universe today is dominated by 72% of dark energy [13,14].

$\bar{\beta}$	Ω_ϕ	Ω_{GB}	Ω_m	Ω_r
11	0.529	0.157	0.310	0.003
12	0.447	0.276	0.273	0.002
13	0.346	0.456	0.196	0.001

Table 2. The values of each density parameter in Gauss-Bonnet model with vector field for variation of $\bar{\beta}$ at $N = 21$

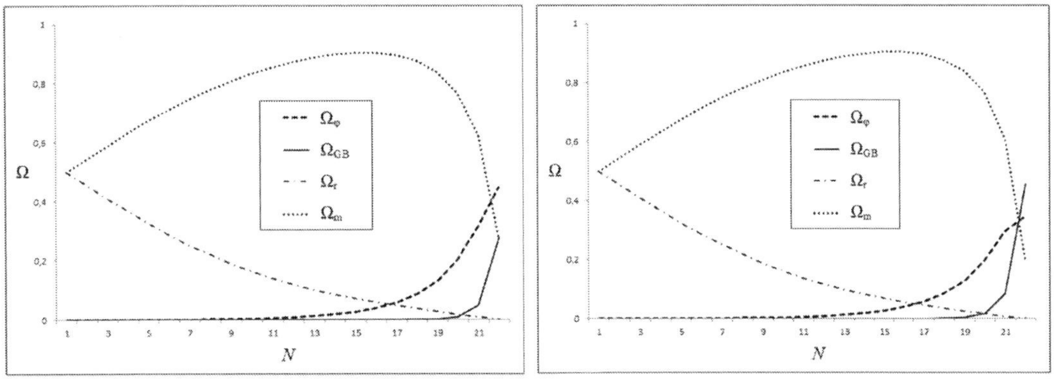

(a) (b)

Figure 2. The evolution of variables Ω_ϕ, Ω_{GB}, Ω_r, and Ω_m with initial condition $x_1 = 0.01$, $x_2 = 0.01$ $x_3 = 10^{-6}$, and $x_4 = 0.707$ for : (a) $\bar{\beta} = 12$, $\mu = 4$, and $\lambda = 23$. (b) $\bar{\beta} = 13$, $\mu = 4$, and $\lambda = 23$

4. Conclusion

We studied The evolution of density parameter of matter, radiation, scalar field and Gauss-Bonnet term on Gauss-Bonnet Model with vector field. Parameter $\bar{\beta}$ which contributed by vector field give influence to the density parameters of the model. Variations of $\bar{\beta}$ give change of density parameters specially matter and Gauss-Bonnet term. The model that was analyzed represent the condition of the present universe today which dominated by dark energy. Scalar field and Gauss-Bonnet term in the model dominate the universe as dark energy in Gauss-Bonnet model with 72.3%.

5. Acknowledgement

This work was supported by "Riset Inovasi KK ITB 2016" and "Riset desentralisasi ITB 2016". AS would like to thank Ministry of Research Technology and Higher Education of the Republic of Indonesia for "Fresh Graduate' Scholarship.

References

[1] A. G. Riess, et. al., Astron J. **116**, 1009 (1998); Astron. J. **117**, 707 (1999).
[2] S.Perlmutter et. al., Nature **391**, 51 (1998).
[3] M. Kowalski, et. al., Astrophys. Journal, **686**, p.749 (2008), arXiv:0804.4142
[4] M. Hicken et al., Astrophys. J. **700**, 1097 (2009) [arXiv:0901.4804 [astro-ph.CO]].
[5] E. Komatsu et al. [WMAP Collaboration], Astrophys. J. Suppl. **180**, 330 (2009) arXiv:0803.0547 [astro-ph].
[6] W. J. Percival et al., Mon. Not. Roy. Astron. Soc. **401**, 2148 (2010); arXiv:0907.1660 [astro-ph.CO]
[7] S. Nojiri, S. D. Odintsov, and M. Sasaki, Gauss-Bonnet dark energy, Phys. Rev. **D 71** (2005), 123509.
[8] B. M. N. Carter and I. P. Neupane, Towards inflation and dark energy cosmologies from modified Gauss-Bonnet theory, JCAP 0606 (2006), 004.
[9] L. Amendola, C. Charmousis, and S. C. Davis, Constraints on Gauss-Bonnet gravity in dark energy cosmologies, JCAP 0612 (2006), 020.
[10] Y. Fujii and K. Maeda, The Scalar-Tensor Theory of Gravitation, Cambridge University Press (2003).
[11] Arianto, F. P. Zen, Triyanta, and B. E. Gunara,Attractor Solutions in Lorentz Violating Scalar-Vector-Tensor Theory, Phys. Rev. **D 77**, 123517 (2008).
[12] F.P. Zen et al.,Cosmological Evolution of Interacting Dark Energy in Lorentz Violation, The European Phys. Journal C, Vol. 63, No. 3, (2009), pp. 477-490 [arXiv:0809.3847]
[13] N. Jarosik et al., Seven-Year Wilkinson Microwave Anisotropy Probe (WMAP) Observations: Sky Maps, Systematic Errors, and Basic Results, Astrophys. J. Suppl. **192** (2011) 14 [arXiv:1001.4744].

IOP Conf. Series: Journal of Physics: Conf. Series **1127** (2019) 012022 doi:10.1088/1742-6596/1127/1/012022

[14] M. Kowalski et al., Improved Cosmological Constraints from New, Old and Combined Supernova Datasets, Astrophys. **J. 686** (2008) 749 [arXiv:0804.4142].

IOP Conf. Series: Journal of Physics: Conf. Series **1127** (2019) 012023 doi:10.1088/1742-6596/1127/1/012023

Group theory description of third harmonic generation in diamond and zincblende lattice

B Nurjanati [1], H Alatas [1, 2] and H Hardhienata [1, 2]

[1] Theoretical Physics Division, Department of Physics, Bogor Agricultural University, INDONESIA.
[2] Research Cluster for Dynamics and Modelling of Complex Systems, Faculty of Mathematics and Natural Sciences, Bogor Agricultutal University, INDONESIA.

corresponding author: hendradi@ipb.ac.id

Abstract. We provide a description that shows the importance in finding a susceptibility tensor in nonlinear optics and apply group theory to derive a fourth rank tensor of O_h point group related to the bulk symmetry of monoatomic diamond semiconductor such as Silicon. It is well known that the fourth rank tensor is closely related to the third order nonlinear susceptibility tensor in nonlinear optics through Neumann's principle and is crucial to understand third harmonic generation (THG) and electric field induced second harmonic generation (EFISH). Using a general Rodriguez rotation, we were able to obtain a three-dimmensional rotation matrix required to perform the symmetry operations, whereas their reflection matrixes are derived using a Householder fromula. We also show how a T_d point group corresponding to zincblende can be obtained by breaking the symmetry of the O_h tensor rather than repeating the tedious procedure when obtaining the O_h tensor. Assuming a driven frequency far from the resonance frequency, we can apply Kleinman symmetry and demonstrate that the number of independent components in the O_h and T_d fourth rank tensor is reduced from four to just two.

1. Introduction

We live in a world where the system's response towards perturbations is often nonlinear. Nonlinearity can be found in many fields of science, such as the generation of higher harmonics radiation. It is produced by a nonlinear polarization source that is generated inside the material and can be treated both classically or quantum mechanically using perturbation theory introduced by Amstrong, Blombergen, Ducuing, and Pershan (ABDP) [1]. Although the quantum mechanical model should be valid for a wider phenomenon, such as nonlinearity due to excitons in semiconductors [2], the physics is difficult to visualize. Thus, classically phenomenological models pioneered by John Sipe [3] through a polarizable charged sheet using Green's theory are therefore still maintained but can contain up to 10 parameters to interpret the second harmonic generation (SHG) intensity from nonvicinal semiconductors [4]. The understanding that nonlinearity can be highly simplified through a classically phenomenological model came in 2002 through the work of D. E. Aspnes Group who introduced the simplified bond hyperpolarizability model (SBHM) assuming SHG radiation produced by anharmonic charge oscillation along atomic bonds [5]. It was later extended to investigate relative dipole and quadrupole contribution [6] and to cope for third harmonic generation (THG) [7].

It was soon realized that the validity of the higher rank tensor obtained in SBHM by direct product of bond vectors hinges on comparison with the tensor obtained by group theory, since the latter depends only on the symmetry of the material which must be geometrically correct. This link was established through the work of Kurt Hingerl's group in 2013 where one us (Hardhienata) was involved [8]. It was shown that the tensor obtained in SBHM can be related to group theory (GT) with good agreement, especially if Kleinman Symmetry (KS) is applied [9]. They proved later on that KS is indeed an inherent property of SBHM [10]. The model was recently adapted to include THG and EFISH in diamond semiconductors [11], arbitrary input polarization angle to identify bulk quadrupoles [12], SHG from zincblende structures [13], and the ability to determine vicinality and absorption in semiconductors using nonlinear optics particularly SHG [13].

Furthermore, group theory (GT) usually used to simplify the computational effort in band stucture calculation with highly computational electronic. Macroscopically the type of linear and nonlinear optical source from within the material depends on the band structure and symmetry of the semiconductor [14], the latter is expressed through the macroscopic nonlinear susceptibility. The mathematical procedure to reduce the tensor components based on symmetry property is group theory and plays a vital role in the investigation of solid state physics both classically and quantum mechanically [15], which is an important discipline that was developed in the 20th century [16].

The optical response of the materials can be described by the relation between polarization and electric field. However, if the electric field used is quite large, as is the case for Femto or Pico second lasers, the polarization response is nonlinear, and the relation between the polarization and electric field comprises of a linear and nonlinear terms:

$$P_i = \chi_{ij}{}^{(1)} E_j + \chi_{ijk}{}^{(2)} E_j E_k + \chi_{ijkl}{}^{(3)} E_j E_k E_l \ldots \quad (1)$$

Equation (1) applies for electric dipoles and can be derived from a classical force equation assuming damped charge forced anharmonic oscillation solved using the Rayleigh-Schrödinger perturbation theory [17]. The magnitude of $\chi^{(1)}$ is in general function of the frequency which is inferable through familiar linear dispersional relations and is a consequence of causality (e.g.) Kramers-Krönig relation. Here, $\chi^{(2)}$ and $\chi^{(3)}$ is the second-, third-order susceptibility tensor (also known as the third and fourth rank tensor, respectively). The susceptibility tensor contains many information about the symmetry of nonlinearly optical materials that are crucial in understanding the origin of nonlinear radiation. If the potential is centrosymmetric as is the case for diamond cubic systems such as Silicon, the second order nonlinear susceptibility in equation (1) vanishes (zero) inside the bulk due to parity symmetry. However, the odd order susceptibilities including third order (or fourth rank tensor) are still allowed and will be analyzed using GT.

2. Methodology

2.1. Cartesian coordinate rotation

To obtain the correct tensor that represents a given symmetry, we apply certain matrix transformational operation on the most general tensor with the idea that these operations should leave the geometry of the unchanged symmetry. The amount of allowed symmetry operation depends on how high the symmetry of the crystal is. Generally, if a crystal structure has high symmetry, such as the case of a silicon bulk, then the number of requirely transformational operation becomes higher. Due to symmetry, applying certain rotational and mirror operation should invariantly leave the crystal geometry corresponding to certain point group, therefore the tensor before and after the operation should be the same and a mathematical consistency requires some elements to be zero. Generally, the applied transformational operation that is used to obtain the three-dimensional rotation matrix of the materials consists of rotation of the Cartesian coordinates [18].

To find the identity matrix E (derived from the German word *Einheitsmatrix*), we use 360^0 or 2π for θ. Because interestingly, crystals in nature tend to maintain an efficient periodicity without free or blank spaces in between they retain such symmetry, so that it is invariant through certain rotation only. In other words, because the crystals must completely be filled with the periodic repetition of the basic building block, only two-, three-, four-, and six-fold rotations are possible [8]. These possible rotations are given by the angles of π, 2π, π/2, and π/3. The type of the symmetry axis of the crystals is symbolized by the rules of the Schönflies notation, which for rotation is symbolized by C_n, where C is the rotation and n denotes the n-fold rotation. For a two-fold rotation, the angle is 180 degrees so that $\frac{2\pi}{n}$ [18].

2.2. *Rodriguez rotation*

Because the bulk symmetry is three dimensional, the rotaion of the Cartesian coordinate cannot be used directly, but it must be combined with Rodriguez rotation. It is a formulation that is associated with the parameter and Euler-Rodriguez formula that applies to a general three-dimensional rotation in th bulk [19]. The combination of rotation about the Cartesian coordinates and Rodriguez equation gives the correct representation of the bulk tensor.

The Rodriguez rotation satisfies $\boldsymbol{v} = \boldsymbol{v}_{=} + \boldsymbol{v}_{\perp}$. Here, $\boldsymbol{v}$ is the vector contained in the rotation elements of the group. This indicates that the each vector element within this group contains a parallel and perpendicular vector. Using the identity rule from calculus vectors and allowing rotation $\boldsymbol{v}_{rot} = \boldsymbol{v}_{rot\,//} + \boldsymbol{v}_{rot\perp}$ and $\boldsymbol{v}_{rot\perp} = (\cos\theta)\boldsymbol{v}_{\perp} + (\sin\theta)(\boldsymbol{n}\times\boldsymbol{v}_{\perp})$ we obtained the equation:

$$\begin{pmatrix} (\boldsymbol{n}\times\boldsymbol{v})_x \\ (\boldsymbol{n}\times\boldsymbol{v})_y \\ (\boldsymbol{n}\times\boldsymbol{v})_z \end{pmatrix} = \begin{pmatrix} n_y v_z - n_z v_y \\ n_z v_x - n_x v_z \\ n_x v_y - n_y v_x \end{pmatrix} = NV \tag{2}$$

Because $\boldsymbol{v}_{rot} = \boldsymbol{v}R$, we obtained the following equation:

$$R = I + \sin\theta(\boldsymbol{n}) + (1-\cos\theta)\boldsymbol{n}^2 \tag{3}$$

where, R is the so called Rodriguez rotation matrix, and I is the identity element matrix. Then we can get the new rotation transformation [20]:

$$R_{(\theta)} = \begin{pmatrix} n_1^2(1-\cos\theta)+\cos\theta & n_1n_2(1-\cos\theta)-n_3\sin\theta & n_1n_3(1-\cos\theta)+n_2\sin\theta \\ n_1n_2(1-\cos\theta)-n_3\sin\theta & n_2^2(1-\cos\theta)+\cos\theta & n_2n_3(1-\cos\theta)-n_1\sin\theta \\ n_1n_3(1-\cos\theta)-n_2\sin\theta & n_2n_3(1-\cos\theta)+n_1\sin\theta & n_3^2(1-\cos\theta)+\cos\theta \end{pmatrix} \tag{4}$$

where, $\boldsymbol{n}$ is the unit vector and satisfies $n_1^2 + n_2^2 + n_3^2 = 1$.

2.3. *Reflection operation*

We now describe a reflection operation which is as important as the rotational operation discussed beforehand. Generally, the reflection operation can be obtained from the called Householder matrix, the matrix is usually called by the various names: Householder formula, Transformation of Householder, Householder reflection, or elementary reflector. Householder transformation is a reflection through orthogonal plane to a given vector unit, such as vector unit v and expressed in the form of a matrix [21]. This reflection can be expressed as follows:

$$Hv = \boldsymbol{v} - 2\boldsymbol{v}\boldsymbol{v}_{//} \tag{5}$$

Because the preffered H matrix is a single matrix that can transform to all of the normal vector unit $\boldsymbol{v}$, then the normal vector $\boldsymbol{v}$ that is parallel must be projected on the x-axis, therefore:

$$\boldsymbol{v}_{//} = \frac{\boldsymbol{x}}{\|\boldsymbol{x}\|}\|\boldsymbol{v}\|\cos\theta \tag{6}$$

By eliminating $\cos\theta$ in equation (6), we found the new reflection matrix of H. the form of the Householder matrix is:

$$H = \begin{pmatrix} 1-(n_1)^2 & -2n_1n_2 & -2n_1n_3 \\ -2n_2n_2 & 1-(n_2)^2 & -2n_2n_3 \\ -2n_1n_3 & -2n_2n_3 & 1-(n_3)^2 \end{pmatrix} \tag{7}$$

3. Result

The bulk fourth rank tensor related to the Third Harmonic Generation (THG) from a dimond cubic crystal, such as Silicon, can be produced by applying the following procedure: apply rotation of $\pi/3$ about the z-axis, apply rotation of $\pi/2$ about the center side of the cube to another center side of the cube where the axis also passes through the center of cube, then apply the $\pi/4$ rotation. Afterwards apply the before mentioned angle of rotation to the combined Cartesian coordinate and Rodriguez rotation, inverse the identity matrix, then apply the reflection to diagonal and horizontal plane. When all the procedures are implemented accordingly, the next step is combining the rotation operation with reflection operation through the equation:

$$S_4 = \sigma_h C_4 \tag{8}$$

When the procedure is completed a total of 16 matrix elements of the O_h is obtained. It is the point group of Silicon bulk. The most general form of the fourth rank tensor is contain a total of 81 components. To simplify the general tensor from is transformed using the corresponding rotation and reflection operation that we have found previously. The procedure of transforming the tensor is [19]:

$$d'_{ijkl} = r_{im}r_{jn}r_{ko}r_{lp}d_{mnop} \tag{9}$$

Equation (9) must satisfy Neumann principle which can be stated as: the component of the symetry operation of the group in the crystals must be invariant under transformation [9]. Based on this rule, the entire tensor components after the transformation that are equal to the negative of the corresponding elements before the operation is performed must be zero. Then the fourth rank tensor becomes:

$$\begin{pmatrix} \begin{pmatrix} d_{3333} & 0 & 0 \\ 0 & d_{3322} & 0 \\ 0 & 0 & d_{3322} \end{pmatrix} & \begin{pmatrix} 0 & d_{3232} & 0 \\ d_{3223} & 0 & 0 \\ 0 & 0 & 0 \end{pmatrix} & \begin{pmatrix} 0 & 0 & d_{3232} \\ 0 & 0 & 0 \\ d_{3223} & 0 & 0 \end{pmatrix} \\ \begin{pmatrix} 0 & d_{3223} & 0 \\ d_{3232} & 0 & 0 \\ 0 & 0 & 0 \end{pmatrix} & \begin{pmatrix} d_{3322} & 0 & 0 \\ 0 & d_{3333} & 0 \\ 0 & 0 & d_{3322} \end{pmatrix} & \begin{pmatrix} 0 & 0 & 0 \\ 0 & 0 & d_{3232} \\ 0 & d_{3223} & 0 \end{pmatrix} \\ \begin{pmatrix} 0 & 0 & d_{3223} \\ 0 & 0 & 0 \\ d_{3232} & 0 & 0 \end{pmatrix} & \begin{pmatrix} 0 & 0 & 0 \\ 0 & 0 & d_{3223} \\ 0 & d_{3232} & 0 \end{pmatrix} & \begin{pmatrix} d_{3322} & 0 & 0 \\ 0 & d_{3322} & 0 \\ 0 & 0 & d_{3333} \end{pmatrix} \end{pmatrix} \tag{10}$$

The tensor elements in equation (10) show that the maximum independent component of the bulk that can be owned by O_h point group is four. Assuming that the input frequency is much smaller than natural frequency (e.g. far resonance) of the silicon crystals, then we can use Kleinman symmetry: the index of the each components can be exchanged freely [9]. Application of Kleinman symmetry immediately reduces the amount of independent component from four to just two. According to the SBHM only one independednt component is responsible in the fourth rank tensor to generate THG [13].

Interestingly, when the material is Zincblende, which just diamond cubic, but with two atoms instead such as GaAs, rather than the monoatomic Si the symmetry inside the bulk breaks from O_h to a lower symmetry because now the structure has Ga or As centre surrounded by four As or Ga atoms respectively forming an alternating $Sp3$ atomic orbital (tetrahedral). If this is the case, then from the 48 symmetry elements only 24 remain, and this is called T_d point group. It is understandable then, that T_d is a subgroup of O_h, because it can be obtained by breaking the O_h symmetry. When the symmetry is reduced from 48 to 24 the tensor in equation (10) is the same with the tensor of the T_d point group in fourth rank tensor, but in the third rank tensor the tensor form of the O_h point group and T_d point group is very different. Equation (10) can be formed into reduced tensor, to make it simple, because the reduced tensor has merely 36 component. In the reduced tensor notation, equation (10) becomes:

$$\begin{pmatrix} d_{3333} & d_{3322} & d_{3322} & 0 & 0 & 0 \\ d_{3322} & d_{3333} & d_{3322} & 0 & 0 & 0 \\ d_{3322} & d_{3322} & d_{3333} & 0 & 0 & 0 \\ 0 & 0 & 0 & 2d_{3223}+2d_{3232} & 0 & 0 \\ 0 & 0 & 0 & 0 & 2d_{3223}+2d_{3232} & 0 \\ 0 & 0 & 0 & 0 & 0 & 2d_{3223}+2d_{3232} \end{pmatrix} \tag{11}$$

Another form of the reduced tensor of a T_d and O_h point group is by simplifying the indeces of the tensor component with just two indices referring to the convention in [18]. Then equation (11) becomes:

$$\begin{pmatrix} \begin{pmatrix} d_{11} & 0 & 0 \\ 0 & d_{12} & 0 \\ 0 & 0 & d_{12} \end{pmatrix} & \begin{pmatrix} 0 & \frac{d_{44}}{4} & 0 \\ \frac{d_{44}}{4} & 0 & 0 \\ 0 & 0 & 0 \end{pmatrix} & \begin{pmatrix} 0 & 0 & \frac{d_{44}}{4} \\ 0 & 0 & 0 \\ \frac{d_{44}}{4} & 0 & 0 \end{pmatrix} \\ \begin{pmatrix} 0 & \frac{d_{44}}{4} & 0 \\ \frac{d_{44}}{4} & 0 & 0 \\ 0 & 0 & 0 \end{pmatrix} & \begin{pmatrix} d_{12} & 0 & 0 \\ 0 & d_{11} & 0 \\ 0 & 0 & d_{12} \end{pmatrix} & \begin{pmatrix} 0 & 0 & 0 \\ 0 & 0 & \frac{d_{44}}{4} \\ 0 & \frac{d_{44}}{4} & 0 \end{pmatrix} \\ \begin{pmatrix} 0 & 0 & \frac{d_{44}}{4} \\ 0 & 0 & 0 \\ \frac{d_{44}}{4} & 0 & 0 \end{pmatrix} & \begin{pmatrix} 0 & 0 & 0 \\ 0 & 0 & \frac{d_{44}}{4} \\ 0 & \frac{d_{44}}{4} & 0 \end{pmatrix} & \begin{pmatrix} d_{12} & 0 & 0 \\ 0 & d_{12} & 0 \\ 0 & 0 & d_{11} \end{pmatrix} \end{pmatrix} \tag{12}$$

From equation (12), we can see that the maximum independent tensorial component owned by a T_d point group is 2, with the rule that $\frac{d_{44}}{4}$ is d_{12} and d_{12} is the same with the d_{21}. Further work is required to compare this finding with the SBHM.

4. Conclusion

We have derived the corresponding fourth rank tensor describing THG for a diamond cube semiconductor point group using grup theory (GT) which has the O_h point group. We show that four independent tensorial components are necessary to describe THG which are immediately reduced to two if Kleinman symmetry is assumed. We show how the symmetry is broken from O_h to T_d for the case of zincblende structure which is a subgroup of O_h with only 2 independent component.

References

[1] Amstrong J A, Bloembergen N, Ducuing J, and Pershan P S 1962 *J. Phys. Rev.* 127, 1918-1939
[2] Nye J F 1985 *Physical Properties of Crystals* (Oxford: Oxford University Press) p 41
[3] Sipe J E, Moss D J, and van Driel H M 1988 *J. Phys. Rev.* **B** 35, 1129-1141.
[4] Lupke G, Bottomley D J, and van Driel H M 1994 *J. Opt. Soc. Am.* **B.** 11 33
[5] Wang J-F T, Powell G D, Johnson R S, Lucovsky G and Aspnes D E 2002 *J. Vac. Sci. Technol.* **B** 20 1699
[6] Peng H J, Adles E J, Wang J-F T and Aspnes D E 2005 *Phys. Rev.* **B.** 72 205203
[7] Peng H and Aspnes D E 2004 *Phys. Rev.* **B.** 70 165312
[8] Jatirian-Foltides E S, Escobedo-Alatorre J J, Marquez-Aguilar P A and Hardhienata H 2016 *J. Rev. Mex. Fis.* E 62 (2016) 5-13
[9] Hardhienata H et all 2016. *J. Opt. Soc. Am.* B 33(2), 33 195
[10] Alejo-Molina A, Hardhienata H and Hingerl K 2014 *J. Opt. Soc Am.* **B.** 31 526
[11] Alejo-Molina A, Hingerl K and Hardhienata H 2015 *J. Opt. Soc. Am.* **B** 232436
[12] Reitbock C, Stifter, Alejo-Molina A, Hinger K, Hardhienata H *J. Opt.* 18 (2016) 035501
[13] Hardhienata H, Prylepa A, Stiffer D and Hinger K 2013 *J. Phy* 423, 012046
[14] Yu Y P, Cardona M 2010 *Fundamentals of Semiconductors: Physics and Material Properties Fourth Edition* (Berlin: Springer) p 17-61
[15] Dresselhaus M S 2008 *Group Theory: Application to the Physics of Condensed Matter* (Berlin: Springer) p 3-70
[16] Mulyadi 2004 *Klasifikasi Hadron dan Meson sebagai Representasi Uniter pada Sistem Partikel Elementer* (Depok: Universitas Indonesia) p 1
[17] Boyd W R 2008 *Nonlinear Optics: Third Edition* (Amsterdam: Elsevier) p 1-204
[18] Powell R C 2010 *Symmetry, Group Theory, and the Physical Properties of Crystals* (Berlin: Springer) p 25-77
[19] Hardhienata H 2016 *Tutorial Optika Nonlinier dari sisi Fisika Klasik* (Bogor: IPB Press) p 81-102
[20] Eminger C 2013 *Transformation of Third-rank Tensor* (Linz: Johannes Kepler University) p 19
[21] Householder S 1970 *The Numerical Treatment of a Single Nonlinear Equation* (US: Mc Graw-Hill Book Company)

Acknowledgements

H. H would like to thank Adalberto Alejo-Molina for valuable disscussion on group theory. This research was funded by the "Penelitian Unggulan Perguruan Tinggi 2017 No: 011/SP2H/LT/DRPM/IV/2017".

IOP Conf. Series: Journal of Physics: Conf. Series **1127** (2019) 012024 doi:10.1088/1742-6596/1127/1/012024

Preliminary Design Study of Small Power 30-50 MWt Experiment Power Reactor based on High Temperature Pebble Bed Gas Cooled Reactor Technology

Dwi Irwanto[1*], Sidik Permana[1], Asril Pramutadi[1] and Syeilendra Pramuditya[1]

[1]Nuclear Physics and Biophysics Research Group, Faculty of Mathematics and Natural Sciences, Institut Teknologi Bandung, Jl. Ganesha No 10, Bandung, 40132, Indonesia

*Email: dirwanto@fi.itb.ac.id

Abstract. Pebble Bed Reactor (PBR) is one of the most promising nuclear power plant. It has advantages in terms of high thermal efficiency, economic and its safety features. To build an experimental power reactor based on PBR technology, it is important to understand all the aspects related to the neutronic, thermal hydraulic and safety features of this reactor type. In the present study, a set of calculations have been performed to find characteristics of selected parameters of the PBR design that significantly contributed to the neutronic aspect, such as number of uranium enrichment, pebble fuel speed, fuel loading scheme and burnup. Calculation and analysis was performed for 30 MWt and 50 MWt power.

1. Introduction

Recently, nuclear power generation is an established part of the world's electricity mix providing a significant portion of world electricity. Continuous improvement to the reactor designs resulting Generation IV reactors (Gen IV) as a set of theoretical nuclear reactor designs that currently under intensive research in the world. Many reactor types were considered initially but the list was downsized to focus on the most promising technologies. These systems offer significant advances in safety and reliability, proliferation resistance, sustainability and economics.

The Very High Temperature Reactor concept utilizes a graphite-moderated core with outlet temperature of around 1000 °C is also one of the Gen-IV reactor that being researched. The high temperatures enable applications such as hydrogen production, desalination of sea water and semo other industrial processes. The reactor core can be either a prismatic-block or a pebble bed reactor design. The Pebble Bed Reactor (PBR) has three fuel loading scheme: Multipass, Once Through Then Out and Peu a Peu or accumulative fuel loading scheme which have good neutronic aspect for both using uranium or thorium fuel [1-3].

Pebble bed reactor considered as a safety system because of the low power density of the reactor, which means that the amount of energy and heat produced is volumetrically low. The use of large amount of graphite as a moderator and reflector make this system has high heat capacity. Moreover, it is expected that natural mechanisms to remove the heat such as conductive and radiative heat transfer will works even if no convective core cooling is provided. This becomes important because the

IOP Conf. Series: Journal of Physics: Conf. Series **1127** (2019) 012024 doi:10.1088/1742-6596/1127/1/012024

temperature that is reached in a case of loss of coolant accident is far below the fuel damage threshold temperature and it takes long time to reach the peak temperature. So, that fuel damage will not occur and the core will not melt even in a severe accident [4].

Indonesia is now considering to build an Experiment Power Reactor based on PBR technology. In order to master this technology, it is important to understand all the aspects related to the neutronic, thermal hydraulic and safety features of this reactor type. The purpose of the present study is to analyze characteristics of selected parameters of the PBR design that significantly contributed to the neutronic aspect, such as number of uranium enrichment, pebble fuel speed, fuel loading scheme and burnup. Calculation and analysis are performed for 30 MWt and 50 MWt power.

2. Calculation Methods and Model

Pebble Bed Reactor has unique design and characteristic compared to other types of reactor. To accurately calculate Pebble Bed Reactor system, one have to consider double heterogeneity of PBR. The first heterogeneity is the randomly distributed coated fuel particle inside a pebble fuel ball. This pebble fuel balls itself are then randomly distributed inside the reactor core, that make it the second heterogeneity. A suitable computational method to simulate this system is the continuous energy Monte Carlo Method. This method can deal with reactor systems which have double heterogeneous structure. In Monte Carlo simulation of particle transport in the region where spherical fuel element exists, a particle enters a spherical fuel element, penetrates it and enters the next spherical fuel element. The whole particle transport history is made up of these basic processes. Therefore, it is not necessary to know the locations of all spherical fuel elements at the same time in the calculation. The position of the fuel particles itself is determined probabilistically from the Nearest Neighbor Distribution (NDD) in the process of the random walk. Those methods are implemented in the MVP/MVP-BURN code with JENDL 3.3 as nuclear data library which are used in this research. [5-7]

A reference design based on HTR-10 are implemented in the calculation [8]. In the present step, simplification to the actual reactor was done by created a cylindrical core loaded by pebble fuel ball. Figure 1 shows calculation model of the small Pebble Bed Reactor.

Figure 1. Calculation Model

Specification of the reactor core design is shown in Table 1. The core is 197 cm high with 90 cm radius. 180 cm of it height is filled with pebble fuel balls with homogen uranium enrichment. Parametric survey was performed for pebble bed fuel with uranium enrichment between 7% to 20%.

IOP Conf. Series: Journal of Physics: Conf. Series **1127** (2019) 012024 doi:10.1088/1742-6596/1127/1/012024

Table 1. Reactor Design Specification

Design Specification		
Reactor Power	30 50	MWth
Fuel	TRISO	
Core radius	90	cm
Core Height	197	cm
Reflector width	100	cm
Uranium enrichment	7-20	%
U loading	5	gram

Number of uranium loaded per one ball set to be constant at 5 grams for each ball and enrichment. TRISO fuel configuration is used and its detail configuration including material and density of each layer is shown in Table 2.

Table 2. Coated Fuel Particle Specification

Coated Fuel Particle		
Fuel Kernel		
✧ Material	UO_2	
✧ Radius of the kernel	0.250	mm
✧ Density	10.4	g/cm^3
Coatings		
➢ First Buffer Layer (PyC)		
✧ Thickness	0.09	mm
✧ Density	1.1	g/cm^3
➢ Second Layer (PyC)		
✧ Thickness	0.04	mm
✧ Density	1.9	g/cm^3
➢ Third Layer (SiC)		
✧ Thickness	0.035	mm
✧ Density	3.18	g/cm^3
➢ Forth Layer (PyC)		
✧ Thickness	0.04	mm
✧ Density	1.9	g/cm^3

3. Calculation Results and Analysis

Calculation are performed to investigate important parameters that affect neutronic characteristics of the reactor design. Uranium enrichment are varying from 7-20% for each reactor configuration with five bulk fuel loaded in the reactor core. Uranium enrichment effects to criticality of the reactor then examined for different approximated pebble fuel speed of 0.5, 0.75 and 1.0 cm/day through the reactor core to see its effects.

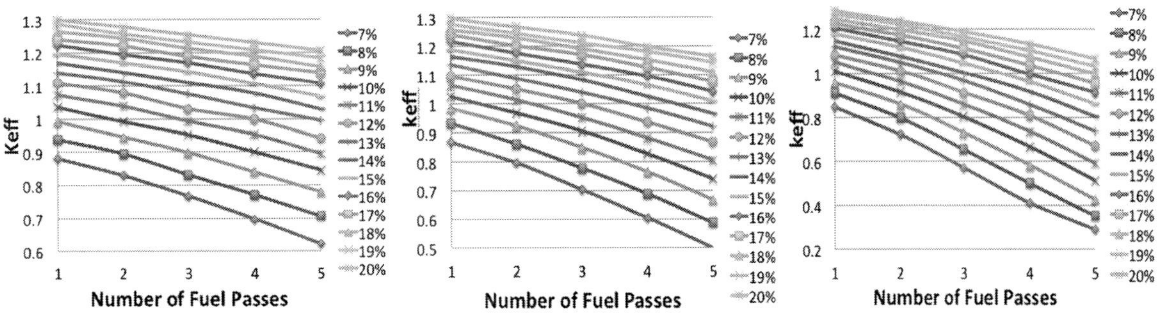

Figure 2. Characteristic of pebble fuel speed (1.0, 0.75 and 0.5 cm/day) with five passes fuel loading scheme and 7-20% uranium enrichment needed for 30 MWth power

Figure 2-3 shows characteristic of several aspect that contribute to neutronic parameters: number of uranium loaded to the pebble fuel and pebble fuel speed with five passes fuel loading scheme.

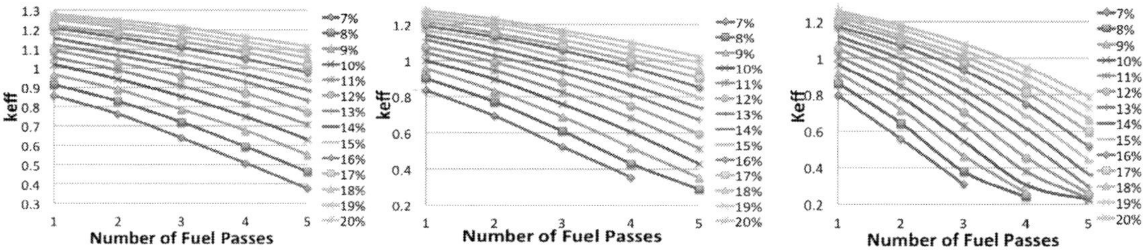

Figure 3. Characteristic of pebble fuel speed (1.0, 0.75 and 0.5 cm/day) with five passes fuel loading scheme and 7-20% uranium enrichment needed for 50 MWth power

From series of performed calculations, one could understand that for five passes fuel loading with 1.0 cm/day fuel ball speed, minimum uranium required is 14% for 30 MWth power reactor. For the same configuration of 30 MWt power and five fuel ball passes, if fuel ball speed decreased to 0.75 cm/day then minimum uranium enrichment required increased to 15%. If fuel ball speed through reactor core further decreased to 0.5 cm/day, one would need 19% of uranium enrichment for the same 30 MWt power.

With the same scheme for 50 MWt, fuel speed of 0.5 cm/day had to have more than 20% of uranium enrichment. For 0.75 and 1.0 cm/day fuel ball speed, it is required 20% and 17% uranium enrichment, respectively.

Furthermore, it is also important to examined number of burnup for each reactor configuration with minimum uranium enrichment obtained from the previous analysis. Figure 4-5 shows the results of burnup and number of fissile material (U-235) used for each reactor configuration.

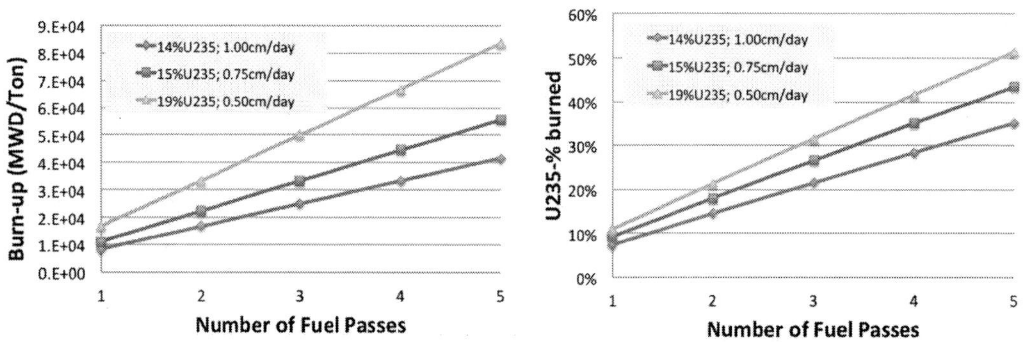

Figure 4. Effect of fuel speed (0.5, 0.75 and 1.0 cm/day), uranium enrichment for five fuel passes loading scheme of 30 MWt power

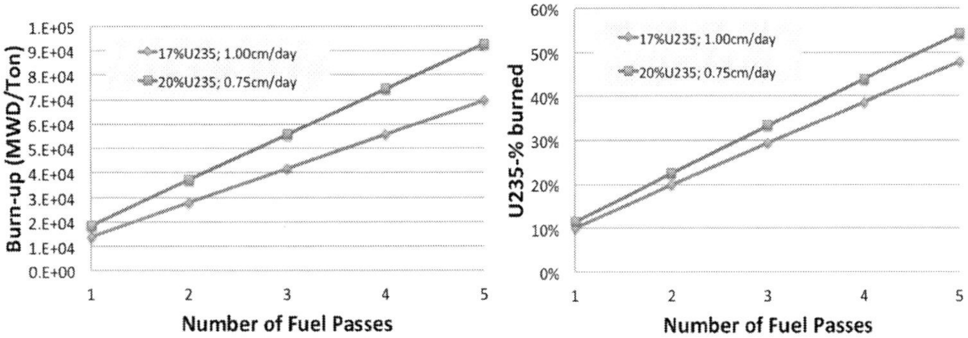

Figure 5. Effect of fuel speed (0.5, 0.75 and 1.0 cm/day), uranium enrichment for five fuel passes loading scheme of 50 MWt power

From performed calculation for different fuel speed of 1.0, 0.75 and 0.5 cm/day and minimum uranium required for each reactor configuration with 30 MWt power, one could get burnup of 41.7, 55.6 and 83.5 MWD/ton, respectively. Each of these reactor configurations also will burn 35.1, 43.4 and 51.4% U-235 as fissile material. For 50 MWth power with 1.0 cm/day speed and 20% uranium enrichment, burnup will be 69.5 MWD/ton and 47.8% fissile burned; meanwhile for 0.75 cm/day and 17% uranium enrichment, it will be 92.7 MWD/ton and 54.2% depleted U-235.

4. Conclusion

Parametric survey and design for small Pebble Bed Reactor have been performed in the present study for 30 and 50 MWt power by considering fuel speed and uranium enrichment for five passes loading fuel through the reactor core. Based on parametric surveys performed in the present study, considering uranium enrichment and fuel economics, configuration of reactor core with 15% enrichment and 0.75cm/day for 30MWt; 20% enrichment and 0.75cm/day for 50MWt have relatively good characteristics. From series of performed calculations, in general one could understand that faster fuel ball speed will required lower uranium enrichment but it will have relatively poor fuel economic. At the other hand, from the calculation it is shown that slower fuel ball speed will required higher uranium enrichment and resulting in relatively higher fuel economic for the same number of fuel passes.

Acknowledgment

This work has been supported by Penelitian Unggulan Perguruan Tinggi (PUPT) Program-Desentralisasi-RISTEKDIKTI 2016.

References

[1] D. Irwanto and T.Obara, 2012 *J. Nucl. Sci. Technol*, vol. 49, No.2, p.222-229
[2] D. Irwanto and T.Obara, 2011 *J. Nucl. Sci. Technol*, vol. 48, No.11, p. 1385-1395
[3] D. Irwanto, T. Obara, K. Yukitaka and I. Yamanaka, 2010 *Trans Am Nucl Soc* 103 p.791
[4] D. Irwanto and T. Obara, 2013 *Ann. Nucl. Energy*, vol. 60, p. 383-395
[5] Y. Nagaya, et al, 2004 Japan Atomic Energy Agency
[6] K. Okumura, et al, 2005 Japan Atomic Energy Agency
[7] K.Shibata, et al, 2002 *J. Nucl. Sci. Technol*, Vol39, No.11, p. 1125-1136
[8] International Atomic Energy Agency, *Evaluation of High Temperature Gas Cooled Reactor Performance: Benchmark Analysis Related to Initial Testing of the HTTR and HTR-10* (2003)

IOP Conf. Series: Journal of Physics: Conf. Series **1127** (2019) 012025 doi:10.1088/1742-6596/1127/1/012025

Study on SMART Reactor Design Using ThO_2-UO_2 Fuel

Rahmi N. Ramdhani[1], Abdul Waris[2,*], Widayani[2] and Rizal Kurniadi[2]

[1]Department of Physics, Faculty of Mathematics and Natural Sciences,
Institut Teknologi Bandung,
[2]Nuclear Physics & Biophysics Research Division, Department of Physics,
Faculty of Mathematics and Natural Sciences,
Institut Teknologi Bandung,
*E-mail: awaris@fi.itb.ac.id

Abstract. There are two heterogenous core design options for thorium based fuel cycle in PWRs. One of them is Whole Assembly Seed and Blanket (WASB) design. In WASB design, seed units and blanket units separately occupy one full-size PWR assembly each, and the assemblies are arranged in the core in a modified checkerboard array. In this study, the seed units was loaded with UO_2 fuel rods and the blanket units was loaded with ThO_2-UO_2 . The configuration for each assemblies in both seed and blanket units were based on 17x17 KOFA (Korean Standard Fuel Assembly). Neutronics calculation was performed by using PIJ and CITATION modules of SRAC 2006 code with JENDL 3.3 as nuclear data library. The calculation showed that for the same fuel material configuration in each asssembly uints the WASB-I core loading pattern produce a better power distribution than WASB-II core loading pattern. The criticality condition for both core loading patterns can be achieved by loading 4.95 wt.% of enriched-U in seed fuel assemblies while blanket fuel assemblies were loaded by 20 % UO_2 (15 wt.% of enriched-U) and 80% ThO_2.

1. Introduction

One of the main issues that concerns the world is the increasing need of energy. Some studies predict that by 2020 the energy consumption worldwide will be about 48% higher than in 2011 [1]. Population growth has always been and will remain one of the key drivers of energy demand, along with economic and social development. The challenges of energy consumption growth make nuclear power a prominent major energy source for the next several decades according to the projections made by the International Atomic Energy Agency (IAEA). The IAEA predicts two scenarios concerning the growth of nuclear power, a low and a high. The low projection assumes that all nuclear capacity which is currently under construction or firmly in the development pipeline, gets completed and attached to the grid, but no other capacity is added. It states that, there would be growth in capacity from 370 GWe at the end of 2006 to 447 GWe by 2030. This means a growth in the nuclear capacity of 20.8% [2].

On the other hand the growth in the nuclear consumption also arises some questions about the viability to accomplish the expectations with the current fuel reserves, to respect the non-proliferation treaty, to reduce the radiotoxicity of the waste and to maintain and enhance the safe and reliable operation. One of the solutions to face these issues is the use of thorium as a fuel.

Korea Atomic Energy Research Institute (KAERI) has been developing a small-sized integral reactor called SMART (System integrated Modular Advanced ReacTor) for a dual purpose: seawater desalination and electricity generation [3,4]. SMART has been developed as a small-sized integral reactor, hence most of its major components is housed within the reactor vessel. Major components including 4 main coolant pumps, 12 steam generator cassettes, a self-pressurizer as well as active core are placed in a single reactor vessel as shown in Fig.1.

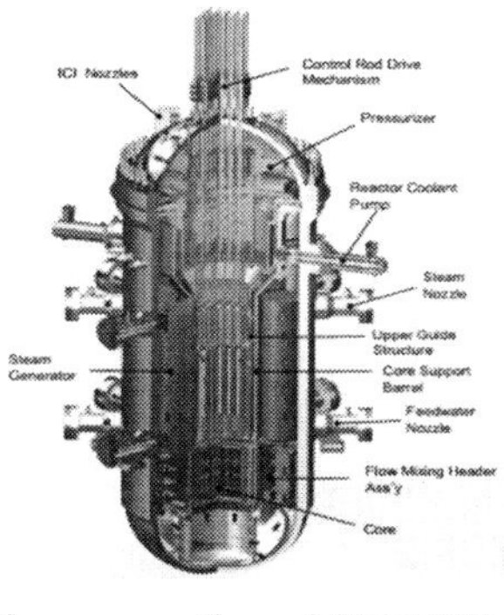

Figure 1. Vertical cross section of SMART reactor vessel [5]

SMART's core consists of 57 FAs with a 17 x 17 fuel rod array. The fuel assembly is based on the design of KOFA (Korean Standard Fuel Assembly) that was designed by KAERI/Siemens-KWU [6]. In this study, the SMART's core configuration was developed based on Whole Assembly Seed and Blanket Concept. The nuclear calculation for this core configuration was performed by using CITATION and JENDL3.3 as its nuclear data library.

2. Nuclear Core Design

The goal of this study is the search for an advanced heterogenous core design for thorium based fuel cycle in SMART reactor. There are two major heterogenous PWR core design: 1) the Seed-Blanket Unit (SBU)/Radkowsky Thorium Fuel (RTF) which employs a seed-blanket unit that is a one-for-one replacement for a conventional PWR fuel assembly; and 2) the whole assembly seed and blanket (WASB) where the seed units and blanket units separately occupy one full-size PWR assembly each, and the assemblies are arranged in the core in a modified checkerboard array [7].

In this study, the SMART's core design was developed based on WASB concept. The WASB concept was developed as an answer to the drawback from SBU concpet[8]. In SBU concept, the fuel assembly has two different regions which makes design, fabrication and refuelling process more complex. Therefore, the WASB concept was developed to simplify refueling and manufacture process because the core is loaded with separate seed and blanket fuel assembly.

The fuel assembly for the seed units was loaded with UO_2 fuel rods and the blanket units was loaded with ThO_2-UO_2. Other than normal fuel rods, each FAs also consist of shim rods as its burnable absorber rods. Burnable absorber material in shim rods are B_4C in Al_2O_3.

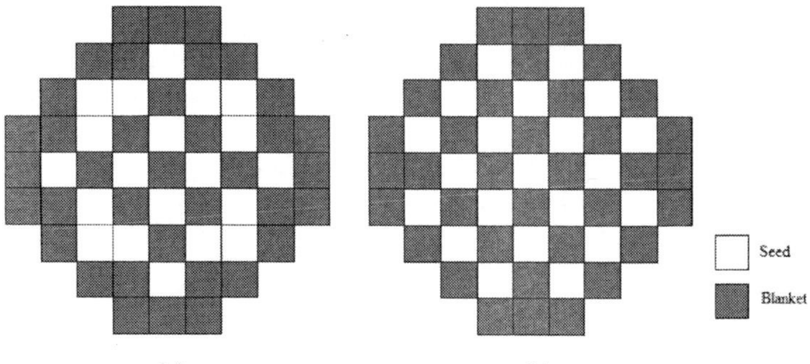

(a) (b)

Figure 2. SMART core loading pattern based on WASB concept a) WASB-1, b) WASB-II

In Fig. 2, SMART core loading pattern based on WASB concept is presented. The WASB core loading pattern is based on the general rule that more energy is extracted from the fuel if the ratio of peak-to-average power in the core is as low as possible [7]. In this study we designed two different core loading pattern with 1:2 seed and blanket units ratio. This arrangements were made to analyzed the effect of seed-blanket units ratio and their location in core performance.

Table 1. SMART major parameters

Parameter	Value
Thermal Output	330 MWt
Electric Output	100 Mwe
Refueling Cycle	3 years
Active Length	200 cm
Cladding Material	Zircaloy-4
Cladding Thickness	0.57 mm
Pellet Diameter	8.05±0.01 mm
Pin Pitch	1.26 cm
Fuel rod to fuel rod	21.10 x 21.10

3. Results and Discussion

The neutronic behaviour of the two core design was studied on different fuel optimization. As seen from Table. 2, there were seven different fuel optimization. The seed unit were loaded with UO_2 fuel with 4.95 wt% enrichment, while the blanket unit were loaded with ThO_2-UO_2 with different fuel fraction and UO_2 enrichment that varied from 10 wt% to 15wt%.

Table 2. Fuel material composition for seed and blanket units

	Fuel material composition	
	Seed	Blanket
I		80% ThO_2 + 20% UO_2 (10 wt%)
II		80% ThO_2 + 20% UO_2 (12 wt%)
III		80% ThO_2 + 20% UO_2 (13 wt%)
IV	UO2 (4.95 wt%)	80% ThO_2 + 20% UO_2 (15 wt%)
V		75% ThO_2 + 25% UO_2 (10 wt%)
VI		75% ThO_2 + 25% UO_2 (12 wt%)
VII		75% ThO_2 + 25% UO_2 (15 wt%)

The multiplication factor of WASB-II core design is smaller than in WASB-1 for the same fuel material configuration. For example when blanket units were loaded with 80% of ThO_2-UO_2 and 20% UO_2 (10 wt% of U-235) and seed units were loaded with 4.95 wt% UO2, the k-eff of WASB-I core was 1.078875 but the k-eff of WASB-II was 1.073251.

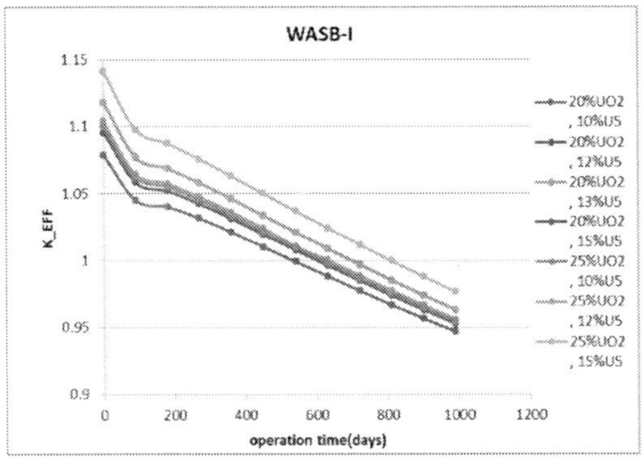

Figure 3. Effective multiplication factors (k_{eff}) of WASB-I core design

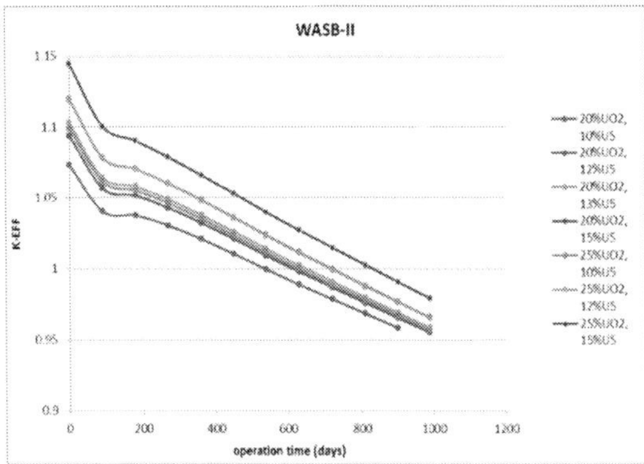

Figure 4. Effective multiplication factors (k_{eff}) of WASB-II core design

Fig. 5, 6, 7 and 8 show the power distribution at the beginning of life in X and Y direction for each core design. For WASB-I core loading pattern, the fuel material composition that gave us the lowest peak-to-average power distribution was the 7th optimization. The seed units were loaded with UO_2 fuel rods (4.95 wt% of U-235), while the blanket units were loaded with 75% ThO_2 and 25% UO_2 (15 wt% of U-235). In WASB-II core loading pattern the fuel material composition that gave us the lowest peak-to-average power distribution was when the blanket units were loaded with 75% ThO_2 and 25% UO_2 (12 wt% of U-235). Therefore if we compare the power distribution in these two core loading pattern, the WASB-I core loading pattern gave us the better power distribution with smaller peak-to-average power distribution than in WASB-II core loading pattern.

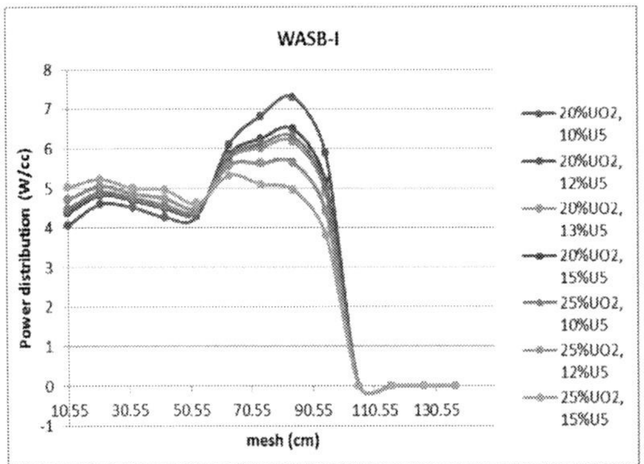

Figure 5. Power distribution in X direction at BOL (Beginning of Life) in WASB-I core design

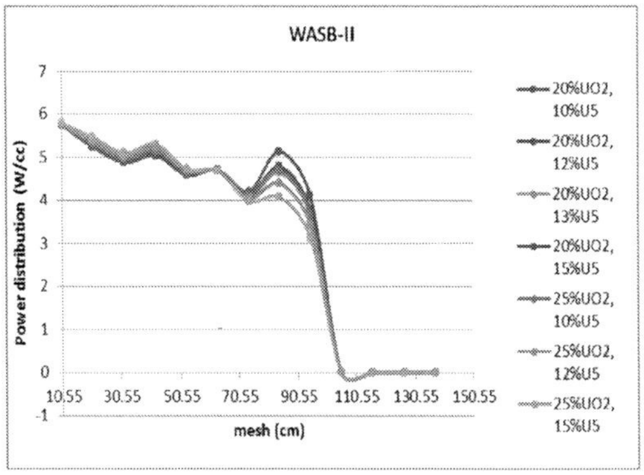

Figure 6. Power distribution in X direction at BOL (Beginning of Life) in WASB-II core design

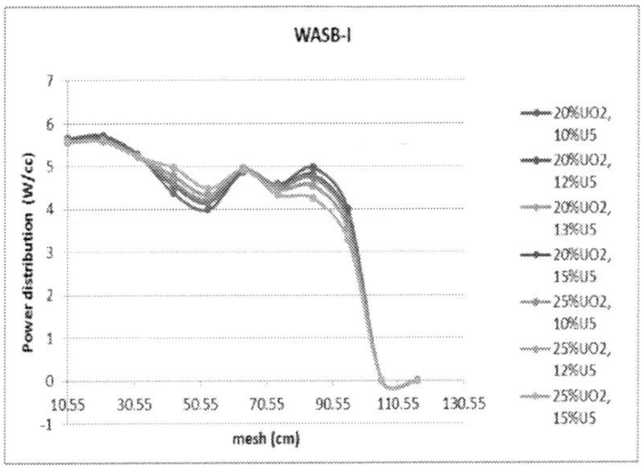

Figure 7. Power distribution in Y direction at BOL (Beginning of Life) in WASB-I core design

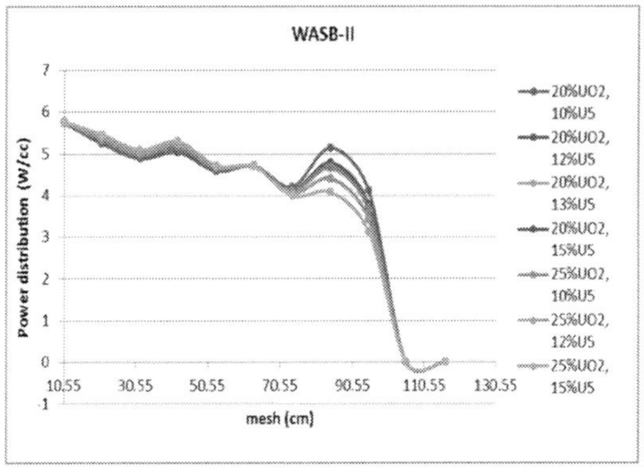

Figure 8. Power distribution in Y direction at BOL (Beginning of Life) in WASB-II core design

4.

5. Conclusion

The study of SMART core design using ThO_2-UO_2 fuel with Whole Assembly Seed and Blanket (WASB) concept are presented. We develop two SMART core loading patterns based on WASB concept with the same seed-blanket units ratio but different configuration. For the same fuel material compositions, the multiplication factor of WASB-II core design was smaller than WASB-I. In WASB-I core loading pattern, the fuel material composition that gave us the lowest peak-to-average power distribution was when the seed units were loaded with UO_2 fuel rods with 4.95 wt.% of enriched-U, while the blanket units were loaded with 75% ThO_2 and 25% UO_2 (with 15 wt% of enriched-U). But in WASB-II core loading pattern, the fuel material composition that gave us the lowest peak-to-average power distribution was when the blanket units were loaded with 75% ThO_2 and 25% UO_2 (12 wt% of enriched-U). Therefore if we compare the power distribution in these two core loading pattern, the WASB-I core loading pattern gave us the better power distribution with smaller peak-to-average power distribution than in WASB-II core loading pattern

6. References

5.

[1] World Energy Council 2013. World Energy Resources: 2013 Survey.

[2] International Atomic Energy Association (IAEA) 2007. Nuclear Power Worldwide: Status and Outlook. Sciencedaily

[3] International Atomic Energy Association (IAEA) 2005 Optimization of the Coupling of Nuclear Reactors and Desalination Systems, Final Report of a Coordinated Research Project (1999–2003) *IAEA-TECDOC 1444*

[4] Seo J K 1997 Advanced Integral Reactor (SMART) for Nuclear Desalination. *IAEA-SM-347/40*. Syposium of Desalination of Seawater with Nuclear energy. Taejon Korea May 26-30 1997

[5] Jung Jeong Seong, Shim Hyeong Jin, Kim Chang Hyo 2001 Monte Carlo Depletion Analysis of SMART core by MCNAP Code. Spring Conference Proceedings Korea Nuclear Society. Seoul National University.

[6] Lee C C, Hwang D H, Zee S Q, and Chang M H 2003 Nuclear and Thermal Hydraulic Design Characteristics of the SMART Core. International Confrence on Global Environment and Advanced Nuclear Power Plants Kyoto Japan

[7] Todosow M, dan Kazimi M 2004 Optimization of Heterogenous Utilization of Thorium in PWRs to Enchance Ploriferation resistance and Reduce Waste. Bookheaven National Library New York.

[8] Wang, Dean 2003 Optimization of a Seed and Blanket Thorium-Uranium Fuel Cyce for Pressurized Water Reactors. Dissertation of Doctor of Philosophy Massachusetts Institute of Technology.

IOP Conf. Series: Journal of Physics: Conf. Series **1127** (2019) 012027 doi:10.1088/1742-6596/1127/1/012027

Design of AL-13 Block Cipher Algorithm Based On Extended Feistel Network

Mohamad Ali Sadikin[1], Bety Hayat Susanti[2]

[1]Faculty of Engineering, National Crypto Institute, Bogor, Indonesia
[2] Faculty of Engineering, National Crypto Institute, Bogor, Indonesia
[1]mohamadalisadikin@gmail.com, [2] bety.hayat@stsn-nci.ac.id

Abstract. In this paper, we design a new block cipher algorithm called AL-13. The structure of AL-13 is simple. We use Extended Feistel Network Type II with input-output 128 bits and key input 256 bits. We also use Dragon's key schedule, binary addition, modulus addition, and XOR to give diffusion, and two functions F and N for confusion. We conducted Strict Avalanche Criterion (SAC) tests to determine whether AL-13 satisfy cryptographic measurement parameters.

1. Introduction

Symmetrical cryptosystems are an encryption system in which the sender and receiver share a single, common key that is used to encrypt and decrypt a message. There are two kinds of it, i.e. block cipher and stream cipher [1]. In block ciphers, the plaintext is divided into block of a fixed length, which are then encrypted into blocks of ciphertext of the same length using the same key [2]. There are seven criteria for evaluating block ciphers, which estimates the security level, the key size, throughput, block size, complexity cryptographic mapping, additional data, and error propagation [3]. Whereas according to Shannon, there are two principles that are often used in encryption block cipher, confusion and diffusion. It is a difficult challenge for the designers to design a strong and secure block cipher algorithm that fulfill the criteria, and have principles of confusion and diffusion [4].

In this paper, we present a 128-bit block cipher called AL-13 which is designed using Feistel structure [5] and is considered to have the principles of confusion and diffusion [3]. Principles of confusion and diffusion [6] that already exist can be combined or repeated many times with different techniques and different combinations to get stronger security.

2. AL-13 BLOCK CIPHER ALGORITHM

AL-13 Cipher Algorithm is a block cipher algorithm with a 256-bit key input and 128-bit plaintext. AL-13 algorithm is designed based on Extended Feistel Network type II structure. The number of rounds is 16-round, where in each round there are two distinct functions, namely F and N function. F and N function using the component operations of addition, XOR, delta, and SUM. We use Dragon's key schedule for AL-13's key schedule [7].

2.1. Notations

$\oplus$	bitwise exclusive-OR operation
$\boxplus$	modulo addition operation
$Delta$	golden number $(\sqrt{5}-1)*2^{31}$
SUM	initial value
$\|\|$	concatenation of two operands

2.2. Key Scheduling
Key scheduling process of AL-13 algorithm based on Dragon's cipher algorithm with key input length is 256-bit. For more details, see in [7].

2.3. Encryption and Decryption Scheme
Encryption/decryption scheme of AL-13 algorithm is to process 128-bit plaintext into 128-bit ciphertext and vice versa. The following is a description of encryption and decryption of AL-13 algorithm process.

1. Partition input plaintext/ciphertext
   At this stage, 128-bit plaintext/ciphertext input is partitioned into four subblock 32-bit (A_0, B_0, C_0, D_0).
2. Transformation round
   The next stage is the transformation round with the same operation for each round. We use as many as 16 rounds of iteration.
3. The formation of a 128-bit ciphertext/plaintext
   128 bit ciphertext/plaintext formed from the merger of subblock $A_{16}, B_{16}, C_{16}, D_{16}$ by 32 bits.

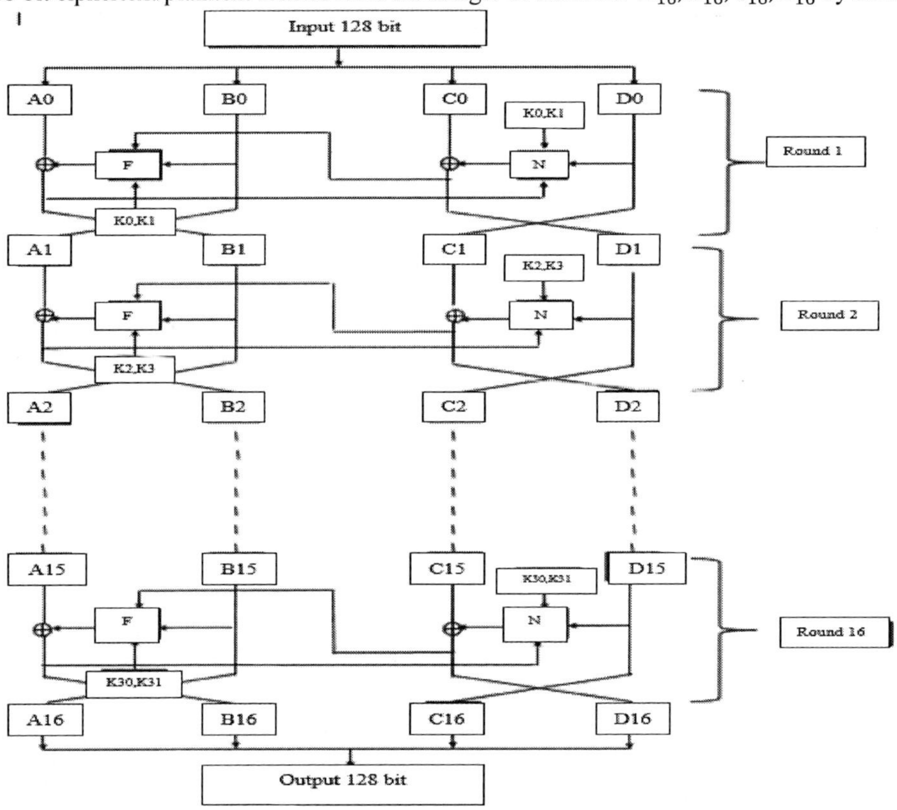

Figure 1. Structure of Encryption/Decryption of AL-13 Algorithm.

2.4. F-Function
In the round process, the subblock of plaintext goes into the second F function with the following mechanism: Subblock input of C and B goes into F function. 32-bit subblock of B XORed by subkeys Rk_0^n. Thereafter, we do the addition modulo of these results with subblock C which is then XORed with subkey Rk_1^n. Later, we do the addition modulo of those results with delta and SUM. Afterthat, we XOR the results with subblock C to produce output.

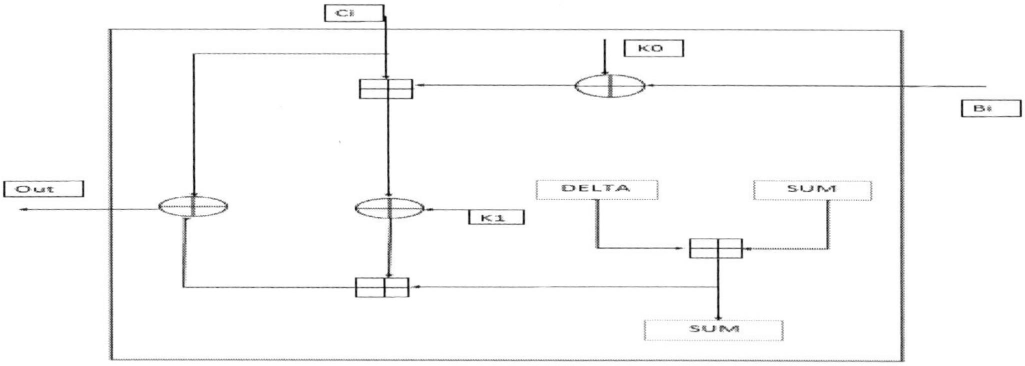

Figure 2. F-Function Structure of AL-13 Block Cipher Algorithm.

2.5. N-Function

In the round process, the subblock of plaintext goes into the second F function with the following mechanism: Subblock input of A and D goes into N function. We do the addition modulo 32-bit subblock of D with subkeys Rk_1^n. Hereafter, the result is XORed with subblock A (result = X) which is then modulo added with delta and SUM. Later, we do XOR subblock of A with subkeys Rk_0^n. Afterthat, we XOR the results with X to get the output.

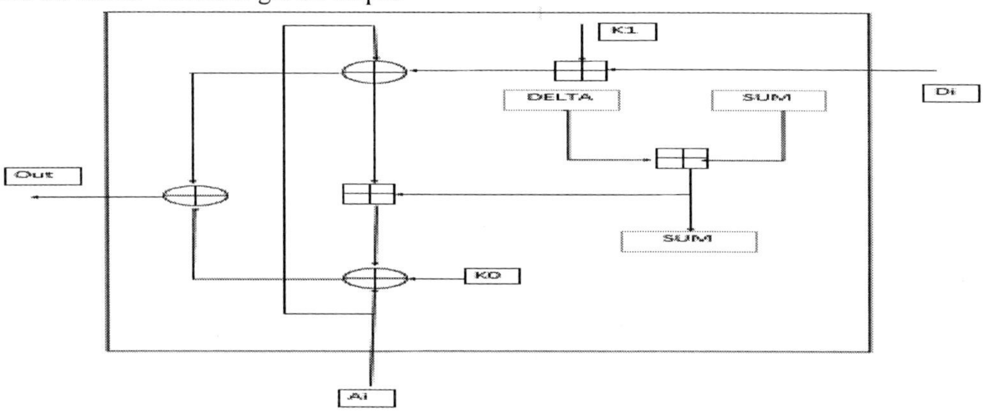

Figure 3. N-Function Structure of AL-13 Block Cipher Algorithm.

3. Rational Design and Implementation of AL-13 Algorithm

3.1. Rational Design of AL-13 Algorithm

3.1.1. The Use of F and N Function

In the AL-13 algorithm there are F function and N function that implements the concept of confusion and diffusion.

1. Confusion

The F function on the AL-13 algorithm fulfill confusion requirement. The results of XOR operation between K_0 and Delta is added with SUM. The goal is to obscure the incoming subblock, so that the adversary will have difficulty to find the key.

Confusion on the N function of AL-13 algorithm generated from the XOR operation conducted by K_0 with K_1 and addition operation between Delta and SUM. The goal is the same with the goal of F function.

a. Delta
Delta is the golden number which is equal to (√5-1) * 2 ^ 31. Different delta is used in every round and so there are no multiplication of bits that do not change.
b. SUM
We use zero initial value for SUM. The SUM value will change every round due to the influence of the addition with delta. Thus, the value is always different in each round.
2. Diffusion
Diffusion generated from operations of addition and XOR which is intended to reduce the relationship between input and output. So, we can prevent attempts to guess the key.
3. The combination of addition and XOR operations
Couple addition and XOR operations do not satisfy the nature of associative and distributive properties that can not be converted into a simpler form to simplify the process of cryptanalysis.

3.1.2. The Use of Key Scheduling Based on Dragon's Algorithm
In the key scheduling based on Dragon's algorithm, there are combination of confusion and diffusion to eliminate the relationship between key input and subkey for encryption. Thus, it will be a strong parameter for the algorithm [7].

3.1.3. The Use of Extended Feistel Network Type II
AL-13 algorithm using Extended Feistel Network type II. From the test results of SAC in Extended Feistel Network type I, Extended Feistel Network type II, and Extended Feistel Network type III, we obtained that Extended Feistel Network type II and Extended Feistel Network type III fulfill the criteria of avalanche probability, i.e. 0.5 [8].

3.2. Implementation of Encryption and Decryption Process of AL-13 Algorithm

Figure 4. Implementation of Encryption the AL-13 Algorithm on C++.

Figure 5. Implementation of Decryption the AL-13 Algorithm on C++.

Figure 4 and 5 shows the implementation of encryption and decryption process of AL-13 algorithm in C++ progamming. The implementation is used to proof that the result of encryption and decryption process in AL-13 algorithm is equal between in C++ programming and based on paper and pencil.

4. Strict Avalanche Criterion (SAC) Test on AL-13 Algorithm

In this section, we discuss the analysis results of SAC test on AL-13 algorithm with plaintext as independent variable and key as independent variable. We refer to [9] for the SAC test.

4.1. Analysis Results of SAC Test on AL-13 Algorithm with Independent Variable of Plaintext

Based on data from SAC test results of AL-13 algorithm with the plaintext to be treated as independent variables showed that the AL-13 algorithm passed SAC with minimum and maximum values as shown in Table 1.

Table 1. Results of SAC test on AL-13 algorithm with plaintext as independent variable

SAC value	*Vector Unit*	*Avalanche Vector*	*SAC Value (%)*	*Relative Error*	*Description*
Min	98	101	49,23629	0,0152742	Passed
Max	100	94	50,7454	0,014908	Passed

According to Table 1, the largest error value is 0,0152742 and obtained by

$\varepsilon_s = \max_{1\leq i,j\leq n}|2k_{SAC}(i,j) - 1| = |2 \cdot 0{,}4923629 - 1| = 0{,}0152742$

Interval value of SAC $= \frac{1}{2}(1 - \varepsilon_s) \leq k_{SAC}(i,j) \leq \frac{1}{2}(1 + \varepsilon_s)$

$= \frac{1}{2}(1 - 0{,}0152742) \leq k_{SAC}(i,j) \leq \frac{1}{2}(1 + 0{,}0152742)$

$= 0{,}4923629 \leq k_{SAC}(i,j) \leq 0{,}5076371$

Interval value of SAC (%)= $49{,}23629\% \leq k_{SAC}(i,j) \leq 50{,}76371\%$.

Table 1 shows that AL-13 algorithm has good diffusion properties as indicated by the largest error value that less than 2%. The value of the largest error occurs at different bit positions (unit vector) into bit position 98 and the avalanche vector to 101, meaning that when the input plaintext bits to 98 changed then output bits to 101 will be changed with a probability of 49,23629%.

In general, it can be said that changing one bit plaintext will cause changes in output bits with a probability of 50% with the largest relative error value is 0,0152742. So, it can be stated that the AL-13 algorithm has good diffusion properties.

4.2. Analysis Results of SAC Test on AL-13 Algorithm with Independent Variable of Key

The SAC test results of AL-13 algorithm with keys are treated as independent variables showed that AL-13 algorithm passed SAC test with minimum and maximum values as shown in Table 2.

Table 2. Results of SAC test on AL-13 Algorithm with key as independent variable

SAC value	*Vector Unit*	*Avalanche Vector*	*SAC value (%)*	*Relative Error*	*Description*
Min	98	46	49,26223	0,014755	Passed
Maks	121	11	50,72099	0,01442	Passed

According to Table 2, the largest error value is 0,014755 and obtained by

$\varepsilon_s = \max_{1\leq i,j\leq n}|2k_{SAC}(i,j) - 1| = |2 \cdot 0{,}4926223 - 1| = 0{,}014755$

Interval value of SAC $= \frac{1}{2}(1 - \varepsilon_s) \leq k_{SAC}(i,j) \leq \frac{1}{2}(1 + \varepsilon_s)$

$= \frac{1}{2}(1 - 0{,}014755) \leq k_{SAC}(i,j) \leq \frac{1}{2}(1 + 0{,}014755)$

$= 0{,}4926223 \leq k_{SAC}(i,j) \leq 0{,}5073775$

Interval value of SAC (%)= $49{,}26223\% \leq k_{SAC}(i,j) \leq 50{,}73775\%$.

Table 2 shows that AL-13 algorithm has good confusion properties as indicated by the largest error value that less than 2%. The value of the largest error occurs at different bit positions (unit vector) into bit position 98 and the avalanche vector to 46, meaning that when the input palintext bits to 98 changed then output bits to 46 will be changed with a probability of 49,26223%. In general, it can be said that changing one bit plaintext will cause changes in output bits with a probability of 50% with the largest relative error value is 0,0147554. Thus, it can be stated that AL-13 algorithm has good confusion properties.

IOP Conf. Series: Journal of Physics: Conf. Series **1127** (2019) 012027 doi:10.1088/1742-6596/1127/1/012027

5. Conclusion

In this paper, we proposed a new block cipher algorithm called AL-13 that use Extended Feistel Network type II structure with 256-bit key input, 128-bit plaintext, and 16 total of rounds. F-function and N-function in AL-13 algorithm is used to fulfil the concept of confusion and diffusion. Furthermore, SAC test is used to test AL-13 algorithm. The test results shows that AL-13 algorithm has good diffussion and confussion properties.

References

[1] Girault M, Juniot L and Robshaw M J B 2005 *The Feasibility of On-the-Tag Public Key Cryptography* France Telecom Research and Development

[2] Stallings W 2014 *Cryptography and Network Security: Principle and Practice, Sixth Edition.* (New Jersey: Prentice Hall) p 63-83

[3] Menezes A J, Van Oorschot P C and Vanstone S A 1997 *Handbook of Applied Cryptography* (Boca Raton: CRC press CLC) p 223-224

[4] Kevin S C 2004 *A Block Cipher Cryptosystem UsingWavelet Transforms Over Finite Fields* IEEE Transactions On Signal Processing, Vol. 52, No. 10. p 2975-2991

[5] Youssef A M 1997 *Analysis and Design of Block Cipher* Thesis in Department of Electrical and Computer Engineering, Queen's University

[6] Schneier B 1996 *Applied Cryptography: Protocols, Algorithms, and Source Code in C Second Edition* (New York: John Wiley & Sons Inc) p 442

[7] Chen K., et al 2004 *Dragon : A Fast Word Based Stream Cipher* Int. Conf. on Information Security and Cryptology (ICISC) 2004 LNCS 3506 p 33-50

[8] Ibrahim S Maarof M A and Idris N B 2005 *Avalanche Analysis of Extended Feistel Network* Proccedings of the Postgraduate Annual Research Seminar FSKSM 2005 UTM p 17-18

[9] Vergili I and Yucel M D 2001 *Avalanche and Bit Independence Properties for the Ensembles of Randomly Chosen n x n S-boxes* Turk J Elec Engin, vol. IX p 50-62

IOP Conf. Series: Journal of Physics: Conf. Series **1127** (2019) 012028 doi:10.1088/1742-6596/1127/1/012028

Heat transfer analysis for Casson fluid flow over stretching sheet with Newtonian heating and viscous dissipation

K Ahmad[1], Z Wahid[1] and Z Hanouf[2]

[1]Department of Science in Engineering, Kulliyah of Engineering, 50728 Gombak, Kuala Lumpur, Malaysia
[2]Department of Mechanical Engineering, Kulliyah of Engineering, 50728 Gombak, Kuala Lumpur, Malaysia

Abstract. A steady two-dimensional boundary layer flow and heat transfer of a Casson fluid over a stretched plate with Newtonian heating boundary condition is investigated. The governing partial differential equations are first reduced to nonlinear ordinary differential equations by using suitable similarity transformations before being solved numerically using an implicit finite difference scheme. The features of the flow and heat transfer characteristics for some embedded parameters, such as Casson fluid parameter β, Eckert number *Ec*, Prandtl number *Pr* and conjugate Newtonian heating parameter γ, are analyzed and discussed.

1. Introduction

Due to great application in industry and manufacturing processes, flow induced by stretching sheet is important as the final desired output is very much influenced by the rate of stretching and heat transfer at the surface. Plastic and metal extrusions, production of polymer films or thin sheet and glass-fiber production are some examples where the stretching process takes place. In industry, polymer sheets and filaments are manufactured by continuous extrusion of the polymer from a die to a windup roller, which is located a finite distance away. The thin polymer sheet constitutes a continuously moving surface with a non-uniform velocity through an ambient fluid [1]. Crane [2] provided the closed form solution for steady two-dimensional flow of an incompressible viscous fluid generated by stretching surface. Later, his work was extended by other researchers for various velocity and thermal boundary conditions to both Newtonian/ non-Newtonian flow, along with several physical effects (see [3]-[10]).

Most fluids are non-Newtonian in nature, eg. honey, paint, slurry, ketchup, muds, glues, lubricants and others. Hence, it should be expected that non-Newtonian fluids are highly non-linear and much more complicated compared with Newtonian fluids. Due to the complexity, there is no single constitutive equation exhibiting all properties. As such, there are few models available in the literature to cater to the characteristics which differ from one another. However, due to promising industry application, non-Newtonian fluid has been the interest of researchers looking to further explore this type of fluid, one of them being Casson fluid. Casson fluid is classified as the most popular non-Newtonian fluid, and has several applications in food processing, metallurgy, drilling operations and bio-engineering operations [11]. Casson fluid was introduced by Casson in 1959 for the prediction of the flow behavior of pigment-oil suspensions [12]. According to Dash et al. [13] and Fung [14], human blood can also be treated as Casson fluid due to the presence of several substances such as protein, fibrinogen, globulin in aqueos base plasma and red blood cells. Shawky [15] studied MHD Casson fluid flow with heat and mass transfer through a porous medium over a stretching sheet, Shateyi and Marewo [16] investigated the hydromagnetic boundary layer flow, heat and mass transfer of Casson fluid near a stagnation point over a stretching surface in the presence of thermal radiation, viscous dissipation and chemical reaction and Chenna and Shankar [17] dealt with Casson fluid flow over a nonlinearly stretching sheet with the effects of radiation and heat source/sink. Very recently, Kumar and Gangadhar [18] considered MHD stagnation point flow of Casson fluid and heat and mass transfer over a stretching sheet with the effects of radiation, chemical reaction and thermal slip and Medikare et al. [19] analyzed MHD stagnation point flow of a Casson fluid over a nonlinearly stretching sheet with viscous dissipation. However, all the above mentioned studies have not taken into consideration the Newtonian heating effect. Das et al. [20] studied the influence of Newtonian heating on heat and mass transfer in unsteady hydromagnetic flow of Casson fluid past a vertical plate in the presence of radiation and chemical reaction, Jain [21] analyzed Casson fluid flow past an exponentially accelerated infinite vertical plate with Newtonian heating in the presence of radiation and very recently, Ullah et al. [22] examined the slip effect on MHD free convective flow of Casson fluid over a nonlinearly stretching sheet saturated in porous medium with Newtonian heating. Hence, motivated by the above works, we would like to study the effects of Newtonian heating and viscous dissipation towards Casson fluid flow moving over a stretching sheet. To the best knowledge of the authors, such a study has never been conducted before.

2. Analysis

Consider a steady two-dimensional Casson fluid flow along a stretched plate of ambient temperature T_∞. The plate is stretched with velocity $u_w = ax$ $(a > 0)$ and heated due to Newtonian heating. Here, the x –axis is measured along the stretched plate and the y – axis is measured in the direction normal to it. Under the Boussinesq and boundary layer approximations, the governing boundary layer equations are

$$\frac{\partial u}{\partial x} + \frac{\partial v}{\partial y} = 0, \tag{1}$$

$$u\frac{\partial u}{\partial x} + v\frac{\partial u}{\partial y} = \upsilon\left(1 + \frac{1}{\beta}\right)\frac{\partial^2 u}{\partial y^2}, \tag{2}$$

$$u\frac{\partial T}{\partial x} + v\frac{\partial T}{\partial y} = \frac{k}{\rho c_p}\frac{\partial^2 T}{\partial y^2} + \frac{\mu}{\rho c_p}\left(\frac{\partial u}{\partial y}\right)^2, \tag{3}$$

where u and v are the velocity components along the x and y directions, respectively. υ is the kinematic fluid viscosity, β is the non-Newtonian (Casson) fluid parameter, T is the fluid temperature, k is the thermal conductivity, ρ is the density of the fluid, c_p is the specific heat at constant pressure and μ is the dynamic viscosity. The flow is subjected to the boundary conditions:

$$u = u_w, \quad v = 0, \quad \frac{\partial T}{\partial y} = -h_s T \ (\mathrm{NH}) \quad \text{at} \quad y = 0,$$

$$u \to 0, \quad T \to T_\infty \quad \text{as} \quad y \to \infty, \tag{4}$$

where h_s is the heat transfer parameter for Newtonian heating. Using the following transformation:

$$\psi = \sqrt{a\upsilon}\, x\, f(\eta), \quad \eta = y\sqrt{\frac{a}{\upsilon}}, \quad \theta(\eta) = \frac{T - T_\infty}{T_\infty}, \tag{5}$$

where ψ is defined in usual fashion, i.e $u = \partial\psi/\partial y$ and $v = -\partial\psi/\partial x$, and η is the similarity variable, Equation (1) is automatically satisfied and Equations (2) - (4) reduced to

$$\left(1 + \frac{1}{\beta}\right) f''' + f f'' - f'^2 = 0, \tag{6}$$

$$\theta'' + Ec\, Pr\, f''^2 + Pr\, f\, \theta' = 0, \tag{7}$$

along with boundary conditions

$$\begin{aligned} &f(\eta)=0, \quad f'(\eta)=1, \quad \theta'(\eta)=-\gamma[1+\theta(0)] \quad \text{at} \quad \eta=0, \\ &f'(\eta)\to 0, \quad \theta(\eta)\to 0 \quad \text{as} \quad \eta\to\infty, \end{aligned} \tag{8}$$

where prime denotes differentiation with respect to η, $Ec = \dfrac{u_w^{\,2}}{c_p T_\infty}$ is the Eckert number, $Pr = \dfrac{\upsilon}{\alpha}$ is the Prandtl number and $\gamma = h_s\sqrt{\dfrac{\upsilon}{a}}$ is the conjugate parameter for Newtonian heating.

3. Results and Discussion

Equations (6) and (7), along with boundary conditions (8), have been solved numerically for several values of Casson fluid parameter β, Eckert number Ec, Prandtl number Pr and conjugate Newtonian heating parameter γ. We opt to use an implicit finite-difference method, namely the Keller-box method, which involves four steps [23]:

(i) Reduce Equations (6) and (7) to a first-order system;
(ii) Write the difference equations using central differences;
(iii) Linearize the resulting algebraic equations by Newton's method and write them in matrix-vector form; and,
(iv) Solve the linear system obtained by using a block-tridiagonal elimination technique.

$\Delta\eta$ chosen was 0.02 and found to be satisfactory to achieve convergence of 10^{-5}. The numerical method used here has been proven to be unconditionally stable and the most flexible of the common methods, being easily adaptable to solve equations of any order [23].

The effect of Ecker number along with Newtonian heating when $Pr = 7$ towards the heat transfer is depicted in Figure 1. It is seen that as the Eckert number increases, the heat transfer rate on the surface also increases. On top of that, Casson fluid flow is found to increase the heat transfer as compared with Newtonian flow ($\beta = \infty$). As the surface is being heated by Newtonian heating, it is expected that as γ increases, surface temperature will also increase.

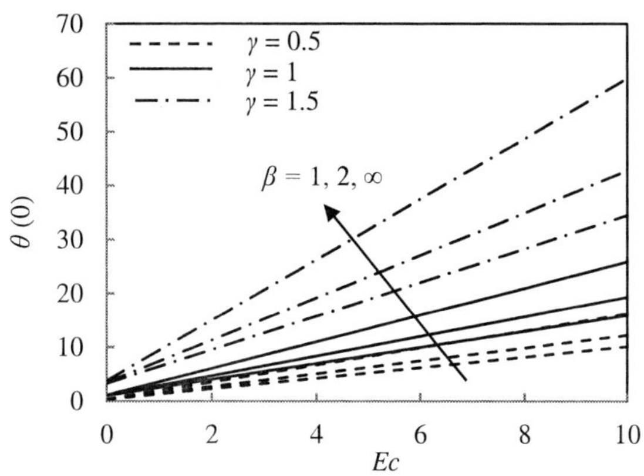

Figure 1. Variations of θ (0) for various values of β, Ec and γ when $Pr = 7$.

Figure 2 depicts the surface temperature when Ec = 2 and β = 1 for various values of γ. It is noted that at fixed γ, the solution for the surface temperature started at certain value of Pr, namely Pr_c which is Pr_c = 1.99, 6.96 and 15.01 for γ = 1, 2 and 3, respectively. At this point, the surface temperature is infinitely large and decreases monotonically as $Pr > Pr_c$. No solution is found when $Pr < Pr_c$. However, the range of Pr_c can be increased by decreasing γ.

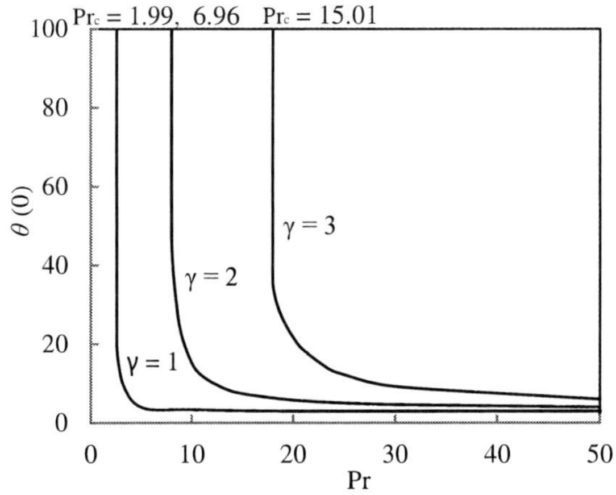

Figure 2. Variations of θ (0) for various values of γ and Pr when β = 1 and Ec = 2.

The temperature profiles for various values of the Ec numbers and γ when β = 1 and Pr = 50 are plotted in Figure 3. It is seen that an increment in the Eckert number will result in an increment of the surface temperature as well. This is in line with the results obtained in Figure 1. Furthermore, the thermal boundary layer thickness is found to be thicker for a higher Eckert number. As the plate is subjected to Newtonian heating, it is expected that the surface temperature is higher when $\gamma = 2$ when compared with $\gamma = 1$.

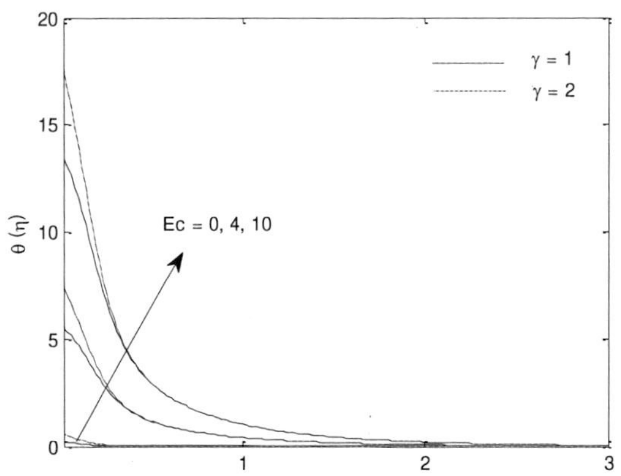

Figure 3. Temperature profile when β = 1 and Pr = 50 for several values of Ec and γ.

4. References

[1] Takhar H S, Chamka A J and Nath G 2001 *Acta Mech.* **146** 59
[2] Crane L J 1970 *J. Appl. Math. Phys.* **21** 645
[3] Chakrabarti A and Gupta A S 1979 *Quarterly of Applied Mathematics* **37** 73.
[4] Surma Devi C D and Nath G 1990 *Indian journal of technology* **28** 93
[5] Lawrence P S and Nageswara B R 1997 *Journal of Applied Mathematics and Mechanics* **77** 317
[6] Kechil S A 2009 *Journal of Porous Media* **12**
[7] Vidyasagar G, Reddy B A P, Ramana B and Sugunamma V 2012 *Advances in Applied Science Research* **3** 4016
[8] Ahmad K, Ishak A and Nazar R 2013 *Mathematical Problems in Engineering* **2013**, Article ID 257161
[9] Heydari M, Loghmani G B and Dehghan A A 2014 *Journal of Applied Fluid Mechanics* **7** 473
[10] Ahmad K, Hanouf Z and Ishak A 2016 *AIP Advances* **6**: 035024-7
[11] Mahanta G and Shaw S 2015 *Alexandria Engineering Journal* **54** 653
[12] Casson N A 1959 *Flow Equation for the Pigment Oil Suspensions of the Printing Ink Type* (New York: Pergamon)
[13] Dash R K, Mehta K N and Jayaraman G 1996 *International Journal Engineering Science* **34** 1145
[14] Fung Y C 1984 *Biodynamics Circulation* (New York: Springer-Verlag)
[15] Shawky H M 2012 *Journal of Porous Media* **15**
[16] Shateyi S and Marewo G T 2014 *Proceedings of the 7th International Conference on Finite Diferences, Finite Elements, Finite Volumes, Boundary Elements (Gdansk, Poland)*
[17] Sumalatha, Chenna and Bandari S 2015 *World Journal of Mechanics* **5** 257
[18] Kumar P S and Gangadhar K 2015 *Advances in Applied Science Research* **6** 205
[19] Medikare M, Joga S and Chidem K 2016 *American Journal of Computational Mathematics* **6** 37
[20] Das M, Mahato R and Nandkeolyar R 2015 *Alexandria Engineering Journal* **54** 871
[21] Jain P 2015 *International Journal of Engineering Research and General Science* **3** 873
[22] Ullah I, Shafie S and Khan I 2016 *Journal of King Saud University-Science* (In Press- Corrected Proof)
[23] Cebeci T and Bradshaw P 1988 *Physical and Computational Aspects of Convective Heat Transfer* (New York: Springer)

Acknowledgment
This work is supported by FRGS grant (FRGS15-228-0469) from the Ministry of Higher Education Malaysia.

Electronic Structure and Impurity Level Location of Rare-Earth (Ce, Pr, Nd, Dy) Doped GaN : GGA Approximation

Muhammad Y.H. Widianto[a] **and Acep Purqon**[b]

[a]Informatics Engineering, Del Institute of Technology, Jl. Sisingamangaraja Toba Samosir Sumatera Utara 22381, Indonesia

[b]Physics of Earth and Complex Systems, Institute of Technology Bandung, Jl Ganesha 10 Bandung 40132, Indonesia

E-mail: yusuf.widianto@del.ac.id

Abstract. Electronic structure and impurity level location of rare-earth (RE) doped GaN have been investigated. Based on density functional theory GGA Approximation, the electronic structure of GaN and GaN:RE are calculated in hexagonal structure. The lattice parameter, bond length of Ga-N and the tetrahedral angle of N-Ga-N show the 3% change from experimental results. When we substitute Ga to RE in the GaN structure, the bond length changes 12% longer with the equilibrium bond length of RE-N varies between 2.1382Å - 2.2428Å . The ionic radii of RE larger than Ga, it will affect to tetrahedral angle which vary from $112.68° - 113.60°$. In the electronic properties calculation, the pristine GaN has direct band gap 2.58 eV. The impurity energy level of Ce, Pr, and Nd are detected in the band gap while Ce level lies near conduction band minimum and the impurity level is contributed by 4f electron states. By this, the 4f energy can explain the luminescence process from rare-earth atoms to host materials. The band structure of GaN:RE is expected to allow optical transition in IR-UV emission range. GaN:RE has high potential to be applied to lower energy of light emitting diodes and energy-saving devices.

Keywords: GaN, Rare-Earth, electronic structure, GGA approximation

1. Introduction

Semiconductor III-V has been used for high performance light emitting devices. This is the most recent revolutionary lighting advancement for saving and lower energy. At the room temperature, III-nitrides materials have the direct transition with energy band gap ranging 1.9-6.2 eV for InN, GaN, and AlN[1][2]. Blue light-emitting device was found by Nakamura et. al. using nitride materials[3][4]. Magnesium as an impurity transition contributed to blue light in the p-semiconductor layer, this performance has high efficiency several times than SiC LED[5]. The significant progress is developed from blue-UV spectra of InGaN mixed with a yellow phosphor to get white light. The light is generated within active layer of InGaN due to pair recombination e-h injected impurity. Addition impurities materials on the host material makes changes to the band's energy.

Rare-earth (RE) has unique properties since strong visible wavelength was explored when GaN doped RE[6][7]. The energy level 4f has high energy that the position is relative to valence band and conduction band of host materials[8][9]. Energy of the f shell RE contributed to narrowing band energy, this is detected on RE doped which has high peaks spectral[10]. For example Eu atom has 6 electron in f shell, these produce red emission when doped in GaN[11]. GaN:Ce and GaN:Pr clearly has red emission with the number of electron in f shell smaller [12][13]. Another color has found in GaN:Tm (blue) and GaN:Eu (green) [14]. Sharp peak energy was shown in the transition $^5D_0 \rightarrow^7 F_2$ by RE ions doped which produce IR spectrum[15]. It indicates that valence electron of RE excited to host materials.

Basic understanding of electronic structure of RE doped GaN is needed to explain the wide range energy band. In this paper, investigation the electronic properties of GaN doped RE is clarified based on DFT calculations. We study RE elements including Ce, Pr, Nd, Dy to complete basic understanding impurity level location to host materials. In Sec II. the computational model which is used will be presented. In Sec. III. analysis of structure and electronic behavior of doped RE are also studied to understand the energy structure of impurity. Finally, the summary will be presented in Sec. IV.

2. Computational Model

PHASE0 is used to calculate electronic structure of pristine GaN and GaN:RE[16]. GGA Approximation was applied for exchange-correlation energy, E_{xc}, to give rather accurate result for the systems[17]. The calculations are carefully calculated with Ultrasoft pseudopotential, which the cut-off energy for wave functions is 25 Ry and cut-off energy for charge density is 225 Ry. The structure evolution is set with atomic force 0.1^{-3} Hartree/Bohr. K-points generate $2 \times 2 \times 2$ mesh generation for geometry optimization. Convergence of total energy is carefully examined to the cut-off energy and k-point sampling. Both GaN and GaN:RE are modeled in hexagonal structure 32 atoms, the most stable in bulk structure. First, optimization of unit cell hexagonal structures are performed with four atoms per hexagonal unit cell. Group III elements are placed on (0,0,0) and $(0.6667, 0.3333, 0.5)$. Nitrogens is placed on (0,0,u) and $(0.6667, 0.3333, u + 0.5)$, which u is internal parameter. This calculation uses vectors **a**=(a,0,0), **b**=$(0.5a, 0.866a, 0)$, and **c**=(0,0,c), which a and c are lattice parameter of hexagonal structure[18]. These parameters are related by $(\frac{c}{a})^2 = \frac{1}{u} = \frac{8}{3}$.

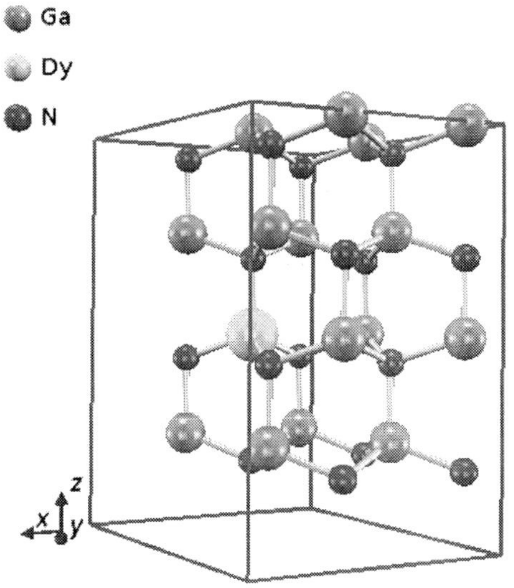

Figure 1. Model of hexagonal structure in supercell 32 atoms of $Ga_{15}N_{16}Dy$. The RE has clearly four bonds of N atoms in the tetrahedral angles.

The optimal structure of hexagonal can be achieved with the unit cell volume parameter V. The independent parameters V, $\frac{c}{a}$, and u should be optimized as done by Stampfl et. al.[19]. In the first step, the ideal parameters of hexagonal structure are used and combined with varying lattice parameter of a to gain equilibrium value of a. Next, vary the ratio of $\frac{c}{a}$ while remain keeping the ideal of u and the last lattice constant of a. For the second time, vary parameter of a with control the ratio of $\frac{c}{a}$ to gain equilibrium of a. With keep retain ratio of $\frac{c}{a}$ and the last value of a, find the equilibrium of u by varying them. Finally, calculate the unit cell volume V hexagonal structure that consist of two hexagonal closest packed sublattices. RE atom substitute Ga site as illustrated in Fig. 1. Experimental result and theoretical studies have found that RE tend to occupy cation site, by binding four N atoms[20]. The composition RE on GaN structure written as $Ga_{1-x}RE_xN$ $(0 \leq x \leq 1)$, x is the RE dopant concentration.

3. Result and Discussion

On the last study, we clearly explain the geometry structure of pristine GaN and GaN:RE in hexagonal structure. The lattice parameter, bond length, and tetrahedral angles have slightly changed on the GaN case. When RE is substituted Ga site, the N atoms can bind strongly and keep the position stability of the RE, thus affecting only the crystals conditions around the RE. The bond length of RE-N is longer than Ga-N because the RE radii is larger than Ga[21]. The average bond length of Ce-N, Pr-N, Nd-N, and Dy-N are sequentially 2.206Å, 2.201Å, 2.192Å, and 2.165Å. By the average of Ga-N bond length is 1.9266 Å , the bond length ratio of RE-N:Ga-N is 1:0.879. The tetrahedral angles has slightly changed in the GaN:RE structure with the largest deviation from a stable tetrahedral angle is N-Ce-N with 113.6°. The ratio tetrahedral angle GaN:Ce with GaN is 1:0.961. This ratio shows that RE is stable in the hexagonal structure of GaN.

Energy from RE affects the electronic structures of host materials. It may be used to determine the narrowing band gaps near the IR-visible light. For pristine GaN the fermi energy E_f is 0.189 eV and the highest occupied band energy (vbm) of GaN is detected 0.085 eV, the electrons fill the lowest energy level. Band gap energy for pristine GaN shows 2.58 eV, it is 19.38% smaller than experimental result. Band structure is also found in direct band gap obtained with CBM and VBM lie on $\Gamma - \Gamma$ points[18]. It is because DFT calculation of GGA approximation underestimate. Calculation with Hubbard U provides accurate results by considering Coulomb interaction.

Energy transfer is obtained when the excitation of the impurity becomes unoccupied, potentially use 4f energy. The density of states (DOS) for GaN:Ce, GaN:Pr, GaN:Nd, and GaN:Dy are shown in the Fig.2, which E_f was set to be 0.0 eV for convenience. In this work, the ion charge of RE element is zero. The impurity energy of RE was found between conduction band and valence band on the GaN:Eu, GaN:Er, and GaN:Tm[20][21]. The conduction and valence energy of GaN:Ce was shown in the straight black line which impurity of f shell lies near conduction. GaN:Pr was shown in red dash line that indicates the position of valence energy move closest to impurity of f shell Pr. When the number of RE atom is larger, the position of valence approach to impurity and stay away from conduction, indicated by GaN:Dy which has the number of electron $4f^9$. When the RE ions are given charge, Ce^{+3}, Pr^{+3}, that can trap holes from the valence band and may act as a hole donor in GaN. The position of Pr^{+3} energy below the conduction band[9]. In the band structure calculations, the valence band maximum of GaN:Ce, GaN:Pr, and GaN:Nd ware detected in the K-point and conduction band minimum in Γ. The band structure will be discussed later.

4. Summary

The optimum structure of pristine GaN has slightly changed 2% from experimental result. When the pristine GaN structure was doped with RE, the lattice parameter, bond length and

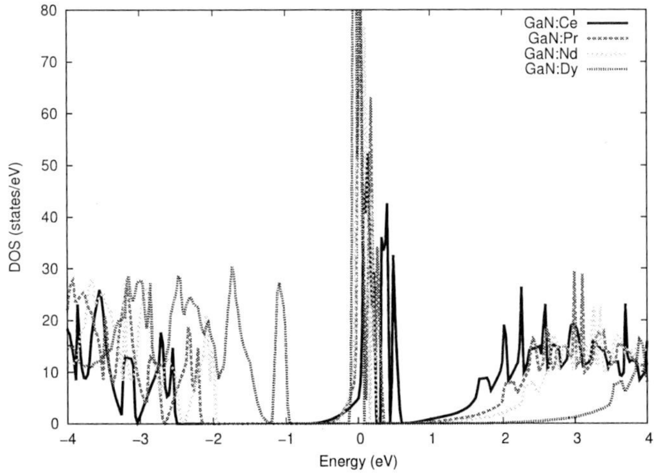

Figure 2. The DOS of GaN:RE which denoted by line colors GaN:Ce (black), GaN:Pr (red), GaN:Nd (green), and GaN:Dy (blue). The impurity of RE was clearly found between conduction dan valence band of host materials.

tetrahedral angles are stable in the hexagonal structure of GaN:RE, which four N atoms bond around RE atom in tetrahedral angles. In the calculation of electronic properties, the GaN has direct band gap 2.58 eV. It is smaller than experimental result (3.4eV). That calculation is required Coulomb interaction in the calculation to get high performance which study later. When the RE atom was substituted in the Ga site, the impurity energy was found between conduction band and valence band of host material. With this scheme, it can be started to explain optical and electronic properties RE doped GaN. Charged ions of RE is interested when placed in the host materials [9].

5. Acknowledgements

MW thanks to FMIPA ITB for financial support through the research. GRID LIPI High Performance Computing for numerical calculations were performed. Finally, MW thanks to MS for all encouragement and great support. Lets maintain this enthusiasm last forever. So, will you marry me?

6. References

[1] I. Vurgaftman, J. R. Meyer, L. R. Ram-Mohan, Journal of Applied Physics 89 (11) (2001) 5815–5875.
[2] I. Vurgaftman, J. R. Meyer, Journal of Applied Physics 94 (6) (2003) 3675–3696.
[3] S. Nakamura, T. Mukai, M. Senoh, Applied Physics Letters 64 (13) (1994) 1687–1689.
[4] P. Fischer, J. Christen, S. Nakamura, Japanese Journal of Applied Physics 39 (2B) (2000) L129.
[5] C. H. Park, C. Byoung-Ho, L. Keun-Ho, C. K. J., Phys. Rev. B 49 (1994) 4485–4493.
[6] A. J. Steckl, J. C. Heikenfeld, D.-S. Lee, M. J. Garter, C. C. Baker, Y. Wang, R. Jone, IEEE Journal of Selected Topics in Quantum Electronics 8 (4) (2002) 749–766.
[7] D. S. Lee, A. J. Steckl, Applied Physics Letters 82 (1) (2003) 55–57.
[8] K. C. Mishra, V. Eyert, P. C. Schmidt, Zeitschrift fr Physikalische Chemie 221 (2009) 1663–1676.
[9] P. Dorenbos, E. van der Kolk, Applied Physics Letters 89 (6) (2006) 061122.
[10] V. Kumar, A. F. Khan, S. Chawla, Journal of Physics D: Applied Physics 46 (36) (2013) 365101.
[11] S. Goumri-Said, M. B. Kanoun, Journal of Physics D: Applied Physics 41 (3) (2008) 035004.
[12] A. Majid, A. Ali, Journal of Physics D: Applied Physics 42 (4) (2009) 045412.
[13] A. Majid, A. Dar, J. Zhu, Journal of Magnetism and Magnetic Materials 374 (2015) 676 – 679.
[14] A. Steckl, J. Heikenfeld, D. Lee, M. Garter, Materials Science and Engineering: B 81 (13) (2001) 97 – 101.
[15] A. R. Zanatta, Journal of Physics D: Applied Physics 42 (2) (2009) 025109.

[16] C. for Research on Innovative Simulation Software (CISS), the Institute of Industrial Science (IIS), the University of Tokyo, First-principles electronic structure calculation program phase/0.
URL https://azuma.nims.go.jp/software/phase
[17] J. P. Perdew, K. Burke, M. Ernzerhof, Phys. Rev. Lett. 77 (1996) 3865–3868.
[18] D. Andiwijayakusuma, M. Saito, A. Purqon, Journal of Physics: Conference Series 739 (1) (2016) 012027.
[19] C. Stampfl, C. G. Van de Walle, Phys. Rev. B 59 (1999) 5521–5535.
[20] S. Sanna, W. G. Schmidt, T. Frauenheim, U. Gerstmann, Phys. Rev. B 80 (2009) 104120.
[21] M. Widianto, A. Purqon, Journal of Physics: Conference Series 877 (1) (2017) 012027.

Controlling polypropylene rheological properties by promoting organic peroxide during extrusion with improved properties for automotive applications

B Herlambang[1] and S Bramantika[1]

[1]PT. Chandra Asri Petrochemical, Tbk, Cilegon 42447, Indonesia

E-mail: GroupRND@capcx.com

Abstract. Excellent stability and process-ability of Controlled-Rheology Polypropylene (CRPP) was mandatory parameters especially for automotive application. Selected organic peroxide which was used to control the rheology of PP should perform good stability in physical properties due to pre-process of compounding for automotive application. The common organic peroxide that widely used is 2,5-dimethyl-2,5-di-tert-butylperoxyhexane (DHBP). However, the problems that usually encountered are Melt Flow Index (MFI) stability, color stability, initial color, izod impact properties, and the odor which come from decomposition process. Research and Development (RND) team of PT Chandra Asri Petrochemical, Tbk (CAP) did the study to change the type of organic peroxide to 3,6,9-triethyl-3,6,9,-trimethyl-1,4,7-triperoxonane (TETMTPA) which was intended to improve these inferiorities when using DHBP. The results indicate similar MFI stability, similar color stability, better initial color, higher izod impact properties, and less odor when using TETMTPA than DHBP. This superiority is very applicable particularly for automotive application. Detailed analysis about volatile decomposition product from both peroxide indicated CRPP with TETMTPA has lower volatile compound which result lower odor level than CRPP with DHBP.

Keywords: rheology, Melt Flow Index, DHBP, TETMTPA, odor

1. Introduction

Polypropylene (PP) produced in conventional reactors by using fourth generation of Ziegler Natta (ZN) catalyst system presents relatively high Molecular Weight (MW) and broad Molecular Weight Distribution (MWD). These features generate high melt viscosity (low MFI) which makes them unsuitable for commercial end-use such as fiber spinning, blown film, and injection-molded thin walled product. The MWD largely determines the rheological properties of PP melts and many physical properties. Therefore, this parameter must be controlled to improve the material response during processing and to achieve the diversity in polymer grades suitable for the different applications of PP. Controlling MWD of PP can be done economically by post-reactor operation such as reactive extrusion. This technique incorporate peroxide to induce controllable degradation of PP chains which efficiently result in polymer with tailor-made properties [1].

Organic peroxide is widely used to control the rheological properties of PP as well as polyethylene (PE). PE is an easily crosslinking polymer when organic peroxide is promoted, contrary with isotactic polypropylene (iPP) which is considered until very recently as a non-crosslinking-able polymer due to fact that the β-scission degradation process predominated over the crosslinking mechanism [2]. Therefore, organic peroxide acts as cracking agent in this PP case. The modification of molecular structure of PP during reactive extrusion after polymerization by the action of peroxide-generated radicals is known as Controlled-Rheology Polypropylene (CRPP) [3-6]. When PP resin and peroxide is melt-extruded, the peroxide thermally decomposes to produce peroxy radicals that attack the polymeric chain then decrease the molecular weight. Statistically, the larger molecular weight chains are cleaved more than low molecular weight chains, resulting in the overall effect of a decrease in the average molecular weight and a narrowing of the MWD [6-7]. This process converts the low MFI commodity resin to polymer with higher MFI (up to 100 times the original value) [8].

CRPP has good process-ability because of less elasticity and less shear sensitivity. CRPP also has less part warpage and better physical properties such as heat deflection temperature, and accurate dimensional stability [9-10]. These all properties are required for automotive applications.

It is necessary to have sufficient stabilization in order that direct recycling PP –which is commonly occurs at pre-process of compounding at automotive application - presents no problem such as significant change in molar mass, alteration to the polymer chains or morphology [11]. To study this behavior, the PP rheological properties were determined by multiple extrusions.

Some attention should be addressed to the peroxide especially on decomposition substances and the performance of end product. The decomposition substance of organic peroxide is the item that needs to be considered especially related to the acceptance of end product especially for the application requiring stringent requirements such as for food contact, automotive application, etc. It is not advisable if the peroxide decompose to be high level of volatiles compound which remaining in the end product because it give rise to odor and taste problem or even worse if toxic. So, the present work aims to compare the odor level on the use of organic peroxide at PP production.

2. Methods

2.1. Materials

Commercial PP Trilene® BI9.0GA was used and supplied by our PP production of PT Chandra Asri Petrochemical at polymer plant site. The physical properties of the product are shown in table 1.

Table 1. Physical properties of PP Trilene® BI9.0GA

Physical Properties	Test Method	Unit	Value
Melt Flow Index (230 °C / 2.16 kg)	ASTM D 1238	g/10 min	9
Density	ASTM D 792	g/cm^3	0.9
Tensile Yield Strength (@50 mm/min)	ASTM D 638	MPa	27
Hardness, Rockwell	ASTM D 785	R-Scale	80
Melting Temperature	ASTM D 3418	°C	163

The organic peroxide is 2,5-dimethyl-2,5-di-tert-butylperoxyhexane (DHBP) and 3,6,9-triethyl-3,6,9,-trimethyl-1,4,7-triperoxonane (TETMTPA) were supplied from our additive vendor. DHBP is the organic peroxide which is commonly used in production of CRPP and polymer grafting. DHBP is our existing cracking agent in production of CRPP. TETMTPA is new generation of organic peroxide that claimed generates less volatile, improved safety and organoleptic profile, and food contact approved. The properties comparison of DHBP and TETMTPA are shown in table 2.

IOP Conf. Series: Journal of Physics: Conf. Series **1127** (2019) 012030 doi:10.1088/1742-6596/1127/1/012030

Table 2. Physical properties of DHBP vs. TETMTPA

Properties	Unit	DHBP	TETMTPA
Molecular weight	g/mol	290	264
Density, 20°C	kg/m^3	870	875
Viscosity, 20°C	MPa.s	6.4	5
Active oxygen	%	10.14	7.3
SADT*	°C	80	110

*Self-Accelerating Decomposition Temperature

2.2. Reactive Extrusion Process

For extrusion, a twin screw extruder at PP production plant from Kobe Steel was used. The normal capacity is 20 ton/hour with screw speed was fixed at two fixed speed at 500 rpm and 375 rpm. During production, initial temperature of extruder is 100°C and increase to peak point at 250°C. Prior to extruder, feeding facility for the liquid peroxide consists of pump and tank. The liquid peroxide was diluted by mineral oil then fed to the feed zone of the extruder at the predetermined rate by the liquid additive feed pump which is diaphragm type. The concentration of peroxide was based on target of MFI at 33. Multiple extrusions test with single screw laboratory mini-extruder was done to analyze for the stability of CRPP's properties. Each polymer resin from virgin toward multiple extrusions (1st, 3rd, and 5th) has to be analyzed for some parameters like MFI, YI, izod impact, and odor level.

2.3. Melt Flow Index (MFI)

MFI values are measured based on the reference: ASTM D1238 using Tinius-Olsen Melt Indexer. This test method covers the determination of the rate of extrusion of molten thermoplastic resins using an extrusion plastometer. After a specified preheating time, resin is extruded through a die with a specified length and orifice diameter under prescribed condition of temperature, load, and piston position in the barrel. For PP, the temperature is at 230°C and the load is 2.16 kg. MFI value is in grams per 10 min. Longer polymer backbone chain needs higher energy used to flow which indicates the material is more viscous (low MFI), and vice versa. The test is done to determine successfulness of peroxide to increase MFI.

2.4. Yellowness Index (YI)

Visually, yellowness is associated with scorching, soiling, and general product degradation by light, chemical exposure, and processing. Yellowness Index is a number calculated from spectrophotometric data that describes the change in color of a test sample from clear or white toward yellow. Higher value of YI indicates more yellow and lower value of YI indicates less yellow or whiter. We use Nippon Denshoku Colorimeter to calculate YI with ASTM E313 as reference. Higher YI indicates more yellow color.

2.5. Izod Impact

The test is done to determine the resistance of materials from impact load. The test uses "standardize" pendulum-type hammers, mounted in "standardize" machine, in breaking standard specimen with one pendulum swing. Automotive application will need higher izod impact due to resist impact load in case of collision. The unit of izod impact is in J/m or kg-cm/cm.

2.6. Odor Level

We use organoleptic test to determine odor level of sample. Organoleptic testing refers to subjective evaluations that need individual experience by their sensing (including taste, sight, smell, and touch). The odor level is determined by selected tester or panelist and can be helped by standard sample. The sensing of tester or panelist is reevaluated periodically to ensure the consistency of testing activity. The standardize PP resin with specific odor level is used as reference to help the tester to determine the odor level of the sample. The standard consists of five level of odor with scale 5 is the highest and 1 is the lowest level of odor.

3. Results

In order to convert MFI from 9 to 33, based on our historical data, we used about 300 ppm concentration of DHBP as our existing organic peroxide. TETMTPA as new organic peroxide at first also use same dosage with DHBP but cannot achieve the MFI target. Then slowly increase the dosage till we can achieve the MFI target by 400 ppm of TETMTPA. TETMTPA need higher dosage since it has lower active oxygen which will lead to generate lower peroxy radicals. The result analysis as comparison of multiple extrusions of those two types of peroxides is shown in Figure 1.

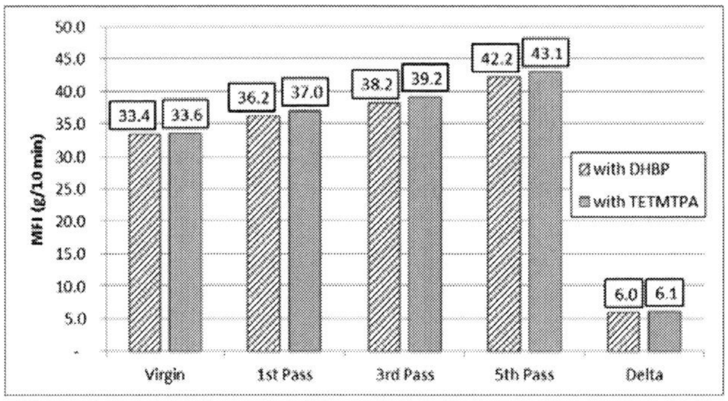

Figure 1. MFI properties of PP end-target with 33 MFI

We conduct multiple extrusions to see MFI stability. After done every pass, MFI value will increase since the polymer will slightly degrade due to heat in extrusion process. Degradation followed by chain scission that generates lower polymer chain length which cause higher MFI value. Good MFI stability will be indicated by smaller delta value of 1st pass and n-th pass, in this activity we multiple extrusion till 5th times (5th pass). Delta value of MFI for CRPP with TETMTPA and DHBP is similar which indicate that MFI stability of CRPP with new TETMTPA is similar than that of existing DHBP.

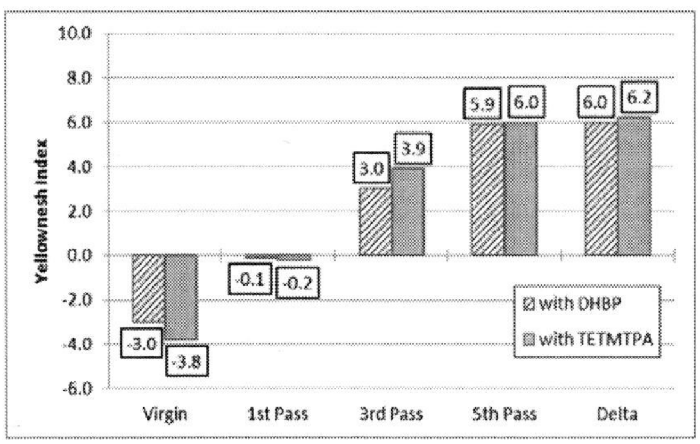

Figure 2. YI of PP end-target with 33 MFI

Initial color of CRPP as shown at virgin YI of figure 2 shows that CRPP with TETMTPA is whiter than CRPP with DHBP. It is caused by the degradation of polypropylene with cyclic ketone peroxides (TETMTPA) results in less yellowing than comparable processes employing their noncyclic ketone peroxides (DHBP) [13]. However, multiples extrusion test indicates those two TETMTPA and DHBP gave similar delta value meaning that they gave similar color stability.

Table 3. Odor test by Panel of PP end-target with 33 MFI

Person	CRPP with DHBP	CRPP with TETMTPA
Man-1	4	2
Man-2	4	3
Man-3	4	2
Man-4	4	2
Man-5	4	2

As shown at table 3, odor test by panel (5 panelists) is done and give the result that CRPP with TETMTPA was lower odor level than CRPP with DHBP. The statement was supported by result analysis of volatile decomposition from DHBP and TETMTPA (data not shown due to confidentiality). Apart from methane and ethane (C1 + C2), both peroxides produce different decomposition product. Volatile like t-butanol and acetone will partially remain in the final product and give rise to odor and taste problem [12]. This is undesirable, especially for certain end uses such as for the manufacture of articles or component part for the automotive industry [13].

According to decomposition mechanism of DHBP [14], DHBP will undergo homolytic reaction producing two molecules of tert-butyl and one alkoxy radical. The further reaction, radical chain continues by hydrogen abstraction (α-decomposition) from the substrate forming tert-butyl alcohol (tert-butanol) and alkyl radical. The other decomposition of tert-butyl by β-decomposition simultaneously forms acetone and methane radical, which decomposes further to form methane and alkyl radical [14]. The following compounds generated during decomposition of DHBP, such as t-butanol, acetone and methane, will give a strong odor level. Thus, CRPP with TETMTPA had significantly lower odor level than CRPP with DHBP.

TETMTPA also gives higher izod impact properties than DHBP, 406 vs 183 J/m. This higher value will give high resistance due to impact load when use for automotive application. Higher value of izod impact with balance with other properties is the desired requirement especially for automotive application.

4. Conclusions

With temperature profile at extruder from 100°C to 250°C as the operating condition, it was concluded that comparison using TETMTPA and DHBP as rheology controller to achieve similar MFI target for CRPP production resulting some points:

- TETMTPA gives similar MFI stability and color stability than DHBP
- TETMTPA gives better initial color to DHBP
- TETMTPA generate less odor than DHBP which was indicated by odor test and detailed analysis of volatile formation in CRPP.
- TETMTPA gives higher izod impact properties than DHBP

Therefore all those positive points are the improved properties required for automotive applications.

Acknowledgments

The author would like to thank RND team and PP production team of PT. Chandra Asri Petrochemical Tbk for their great support that allows this study possible.

References

[1] Salazar A and Rodriguez J 2012 Fracture behaviour of controlled-rheology polypropylenes In rheology Intech

[2] Bouhelal S, Cagiao M E, Benachour D and Calleja F J B 2007 Structure modification of isotactic Polypropylene through Chemical Crosslinking: Toughening Mechanism J. Appl. Polym. Sci. 103 2968–2976

[3] Fernanda M, Coutinho B, Marisa C and Rocha G 1991 Polypropylene controlled degradation J.C. Salamon (Ed.), Concise Polymeric Encyclopedia, first ed, CRC Press p 6588–6594

[4] Niki E, Decker C and Mayo F R 1973 Aging and degradation of polyolefins. Part I: Peroxide-initiated oxidations of atactic polypropylene J. Polym. Sci.: Polym. Chem. 11 2813–2845

[5] Ebner K and White J L 1994 Peroxide induce and thermal degradation of polypropylene Int. Polym. Proc. IX (3) 233–239

[6] Ryu S H, Gogos C G and Xanthos M 1997/1992 Parameters affecting process efficiency of peroxide initiade controlled degradation of polypropylene Adv. Polym. Tech. 11 121–131

[7] Berzin F, Vergnes B and Delamare L 2001 Rheological behaviour of controlled-rheology polypropylenes obtainded by peroxide-promoted degradation during extrution: Comparison between homopolymer and copolymer J. Appl. Polym. Sci. 80 1243–1252

[8] Azizi H, Ghasemi I and Karrabi M 2008 Controlled-peroxide degradation of polypropylene: Rheological properties and prediction of MWD from rheological data Polymer Testing 27 548–554

[9] Azizi H and Ghasemi I 2005 Investigation on influence of peroxide-degradation process on rheological properties of polypropylene Iranian Polym J. 14 465–471

[10] Baik J J and Tzoganakis C 1998 A study of extrudate distortion in controlled-rheology polypropylene, Polym. Eng. Sci., 38 274–281

[11] Da Costa H M, Ramos V D, Rocha M C G 2005 Rheological properties of propylene during multiple extrusion Polymer Testing 24 86–93

[12] Technical Bulletin 2010 Trigonox®301, An innovation for controlled rheology polypropylene production, Akzo Nobel p 1–12

[13] Fink F K 2013 Reactive polymers fundamental and applications Second ed, Elsevier 411–424

[14] Das M and Shu C M 2016 A green approach towards adoption of chemical reaction model on 2,5-dimethyl-2,5-di-(tert-butylperoxy)hexane decomposition by differential isoconversional kinetic analysis J. Hazardous Materials 301 222–232

IOP Conf. Series: Journal of Physics: Conf. Series **1127** (2019) 012031 doi:10.1088/1742-6596/1127/1/012031

Thermophysical parameters of inorganic PCM$CaCl_2.6H_2O$ added by $BaSO_4$derived from T-history measurement as the latent TES

Nissa Sukmawati,[1]Alfriska O Silalahi,[1] I.M. Sutjahja,[1,*)] D. Kurnia,[1]S. Wonorahardjo[2]

[1]Dept. of Physics, Faculty of Mathematics and Natural Sciences, InstitutTeknologi Bandung, Indonesia
[2]Building Technology Research Division, School of Architecture, Planning and Policy Development, InstitutTeknologi Bandung, Indonesia
Jl. Ganesha No. 10, Bandung 40132

[*)]inge@fi.itb.ac.id

Abstract. The thermophysical parameters of inorganic phase change material (PCM) of $CaCl_2.6H_2O$ added by 1%wt $BaSO_4$ as the nucleating agenthas been studied by analyzing the temperature vs time data during liquid-solid phase transition (solidification process) based on the original T-history method and its modification form. The PCM sample and water as reference material are placed in a cylindrical tube with chosen dimensions to fulfill the criterion that the Biot number Bi is less than 0.1. To extract the values of mean specific heat capacities of solid and liquid PCM and the heat of fusion related to liquid-solid phase change, hot liquid PCM in the tube with temperature above its melting point was suddenly exposed to a cool atmosphere to allow the gradual solidification process. We comparedthe results with reference data, both from direct measurements of Differential Scanning Calorimetry(DSC) andindirect measurement of the same method.

1. Introduction

In this era the problem of energy crisis is still been an interested issues to be solved.Thermal energy storage (TES) is an alternative technology to be considered, because it can stored the energy to be used at a later time for heating and cooling applications [1-2]. TES are used particularly in buildings as an heat exchange system to keep the thermal comfort of occupants, due to the fact that it can help the balance between energy demand and supply [3-4]. They can also reduce the electrical energy consumption and its peak demand, as well as to reduce CO_2 emissions that mainly contribute to the local and global warming.

Compared to brick, wood, concrete, and glass as the conventional examples of sensible TES, TES based on phase change material (PCM) is one of the feasible ways for energy conservation in the future. In this case PCM as the medium of latent TES can store relatively large heat at temperatures around its phase transition (solid-liquid), and releasing the absorbed heat back to the environment when the ambient temperature is lower than the temperature of the material. The performance of PCM is mainly determined by its thermophysical parameters that consist of liquid $(c_{p,l})$ and solid $(c_{p,s})$ specific heats and the heat fusion (h_m). For building application, inorganic PCM of $CaCl_2.6H_2O$ with melting temperature around 29°C and relatively high heat of fusion is very good candidate [5].In this paper the thermophysical parameters of $CaCl_2.6H_2O$ added by 0.1%wt $BaSO_4$ as

IOP Conf. Series: Journal of Physics: Conf. Series **1127** (2019) 012031 doi:10.1088/1742-6596/1127/1/012031

the nucleating agent are determined by means of T-history method proposed firstly by Yinping et al. [6]. The data analysis follow the method by Yinping and its modification form by Hong et al. [7]. The results are then comparedwith references data, both resulted from direct (DSC) measurement and indirect measurement of the same method.

Based on T-history method, during the lowering of temperature in cool air environment below the melting temperature, $CaCl_2.6H_2O$ will experience the sensible and latent heats release, while water as reference material will just experience the sensible heat release. The energy equations that related to specific heat capacities of liquid and solid PCM as well as the heat of fusion related to liquid-solid phase changeare described in the followingequations (i.e.first column of table 1), while the corresponding energy equations for water as the reference material are described in the second column of the same table.In other side the phase boundaries of PCM and water that consist of period, temperature and time ranges, and area calculation for further analysis of the dataare described in table 2.

Table 1. Energy equations for PCM and water during the lowering of temperature in cool air environment.

Energy equations for PCM	Energy equations for water (reference)
$\left(m_{t,p}c_{p,t}+m_pc_{p,l}\right)\left(T_0-T_s\right)=hA_cA_1$	$\left(m_{t,w}c_{p,t}+m_wc_{p,w}\right)\left(T_0-T_s\right)=hA_cA_1'$
$\left(m_{t,p}c_{p,t}+m_p\dfrac{c_{p,l}+c_{p,s}}{2}\right)\left(T_m-T_i\right)+m_ph_m=hA_cA_2$	$\left(m_{t,w}c_{p,t}+m_wc_{p,w}\right)\left(T_m-T_i\right)=hA_cA_2'$
$\left(m_{t,p}c_{p,t}+m_pc_{p,s}\right)\left(T_i-T_f\right)=hA_cA_3$	$\left(m_{t,w}c_{p,t}+m_wc_{p,w}\right)\left(T_i-T_f\right)=hA_cA_3'$

In those above equations, the masses of the PCM, water and tube used for PCM and water are each denoted as m_p, m_w and $m_{t,p}$, $m_{t,w}$; the liquid and solid specific heats of PCM as $c_{p,l}$ and $c_{p,s}$; the water and tube specific heatsas $c_{p,w}$ and $c_{p,t}$; andthe heat of fusion of PCM as h_m. In addition, A_c is the convective heat-transfer area of the tube, while two set values of $\{A_1, A_2, A_3\}$ for PCM and $\{A_1', A_2', A_3'\}$ for water correspond to the area below the curve of PCM'sand water's temperatures towards air environment, with the full expression given in the last column of table 2.

Table 2. Period, temperature and time ranges, and area of each period used for analysis of T-history curves of PCM and water. The time-dependent temperature values of PCM and water are denoted by T_p and T_w, respectively.

	Period	Temperature Range	Range of time	Area
PCM	Sensible liquid	T_0-T_s	t_0-t_1	$A_1=\int_{t_0}^{t_1}\left(T_P-T_a\right)dt$
	Phase change	$T_s-T_m-T_i$	t_1-t_2	$A_2=\int_{t_1}^{t_2}\left(T_P-T_a\right)dt$
	Sensible solid	T_i-T_f	t_2-t_3	$A_3=\int_{t_2}^{t_3}\left(T_P-T_a\right)dt$
Water	Sensible liquid	T_0-T_s	$t_0'-t_2'$	$A_1'=\int_{t_0'}^{t_2'}\left(T_w-T_a\right)dt$
	Sensible liquid	T_m-T_i	$t_1'-t_3'$	$A_2'=\int_{t_1'}^{t_3'}\left(T_w-T_a\right)dt$
	Sensible liquid	T_i-T_f	$t_3'-t_4'$	$A_3'=\int_{t_3'}^{t_4'}\left(T_w-T_a\right)dt$

Finally, the thermophysical parameters of PCM consist of the liquid and solid specific heats ($c_{p,l}$ and $c_{p,s}$) and the heat of fusion (h_m) derived from this method are expressed as,

$$c_{p,l} = \frac{m_{t,w}c_{p,t} + m_w c_{p,w}}{m_p}\frac{A_1}{A_1'} - \frac{m_{t,p}}{m_p}c_{p,t} \quad (1)$$

$$c_{p,s} = \frac{m_{t,w}c_{p,t} + m_w c_{p,w}}{m_p}\frac{A_3}{A_3'} - \frac{m_{t,p}}{m_p}c_{p,t} \quad (2)$$

$$h_m = \frac{m_{t,w}c_{p,t} + m_w c_{p,w}}{m_p}\frac{A_2}{A_2'}(T_m - T_i) - \left(\frac{m_{t,p}}{m_p}c_{p,t} + \frac{c_{p,l} + c_{p,s}}{2}\right)(T_m - T_i) \quad (3)$$

2. Methods

The salt hydrate $CaCl_2.6H_2O$ as inorganic PCM and nucleator $BaSO_4$ were purchased from Sigma Aldrich. About 20gr $CaCl_2.6H_2O$ added with 0.1%wt $BaSO_4$ as nucleating agent and 20gr reference sample (water) were placed in the separate tube. The tube dimensions were chosen to ensure the one-dimensional heat transfer along the length of the tube as measured by the Biot number Bi ($=hR/2k$, with h is the natural convective heat-transfer coefficient of air outside the tube, R is the radius of the tube, and k is the thermal conductivity of PCM) is smaller than 0.1 to apply the lumped capacitance method [6]. Each tube was equipped with the temperature sensors (T-type thermocouple with diameter about 1mm) that integrated to the data logger (Applent AT4508A from Applent Instruments Inc.) and computer.

For T-history measurement, the test tube containing molten PCM was suddenly put into the cool air environment at a temperature lower than the melting temperature of PCM (T_m) so that solidification can be initiated.The temperature vs time data were then recorded automatically by data logger connected to a computer.

3. Results and Discussion

The typical T-history curves that show the graphs of temperature vs time of PCM $CaCl_2.6H_2O$+0.1%wt $BaSO_4$ and water as the reference is typically shown in figure1. From this figure, upon coolingin cool air $CaCl_2.6H_2O$ experience the temperature change and liquid to solid phase transition, with supercooling and solidification temperature are each denoted by T_s and T_m, while the difference between them isdenoted as the supercooling degree ($\Delta T_m = T_s - T_m$). The average and standard deviation values of T_s, T_m and ΔT_m from 5 set of data experiments are $T_s \cong (24.58 \pm 0.77)^\circ C$, $T_m \cong (26.34 \pm 0.44)^\circ C$ and $\Delta T_m \cong (1.76 \pm 0.58)^\circ C$, respectively.We note that the resulted solidification temperature (heat release) is lower than the reported values [8] of melting temperature (heat absorption) based on exothermic mode of DSC measurement, which commonly occur in other kinds of material. The relatively small value of supercooling degree mark the effectiveness of $BaSO_4$ as the nucleating agent for the crystallization process of $CaCl_2.6H_2O$ as reported before in Ref. [9].

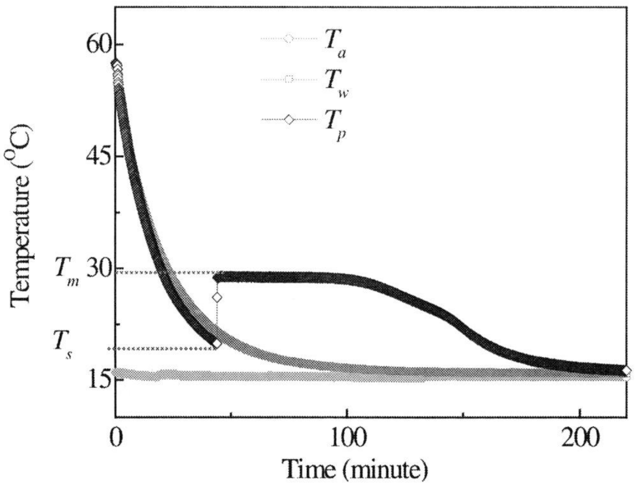

Figure 1. The typical T-history curves of water and $CaCl_2.6H_2O$ +0.1%wt $BaSO_4$ in cool airbath.

For further analysis of the data, we show in figure 2 the area under the curve for PCM and water for each phase boundaries along with the temperature derivative curve of PCM (figure 2(a)). We note that during the phase transition the temperature keeps constant in the latent heat release (denoted by T_m in figure 1), but decreases exponentially in cooling process accompanying sensible heat alone. Thus, following the procedure introduced by Hong et al. [7], the end of phase transition process was determined from the inflection point of the first derivative of T-history curve (figure2(a)). In addition, $t_3$that mark the transformation of overall PCM to become solid was determined from the difference in the trend of temperature derivative for low and high temperature part just after the inflection point.

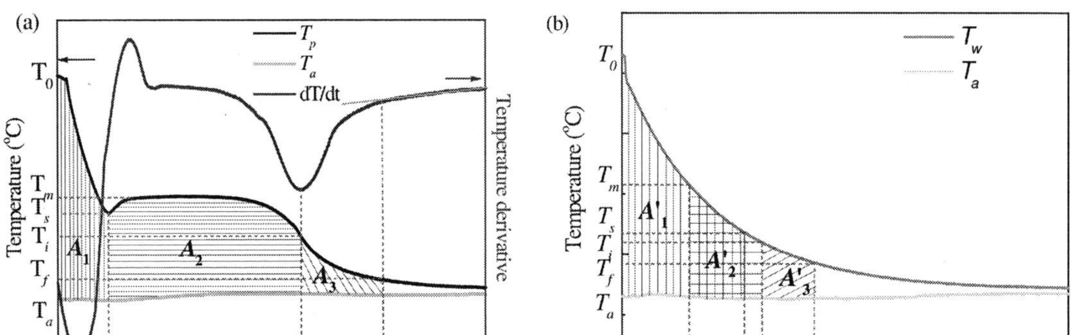

Figure2. Illustration for the analysis of T-history curve of:(a) $CaCl_2.6H_2O$+0.1%wt $BaSO_4$ and (b) water, that show the phase boundaries for sensible liquid, liquid to solid phase change, and sensible solid phases as well as the area under the curve.

The results for specific heats of the liquid and solid PCM and the heat of fusion based on equation (1), (2) and (3) are shown in table 3.

Table 3. The values of liquid specific heat, solid specific heat and the heat of fusion of $CaCl_2.6H_2O$+0.1%wt $BaSO_4$.

Method	Ref	No. of Iteration	$c_{p,l}$[kJ/kg.K]	$c_{p,s}$ [kJ/kg.K]	h_m [kJ/kg]
TH		1	2.36	2.12	188
		2	2.13	2.07	186
		3	2.18	2.49	174
		4	2.33	2.31	186
		5	2.70	1.84	188
		Average ± STDEV	2.34 ± 0.22	2.17 ± 0.25	184 ± 6
TH	[8]	-	2.2	1.77	210
DSC	[8]	-	1.9	1.6	212
DSC	[5]	-	-	-	190.8
DTA	[10]	-	-	-	190.8
DSC	[11]	-	-	-	160
DSC	[12]	-	2.1	1.4	140

We note that the resulted average values of $c_{p,l}$, $c_{p,s}$ and $h_m$are in close agreement with those obtained from the same T-history method (TH), DSC, and DTA. In particular, $c_{p,l}$ is higher than $c_{p,s}$. The slight difference might due to additional dopant of $BaSO_4$ that more or less give the influence to the resulted values.

4. Conclusion

We have determined the thermophysical parameters of inorganic PCM $CaCl_2.6H_2O$+0.1%wt $BaSO_4$ as the nucleator by means of T-history method. Analysis of the data have led to the values of mean specific heat of the liquid,$c_{pl}\cong(2.34\pm0.22)$kJ/kg.K; mean specific heat of the solid,$c_{ps}\cong(2.17\pm0.25)$kJ/kg.K; and the heat of fusion, $h_m\cong(184\pm6)$kJ/kg. These values are close to the references data, both obtained from the same method or Differential Scanning Calorimetry (DSC) measurement.The difference might be due to the fact of additional $BaSO_4$ as the nucleator agent.

Acknowledgement
This research is financially supportedby Kementrian Riset Teknologi dan Pendidikan Tinggi Indonesia through the Desentralisasi 2016 research program under contract number: 583p/I1.C01/PL/2016.

References

[1] Dincer I and Rosen M 2010 *Energy Storage System and Aplications* (Ontario Canada: A John Wiley and Sons, Ltd., Publication).

[2] Harald M and Cabeza L F 2008 *Heat and Cold Storage with PCM an Up to Date Introduction Into Basics and Applications* (Berlin, Germany: Springer).

[3] Tyagi V V and Buddhi D 2007 PCM thermal storage in buildings: A state of art *Renewable and Sustainable Energy Reviews***11** 1146–1166.

[4] Ostermana E, Tyagi V V, Butalaa V, Rahimb N A and Stritih U 2012 Review of PCM based cooling technologies for buildings *Energy and Buildings***49** 37–49.

[5] Lane G A 1980 Low temperature heat storage with phase change materials*Int. J. Ambient Energy***1** 155–68.

[6] Yinping Z, Yi J and Yi J1999A simple method,the T–history method of determining the heat of fusion, specific heat and thermal conductivity of phase-change materials*Meas. Sci. Technol.***10**201–5.

[7] Hong H, Kimb S K and Kim YS 2004 Accuracy improvement of T-history method for measuring heat of fusion of various materials *Int. Journal of Refrigeration* **27** 360–6.

[8] Hasan A, McCormack S J, Huang M J and Norton B2014 Characterization of phase change materials for thermal control of photovoltaics using Differential Scanning Calorimetry and Temperature History Method *Energy Conversion and Management* **81** 322–329.

[9] Lane G A1992 Phase change materials for energy storage nucleation to prevent supercooling *Solar energi materials and solar cells* **27** 135-60

[10] Telkes M 1975 Thermal storage for solar heating and cooling, *Proceedings of the Workshop on Solar Energy Storage Subsystems for the Heating and Cooling of Buildings* Charlottesville (Virginia, USA)

[11] Abhat A 1983 Low temperature latent heat thermal energy storage: heat storage materials *Solar Energy* **30** 313–32

[12] Tyagi VV and Buddhi D 2008 Thermal cycle testing of calcium chloride hexahydrate as apossible PCM for latent heat storage *Sol Energy Mater Sol Cells* **92(8)** 891–9.

The Effect of Various Sintering Condition to the Electrical Characteristics of Fe_2TiO_5 Ceramics for NTC Thermistor

Wiendartun [1*], Risdiana[2], Fitrilawati[2], R.E.Siregar[2]

[1] Department of Physics Education, Indonesia University of Education Jl. Setiabudhi 229, Bandung 40514, Indonesia
[2] Department of Physics, Padjadjaran University Jl. Raya Bandung – Sumedang km. 21 Jatinangor, Sumedang 45363, Indonesia

*Corresponding author: wien@upi.edu

Abstract. Iron Titanium Oxide (Fe_2TiO_5) ceramics for negative temperature coefficient (NTC) thermistor have been synthesized in various sintering condition in order to know the effect of sintering condition, namely, temperature, time and atmosphere condition, on the characteristic especially the electrical characteristic of Fe_2TiO_5 ceramics. Temperature conditions are adjusted between 1200 and 1300°C, for 2 and 8 hours in furnace with atmospheric air and oxygen (O_2) gas. For Fe_2TiO_5 ceramics sintered in the air, the effect of sintering temperature on the grain size was clearly seen. With increasing sintering temperature and sintering time, grain size of Fe_2TiO_5 ceramics increased. However, the effect of atmospheric O_2 gas to their grain size is not clearly seen although it has reduced tendency. XRD analysis showed that all ceramics Fe_2TiO_5 are single phase of orthorhombic structure. From electrical resistances data that was measured at temperature 40 – 250°C, it is found that the value of thermistor constant (B), activation energy (E_a) and room temperature resistance (R_{RT}) decreases with increasing sintering temperature and sintering time. But the value of B, E_a, and R_{RT} increases with atmospheric O_2 gas condition.

1. Introduction

NTC thermistor has been widely used around the world today due to its capability in various fields of electronics, such as thermometer, electric current limiter, water flow sensor, and pressure sensor [1-2]. Fe_2TiO_5 is one of semiconductor ceramics used as based material for main components fabrication of NTC thermistor as temperature sensor. The composition of mineral Fe_2TiO_5 belongs to pseudobrokyte group where the general formula of this compound is X_2YO_5 with octahedral in both side, X and Y. Generally, Fe_2TiO_5 mostly used for gas sensor, nonlinear optic, magnetic, catalyst, and microelectronics [3-4]. Since Fe_2TiO_5 actually has semi-conductivity, it is very capable to use Fe_2TiO_5 for NTC thermistor as based material. Thermistor constant B is a quantity which determine typical characteristic of thermistor corresponding to electrical resistance changes with temperature. The larger thermistor constant leads the better thermistor quality. Many works and studies have been conducted by researchers to enhance thermistor constant B and thermistor sensitivity α. They have studied the effect of sintering time to typical characteristic of Fe_2TiO_5-based ceramics. The research of NTC thermistor for high temperature has been reported [5]. However, the reports of NTC thermistor in high temperature fabricated by Fe_2TiO_5-based ceramics have not been excessively reported so far.

Recently, electrical characterizations of Fe_2TiO_5-based ceramics for NTC thermistor with concentration 0, 0.4, and 1.0 % mole of each MnO_2 dopant sintered at 1100°C for 90 minutes in air have been carried out [6]. Through that research, it is found that thermistor constant (B) of Fe_2TiO_5-based ceramics increased along with increasing of MnO_2 dopant concentration. The obtained value of constant thermistor is 5018–6476 K. It significantly shows that thermistor can be used in accordance with market need which is ≥ 2000 K [7-10]. Beside dopant concentration, in order to obtain optimum condition, sintering parameter (temperature, time, and atmosphere) should be varied since they will give effect to electrical characteristic of obtained ceramics. Therefore, the aim of this research is to investigate the effect of sintering parameter on their electrical characteristic of Fe_2TiO_5-based ceramics for NTC thermistor.

2. Materials and Method

Fe_2TiO_5 thermistor ceramic was prepared by using commercial powder of Fe_3O_3 and TiO_2. Mixture of Fe_2O_3 and TiO_2 with composition of 50:50 in mole %, respectively, to form pellets, calcinated at 700°C for 2 hours. In order to form pellets, pressed powder, the homogeneous mixture of Fe_2O_3 and TiO_2 was pressed at 4.10^7 kg/m^2. The raw pellets were sintered at 1100-1300°C for 2-8 hours in air and oxygen (O_2) atmosphere. Pictures of sintered pellets were taken to know their visual appearance. The crystal structure of the sintered pellets analyzed with x-ray diffraction (XRD) using Kα radiation at 40 kV in voltage and 25 mA in current. The pellets ceramic were investigated by SEM to evaluate their microstructure (morphology). The resistance was measured at various temperatures from 40 to 250°C in steps of 10°C using a digital multimeter and a laboratory made chamber equipped with a digital temperature controller. Both surfaces of sintered pellets had been coated by conductive silver paste colloid silver solution and heated at 600°C for 10 minutes in advance. Thermistor constant (B) was derived from Ln resistivity vs. *1/T* curve where B is the gradient of the curve based on equation 1 [11-13]:

$$R = R_o \exp(B/T) \tag{1}$$

Where, R is the electrical resistance, R_o is a constant or the resistant at the infinite temperature, B is the thermistor constant and T is the temperature in Kelvin and k is the Boltzmann constant. Room temperature resistance (R_{RT}) was determined as the electrical resistance at room temperature (25°C).

From the value of B, the activation energy (E_a) and sensitivity (α) were calculated using equation 2 and 3 [14-16].

$$E_a = B.k \tag{2}$$

$$\alpha = B/T^2 \tag{3}$$

3. Results and Discussion

The XRD profiles of Fe_2TiO_5 pellet ceramics with various sintering condition is shown in Figure 1 the profiles are similar. The XRD profiles show that the structure of pellets ceramics is orthorhombic after being compared to the XRD standard profile of Fe_2TiO_5 from JCPDS No.01-070-2728. No peaks from second phases observed. It may be due to the small concentration of impurities which is smaller than the precision limit of the x-ray diffractometer used. The XRD data indicates that the synthesis of the Fe_2TiO_5 pellets has been well prepared from Fe_2O_3 and TiO_2 powder in various sintering condition.

IOP Conf. Series: Journal of Physics: Conf. Series **1127** (2019) 012032 doi:10.1088/1742-6596/1127/1/012032

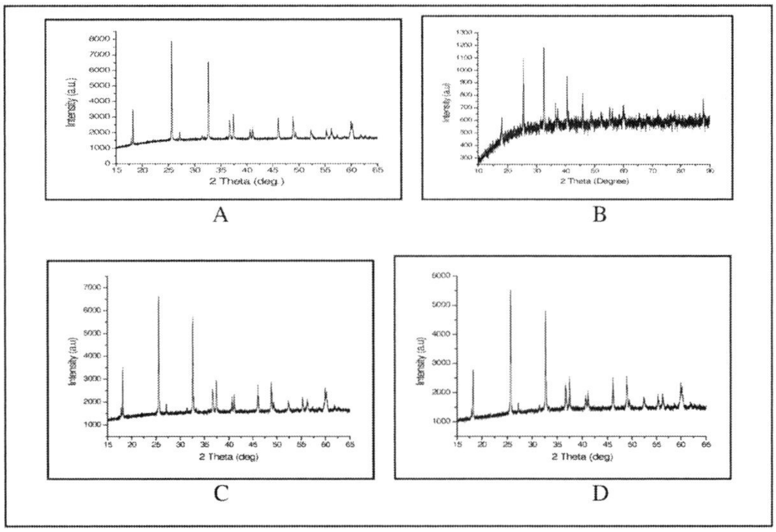

Figure 1. X-ray diffraction patterns of Fe_2TiO_5 ceramic sintered at: (A) 1200°C for 2 hours in air, (B) 1300°C for 2 hours in air, (C) 1200°C for 8 hours in air, and (D) 1200°C for 2 hours in oxygen gas.

Figure 2. Microstructure of Fe_2TiO_5 ceramic sintered at: (A) 1200°C for 2 hours in air, (B) 1300°C for 2 hours in air, (C) 1200°C for 8 hour in air, and (D) 1200°C for 2 hours in oxygen gas.

Microstructure of Fe_2TiO_5 with various sintering condition is shown in Figure 2. Figure 2A and 2B show morphology of Fe_2TiO_5 ceramics with sintered temperature variation. They clearly exhibit the effect of sintering temperature on the grain size. The grain size becomes larger following the increase of the sintering temperature. This is a consequence of the higher mobility of ions at the higher sintering temperature. The higher the mobility of ions is, the larger the grain growth. The grain size calculated by using of the intercept method is found to be 6.04μm and 26.12μm, respectively. Figure 2A and 2C show morphology of Fe_2TiO_5 ceramics with sintered time variation. They clearly exhibit the longer of sintering time on the grain size. The grain size becomes larger following the increase of the sintering time. This is a consequence of the higher mobility of ions at the higher sintering time. The higher the mobility of ions is, the larger the grain growth. The grain size calculated by using of the intercept method is found to be 6.04μm and 8.75μm, respectively. Figure 2A and 2D show morphology of Fe_2TiO_5 ceramics with sintered atmosphere condition. From Figure 2A and 2D are seen that the grain size of the ceramic sintered in oxygen gas is much smaller than that of the ceramic sintered in air. This situation can be explained as follow, relatively poor oxygen for sample sintered in air, make the ceramic could not be well synthesized. Some of Fe_2O_3 and TiO_2 segregated at grain boundaries and inhibited grain growth, producing small grains. The grain size calculated by using of the intercept method is found to be 5.10μm and 6.04μm for sintered in oxygen gas and air, respectively. Sintering atmosphere gases also influence the electrical properties of the pellets Fe_2TiO_5. In an atmosphere which oxygen-rich grain growth does not work properly causing small grain size. Otherwise, in air atmosphere, grains growth better [17-20].

Electrical characteristics of Fe_2TiO_5 ceramic with various sintering condition is shown in Table 1 and Figure 3. As shown in Figure 3 and Table 1, it has found that sintering parameter i.e. temperature, time, and atmosphere has reduced thermistor constant, activation energy, and electrical resistance at room temperature and enlarged the grain sizes. Electrical resistance decreasing mechanism occurs through several ways, such as, growing of grain size due to difference of sintering parameter. The larger grain size result a few grain boundaries. The reduced grain boundaries caused by increase of ion mobility at higher temperature and longer time during sintering process will cause scattering and increasing of mean free path. Therefore, electrical resistance of Fe_2TiO_5 ceramics decreases. The obtained constant thermistor (B) is higher than usual market need (2000 K). It is 4167-6149 K. Generally, the fabricated Fe_2TiO_5 ceramics in this research are still in market price range. Based on the highest constant thermistor B, Fe_2TiO_5 ceramic which is the best choice for main component of thermistor, is sintered at 1200°C for 2 hour in oxygen gas.

Table 1. Electrical characteristic data of Fe_2TiO_5 ceramics.

Temperature/Time/Atmosphere (^{0}C/hour)	B (K)	E_a (eV)	R_{rt} (MΩ)
1100/2/Air	5964	0.51	17.40
1200/2/Air	5879	0.50	13.20
1300/2/Air	4167	0.35	2.53
1200/8/Air	5770	0.49	7.44
1200/2/O_2	6149	0.53	26.54

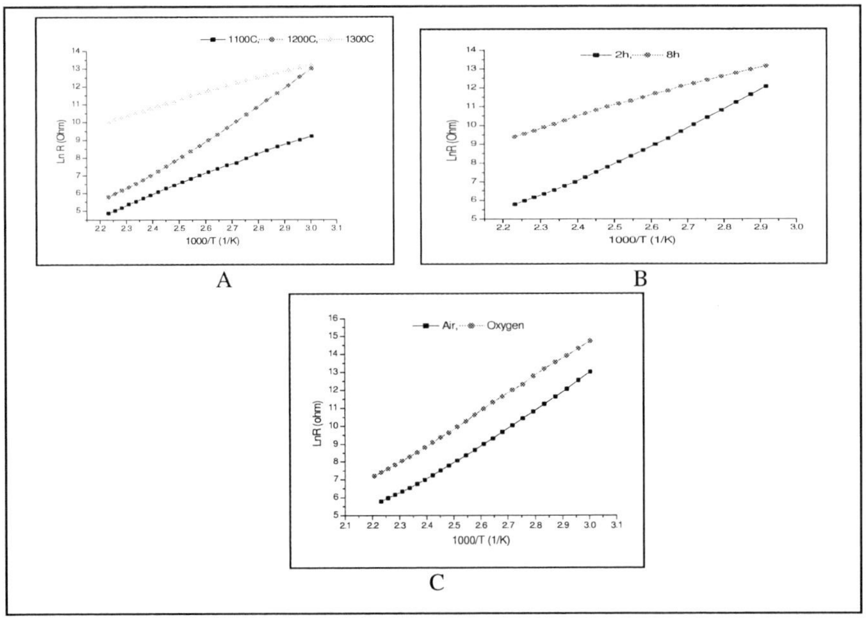

Figure 3. Electrical characteristic of Fe_2TiO_5 ceramic sintered at: (A) 1100°C, 1200°C, and 1300°C for 2 hours in air, (B) 1200°C in air for 2 hours and 8 hours, and (C) 1200°C for 2 hours in air and oxygen gas.

4. Conclusions

Fe_2TiO_5-based ceramics for NTC thermistor prepared from imported Fe_2O_3 and TiO_2 have been successfully fabricated at 1100-1300°C sintering temperature. XRD data show that Fe_2TiO_5 ceramics sintered in all sintering parameter (temperature, time, and atmosphere) have an orthorhombic crystal structure. From morphology data of Fe_2TiO_5 ceramics, it has been investigated that higher temperature, longer time, and change of sintering atmosphere lead to decrease of thermistor constant (B), activation energy (E_a), and electrical resistance at room temperature (R_{RT}). Electrical characteristic data show that Fe_2TiO_5 ceramics have good electrical characteristic for NTC thermistor purpose with thermistor constant (B = 4167-6149 K) and electrical resistivity at room temperature (R_{RT} = 2.53-26.54 MΩ). The values of B and RRT have met a market demand. Sintering temperature, sintering time, and sintering atmosphere is the parameter that can be used to control electrical characteristic of Fe_2TiO_5 ceramic as required.

Acknowledgements

The authors wish to acknowledge their deep gratitude to Directorate General of Higher Education (DIKTI), Ministry of National Education of Indonesian Government for financial support under *Sandwich* program with contract No: B-10944/ Kemsetneg/ Setmen/KTLN/KL.00/ 06/ 2013, dated June 14, 2013 and Hibah Disertasi Doktor program with contract No.3462/UN40/PL/2013, June 19, 2013. It is a pleasure to Koike Laboratory, Department of Applied Physics of Engineering Tohoku University Japan and thanks to Physics of Materials Group, Nuclear Technology Center of Materials and Radiometry BATAN Bandung.

References

[1] R. Metz 2000 *J. Mater. Sci.* **35** 4705–11
[2] M. Vakiv, O. Shpotyuk, O. Mrooz, Hadzaman 2001 *J. Eur. Ceram. Soc.* **21** 1783–85
[3] Wiendartun, D.G. Syarif 2012 *J.Mater. Sci. Research.* **1**(3) 70–75
[4] A. Feltz, W. Polzl 2000 *J. Eur. Ceram. Soc.* **20** 2352–66
[5] G. D. C. Csete de Gyorgyfalva, I. M. Reaney 2001 *J. Eur. Ceram. Soc.* **21** 2145–48
[6] Wiendartun, Waslaluddin, D.G. Syarif 2013 *J. Aust. Cer. Soc.* **49**[2] 141–147
[7] Wiendartun, Risdiana, Fitrilawati, R.E. Siregar 2015 *Mater. Sci. Forum* **827** 262–5
[8] M.N. Muralidharan, E.K. Sunny, K.R. Dayas, A. Seema∗, P.R. Rohini 2012 *Ceram. Int.* **38** 6481–86
[9] R.N. Jadhav, S.N. Mathad, Vijaya Puri 2012 *Ceram. Int.* **38** 6481–86
[10] K. Park, J.K. Lee 2009 *J Alloy Compd.* **475** 513–7
[11] K. Park and D. Y. Bang, J. G. Kim, J. Y. Kim, C. H. Lee, B. H. Choi 2002 J. *Korean Phys. Soc.* **41** 251–6
[12] Ming-Long Liu, DeAn Yang, Yuan-Fang Qu 2010 *J. Alloys. Compd.* **508** 559–564
[13] Huimin Zhang, Aimin Chang, Changwen Peng 2011 *Microelectron. Eng.* **88** 2934–40
[14] A.Veres, J.G.Noudem, O.Perez, S. Fourrez, G. Bailleul 2007 J.*Eur. Ceram. Soc.* **27** 3873–76
[15] Antonio Feteira 2009 *J. Am. Ceram. Soc.* **92** [5] 967-983
[16] N.I. Aljuraide, M.A.A. Mousa, M. Hessien, Mohsen Qhatani and Ameen Ashour 2011 *Int. J. Nanoparticles.* **4** [1]
[17] Xue-Chong Ren, Wei-Li Nie, Yang Bai, and Li-Jie Qiao 2015 *Eur. Phys. J. B.* **88** 232
[18] Chengjian Ma, Yunfei Liu, Yinong Lu, Hao Qian 2015 *J.Alloys. Compd.* **650** 931–5.
[19] Zhuo Wang, Chun Wang, Yongfei Wen & Liangliang Zhang 2016 *Ferroelectrics.* **492** 126-133
[20] Nafiseh Shahgholi, Keivan Asadian *Micro& Nano Letters.* **11** [3], 174–8

IOP Conf. Series: Journal of Physics: Conf. Series **1127** (2019) 012033 doi:10.1088/1742-6596/1127/1/012033

Estimation of Suspended Sediment Concentration by Using Mobile ADCP Instrument

Angga Dwinovantyo[1], **Henry M. Manik**[2]*, **Tri Prartono**[2], **Susilohadi**[3]

[1]Graduate School of Marine Technology, PMDSU Program Batch II, Bogor Agricultural University, Bogor 16680, Indonesia
[2]Department of Marine Science and Technology, Faculty of Fisheries and Marine Science, Bogor Agricultural University, Bogor 16680, Indonesia
[3]Centre for Research and Development of Marine Geology, Bandung, Indonesia

Correspondence Author: henrymanik@ipb.ac.id

Abstract. The measurement of suspended sediment concentration (SSC) by using Acoustic Doppler Current Profiler (ADCP) instrument still not widely used in Indonesia. The application of this technology uses sound wave by four direction transducers placed in the vessel, and this ADCP measured the intensity of backscatter from sediment particles. This study is equipped with RD Instruments WorkHorse Mariner 307.2 kHz ADCP, and the water sample collected from Lembeh Strait, North Sulawesi by using van Dorn bottle sampler. Beyond the experimental purpose of this installation, the interest in having wide area data at different tidal condition is that it provides the possibility of determining the effect of tidal to the SSC. The losses of sound signal were corrected to get more accurate results. Simple linear regression analysis performed to obtain the estimates of the SSC. Estimation of SSC of ADCP measurement was then compared with SSC on water analysis in the laboratory, and the result showed that insignificantly different by statistical analysis. From this research, it showed that the concentration of suspended sediment ranged between 60-73 mg L^{-1} on location near the Port of Bitung, and 50-62 mg L^{-1} at a location away from the port. The result also observed that the highest concentration is in the region below the water column, and tidal condition was affected the condition of SSC. Based on all analysis, the mobile ADCP can provide a suitable measurement of suspended sediment concentration.

1. Introduction

Acoustical methods for measuring suspended sediment concentration have been extensively used especially in the study of sediment transport; one of them uses the Acoustic Doppler Current Profiler (ADCP) instrument. Not many researchers use ADCP as a scientific echosounder for the detection and quantification of suspended sediment in the water column. The using of ADCP still limited to measure speed and direction of water current in oceanography field. On the other hand, there was some research utilized acoustic signal from ADCP to measure the suspended sediment concentration as a baseline to sediment transport study [1, 2, 3].

The acoustic technology in ADCP can produce acoustic backscatter data. The acoustic wave emitted from ADCP to the water column, and the moving particles scattered the acoustic wave back to the instrument equal to echo intensity (EI). In this case, the average of acoustic backscatter data from four transducers in ADCP was used as a base data. The backscatter value was then converted into suspended sediment concentration based on many processes: transmission losses, conversion factor, calibration, and verification. The echo intensity, EI (counts) was then converted into scattering volume, SV (dB) [4] and with the equation from calibration process; the SV will convert into suspended sediment concentration (mg L^{-1}) [5]. This is determined using the sonar equation, and the calibration was conducted using simple linear equations [6].

There are two types of ADCP, static and mobile. Many studies use static ADCP for estimating SSC, because this method provides time series data. In such cases, static ADCP still cannot cover larger areas, due to only take one-point data at one time [7]. Conversely, in order to improve the spatial resolution of ADCP data, the use of mobile ADCP can be reduced the error, and this is the subject of this paper. The advantages of ADCP as a tool for estimating the suspended sediment concentration was having very broad temporal scales, compared with conventional methods using water sampling.

Mobile ADCP use signal processing software that samples the vertical backscatter intensity profile by calculating moving particle in numerous range-gated sample volumes (bin) along a beam path [8]. Suspended sediment concentration can be computed as the product of the strait's cross-section at the point of measurement. This paper describes a study of suspended sediment concentration in Lembeh Strait at 307.2 kHz frequency of mobile ADCP. The measurements were then compared with direct measurement SSC within 30 samples at different tidal condition. The aims of the present study are as follows: (1) to assess acoustic technology using mobile ADCP to estimating suspended sediment concentration in different site (near and far from the port), (2) to estimates SSC in a large spatial data, and (3) to compare SSC from mobile ADCP with in-situ SSC measurement.

2. Materials and methods

2.1. Research location

This research was conducted in 10-12 April 2016 at Lembeh Strait, North Sulawesi, Indonesia. The research vessel was crossed into three different areas: one location was near the Port of Bitung and two locations were away from the port; shipping lanes and dive spot (Fig. 1).

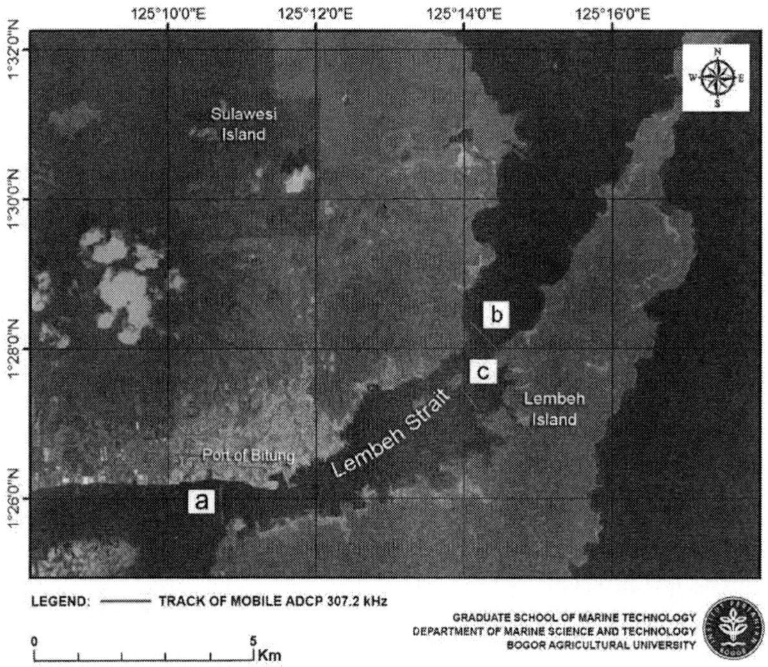

Figure 1. Map of the study area. The red lines correspond to the position of the research vessel in (a) near the port, (b) shipping lanes, and (c) dive spot.

IOP Conf. Series: Journal of Physics: Conf. Series **1127** (2019) 012033 doi:10.1088/1742-6596/1127/1/012033

2.2. *Equipment*
Equipment used in this study was mobile ADCP WorkHorse Mariner by Teledyne RD Instruments and operates at 307.2 kHz, four-beam, convex configuration with a beam angle of 20°. Water samples were collected using van Dorn water sampler from 30 sampling at different site. Other equipment used was Horiba U-50 water quality checker for temperature, salinity, and pH data loggers.

2.3. *Experimental setup*
The strait's cross-section is split into *N* horizontal subsection of equal width and area. The part of the cross-section where mobile ADCP measured backscatter intensity was actually used corresponds to a distance range of 1.5-40 m from the transducer; approximately 30 subsections depend on bathymetry. The setting of mobile ADCP were 100 cm bin size, 60 m maximum profiling depth, transducer placed at 0.65 m depth below surface water, blanking zone 1.5 m below transducer depth. Heading, roll, and pitch data were corrected using gyro. The ship velocity was calculated from position fixes obtained by C-Nav Differential Global Positioning System (DGPS).

2.4. *Data analysis*

2.4.1. Transmission losses and absorption. Sound signal strength from ADCP determined as echo intensity in counts. Before conversion process to suspended sediment concentration, this signal required transmission losses and absorption correction. Transmission losses (dB) were calculated as follows [5]:

$$TL = 20 \log_{10} R + 2\,\alpha R \tag{1}$$

where R is distance between each transducer in ADCP to measured layer or bin (m) and α is sound absorption coefficient (dB m^{-1}). The R can be measured by:

$$R = r + \frac{D}{4} \tag{2}$$

where r is a half size of bin size and D is distance from each bin (m). The parameters were including temperature (T), salinity (S), pH, and material absorbing the sound signal; they were boric acid (f_1) and magnesium sulphate (f_2). Absorption by chemicals where f_1 and f_2 consist at the equation, are two principal components in the ocean. Beside those factors, α also depends on frequency used [9] in ADCP. Absorption coefficient can be calculated using [10] formula as follow:

$$\alpha = 0.106 \frac{f_1 f^2}{f_1^2+f^2} e^{(pH-8)/0.56} + 0.52\left(1+\frac{T}{43}\right)\left(\frac{S}{35}\right)\frac{f_2 f^2}{f_2^2+f^2} e^{-D/6} + 0.00049\, f^2 e^{-\left(\frac{T}{27}+\frac{D}{17}\right)} \tag{3}$$

2.4.2. Counts to dB conversion. Since transmission losses were calculated, echo intensity needed to be corrected. The Doppler-shift frequency in ADCP is not based strictly on the continuous Doppler frequency shift, but is estimated using a discrete phase change on a pulse to another pulse in each transducer. Echo intensity in counts was converted to reverberation level (*RL*). In the sonar equation, all terms are in dB. When measuring with an ADCP, the *RL* comes as stated on equation:

$$RL = K_c\,(E - E_r) \tag{4}$$

where K_c is echo intensity scale for conversion to decibels unit, E is echo intensity received signal strength indicator (RSSI), Er is echo intensity at the reference level (both in counts) [4]. A typical value for Er is 40 counts. Relative acoustic backscatter (*RB*) which is echo level measured at transducer plus two-way transmission losses were calculated as follow:

$$RB = RL + 2\,TL \tag{5}$$

2.4.3. Estimation of suspended sediments concentration. To estimate suspended sediment concentration from *RB* time series data, the following approaches were used including calibration process. By utilizing SSC laboratory result in different layer of depth [11], a simplified equation can be derived for estimating SSC. *RB* values can change due to particle move. Empirical formula was used by simple linear regression by calibration process. From this analysis, the regression line should be correlated direct sample and echo intensity data from ADCP. Simple regression analysis was used to find the regression slope and intercept. These relations are based on sonar equation by:

$$SSC_{ADCP} = 10^{(Intercept\, +\, Slope\, \cdot\, RB)} \tag{6}$$

3. Results and discussion

3.1. Conversion of echo intensity in counts to dB

Calibration curve was built to relate echo intensity values to SSC. Data from direct measurement of suspended sediment concentration were then matched with echo intensity in same layer or bin to obtain information of calibration [12]. It was found that the best-fit values for slope and intercept values are 0.0202 and 1.6121, respectively. Suspended sediment concentration profiles have an acceptable fit of r = 0.9197. Measured echo intensity compared with calculation show a fairly good agreement. Despite that the introduced method is calibrated against detailed field measurement, there might be different coefficient of slope and intercept due to location which have to be considered in different area of survey. First, the effect of grain size distribution at one location will be different to other location and it will affect acoustic signal detected by ADCP [13]. One of limitations of using ADCP to measure suspended sediment concentration is acoustic sensors in ADCP is dependence on sediment properties such as particle size and more sensitive to large particle [14]. Second, measurement of direct sampling was lack of data for calibration because it has only 30 data pairs. This calibration data were then used for estimating suspended sediment concentration (Fig. 2).

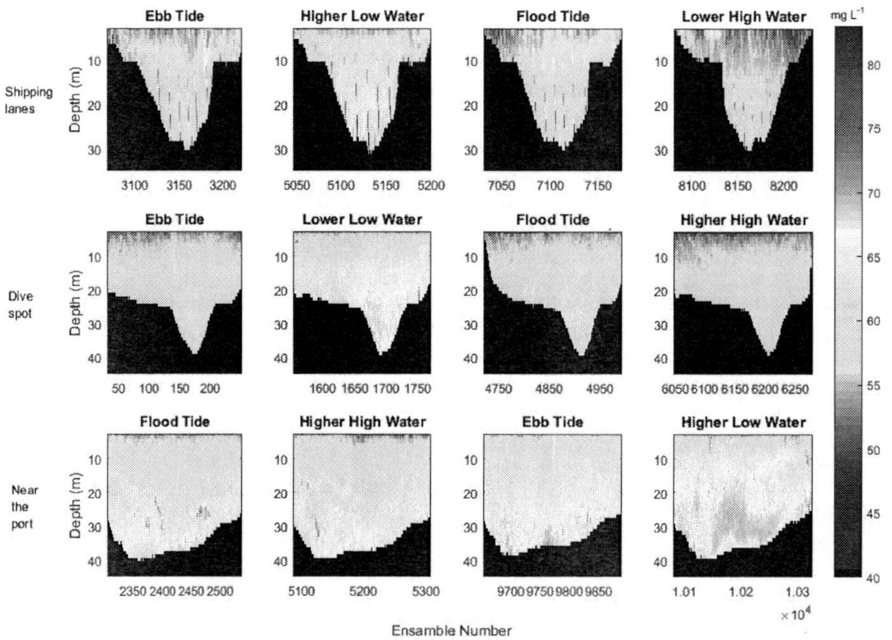

Figure 2. Echogram of suspended sediment concentration at three observation location. Vertical axes indicate water depth in meters, and horizontal axes is ensemble number of track.

For the analysis, we choose an averaged data of the surveyed cross-section and extracted echo intensity data for each tidal condition. The result has been recently investigated with detailed various echo intensity. Because of many factors affecting acoustic signal transmission in natural water column environment such as transmission loss and absorption, raw echo intensity (EI) measured by ADCP were filtered by transmission losses correction to reducing error. Field investigations were performed during both high and low tidal, and also flood and ebb tide cycles in all area. Spatial distributions of averaged suspended sediment concentration were shown in Fig. 2. In echogram, the yellow-orange shades represent higher concentration. Over the cross-section the largest water depth is around 40 m. An important aspect of tidal condition of Lembeh Strait is that current usually flow from the southwest to northeast during ebb tide to low water; and vice versa during flood tide to high water. This area has an average tidal range of 1.2 m and the pattern of tide is a mixed semi-diurnal tide.

Suspended sediment concentration values along Lembeh Strait during the study period ranged from 45 to 80 mg L^{-1}. The range increased during low tide ranged from 10 to 15 mg L^{-1} in all location. The measured SSC was found to be higher in the location near Port of Bitung compared to other location. Comparison between ebb and flood tide in SSC measurement show that the concentration will vary. But during the low water, the SSC values were significantly higher because of large amounts of suspended sediment are alternatingly eroded, resuspended, and deposited [15, 16]. This suggests that suspended sediment variation along Lembeh Strait was affected by tidal condition. During high tide water, the volume of water in Lembeh Strait increases, thus reducing the concentration of suspended sediments. One reason for the higher SSC obtained during low water measurement is probably higher sediment input through northwest current due to inlet from Bitung City.

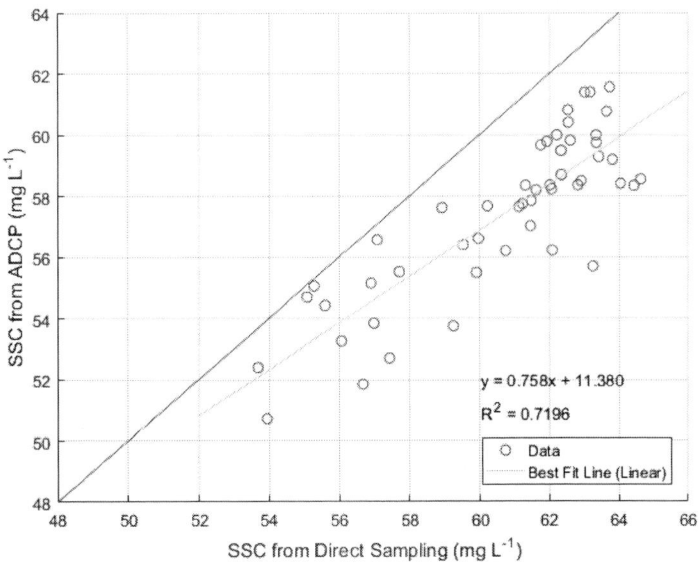

Figure 3. Validation of estimated suspended sediment concentration from ADCP compared to direct sampling.

Figure 3 shows a distribution of suspended sediment concentration derived using the ADCP acoustic intensity and direct measurement. By comparing the 1:1 line with the best fit linear regression line, the ADCP-based SSC estimates are slightly lower than the estimates based on direct measurement. Results of the acoustic method shows a good qualitative agreement with statistical relationships between the SSC and direct SSC measurement method by $r = 0.8483$. This results may be caused by several factors, for example there were differences of particle size distribution at various depths as well as differences in sampling intervals [17].

4. Conclusion

Mobile ADCP is one of promising acoustic instrument that reliable to estimate SSC in a more detailed temporal and spatial resolution. The acoustic parameter was calibrated against the in-situ data sets obtained. The concentration during the low water was higher than high water. Such a measurement method can support the sediment transport study as well as to the calibration process.

Acknowledgements

The authors would like to extend their gratitude to Marine Geological Institute, Indonesia for opportunity to join this research in Lembeh Island. This work also supported by Directorate General of Higher Education for financial support through PMDSU Grant (No. 330/SP2H/LT/DRPM/IX/2016).

References

[1] Gartner J W 2004 Estimating suspended solids concentration from backscatter intensity measured by acoustic Doppler current profiler in San Francisco Bay, California *Marine Geology* 2004 **211** 169-187
[2] Poerbandono, Mayerle R 2004 Assesment of approaches for converting acoustic echo intensity into suspended sediment concentration. *Proceedings 3rd FIG Regional Conference* **2004** 1-13
[3] Rai A K, Kumar A 2015 Continuous measurement of suspended sediment concentration: Technological advancement and future outlook *Measurement* **76** 209-227
[4] Deines K L 1999 Backscatter estimation using broadband acoustic Doppler current profilers. *IEEE* **1999** 249-253
[5] Ghaffari P, et al. 2011 Estimating suspended sediment concentrations using a broadband ADCP in Mahshahr tidal channel *Ocean Science Discussion* **8** 1601-1630
[6] Wall G R, Nystrom E A, Litten S 2006 Use of an ADCP to compute suspended-sediment discharge in the tidal Hudson River, New York. *Geological Survey Scientific Investigations Report* **5055** 1-16
[7] Sentchev A, Yaremchuk M 2016 Monitoring tidal currents with a towed ADCP system *Ocean Dynamics* **66** 119–132
[8] Holdaway G P, et al. 1999 Comparison between ADCP and transmissometer measurements of suspended sediment concentration *Continental Shelf Research* **19** 421-441
[9] Furusawa M, 2015 Effect of noise and absorption on high frequency measurement of acoustic backscatter from fish *International Journal of Oceanography* **2015** 1-11
[10] Ainslie M A, McColm J G 1998 A simplified formula for viscous and chemical absorption in sea water *Journal Acoustic Society America* **103** 1671-1672
[11] American Public Health Association 2012 Standard Methods for the Examination of Water and Wastewater 22nd Ed. **2012**. Washington (US): APHA.
[12] Baranya S, Józsa J 2013 Estimation of suspended sediment concentrations with ADCP in Danube River *Journal of Hydrology and Hydromechanics* **61** 232-240
[13] Wang Y, Gao S 2013 ADCP measurements of suspended sediment flux at the entrance to Jiaozhou Bay, western Yellow Sea *Acta Oceanologica Sinica* **32** 96–103
[14] Voulgaris G, Meyers S T 2004 Temporal variability of hydrodynamics, sediment concentration and sediment settling velocity in a tidal creek *Continental Shelf Research* **24** 1659–1683
[15] Dwinovantyo A, et al. 2017 Estimation of suspended sediment concentration from Acoustic Doppler Current Profiler (ADCP) instrument: A case study of Lembeh Strait, North Sulawesi *IOP Conference Series: Earth Environment Science* **54** 1-8
[16] Hoque M A, Ahad B G, Saleh E 2010 Hydrodynamics and suspended sediment transport at tidal inlets of Salut Mengkabong Lagoon, Sabah, Malaysia *International Journal of Sediment Research* **25** 399-410
[17] Dwinovantyo A, et al. 2017. Quantification and analysis of suspended sediments concentration using mobile and static acoustic Doppler current profiler (ADCP) instruments *Advances in Acoustics and Vibration* **2017** 4890421

Study of galaxy clusters as standard bucket using Chandra X-ray satellite data

A N Indra Putri[1*] and H R T Wulandari[2]

[1]Institut Teknologi Sumatera, South Lampung, Lampung, Indonesia
[2]Astronomy Study Program, Faculty of Mathematics and Natural Sciences, InstitutTeknologi Bandung, Bandung, Indonesia

*Email: annisa.putri@fi.itera.ac.id

Abstract. Clusters of galaxies are the most massive structures in the Universe, confined together in gravitationally bound systems with typical mass in the range of 10^{13} $M_{\odot}$–10^{15} $M_{\odot}$. Clusters of galaxies may be considered as self-similar, meaning that the properties of low mass clusters can be scaled up from the properties of more massive clusters, and vice versa. Clusters of galaxies have been thought to have standard mass fraction of gas, independent of their redshifts or total masses, and therefore proposed to be used as "standard buckets".The purpose of this work is to check the validity of clusters of galaxies as standard buckets by studying larger sample than those analysed by Mantz et al. (2014). The data set employed here consists of Chandra observations of 47 relaxed clusters at redshift $0.069 \leq z \leq 1.063$. The results of this study show that compared to the differential gas fraction ($f_{gas,diff}$), the cumulative gas mass fractions ($f_{gas,2500}$) of the total sample of clusters are less dependent to the total mass M_{2500} and redshift. Relation between gas mass fraction and total mass M_{2500} suggests that massive clusters are more reliable to be used as standard buckets.

1. Introduction

Galaxy clusters are the largest gravitationally bound structuresin the Universe, so that their matter content can be considered sufficiently representative to describe the content of the Universe. The simplest model of structure evolution depends on gravitation prediction that clusters are self-similar, i.e. that clusters are simply scaled up and scaled down versions of each other. Strong self-similarity means that clusters of different masses are identical, scaled versions of each other. Weak self-similarity means that as long as we consider the changing of the density of the Universe, clusters at high redshift are identical to clusters of the same mass at low redshift. Based on these assumpsion and prediction, galaxy clusters are proposed as standard buckets, meaning that gas mass fraction in galaxy clusters is standard, independent of the redshifts and masses. Therefore, clusters of galaxies can be employed to constrain cosmological parameters.

However, various studies show discrepancies regarding at what radius from cluster's center is gas mass fraction expected to be similar from cluster to cluster. Another issue is whether cumulative gas mass fraction ($f_{gas,cum}$) or differential gas mass fraction ($f_{gas,diff}$) is more approximately the same from cluster to cluster. Cumulative gas mass fraction $f_{gas,cum}$ is the fraction of gas enclosed in certain radius,while differential gas mass fraction $f_{gas,diff}$ is measured in a certain shell. For example, Vikhlinin et al. (2009) [1]and Allen at al. (2008)[2] found thatcumulative gas mass fraction $f_{gas,cum}$measured within R_{500} and R_{2500}respectively is constant, whereas Mantz et al. (2014)[3] showed that clusters exhibit similar $f_{gas,diff}$ in the shell range of 0.8 – 1.2 R_{2500}. All the three groups employed data from Chandra.

The goal of this work is to revisit the validity of clusters as standard buckets, by examining gas mass fractionsof a larger sample of clusters compared to Mantz et al. (2014). We begin the procedure by identifyingrelaxed clusters from Chandra X-Ray images. After constructing temperature and density profiles, we proceeded to determining gas mass, total mass, and gas mass fraction of each selected cluster. We then checked how gas mass fraction vary with the radial distance from the cluster center. Finally, we examined whether cumulative ($f_{gas,cum}$) and differential gas mass fractions ($f_{gas,diff}$) are dependent onthe redshifts and total masses ofthe galaxy clusters.

2. Dataset and method

2.1. Dataset

We examined 110 galaxy clusters observed by Chandra X-Ray satellite: 70 clusters are selected from data compiled by Cavagnolo (2008)[4]and 40 clusters are from Mantz et al. (2014). We checked thedegree of relaxation of the clusters. We used softwares CIAO and CALDB to clean and reduce the raw data from Chandra by performing the following steps: data reprocessing, elimination of point source contaminants, and background flares elimination. Blank sky background datasets were tailored to each cleaned observational data.

2.2. Method

Determination of gas mass fraction was carried out in the following steps:

2.2.1 Identification of relaxation states of the galaxy clusters

We adopted relaxation criteria in Mantz et al. (2015) [5]by considering peakness, symmetry, and alignment of the images of the clusters. Peakness is related to surface brightness while symmetry and alignment related to isophotes. Table 1 shows our criteria for the degree of relaxation of galaxy clusters.

Table 1. Relaxation Criteria of Galaxy Clusters.

Criteria	Value	Surface Brightness		Isophotes (n^{th} contour)
		Peakness (log count/pixel)	Central radius (log pixel)	
Very relaxed	5	> 5	≤ 10	Very aligned and symmetric at 1 ≤ n ≤ 5
Relaxed	4	1 ≤ x ≤ 5	30 ≤ y ≤ 10	Rather aligned and symmetric at 1 ≤ n ≤ 5
Unrelaxed	≤ 3	< 1	> 30	Unaligned and asymmetric from n=1

2.2.2 Determining temperature and densityprofile

Temperature and density profile are constructed using deprojection method. We divided the image of a cluster into at least six annuli, each annulus has the same count. We used plasma emission MEKAL (Liedahl et al. (1995)[6]) and WABS models (Morrison andMcCammon, 1983[7]) for spectral analysis. Temperatureprofileswere fitted with the formula adopted from Li et al. (2012)[8]:

$$T(r) = ae^{-(r-b)^2/2c^2} \quad (1)$$

where T denotes temperature, r is radius, a, b, and c are fitting constants. By using deprojectedtemperature profile we could estimate the normalization constant,*norm,* for each region. Then we derived the deprojected electron density (n_e) for each region using the following formula:

$$n_e = \left(\frac{norm * 4\pi D_A{}^2(1+z)^2 10^{14}}{{}^{n_e}/_{n_H} V}\right)^{1/2} \quad (2)$$

Here D_A denotes the angular diameter distance (cm), n_e and n_H (cm^{-3}) are respectively the electron and hydrogen densities, norm is the normalisation constant of each region, and V is the volume of the shell. We fitted the electron density profile using double-β model (Li et al.(2012)):

$$n_e(r) = n_{01}\left[1 + \left(\frac{r}{r_{c1}}\right)^2\right]^{-\frac{3}{2}\beta_1} + n_{02}\left[1 + \left(\frac{r}{r_{c2}}\right)^2\right]^{-\frac{3}{2}\beta_2} \quad (3)$$

where n_{01}, n_{02}, r_{c1}, r_{c2}, $\beta_1$and β_2 are fitting constants.

2.2.3 Determining gas mass fraction

Assuming galaxy clusters are spherically symmetric and are in hydrostatic equilibrium, the total mass within radius *r* can be calculated:

$$M_{\text{total}}(r) = -\frac{kT(r)r}{G\mu m_H}\left(\frac{dln(\rho_{gas})}{d\ln(r)} + \frac{d\ln(T)}{dln(r)}\right) \quad (4)$$

withk denotes Boltzmann constant, G is gravitational constant, μ is mean molecular weight, and m_H is hydrogen mass.Assuming that the electron density in an annulus is constant, gas mass can be determined:

$$M_{gas,annulus} = \frac{4\pi}{3}\rho_{gas,annulus}\int_{r1}^{r2} r'^2 dr' \quad (5)$$

Gas mass fraction of a cluster can be obtained from the relation $f_{gas}(r) = M_{gas}(r) / M_{total}(r)$.

3. Analysis

3.1. Gas massfraction

Using relaxation criteria described in Table 1, we identified that only 57 out of 110 clusters in the sample are classified as relaxed. We further analyzed only 47 galaxy clusters: 21 clusters at low redshifts ($z \lesssim 0.25$) and 26 at high redshifts ($z > 0.26$). We discarded the other 10 clusters that exhibit several problemssuch as too low redshift, high flare background, and too low counts.

We determined cluster total mass enclosed by radius R_{2500} (M_{2500}),where the density is 2500 times the critical density of the Universe at the cluster's redshift using equation (4). Using equation (4) and (5), we can calculate gas mass fraction. Figure 1 shows cumulative ($f_{gas,cum}$) and differential($f_{gas,diff}$) gas mass fraction as a function of M_{2500} (green-red: low redshift, blue-purple: high redshift) and redshift. Also shown are the constants of linear regression, y = a + bx. The errors shown are statistical.

IOP Conf. Series: Journal of Physics: Conf. Series **1127** (2019) 012034 doi:10.1088/1742-6596/1127/1/012034

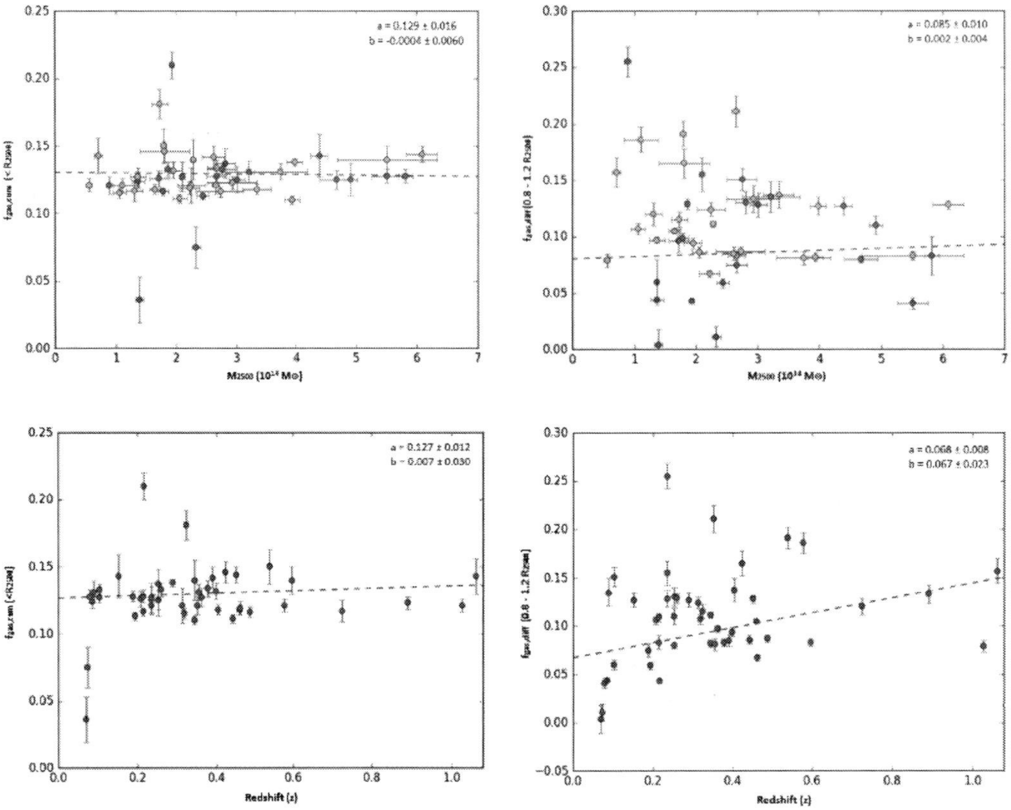

Figure 1. Cumulative (left) and differential (right) gas mass fractions as a function of M_{2500} (top) and redshift (bottom). Dashed-lines indicate the linear regressions, y = a+bx.

It is clear that $f_{gas,cum}$ shows weaker dependencies on M_{2500} and redshift than $f_{gas,diff}$ does. In Figure2, massive clusters (M_{2500}> 2.5 x 10^{14} $M_{\odot}$) exhibit even smaller dispersion in their gas mass fraction. The gas mass fractions of the massive clusters are also less dependent on the redshift compared to the total sample.

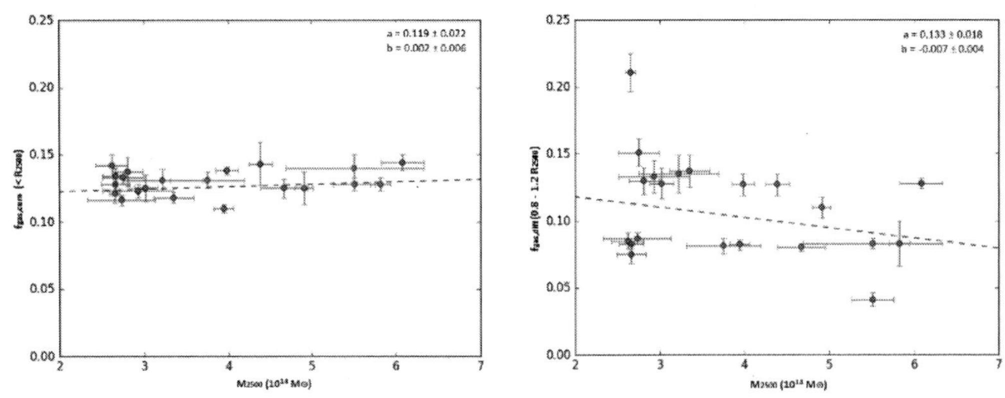

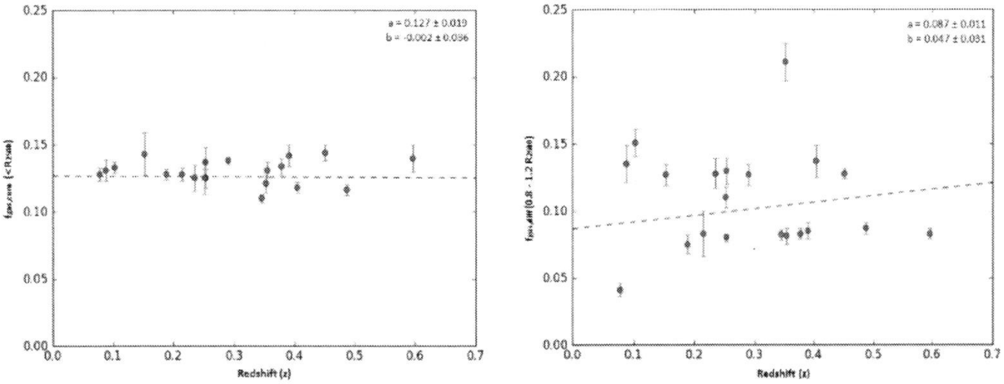

Figure 2. Cumulative (left) and differential (right) gas mass fraction as function M_{2500} (top) and redshift (bottom) for massive clusters only. Dashed-lines indicate linear regressions, y = a+bx.

Figure 3shows our analysis for clusters in Mantz et al. (2014) only. On the left$f_{gas,cum}$is shown as a function ofM_{2500}, whereas on the right is $f_{gas,cum}$ as a function of redshift.We found that $f_{gas,cum}$is less dependent to $M_{2500}$and redshift compared to the relations for all data (Figure 1, left). Mantz et al. (2014) used$f_{gas,diff}$ for analyzing gas mass fraction. We therefore did the same, and compareour results of $f_{gas,diff}$with those obtained by Mantz et al. (2014). Our results show weak dependency of $f_{gas,diff}$on redshift (Figure 4), similar as Mantz et al. (2014) found. Mantz et al. (2014) fitted the relation of $f_{gas,diff}$ and M_{2500} with a power-law, whereas we fitted our results with liniear regression. Mantz et al. (2014) shows more constancy of $f_{gas,diff}$ as a function of M_{2500} than our work does. This might be ascribed to the different methods in the determinations of total mass. Mantz et al. (2014) used three dimensional mass profile model, which adopt *NFWMASS* code fromNulsen et al. (2010), while we assumed hydrostatic equilibrium and used spherically symmetric model.

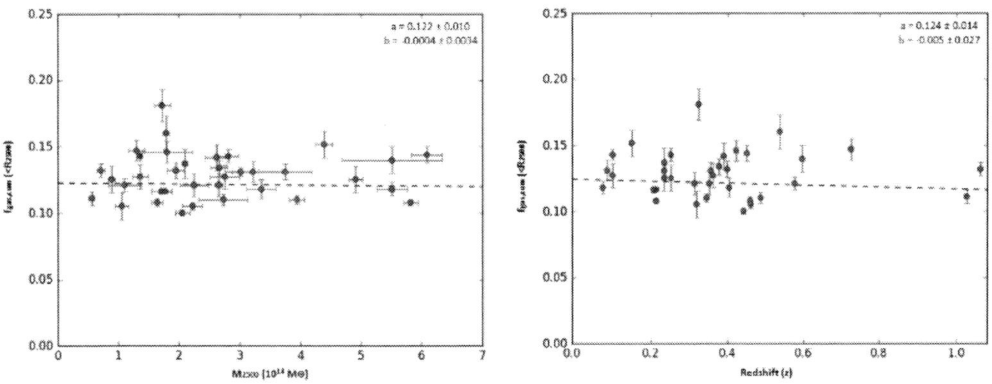

Figure 3. Our results of cumulative gas mass fraction as a function of M_{2500} (left) and redshift (right) for clusters in Mantz (2014) only. Dashed-lines indicate the linear regression, y = a+bx.

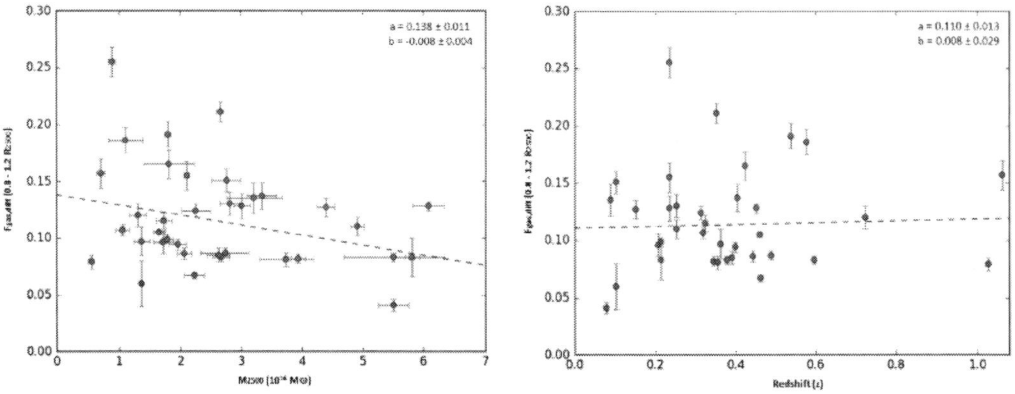

Figure 4. Differential gas mass fraction as a function of M_{2500} (left) and redshift (right) for clusters in Mantz et al. (2014) only. Dashed-linesindicate the linear regression, y = a+bx.

4. Conclusion

We have examined 110 galaxy clusters from Chandra X-Ray data and investigated their relaxation states based on the peakness, symmetry, and alignment of their images. By assuming that the clusters are spherically symmetric and are in hydrostatic equilibrium, we determined the cumulative (within R_{2500}) and differential (in shell 0.8 – 1.2 R_{2500}) gas mass fractions of 47 relaxed clusters. Different from Mantz et al. (2014) results,we found that cumulative gas mass fraction is less dependent on M_{2500} and redshift than differential gas mass fraction is. We also found that massive clusters exhibit smaller dispersion in the gas mass fractions than less massive clusters do. We therefore conclude that massive relaxed clusters are more reliable as standard buckets, and support the use of cumulative gas mass fraction as the measure of gas content in galaxy clusters.

References

[1] Vikhlinin A, Burenim R A, Ebeling H, Forman W R, Hornstrup A, Jones C, Kravtsov A, Murray S S, Nagai D, Quintana H and Voevodkin A 2009 *ApJ.* **692** 1060
[2] Allen S W, Rapetti D A, Schmidt R W, Ebeling H, Morris R G and Fabian A C 2008 *MNRAS.* **383** 879-896
[3] Mantz A B, Allen S W, Morris R G, Rapetti D A, Applegate D E, Kelly P L, von der Linden A and Schmidt R W 2014 *MNRAS.* **440** 2077
[4] Cavagnolo K W 2008*PhD Thesis.* Michigan State Univ
[5] Mantz A B, Allen S W, Morris R G, Schmidt R W, von der Linden A and Urban O 2015 *MNRAS.* **449** 199
[6] Liedahl D A, Osterheld A L and Goldstein W H 1995 *ApJ.* **438** L115
[7] Morrison R and McCammon D 1983 *ApJ.* **270** 119
[8] Li C K, Jia S M, Chen Y, Xiang F, Wang Y S and Zhao H H 2012 *A&A.* **545** A100

IOP Conf. Series: Journal of Physics: Conf. Series **1127** (2019) 012035 doi:10.1088/1742-6596/1127/1/012035

Spectral energy of multisource and multielement of solar energetic particles during the spotless days on solar cycle 24

D Herdiwijaya

Astronomy Research Division and Bosscha Observatory, Institut Teknologi Bandung, Ganesha 10, Bandung, Indonesia 40132

dhani@itb.ac.id

Abstract. We characterized solar energetic particle power-law spectra from selected two sunspotless events with more than 30 consecutive day's length on July 21 to August 20, 2008 (31 days) and July 31 to August 31, 2009 (32 days) in the solar cycle 24. Their spectral of energetic interplanetary multielements of seven charge particles (H, He, C, O, Ne, Si, and Fe) from 15 keV to 0.5 GeV per nucleon have been compiled from onboard instruments of several satellites with different distances from the Sun (GOES, ACE, STEREO, and Voyager). Their spotless spectral energetic particles were also compared with appearances of AR12192, owing very large sunspot area (>2000 millionth of solar hemisphere) without foreshortening effect, on October 19-27, 2014. Our results from six charge elements show that there were intrinsic differences among those events on aspects of flux particles enhancement by double power-law or V-shape. Flux enhancements occurred after 10 MeV for sunspotless days and 20-30 MeV for AR 12192 passage. The effects of very strong magnetic field in AR 12192 influences two mechanisms, flux enrichment of charge particles in lower energy and on the contrary to inhibit streaming of particles in higher energy. It is suggested that interactions of multielements of solar energetic particles with interplanetary magnetic fields and plasma speed play an important role, as seen from the variations of rollover power-law of spectral energy.

1. Introduction

The Sun, as a star, emits wide range energies of photon and plasma particles or Solar Energetic Particles (SEPs). Solar energetic particles constitute a flock of multielement energetic plasma particles, which can be often observed in the near Earth space and heliosphere. They bear information about processes of energy enhancement and turbulence propagation of charged particles in the solar and heliospheric magnetic fields. Many studies concentrated on high energetic solar events related to flare, coronal mass ejection (CME), solar proton events (SPEs), shock, etc. which may abruptly cause relativistic events. Interplanetary SEP observations have already focused to study high-energy particles, plasma physics or space weather. In powerful plasma events, the particle fluxes occupy more than eight orders of magnitude and the energy range extends over more than four orders of magnitude, from MeV to tens of GeV. After more than a half century years, observation results have still remained unknown physical mechanism about plasma interactions and acceleration processes in the interplanetary space [1,2,3,4].

Energetic particles are accelerated near the Sun in at least two distinct locations, namely in solar flares via magnetic reconnection-driven mechanisms so-called impulsive SEPs [5] and at large-scale shock waves driven by fast coronal mass ejections or CMEs known as large gradual SEP events [6].

Their spectra yield a power-law energy spectrum [7]. The power-law index is fully determined by shock properties and is independent of species [8]. The slope of the spectrum of interplanetary fluctuations is important in that it gives us evidence for the dynamics of their evolution. However, the energy spectra often show breaks or rollover power-law at a few to tens of MeV per nucleon and having distinct characteristic in different species of charge particle. Moreover, energy spectra vary from event to event, with their formation, interactions and parameters, depending on the measurement time and location. These SEPs have larger fluences and are responsible for significant disturbances of space weather that can pose serious unrecoverable hazards to technological assets and humans in space. The behavior of the energy spectra in large SEP events is directly indicative of the intrinsic nonthermal radiation threat and has been the subject of many previous studies [9,10].

In this paper, we studied power-law properties of plasma with energy 15 keV–500 MeV per nucleon for seven charged species (H, He, C, O, Ne, Si, and Fe) spectra during at least 30 consecutive sunspotless days cases at July 21 to August 20, 2008 (31 days) and July 31 to August 31, 2009 (32 days) in solar cycle 24. We also compared those very quiet solar activity periods of sunspotless with 9 days passage of the largest area of Jupiter size active region (>2500 msh) on the solar disk, AR 12192 on October 19-27, 2014. All data have already been made publicly available by NASA using interface of The Virtual Energetic Particle Observatory (VEPO) - Multi-Source Spectral Plots (MSSP) of energetic particle (http://vepo.gsfc.nasa.gov/). We have used one hour temporal resolution for ACE [11], GOES 8-11, and STEREO A-B satellites [12], except Voyager 1-2 satellites with 6-hour resolution [13]. At altitude of 6.6 Re, GOES satellite is a geosynchronous satellite, so the spacecraft are inside, but near (when on the day side), the Earth's magnetopause, so GOES energetic particle fluxes are representative of fluxes in the nearest interplanetary medium. ACE satellite orbit has a unique location called Lagrange point, or L1 at altitude $1.5x10^6$ km. This orbit is a gravity neutral point in space, allowing satellite to essentially hover between the Sun and Earth at all times, maintaining a constant view of the Sun and sun-lit side of Earth. Twin satellites STEREO A and B move along a distance of about 1 AU encircling the Sun with opposite directions. Voyager 1 and 2 satellites located at about 24 AU in year 2008 to 2009. For each satellite, it carried several instruments for specific range of energy spectra. The energy spectral points for each source were fitted to a simple power-law relation in logaritmic scale, log (flux) = log (A)+b*log (energy), where A and b are the fitting parameters, which are determined by a linear least squares fit.

2. Distribution of proton speed, density, electric field, and beta ratio

Figure 1 shows proton speed, density, electric field, and beta ratio distribution during two periods of one-month spotless days and passage of the largest area of sunspot group AR 12192 on solar disk. They revealed a similar pattern in density distribution. The last event blew up the fastest average proton speed of (442.32±65.79) km/s with shifted top speed distribution and the lowest density of (4.60±2.07) particles/cm^3. Spotless days of July 31-August 31, 2009 generated the lowest average plasma speed and the largest density of (384.29±65.92) km/s and (5.58±3.10) particles/cm^3, respectively. On the other hand, spotless days on July 21-August 20, 2008 caused (430.41±100.65) km/s and (5.00±3.53) particles/cm^3 for average speed and density, but it poured out up to 30 particles/cm^3. Median of electric field strengths were -0.03 mV/m, -0.07 mV/m for spotless day's period in year 2008 and 2009, and significantly increased to 0.55 mV/m for passage days of AR 12192. Electric field is guided by plasma speed and magnetic field in Z-axis component or Bz (in GSM coordinates). The negative value of Bz denotes southward direction of interplanetary magnetic field. It means that AR 12192 had the strongest positive Bz, and in year 2009 was the smallest variations of electric field with negative Bz. Moreover, influence of strong magnetic field during passage of AR 12192 was also pointed out by minimum value of interplanetary beta ratio with median 1.11, compared with sunspotless days in year 2008 with median 2.35, and year 2009 with median 2.57. Small value of dimensionless parameter beta ratio indicates stronger magnetic filed pressure than gas pressure, and vice-versa. These interplanetary magnetic field environments will subsequently have impact on spectral energetic particles, as described below.

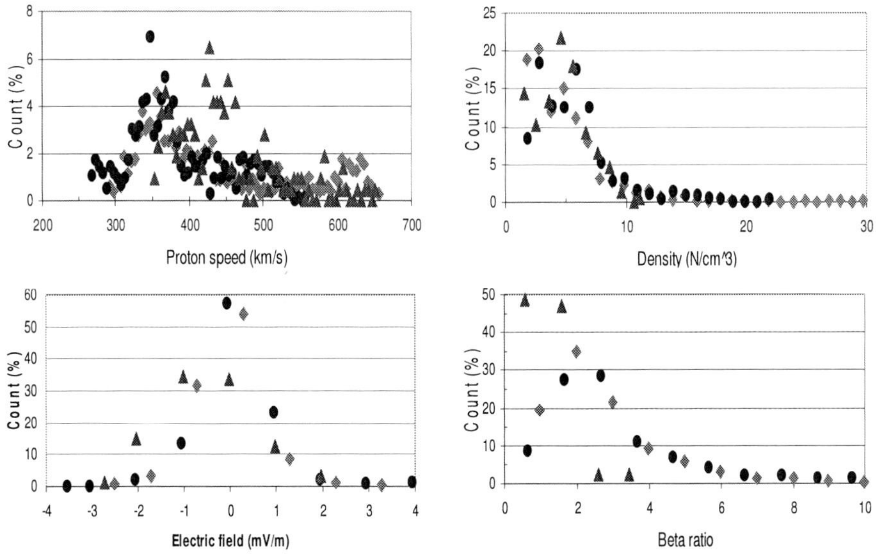

Figure 1. Distribution of proton speed (a; top-left), density (b; top-right), electric field (c; below-left), and beta ratio (d; below-right) during spotless days on July 21-August 20, 2008 (N=792; red diamond), July 31-August 31, 2009 (N=768; black circle), and AR 12192 (N=216; blue triangle).

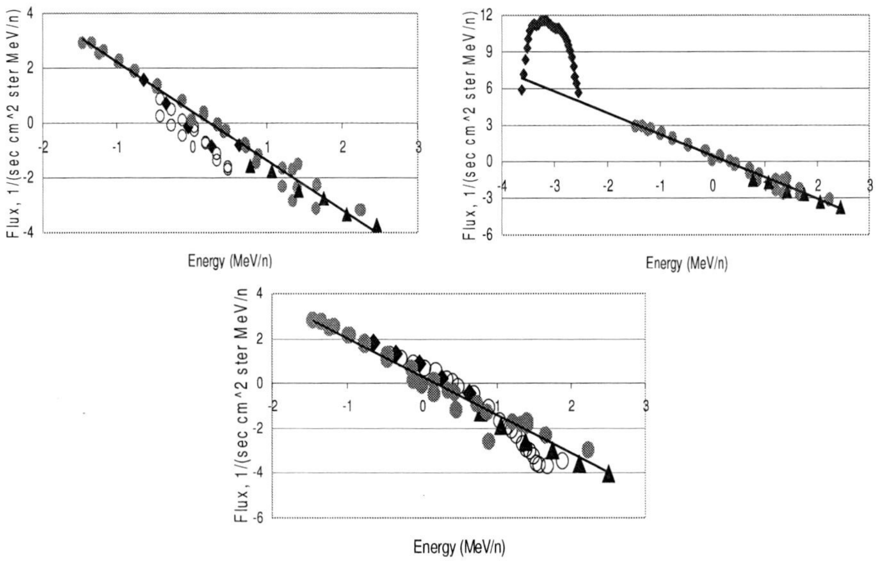

Figure 2. Spectra energy of proton and flux in logaritmic scale during spotless days on July 21-August 20, 2008 (top-left) and July 31-August 31, 2009 (top-right), during passage of the largest active region AR 12192 on October 19-27, 2014 (bottom). Symbols refer to GOES 8-11 (black triangle), Voyager 1-2 (red filled-circle), ACE (blue diamond), STEREO (open circle) satellites.

3. Multisources of proton spectra energy

Figure 2 shows spectra energy for spotless days in years 2008 and 2009 for different satellites. In the range of energy 0.03 – 300 MeV, we can obtain similar parameter slopes of -1.74 and -1.77, respectively. The peak of proton flux (with $4.11\text{x}10^{11}$ particles/(sec cm^2 ster MeV/n)) appeared at very low energy of 619 eV, that observed by ACE satellite in year 2009. See figure 2b.

Homogenous proton spectra energy penetrated across interplanetary region or heliosphere up to 24 AU during spotless days, as detected by Voyager satellites. In the figure 2c, during the passage on the solar disk of the largest area of sunspot group AR 12192 on October 19-27, 2014, gradient reached to -1.95, which was slightly steeper than two periods of sunspotless day. A broken power-law can be seen for energy larger than 10 MeV, which has a slope of -2.1, then rollover with positive gradient.

4. Spectra energy of multielements

4.1. He

During sunspotless days in year 2008 and year 2009, gradients or He (alpha particle) power spectra were -3.31±0.22 and -2.54±0.24 (0.01 MeV<E<10 MeV), respectively. Passage of AR 12192 gave more soft gradient of -2.21±0.11 (0.01 MeV<E<100 MeV). The peak of alpha particle flux took place at energy 1 keV with $2.84\text{x}10^9$ particles/(sec cm^2 ster MeV/n), which were two-folds fewer than proton. See figure 3a. In general, spotless days supply more low energy alpha particles with higher rate through dominant open magnetic field structures. On the other side, alpha particles flux increased for higher energies by changing to positive power-law gradient. This particle injection will shift V-shape of higher energy spectra due to increasing particle flux.

4.2. C

In figure 3b, a similar V-shape power-law can be detected near 10 MeV for all spotless cases, and shifted to about 30 MeV for AR 12192 passages. However, they differed in their gradients. In case of AR 12192 passage, this event produced larger particle flux up to E<30 MeV, then decreased for E>30 MeV. Very large magnetic field strength of AR 12192 clearly shielded outward stream of high energetic particles. In year 2008, energy smaller and larger than 10 MeV contributed to power-law gradient -3.21±0.11 and 0.77±0.08, respectively. In year 2009, their gradients were -2.40±0.12 and 0.78±0.08, respectively. During AR 12192 passage, the turning point occurred with slopes -2.42±0.09 and 0.91±0.13 for smaller and larger than about 30 MeV, respectively. So, very strong magnetic field triggered larger positive gradient or increasing flux of C charge particles.

4.3. O

The turning point of a broken power-law shifted to a larger energy at about 20 MeV for spotless days and about 30 MeV for AR 12192. See figure 3c. Starting for 10 MeV, particle flux during sunspot group passage were gradually smaller than spotless days. In year 2008, power-laws were -2.21±0.07 and 0.46±0.05 for smaller and larger than about 20 MeV. A similar pattern in year 2009, their pairs were obtained -1.44±0.06 and 0.42±0.04. In year 2014 of AR 12192 passage days, gradients were -2.35±0.07 and 0.66±0.06 for smaller and larger than about 30 MeV, respectively. AR 12192 caused larger gradient of growing flux than spotless periods.

4.4. Ne

Shifted energy of V-shape power-law was similar to O element, also particle flux variations, as seen in figure 3d. Their gradient pairs were -1.31±0.17 and 0.57±0.06, -1.59±0.11 and 0.46±0.04, -3.25±0.14 and 0.96±0.17 for year 2008, 2009, and 2014, respectively. Spotless period significantly showed smaller positive gradient for Ne charge particles.

4.5. Si

Shifted energies of rollover power-law at 10 MeV and 20 MeV were similar to C element, also particle flux variations, as seen in figure 3e. Their power-law pairs were -1.70±0.20 and 0.52±0.06, -2.11±0.29 and 0.45±0.06, -2.79±0.15 and 0.74±0.10 for year 2008, 2009, and 2014, respectively.

4.6. Fe

Shifted energies of broken power-law from 10 MeV to 20 MeV were similar to C and Si elements, also particle flux variations, as seen in figure 3.e. Their gradient pairs were -2.90±0.17 and 0.55±0.07, -2.51±0.21 and 0.55±0.06, -2.48±0.09 and 0.54±0.04 for year 2008, 2009, and 2014, respectively. For heavy element, the positive gradients were not depend on those two extreme conditions.

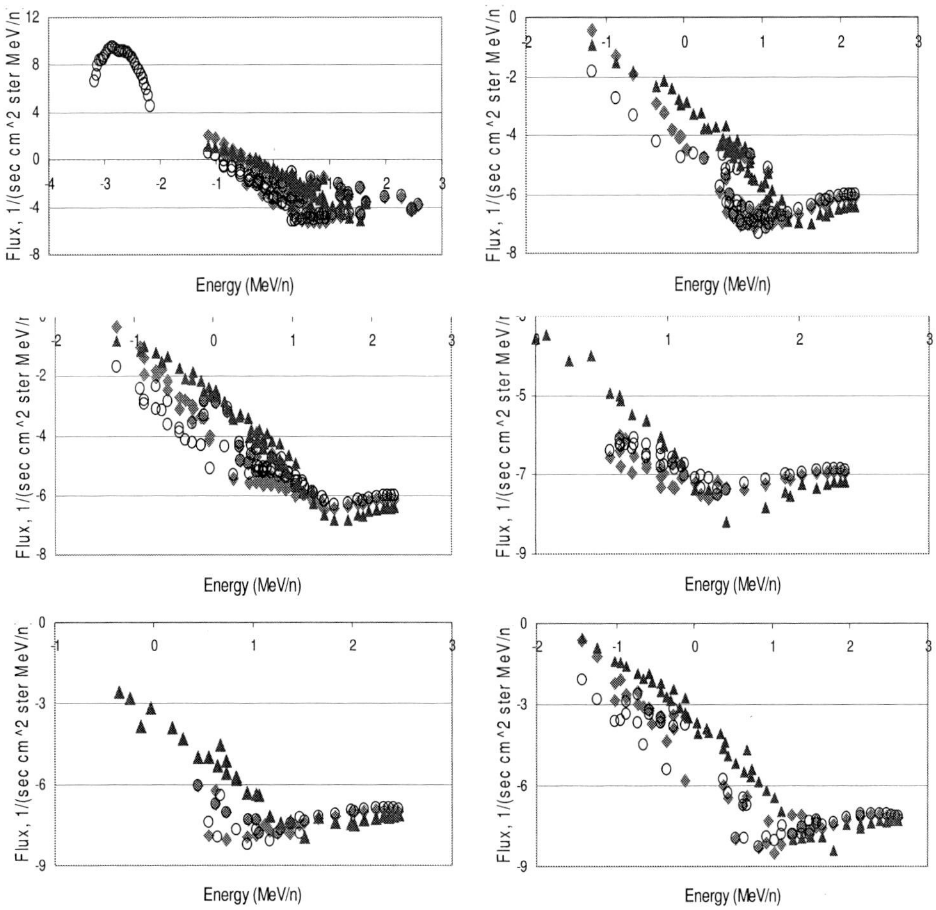

Figure 3. From left to right. Spectra energy of He - alpha particle (a), C (b), O (c), Ne (d), Si (e), and Fe (f) in logaritmic scale during sunspotless days on July 21-August 20, 2008 and July 31-August 31, 2009, during passage of the largest active region AR 12192 on October 19-27, 2014. Symbols refer to sunspotless in year 2008 (red diamond), 2009 (open circle), and AR 12192 (blue triangle)

5. Conclusions

We have clearly showed that there were shifted broken or V-shape power-law for two extreme cases of long sunspotless days and the largest sunspot area passage of AR 12192 for six charge elements. The turning point energy occurred at 10 MeV for sunspotless days and 20-30 MeV for AR 12192 passage. During the sunspotless days, less particle fluxes have been emitted than AR 12192-passage period in lower energy. Then, in higher energy, flux enhancements with positive gradient have been seen for all events, but more fluxes outpouring for sunspotless periods. Strong magnetic field obviously shielded energetic particle streaming.

Acknowledgements: This work has been supported by Institut Teknologi Bandung Research Grant Year 2016. We also acknowledge use of NASA/GSFC's Space Physics Data Facility's OMNIWeb (or CDAWeb or ftp) service, and OMNI data.

References

[1] Forbush S E 1946 Three unusual cosmic-ray increases possibly due to charged particles from the Sun *Phys Rev* **70** 771 (doi: 10.1103/PhysRev.70.771)

[2] Wild J P, Smerd S F, Weiss A A 1963 Solar bursts *Ann Rev Astron Astrophys* **1** 291 (doi: 10.1146/annurev.aa.01.090163.001451)

[3] Miroshnichenko L I, Perez-Peraza J A 2008 Astrophysical aspects in the studies of solar cosmic rays *Int. Journal of Modern Physics* **A 23**(1) 1 (doi:10.1142/S0217751X08037312)

[4] Mewaldt R A, Cohen C M S, Labrador A W, Leske R A, Desai M I, Looper M D, Mazur J E, Selesnick R S, Haggerty D K 2005 Proton, helium, and electron spectra during the large solar particle events of October November 2003 *J Geophys Res* **110** A09S18 (doi: 10.1029/2005JA011038)

[5] Drake J F, Swisdak M, Fermo R 2013 The power-law spectra of energetic particles during multi-island magnetic reconnection *ApJL* **763** L5 (doi: 10.1088/2041-8205/763/1/L5/)

[6] Reames D V 1999 Particle acceleration at the Sun and in the heliosphere *Space Science Reviews* **90** 413 (doi:10.1023/A:1005105831781)

[7] Blandford R D, Ostriker J P 197, Particle acceleration by astrophysical shocks *ApJL* **221** L29 (doi/10.1086/182658)

[8] Lee M A 1983 Coupled hydromagnetic wave excitation and ion acceleration at interplanetary traveling shocks *J Geophys Res* **88** A8 6109 (doi:10.1029/JA088iA08p06109)

[9] Ellison D C, Ramaty R 1985 Shock acceleration of electrons and ions in solar flares *Ap J* **298** 400 (doi/10.1086/163623)

[10] Tylka A J, Cohen C M S, Dietrich W F, et al. 2005 Shock geometry, seed populations, and the origin of variable elemental composition at high energies in large gradual solar particle events Ap J **625** 474 (doi: 10.1086/429384)

[11] Stone E, Frandsen A, Mewaldt R et al. 1998 The Advanced Composition Explorer Space Science Reviews **86** 1 (doi:10.1023/A:1005082526237)

[12] Luhmann J G, Curtis D W, Lin R P, Larson D, Schroeder P et al. 2005 IMPACT: Science goals and firsts with STEREO Advances in Space Research **36** 8 1534 (doi:10.1016/j.asr.2005.03.033)

[13] King J H, Papitashvili N E 2005 Solar wind spatial scales in and comparisons of hourly Wind and ACE plasma and magnetic field data *J Geophys Res* **110** A02104 (doi:10.1029/2004JA010649)

IOP Conf. Series: Journal of Physics: Conf. Series **1127** (2019) 012036 doi:10.1088/1742-6596/1127/1/012036

Relationship of mixed layer depth and sea surface temperature onlongwave radiation flux up in the west of Sumatra waters

M Fadli and M R Putri

Graduate Program of Earth Sciences, Graduate School of Institute Technology Bandung, JalanGanesha No 10, Bandung 40132, Indonesia.

Corresponding author : fadliose07@gmail.com

Abstract. Thermal energywhich stored in upper ocean mixed layer will be released in the form oflongwave radiation flux up (LWUp). LW Up is a longwavethat emitted from sea surface to atmosphere. Variability of upper ocean mixed layerhas influence on ocean thermal capacity and then affected LW Up variation. The areaof interest of this research was west of Sumatra waters. Vertical temperature profiles, sea surface temperature (SST) and LW Updata from 2000 to 2015 were used. Mixed layer depth (MLD) was computedbyThreshold Method based on temperature profiles with 0.31°Ctemperature criterion (ΔT). The results show high LW Up flux occur during shallow MLD and low LW Up occur when MLD become deeper. Correlation between SST and LW Up give a fine value about 0.9 in all station. Good correlation coefficient between MLD and LW Up found in station 1 (-0.81) and 9 (-0.77) with 1 month lag. This results suggest that shallow (deep) MLD followed by high (low) SST so that the result is increase (decrease) LW Up by 1 month lag. Small correlation between MLD and LW Up found in area near equator region. Indian Ocean circulation, westerly wind, and Wyrtki jet are possibly causing the small correlation in equator region.

1. Introduction

Solarenergy reaching earth surface is stored temporarily in upper ocean layer or ocean mixed layer as thermal energy. Thermal energy released mostly as latent heat by evaporation, followed bynet longwaveflux radiation (hereafter netLW) and then as sensible heat by conduction [1]. Net LWconsist of longwave radiated from ocean surface to the atmosphere called LW Up and longwave radiated from atmosphere to ocean surface called LW Down which is compute as the difference between LW Up and LW Down [2]. Many literatures discuss aboutlongwave in form of net LW even thoughLW Up has important role and become a main source for atmosphere to emitted LW Down.

The amount of LW Upemitted depend on ocean surface temperature. If strong diurnal cycle ofsea surface temperature (hereafter SST) in top few meters of the ocean was neglected, LW up depend on mixed layer temperature. Ocean mixed layer is a layer that have homogenous temperature, salinity, and density vertically. This homogenous layer results from active vertical-mixing near the surface, promote by various processes like wind-stirring, wave, turbulence generation by vertical shear or night time convective mixing [3].Mixed layer depth (MLD) variability influence ocean heat capacitythat will affect SST and LW Up.Shallow MLD exhibits a reduced thermal capacity and can hence promote large SST anomaly otherwise deep MLD exhibits an increased thermal capacity so that SST will remain stable to some perturbation [4]. Shallow MLD which cause heating of the atmosphere dominated by surface energy flux (latent and longwave flux) whereas the heating dominated by shortwave absorption within the atmospheric column in the deep MLD condition [5].

Considering the linkages between MLD, SST and LW Up, a preliminary research was conducted. The objective of this research is to gain understanding about the relationship between MLD, SST, and LW Up. The paper is organized as follows : a brief description of research and data in section 2. Section 3 deals with methodology used to compute MLD and cross correlation. Section 4 deals with results and discussion. Conclusions are presented in section 5.

2. Research area and data source

The research area of interest waswest of Sumatra waters from 79°E to 101°E and 16°N and 9°S which divided into 10 study areas (see figure 1). All study areas located in a tropical basin and forced by semi-annually reversing wind system, the monsoon. The surface circulation also reverse semi-annually like wind, however it is not strictly but appears to lead it.

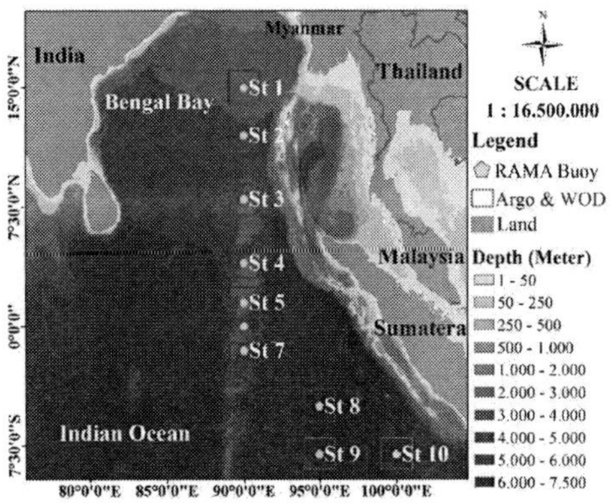

Figure 1. Research area and data positions. Green circle show RAMA buoy data. Red rectangular (2°×2° spacious) show area for argofloat, WOD, SST and LW Up sampled data.

For the purpose of these research, 79,709 observation vertical temperature profiles have been used. Temperature profile from Research Moored Array for African – Asian-Australian Monsoon Analysis and Prediction (RAMA), Argofloat and World Ocean Data (WOD) were sampled with an area 2°x2°where RAMA buoy position as a center. The monthly 1°x1°griddedOptimum Interpolation Sea Surface Temperature (OISST) data are obtainedfrom National Oceanic and Atmospheric Administration (NOAA). Monthly longwave flux Up is obtainedfrom Clouds and Earth's Radiant Energy System Project (CERES) with resolution 1°x1°gridded. All data 2000 through 2015 were used in this research.

3. Methodology

3.1. Mixed Layer Depth

Daily average vertical temperature profiles was used to calculateMLD. MLD was calculated using Threshold Method base on temperature [6]. Temperature criterion used in this research is ΔT=0.31°C.Threshold method give more stable MLD value than gradient method [7].

In threshold method, MLD calculate base on temperature difference (ΔT_{cal}) between temperature in certain depth (T_{depth}) and reference temperature (T_{ref}). ΔT_{cal} is calculated by following equation :

$$\Delta T_{cal} = \left| T_{depth} - T_{ref} \right| \tag{1}$$

then, ΔT_{cal} compared to temperature difference reference(ΔT_{ref}). In this research, ΔT_{ref} =0.31°C is used. If $\Delta T_{cal} \geq \Delta T_{ref}$ then depth in T_{depth} was a mixed layer depth.Reference temperature (T_{ref}) used in this research is temperature at 10 meter from surface.Utilization temperature at 10 meter depth was intended to avoid daily cycle of surface water temperature [6].

3.2. Correlation
Cross correlation was used to calculate relationship between MLD, LW Up and SST with some lag time. Cross correlation coefficientcalculatedby [8] :

$$r_{xy} = \frac{C_{xy}(\tau)}{\sigma_x \sigma_y} \tag{2}$$

where r_{xy} is cross correlation function, $C_{xy}(\tau)$ is cross covariance, σ_x and σ_y are standard deviation of parameter x and$y$. The next step is calculated Z score to figure out if results from a test are valid or not. Z score is calculate by :

$$Z_{score} = r\sqrt{N-1} \tag{3}$$

where$r$ is correlation coefficient and N is number of data. Last step is Z score compared with Z analytic (± 1,96 for 5% two tailed). If Z score greater than Z analytic, it meansthe cross correlation coefficient is valid and the relationship can be accounted.

4. Results

Monthly climatology of mixed layer depth, sea surface temperature and LW Up radiation are shown in figure 2 below.LW Up in east equator zone (St 5 to St 7) has minimum (maximum) value about 469 W/m^2(477 W/m^2) while SST value is 28.7°C (30°C) and MLD value is 70 meter (41 meter). Minimum condition happenedin southwest monsoon (June – September) and maximum in northeastmonsoon (December – March). For north zone, minimum (maximum) value of LW Up about 456 W/m^2 (478 W/m^2) which happen while SST value is 26.3°C (30°C) and MLD value is 80 m (28 m). Minimum condition happened in northeast monsoon while maximum in transition 1 (April – May). For south zone, minimum (maximum) value of LW Up about 460 W/m^2 (475 W/m^2) which happen while SST value are 27°C (29.2°C) and MLD value are 60 m (28 m). Minimum condition happened in southwest monsoon while maximum in transition 1.

From description above, it appears that LW Up maximum occurs while MLD become shallowand LW up minimum occurs in deep MLD condition. Shallow MLD exhibits a decrease thermal capacity so that SST increase more rapidly as a result of solar energy absorption. High SST means that ocean has excess energy and must release it, one of them as LW Up to achieve the equilibrium state. Deep MLD exhibits an increase thermal capacity. It makes ocean need to absorbs more energy to increase SST by 1°C. At this time, SST has decrease and followed by the decreasing of LW up too.

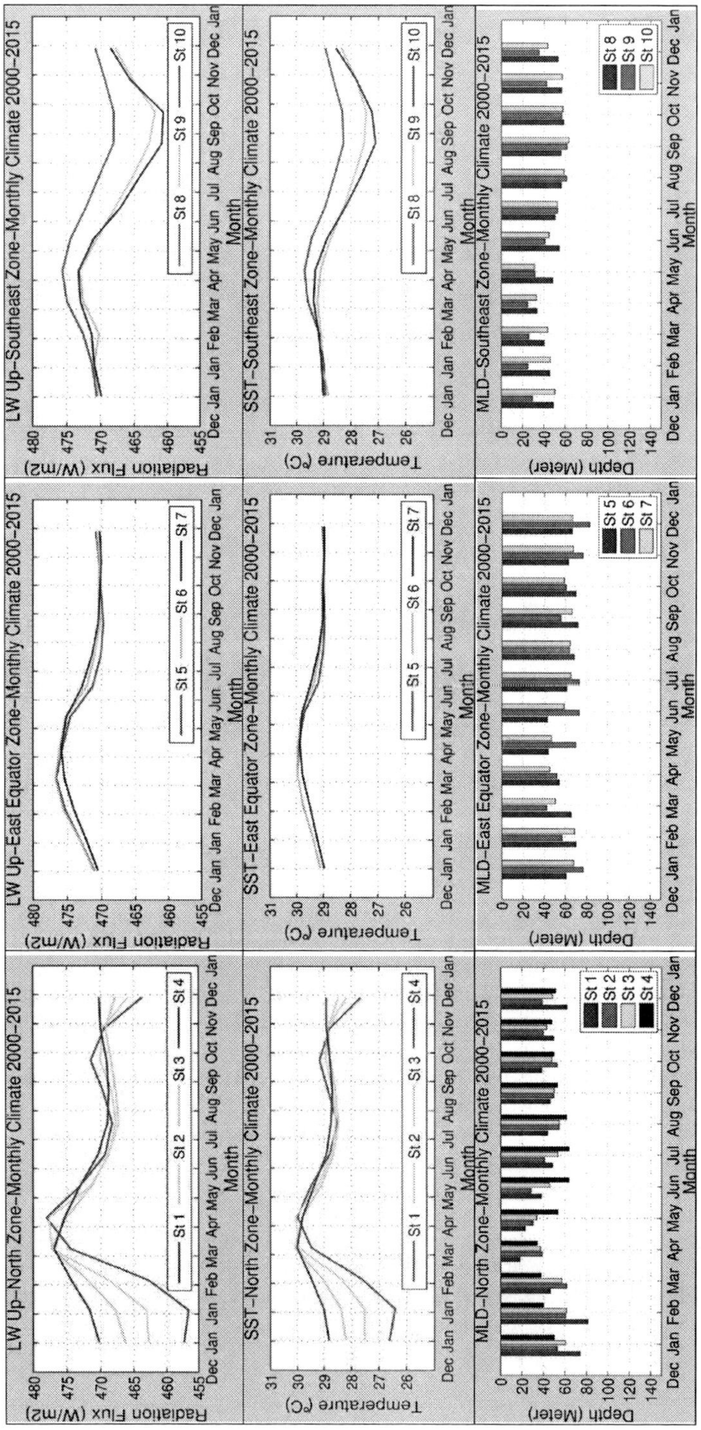

Figure 2. LW Up (above), SST (middle) and MLD (below) monthly climate from 2000 – 2015 in north zone, east equator zone and southeast zone.

Variation of SST and MLD in Indian Ocean depend on monsoon system, water column stability and ocean circulation. MLD and SST in east equator zone mainly affected by equatorial circulation and Wyrtki Jet. Equatorial current that moving eastward throughout the year, will accumulate water inthe eastern equatorial Indian ocean and causing mean MLD in equator zone is deeper compared to north and southeast zone. In transitional season, the Wyrtki jet accumulate more warm water the east so that make MLD become deeper and SST higher during 2 to 3 month. For southeast zone, southwest monsoon lead upwelling along south coast off java and moving westward to Indian Ocean and west coast of Sumatra. Strong wind and cold water weaken the water column stability and cause MLD to become deeper. This condition cause ocean emit less LW Up radiation. Meanwhile in northeast monsoon, weak wind and warm water which carried by equator current and Wyrtki Jet from equator increasing SST and make MLD become shallow. At this time, sun position in the southern hemisphere contribute to increase the SST. This condition lead to increasing of LW up radiation. For north zone, southwest wind that stronger than northeast wind are lead to a deeper MLD while cold water from coastal upwelling in the north of St 1 that move southward in northeast monsoon make water column unstable so that MLD is deeper. Deeper MLD with cold SST weaken LW up radiation. Meanwhile in transitional season, weak wind and warm water input from equator carried by Wyrtki Jet to Bengal Bay lead to shallow MLD and increasing of SST. It makes ocean emit more LW up radiation to the atmosphere.

Table 1. Correlation coefficient between MLD, LW up and SST in 10 stations in Indian Ocean. r = correlation coefficient

Zone	Station	MLD and LW Up			SST and LW up		
		r	Z	Lag (month)	r	Z	Lag (month)
North	St 1	-0.81	-8.06	0	0.97	9.65	0
	St 2	-0.77	-6.11	1	0.97	7.70	0
	St 3	-0.55	-3.30	0	0.92	5.52	0
Equator	St 4	-0.57	-5.44	-1	0.96	9.16	0
	St 5	-0.48	-5.10	1	0.97	10.31	0
	St 6	-0.38	-3.89	-1	0.97	9.94	0
	St 7	-0.45	-5.11	1	0.96	10.90	0
Southeast	St 8	-0.48	-5.58	-1	0.94	10.92	0
	St 9	-0.77	-6.30	-1	0.91	7.45	0
	St 10	-0.64	-4.87	0	0.91	6.93	0

Cross correlation coefficient between SST and LW up shown in Table 1. All station shown strong relationship with coefficient greater than 0.9 and 0 month lag time. Positive coefficient mean that LW up emitted from ocean surface depend on its temperature. High SST make ocean emit LW up and vice versa. Zero month lag time mean that response of temperature change to LW up variation take a short time.

Cross correlation coefficient between MLD and LW up shown in Table 1. The coefficientvaries from -0.38 to -0.81 with Z calculation greater then Z analytic. Negative coefficient mean that deeper MLD followed by decrease of LW flux up and shallow MLD followed by increase of LW up flux. Deep MLD has larger thermal capacity than shallow MLD. It makes ocean must absorb more energy to increase the temperature by 1°C. The result is ocean will emit lesser LW up flux in deep MLD than in shallow MLD condition. Based on Table 1, stations further from equator have higher correlation coefficient than station near equator. It explains that variability of LW Up in equator is not strongly affected by MLD. Equatorial circulation and Wyrtki jet are possibly the causes of this small value. Lag time calculation give negative and positive results with range 1 month. Positive value indicate thatchange of LW Uphappen first before change of MLD while lag time negative mean change of MLD happen first before LW up change. Lag time zero mean that MLD or LW up change followed by other parameter change in short time.

5. Conclusion

From this preliminary research, it can be conclude that LW Up in Indian Ocean has strong relationship with SST in all study area. It mean that warm water emit more longwave to the atmosphere than cold water. LW Up and MLD has strong relationship in area far from east equatorial Indian Ocean while weak correlation seen in east equator zone. SST and MLD which influence LW Up variability are driven by wind induce current that transport warm or cold water into or out of the study area.

Acknowledgement
This result is part of research activity "The Ocean Current System of Indonesian Waters and its effects on Marine Fisheries Production" funded by Research and Innovation, Bandung Institute of Technology 2017.

References

[1] Stewart S H 2008*Introduction to Physical Oceanography*(Texas: Departemen of Oceanography of Texas A & M University) chapter 5 pp 51 -72

[2] Siegel D A and Dickey T D 1986 Variability of net longwave radiation over the eastern north pacific ocean *J. Geophyisical Res.***91(C6)** 7657-66

[3] KeerthiM G, LengaigneM, VialardJ, de Boyer MontegutC and MuraleedharanP M 2012 Interannual variability of the tropical indian ocean mixed layer depth*Clim Dyn***DOI : 10.1007/s00382-012-1295-2**

[4] Shinoda Tand Hendon H H 1998 Mixed layer modeling of intraseasonal variability in the tropical western pacific and indian ocean *Journal of Climate,***11**, 2668-2685

[5] Donohoe A, Frierson D M W and Battisti D S 2013The effect of ocean mixed layer depth on climate in slab ocean aquaplanet experiments *Clim Dyn***DOI : 10.1007/s00382-013-1843-4**.

[6] de Boyer Montègut C, Madec G, Fischer A S, Lazar A and Iudicone D 2004 Mixed layer depth over the global ocean : an examination of profile data and a profile based climatology*J.Geophysical Res.***109** C120031-20

[7] Brainerd K E and Gregg MC1995 Surface mixed and mixing layer depths*Deep Sea Res.***9** 1521-43

[8] EmeryWJ and Thomson RE2001*Data Analysis Methods in Physical Oceanography 2nd* (Amsterdam: Elsevier) p243

Star formation rates in type 1 AGN host galaxies

Irham Taufik Andika[1], Mochamad Ikbal Arifyanto[1], Wolfram Kollatschny[2]

[1]Astronomy Study Program, Institut Teknologi Bandung, Jl. Ganesha No. 10, Bandung 40116, Indonesia
[2]Institut for Astrophysics, University of Goettingen, Friedrich-Hund-Platz 1, Goettingen 37077, Germany

E-mail: irham.andika@students.itb.ac.id

Abstract. We present a study of star formation rates of 3191 type 1 AGN host galaxies at $z < 0.35$, selected from Sloan Digital Sky Survey Data Release 12. We find that low ionization lines, [O II] $\lambda3727$, [N II] $\lambda6548$, [S II] $\lambda\lambda6717, 6731$ luminosities increase with increasing z while intermediate ionization lines, [O III] $\lambda5007$ luminosities does not increase with z, consistent with no evolution in AGN SED. We suggest the increasing luminosity of [O II] is caused by host galaxy contribution. The star formation rates that are estimated from [O II] luminosities show a flatter increase with z than non-AGN galaxies which indicates lower star formation activity across z in type 1 AGN host galaxies.

1. Introduction

The term of active galactic nuclei (AGN) is related to the compact region at the center of galaxy which emits stronger radiation than the rest of the host galaxy. The energy source of AGN is produced by accretion of gas onto supermassive black hole (SMBH; [1]) and then released in the form of radiation, winds, and/or jets. Circumnuclear gas at the center of AGN is photoionized by the energy and produces broad emission lines with typical width thousands of km s^{-1}, and narrow emission lines with typical width hundreds of km $s^{-}1$. The broad emission lines are believed to come from broad line region (BLR) while narrow emission lines are came from narrow line region (NLR). Type 1 AGN is a class of AGN which shows broad and narrow emission lines in their spectra.

There are several evidence that AGN and star formation activity are strongly related [2]. Most of AGN host galaxies in the low redshift universe contain significant amounts of molecular gas and dust. This makes nearby AGN host galaxies have the potential to undergo enhanced star formation. However, studies by [3] and [4] show that AGN activity somehow suppress star formation efficiency in these systems. NLR can be used as a proxy to study the property of gas in AGN host galaxies. It can be described as low-density and low-velocity gas extending from outer AGN torus up to thousands of parsecs along the general direction of ionization cones. This makes NLR emission not only driven by photoionization from accretion disk of AGN but also affected by the star formation activity in the host galaxy.

In this work we will investigate the NLR emission properties across a range of redshift by using new large sample of type 1 AGN from Sloan Digital Sky survey Data Release 12 [5]. Then,

IOP Conf. Series: Journal of Physics: Conf. Series **1127** (2019) 012037 doi:10.1088/1742-6596/1127/1/012037

star formation rates for each AGN host galaxies will be determined from NLR gas luminosity. We will also examine the correlation between star formation rates with redshift and compare the result with previous studies, including the case for non-AGN galaxies. Throughout this paper, we assume a Friedmann-Robertson-Walker cosmology with $\Omega = 0.3, \Lambda = 0.7, H_0 = 70$ km s^{-1} Mpc^{-1}.

2. Data selection and processing

The data are selected from Sloan Digital Survey (SDSS) Data Release (DR) [5] in order to obtain type 1 AGN sample at low redshift (z). Our search criteria are listed below:

(i) The spectrum of the objects is classified as quasi-stellar object or "QSO" by SDSS pipeline.

(ii) $0.02 \leq z \leq 0.35$, to cover the Balmer series (especially Hβ and Hα), optical forbidden narrow lines [O II], [O III], [N II], [S II], and optical Fe II complex in the spectral lines.

(iii) S/N > 10 in the spectrum based on median S/N of whole pixels covered by the spectrum (available in the SDSS database), to improve the accuracy of NLR/BLR decomposition.

A total 9184 spectra pass our criteria. The reduced one-dimensional spectral data are downloaded through SDSS Skyserver (http://skyserver.sdss.org/dr12/en). The data processing throughout this paper were done using custom *Python* codes. We use similar approach which as described by [6] with some modifications in continuum and emission lines fitting procedure. We separated the continuum windows into three parts: around [O II], Hβ, and Hα. For each region, we fit a local power-law continuum ($f_\lambda = A\lambda^{\alpha_\lambda}$), an iron template, and host galaxy spectra to the wavelength region around the line that is not contaminated by the emission line.

We compare our results with the $L_{[\mathrm{O\ III}]}$ measurements of SDSS AGN in [7], who model [O III] using multiple Gauss-Hermite (GH) functions. In the 2657 overlapping objects, the mean difference of our $L_{[\mathrm{O\ III}]}$ to their $L_{[\mathrm{O\ III}]}$ is 0.1 dex with dispersion of 0.08 dex. A total of 4293 objects with acceptable fits (reduced $\chi^2 < 2$) pass these criteria. We note that our rejection criteria may have been too restrictive, and potentially excluded objects with an unusual broad line profile.

We utilize the diagnostic diagram to classify our sample and exclude objects other than AGN [8]. Therefore, our sample now contains 3191 type 1 AGN (T1 AGN) which reside at $42.4 < \log L_{5100} < 45.3$ with median $\log L_{5100} = 43.9$, where L_{5100} is averaged continuum luminosity at rest wavelength 5100 Å.

3. Result and analysis

3.1. Narrow lines luminosity and redshift

In this subsection, we will investigate the NLR emission across a range of redshift. First, we note that SDSS data is likely to miss high luminosity objects in low redshift and low luminosity objects in high redshift. This was expected because the flux limit of SDSS for AGN spectroscopy is $i < 19.1$ and the evolution of AGN luminosity function. To overcome this problem, we group T1 AGN according to their $\log L_{5100}$ into 10 luminosity classes ($L1 - L10$). We thus divide eeach group into several subsets according to their redshift (z) with bins of width 0.03. We only consider the subset that has more than 5 objects to get significant result. Based on this criteria, we only consider $L4 - L7$ luminosity bins which spans more than 50% z range of $0.02 - 0.35$. We will treat each bin separately and then average the measured correlation parameter.

There are four strongest forbidden narrow lines that we measured in T1 AGN spectral range: [O II] λ3727, [O III] λ5007, [N II] λ6548, and [S II] $\lambda\lambda$6717, 6731. We plot the luminosity of these lines against z to find if there is a correlation in Figure 1. We can see overall tendency of increasing L_{NLR} as z increases. To examine this quantitatively, we calculate the Spearman rank

IOP Conf. Series: Journal of Physics: Conf. Series **1127** (2019) 012037 doi:10.1088/1742-6596/1127/1/012037

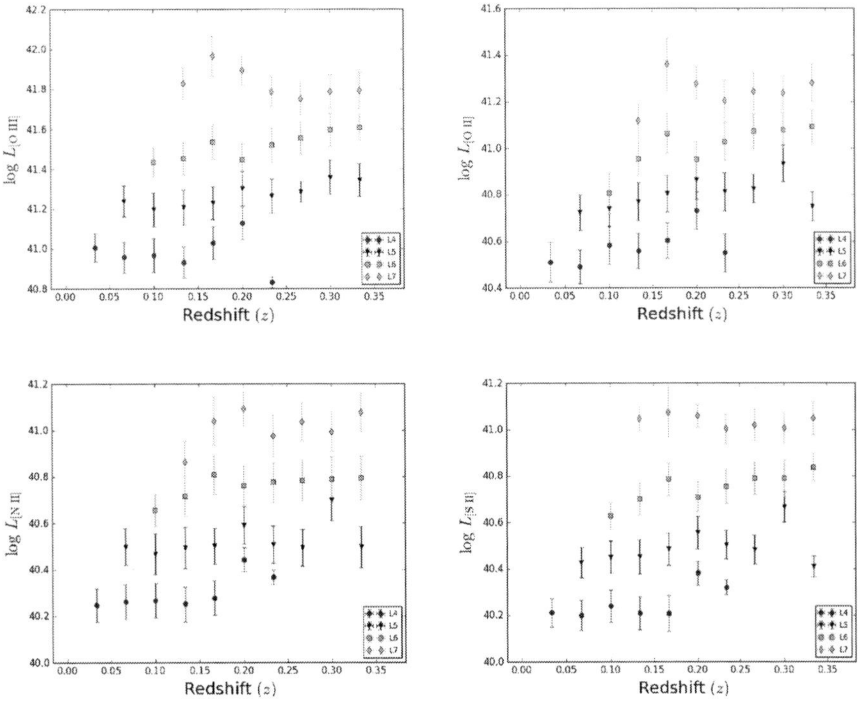

Figure 1. The line luminosity versus redshift of the four narrow lines. In the each luminosity bin, the figure shows a clear increase in luminosity of each line towards higher redshift. The dots associated error bars mark the mean and the standard deviation of λL_λ in each redshift bin.

Table 1. Overall result of the linear regression and the Spearman rank correlation coefficient for the lines luminosity and redshift for the four narrow lines.

Line	Slope (m)	Constant (c)	Std. Error	ρ	P
[O III]	0.129	41.374	0.300	0.250	0.252
[O II]	0.565	40.800	0.339	0.567	0.229
[N II]	0.531	40.552	0.289	0.526	0.260
[S II]	0.387	40.566	0.231	0.342	0.158

correlation coefficient and lope in the Table 1 which is derived from least-squares regression in the form of equation:

$$\log L_{\rm NLR} = {\rm m}z + {\rm c} \quad (1)$$

As shown in Figure 1, the strongest correlation is found in $L_{[\rm O\ II]}$ versus z with Spearman correlation coefficient of $\rho = 0.567$ with $P = 0.299$ (P is the probability of no correlation). A weaker correlation was found in [N II] ($\rho = 0.526$, $P = 0.260$) and [S II] ($\rho = 0.342$, $P = 0.158$). Basically, [O II], [N II], and [S II] have low ionization potential (IP). In the idea of stratified NLR, these lines are located in the outer region of NLR. On the other hand, [O III] which has higher IP and located in the inner region of NLR has weakest correlation ($\rho = 0250$, $P = 0.252$).

The slight increase in the luminosity of these narrow lines might be result of the host galaxy contribution which undergoes enhanced star formation rates (SFR) at higher z. [O III] line that is located closest to the center of AGN will be strongly influenced by AGN radiation and less affected by host galaxy contamination. This was expected as we found the weak correlation in [O III]. The strong corrlations which are found in [O II], [N II], and [S II] emissions show that these lines are more affected by the host galaxy contamination even while they are still driven by AGN photoionization.

The apparent increase in $L_{[\mathrm{O\ II}]}$ as a function of z could be also caused by different factors other than an increase in SFR of typical AGN host galaxy. For example, the 3-arcsec fixed-size aperture that is used by SDSS means the spectra potentially include a larger fraction of the host galaxy gas at high redshift than in lower redshift objects due to the larger projected physical size at higher redshift. More specifically, at $z = 0.02$ the physical size is 1.2 kpc and at $z = 0.35$ the physical size is 14.8 kpc. However, the typical size of entire NLR is probably less than 10 kpc, as found by [4]. This means apparent increase in $L_{[\mathrm{O\ II}]}$ across z can not be caused solely by AGN photoionization.

3.2. Star formation rates and redshift

There are various methods to measure star formation rate (SFR) in galaxies. One of these methods is using [O II] line as reliable tracer. This line becomes particularly important for objects with $z > 0.5$ because the primary SFR tracer of Hα has passed out of the optical bandpass. Unlike Hα, the [O II] line luminosity is not only influenced by electron temperature but also the degree of ionization and metallicity of the gas.

Assuming that [O III] is produced entirely by the AGN and is a good tracer of AGN bolometric luminosity as suggested by [5], we can use the $L_{[\mathrm{O\ III}]}$ to constraint $L_{[\mathrm{O\ II}]}$ contribution from star-forming region in the host galaxy. To estimate the SFR using [O II], we use $L_{[\mathrm{O\ II}]}$ calibration equation from [9] for non-AGN galaxies:

$$\mathrm{SFR}_{[\mathrm{O\ II}]} = 8.4 \times 10^{-41} L_{[\mathrm{O\ II}]}\ \mathrm{M}_\odot\ \mathrm{yr}^{-1} \tag{2}$$

This calibration equation is useful because it simplifies SFR calculation without directly involving assumptions about metallicity and intrinsic extinction that causes additional uncertainty. We first subtracted 10% of the $L_{[\mathrm{O\ III}]}$ from the $L_{[\mathrm{O\ II}]}$ to account the AGN contribution. This way we can estimate the excess of $L_{[\mathrm{O\ II}]}$ contributed by star-forming region of host galaxy and calculate the upper limit SFR of T1 AGN.

We plot the result in Figure 2 which shows clearly that the SFR of T1 AGN increases with increasing z. The figure also shows that higher luminosity objects have higher SFR as expected from higher $L_{[\mathrm{O\ II}]}$ in higher L_{5100}. Our least-squares regression shown in the figure with a slope of m $= 2.02 \pm 0.7$ which is notably consistent with [4] fit (m $= 2.3 \pm 0.4$) but somewhat smaller than the [9] fit (m $= 4.5$).

Study of star formation in AGN host galaxies using [O II] line also has been conducted by [3]. From a review of extant literatures, [3] concludes that SFR in type 1 AGN host galaxies are quiet modest, being no greater than $\sim 1 - 10$ M$_\odot$ yr^{-1}. By using CO measurements, [3] also proved that the star formation efficiencies (SFR per unit gass mass) are also low, suggesting that the AGN somehow inhibits star formation. Another research was conducted by [2] who conducted detailed analysis of the narrow line spectrum of type 1 AGN. [2] also use [O II] line to constrain the ongoing SFR in the host galaxies and compared the measurement with a new set of photoionization models. They found [O II] line can be readily reproduced using conventional AGN parameters. This places strong constraints on any additional star-forming contribution to the [O II] luminosity. Their result also showed that the host galaxies of type 1 AGN evidently experience very modest star formation.

Therefore, we suggest that the smaller power-law slope of SFR we found compared with [9] indicates smaller star formation activity across z in the type 1 AGN host galaxies compared

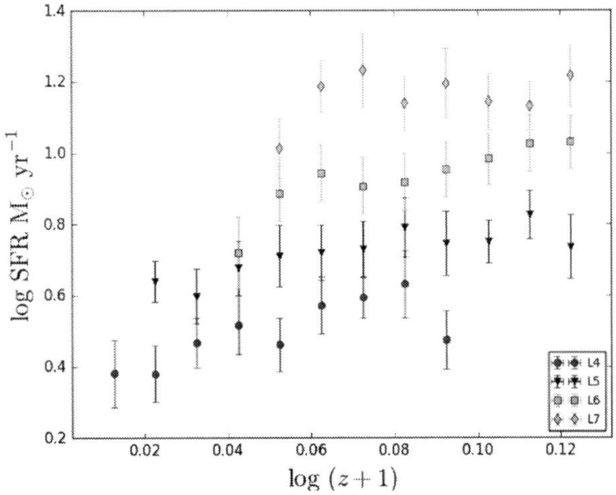

Figure 2. SFR calculated using Equation 2. The trend of increasing SFR towards higher redshift is clearly seen in the figure. The red line shows average linear regression of the luminosity groups.

with normal galaxies, at least up to $z = 0.35$.

4. Conclusion

In this paper, we derive a well-selected sample of type 1 AGN at low redshift using large sample SDSS DR 12 data to get the optical spectra of the sample. Spectral reduction and analysis to optical spectra carried out in order to derive emission lines parameters. We use the measured emission lines parameters to explore correlation and properties of type 1 AGN emission.

The luminosity of forbidden narrow lines increase with redshift. The strongest correlation was found in [O II] line. Because the [O II] line is known to be a reliable SFR estimator in normal galaxies, we interpret this result as evidence for higher SFR in the host galaxies of AGN at higher redshift. We calculate the SFR of T1 AGN sample by using the $L_{[\mathrm{O\ II}]}$ scalling relation and find that the estimated SFR is increasing with redshift. The calculated power-law slope that we found ($\mathrm{m} = 2.02 \pm 0.7$) is consistent with [5] but smaller than non-AGN galaxies ($\mathrm{m} = 4.5$; [9]). Supported by [3] and [2] findings, we suggest that this smaller slope indicates smaller star formation activity across z in type 1 AGN host galaxies compared with normal galaxies, at least up to $z = 0.35$.

References

[1] Netzer H., 2013, The Physics and Evolution of Active Galactic Nuclei
[2] Kim M. Ho Luis C., Im M., 2006, ApJ, 642, 702
[3] Ho Luis C., 2005, ApJ, 629, 680
[4] Tammour A., Gallagher S. C., Richards G., 2015, MNRAS, 448, 3354
[5] Alam S., et al., 2015, ApJS, 219, 12
[6] Shen Y., et al., 2011, ApJS, 194, 45
[7] Stern J., Laor A., 2012b, MNRAS, 426, 2703
[8] Kewley L. J., Dopita M. A., Sutherland R. S., Heisler C. A., Trevena J., 2001, ApJ, 556, 121
[9] Rosa-Gonzlez D., Terlevich E., Terlevich R., 2002, MNRAS, 332, 283

IOP Conf. Series: Journal of Physics: Conf. Series **1127** (2019) 012038 doi:10.1088/1742-6596/1127/1/012038

Temporal Earth Coorbital Types of Asteroid 2016 HO3

B Dermawan

Astronomy Research Division, Faculty of Mathematics and Natural Sciences, Bandung Institute of Technology, Jalan Ganesha 10, Bandung 40132, Indonesia

E-mail: budider@as.itb.ac.id

Abstract. A newly discovered asteroid 2016 HO3 has coorbital type of retrograde (or quasi) satellite with respect to Earth. The discovery enlarges number of known retrograde satellites of Earth. In this work numerical studies have been performed to investigate orbital dynamics of the asteroid by including all planets and the Moon as perturbers. Hypothetical asteroids were also generated to assess its general dynamics. Results show that since several hundreds thousands years ago the asteroid has been trapped in alternating coorbital types of retrograde satellite (RS) and horseshoe (HS). The RS and HS types, respectively, have taken place for some hundreds and some thousands years, and so those for the future. Long term orbital evolution for one million years predicts that about a few hundreds thousands years from the epoch the asteroid will completely escape its coorbital configuration with Earth. Orbit of the asteroid does not follow Kozai resonance state, and by this study it is found that asteroid 2016 HO3 may behave a coorbital pair with asteroid 2000 WN10.

1. Introduction

A coorbital asteroid has almost same orbital period with that of the corresponding celestial object (planet or another asteroid). This is an orbital 1:1 mean-motion resonance. With respect to rotating coordinate system, there are three basic coorbital types of such orbit, i.e. Tadpole or Trojan (T), Horseshoe (HS), and Retrograde (or quasi) Satellite (RS). Moreover, there are transient coorbital types which are a mixture of two out of the three types [1]. It has been known that Earth harbours asteroids whose orbits belong to all the three types. It has been reported elsewhere, e.g. [2], that Earth has at least four RS companions, i.e. 2004 GU9, 2006 FV35, 2013 LX28, and 2014 OL339. A recent discovered RS companion, asteroid 2016 HO3 (Fig. 1), enlarges this number.

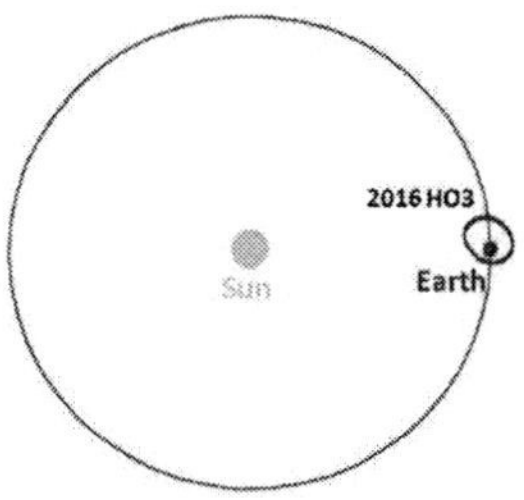

Figure 1. Orbit of a retrograde satellite 2016 HO3 (encircling Earth) in the Sun-Earth rotating coordinate system.

In this work we study orbital dynamics of asteroid 2016 HO3. The dynamics are explored numerically for both short and long term integrations. In more details, three investigations can be probed by the integrations, i.e. the asteroid may show other coorbital types beside RS; the asteroid

IOP Conf. Series: Journal of Physics: Conf. Series **1127** (2019) 012038 doi:10.1088/1742-6596/1127/1/012038

may be trapped in Kozai resonance state; and the asteroid may probably belong to coorbital pair such as the well-known Janus and Epimetheus in Saturnian system.

2. Orbital Elements and Integration Scheme

Orbital elements of asteroid 2016 HO3 (see Table 1) and all planets (including the Moon) are available at the JPL-NASA Solar System Dynamics (`http://ssd.jpl.nasa.gov/horizons.cgi`). Recent available epoch was MJD 57600 (2016 July 31, 00:00:00 UT).

Table 1. Orbital elements of asteroid 2016 HO3 for MJD 57600.

Elements	Value
Semimajor axis (a)	1.001229935 ± 2.76E-9 au
Eccentricity (e)	0.1041429 ± 4.80E-7
Inclination (i)	7.77140 ± 3.62E-5 deg
Longitude of ascending node (Ω)	66.51326 ± 4.33E-5 deg
Argument of pericenter (ω)	307.22765 ± 6.68E-5 deg
Mean anomaly (M)	297.53211 ± 9.70E-5 deg

There are two derived parameters to analyze coorbital motion. First, the relative distance $\alpha = (a - a_E)/a_E$, and second, the libration angle $\theta = \lambda - \lambda_E$, with $\lambda = (\Omega + \omega + M)$; subscript E denotes Earth. Because of the 1:1 mean-motion resonance, value of α should be small. On the other hand, RS companion follows $\theta \approx 0°$, which is a complement of HS type ($\theta \neq 0°$), and T type stays for $\theta \approx \pm 60°$.

Two integration packages were used in this work. *Mercury* package [3] was applied for short term backward and forward integrations of some tens thousands years. The other package, *Swift* [4] ver. 4 (released in year 2012), was employed for long term integrations of one million years, including integrations for hypothetical asteroids. Hypothetical asteroids were generated by combining available uncertainties of the asteroid's six orbital elements to finally yield 3^6 (= 729) objects. The hypothetical objects represent a collection of possible orbits of the asteroid. Step of the integrations were chosen to be 1/100 (short term) and 1/200 (long term) year, which are believed to be sufficient. All planets and the Moon behave perturbers for orbit of the 'massless' asteroid/objects along the integrations.

3. Results and Discussion

3.1. Short term integrations

Figure 2 shows results for short term backward and forward integrations, respectively, of twenty thousands years. It is obvious that asteroid 2016 HO3 has temporal RS and HS coorbital types with relatively short time (several hundreds years) as RS companion.

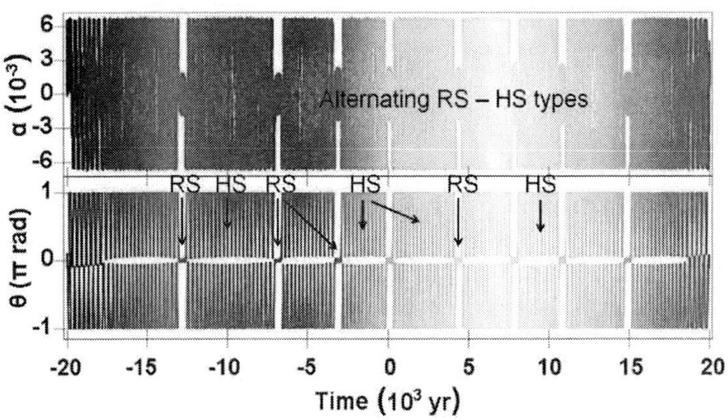

Figure 2. Temporal coorbital RS-HS behavior of asteroid 2016 HO3 along ± twenty thousands years. RS and HS types, respectively, occur in about some hundreds and some thousands years,.

3.2. Long term integrations and Kozai resonance state

Long term orbital integrations show that for several hundreds thousands years from the epoch, orbit of asteroid 2016 HO3 will shrink into inner Earth region (for example, see top panel of Figure 3). More than 60% orbits of the hypothetical asteroids demonstrate roughly the same trend. In line with this, the asteroid will depart from its coorbital configuration after that time because of the cumulative gravitational attractions of the perturbers.

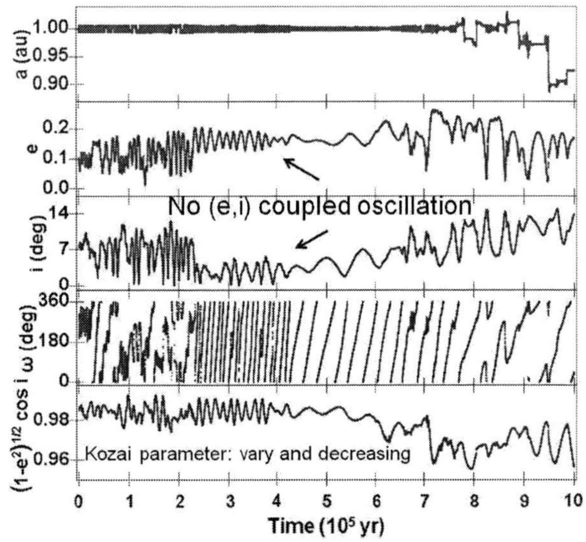

Figure 3. An example of long term integration of asteroid 2016 HO3. No long libration type has beed arised in panel ω (second panel from the bottom).

Regarding long term integrations, it is suitable to investigate Kozai (Kozai-Lidov) resonance state of asteroid 2016 HO3. Orbit of an asteroid is expected to be in Kozai resonance state if the orbit has steady semimajor axis and falls to some criteria at once: i. The Kozai parameter ($(1 - e^2)^{1/2}$ cos i) does not vary appropriately for an order of $10^5 - 10^6$ years; ii. There is a coupled oscillation of e and i; iii. Librations of ω occur at ±60° or 0°/180°. Regarding these criteria it is clear that orbit of asteroid 2016 HO3 does not follow Kozai resonance state. This is in accordance with a new published result [2].

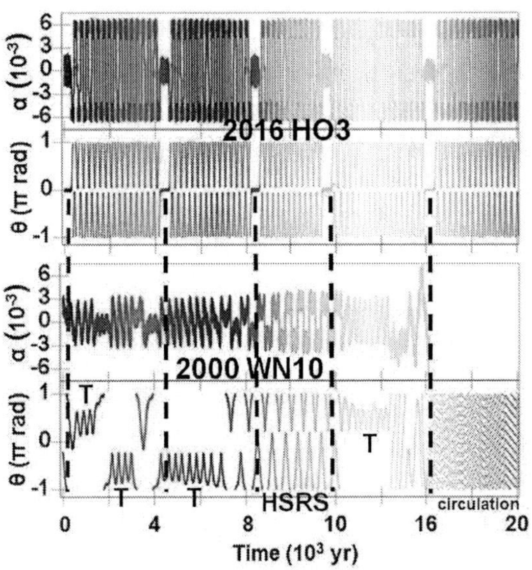

Figure 4. A possible play of two coorbital asteroids as a pair. Each of the change of coorbital type in orbit of 2016 HO3, the corresponding change (marked by vertical dashed lines for clarity) is also happened in orbit of 2000 WN10.

3.3. Earth coorbital pair?
It is interesting to search a coorbital pair among the known Earth coorbital asteroids. A coorbital pair is a pair of two asteroids that show a coupled of two coorbital types. A famous example of this phenomenon is an orbital 'dancing' between Janus and Epimetheus in Saturnian system. Figure 4 illustrates this possibility that the changes of RS-HS types of 2016 HO3 are coincidence in time with those of 2000 WN10 that comprise all coorbital types, including the transient one (a mixture of HSRS). This finding needs a deeper inspection to a bulk of possible orbits of both asteroids.

3.4. New Earth RS-HS coorbital asteroid
By this study we find another (possibly) Earth coorbital asteroid (2016 CA138) that shows alternating RS-HS pattern (Fig. 5) in its orbital dynamics. To our knowledge this may be the fourth asteroid belongs to that pattern of orbit; the other three are 2004 GU9, 2015 SO2, and 2016 HO3.

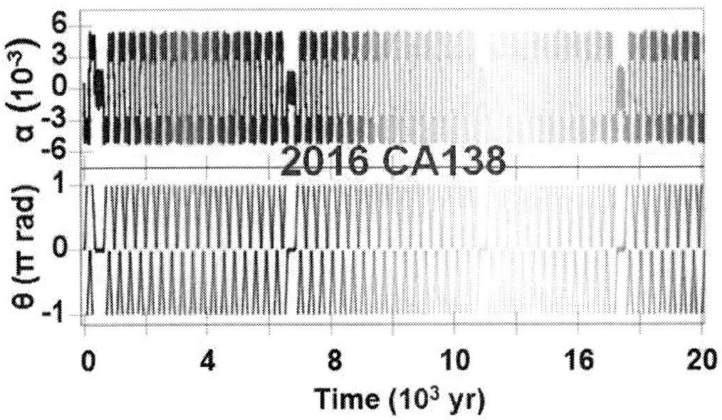

Figure 5. Orbital dynamics of coorbital asteroid 2016 CA138. Alternating RS-HS pattern is clearly shown for at least twenty thousands years.

4. Conclusions
- Asteroid 2016 HO3 will completely escape its coorbital configuration after some hundreds thousands years. The asteroid shows that RS-HS coorbital types occur, respectively, in hundreds and thousands years.
- Orbit of asteroid 2016 HO3 does not in Kozai resonance state.
- Asteroid 2016 HO3 may be a coorbital pair with asteroid 2000 WN10 for at least twenty thousands years.

Acknowledgments
Faculty of Mathematics and Natural Sciences ITB and ICMNS committee are highly acknowledged.

References
[1] Namouni F 1999 *Icarus* **137** 293
[2] de la Fuente Marcos C and de la Fuente Marcos R 2016 *MNRAS* **462** 3441
[3] Chambers J E 1999 *MNRAS* **304** 793
[4] Levison H F and Duncan M J 1994 *Icarus* **108** 18

Study on water mass exchange at Ambon Bay using trajectory model: circulation of one tidal cycle

S Nurfitri and M R Putri

Research Group of Oceanography, Faculty of Earth Sciences and Technology, Bandung Institute of Technology, Jalan Ganesha 10, Bandung 40132, Indonesia

Abstract. Ambon Bay is divided into two parts, Inner Ambon Bay (IAB) and Outer Ambon Bay (OAB) which separated by 10 m narrow canal called sill. It causes IAB and OAB has different water dynamics. Generally, circulation dynamics at Ambon Bay affected by upwelling process in the Banda Sea during southeast monsoon. During that period, tidal upwelling brought high-density water from thermocline layer in OAB to surface and enter the IAB. The mechanism of water mass exchange can be described using a three-dimensional trajectory model which evaluate the origin and movement of water particles. The particle movement during one tidal cycle based on model results are observed. Verification result shows that the model error is 2.32%. After one tidal cycle, particles predominantly move to OAB rather than enter IAB through the western side of the sill.

1. Introduction

Ambon Bay has unique geographical conditions where there is 10 m sill that divides the bay into two parts, Outer Ambon Bay (OAB) and Inner Ambon Bay (IAB) [1]. The sill causes both bays to have different water dynamics. Water circulation in IAB is relatively closed that causes the waste that goes into it would be trapped inside near the sill [2]. One of the factors that affect the circulation in Ambon Bay is upwelling process in the Banda Sea during southeast monsoon. During that period, water mass exchange in Ambon Bay is the most actively occurred [3, 4].

Study on water dynamics of Ambon Bay has been conducted by Wenno and Anderson (1984) [1]. The result stated that the current circulation in the bay predominantly driven by tides. In addition to tides, the current is also driven by density gradient based on the study of Corvianawatie (2014) using three-dimensional hydrodynamic model [5]. The vertical density profile shows that there was a denser water in OAB that moved to IAB during flood tide. When ebb tide, it moved back to OAB, but the small portion of it was left behind in IAB. However, the process of water mass exchange between OAB and IAB could not be explained only by the density profile. The path and source of water mass can be identified using the trajectory model that reviewed the origin and movement of water particles.

2. Methodology

2.1. Trajectory model principle

The principle of the trajectory model is based on the Lagrange approach. In this approach, the measured parameter is the particle displacement, because the position of a particle is a function of time [6]. Particle displacement is formulated as follows [6]:

$$x(t+dt) = x(t) + u \cdot dt$$
$$y(t+dt) = y(t) + v \cdot dt \quad (1)$$
$$z(t+dt) = z(t) + w \cdot dt$$

Where x, y, and z denote the position of a particle in west-east, south-north, and vertical (depth) direction, u, v, and w symbol show particles velocity. The equation above stated that particles position at time $t+dt$ is obtained from the particle position at t plus the particle displacement during dt (time interval).

2.2. Model design and input

Three-dimensional trajectory model used in this study was adapted from Kampf (2010) [7]. It is a baroclinic model which built in Arakawa C grid system [5, 7]. Simulation of particle transport (trajectory) 3D was covered Ambon Bay area at 128.12°-128.24° E and 3.63°-3.68° S. The model area in x and y-direction has a length of 13.98 km and 6.66 km respectively. Model domain is divided into 63 and 30 grid in x and y-direction respectively and 222 m x 222 m of horizontal resolution [5]. The depth was divided into 20 layers with varying of grid size [5]. The simulation was performed using time interval of 60 seconds for 24 hours at neap (August 13, 2008) and spring tide (August 25, 2008).

At the initial condition, 4948 particles are placed in each grid of the model domain. Moreover, for the boundary condition, the closed boundary was applied. If particles reach the mainland, its velocity became zero. On the other hand, if particles move out of the model domain through the open boundary, then specific conditions are applied. If at time step n+1 particle position exceeds the model domain, the position of it will be returned to the position at n.

The input data for 3D (three-dimensional) trajectory model is current velocity i.e. u, v, and w which obtained from 3D hydrodynamics model conducted by Corvianawatie (2014) [5]. The driving forces of the hydrodynamics model were tides and seawater density gradient. Tides input was derived from TMD (Tide Model Driver) which consists of 8 tidal component, i.e. M2, S2, K1, O1, N2, P1, K2, and Q1. Momentum equation used in the model was consist of local and convective acceleration, pressure gradient force, and viscus force, while Coriolis acceleration and river discharge are neglected. In addition to that, the hydrodynamic model is also assimilated with density data from observation at two points every 24 hours [5]. Current velocity every 2 hours obtained from hydrodynamics model was used as input of trajectory model which it was linearly interpolated every simulation time step. The particle movements only drive by advection process, while the diffusion process was neglected.

3. Results and analysis

3.1. Model verification

Model verification was done using a mathematical method. Drifting distance of three particles during 24 hours based on model results are compared with the manual calculation from current velocity. The model error is obtained by calculating the RMSE (Root Mean Square Error) and divided by the total drifting distance of model results as shown in Table 1.

Table 1. Model error based on mathematical verification.

Particle	Initial location	Error (%)
1	128.143° E and 3.676° S	1.44
2	128.183° E and 3.676° S	2.54
3	128.193° E and 3.660° S	2.99

On the other hand, verification results of particle 2 indicate that drifting distance obtained from the model is smaller than calculation. In contrast with particle 2, the drifting distance of particle 3 from the model is larger than manual calculation. Error model of particle 2 is 2.54%, while particle 3 is 2.99%. Based on verification results, it can be concluded that model calculation tends to overestimate the drifting distance with an average error of 2.32%.

3.2. Tides in Ambon Bay

Tide is the dominant driving force of current in Ambon Bay [8]. Based on previous studies, tidal type in Ambon Bay is mixed predominantly semidiurnal with a dominant component of M2. According to Stewart (2008), the period of semidiurnal tidal component M2 is around 12 hours 25.24 minutes [9]. Based on the period, it takes about 24 hours 50.47 minutes to produce 1 semidiurnal tidal cycles (2 times of high water and 2 times of low water) as shown in Figure 1. The graph in Figure 1 shows that in spring condition, it takes around 25 hours to produce one complete tidal cycle. On the contrary, at neap tide, the tidal cycle tend to be diurnal (1 time of high and 1 time of low water) as shown in Figure 1.

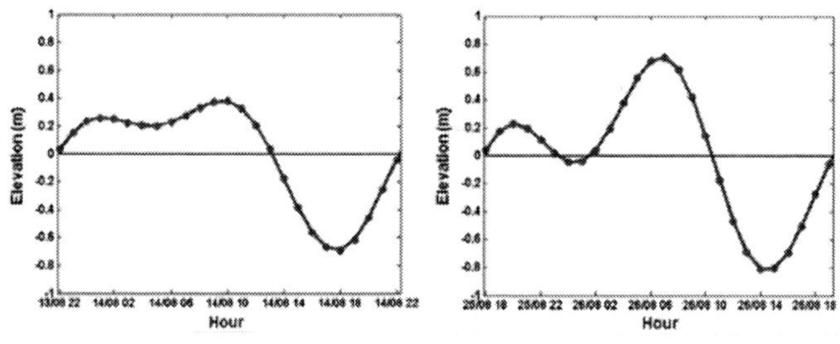

Figure 1. Tides elevation at neap (left) and spring (right) condition obtained from hydrodynamics model.

3.3. Particle trajectory on one tidal cycle

Effect of daily tidal pattern to water mass transport in Ambon Bay can be described by the particle movement during one tidal cycle of neap and spring tide. Simulation result for one neap tidal cycle in August 2008 shows that particles in OAB spread to west and east. The particle from eastern OAB (red) spread to west and east side, while particle from western OAB (orange) dominantly moved to the southwest side. On the contrary, particle in eastern IAB tends to remain in its original position (Figure 2).

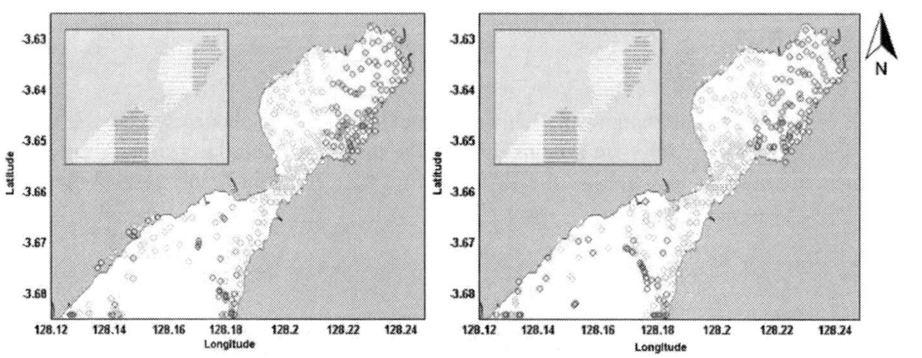

Figure 2. Surface particles distribution after one tidal cycle simulation when neap (left) and spring (right) condition. Initial position of particles shown in red box.

In spring condition, it appears that the particles from eastern OAB (red) spread to west and east side (Figure 2). Similar to them, particles from western OAB (orange) also spread to the west and east side of the bay. Moreover, particles from the sill (green) tend to congregate on the east side which small portion of it entered IAB. Different from OAB, particles from IAB tend to remain inside the bay.

There are some of the particles from western IAB (light blue) which move out to OAB, whereas particles from the eastern side (purple) remain in IAB. These conditions show that circulation in IAB tend to be isolated so that the particles are difficult to pass through the sill.

In order to investigate the circulation pattern in Ambon Bay, the path-lines of five particles were observed. Two particles were released at IAB, two at OAB, and one particle at narrow canal (sill). After one neap tidal cycle, particles at IAB (purple and light blue) move to southwest with average current magnitude of 0.02 m/s and 0.045 m/s respectively. Particles at eastern OAB (red) also move to southwest with an average current magnitude of 0.13 m/s. On the other hand, the particle at sill (green) move to the southwest and then turned back into IAB with average current magnitude of 0.024 m/s (Figure 3).

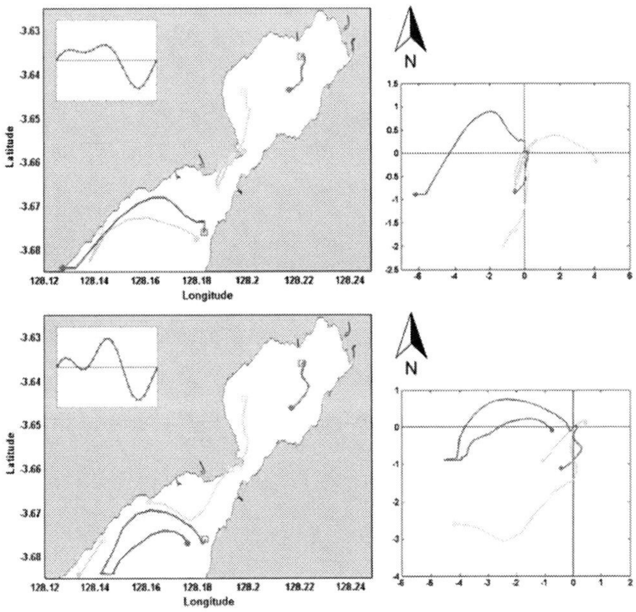

Figure 3. Surface particle path-lines (left) and tracking graph (right) after one tidal cycle simulation when neap (above) and spring (below) condition. Square and dot symbols show initial and final position; x and y-axis show west-east and south-north direction in kilometers.

Different path-lines pattern were found in spring condition. Almost all of five particles move to the southwest on one spring tidal cycle as shown in Figure 3. Based on the graph, four of the particles movement did not create the closed path. The near-closed path was found in particle from eastern OAB (red) which move to the west and then turned back to its starting point. Based on the path-lines pattern, it appears that the entrance and exit way of particles from IAB in through the western side of narrow canal or sill.

Tidal type when the neap condition tends to be diurnal where there is one time of high and low water. Theoretically, these tidal type will produce a tidal stream leading to the bay when flood and coming out the bay when ebb tide [8]. It will cause the particles to move into the bay when flood and out the bay when ebb, so that create a closed path-line. However, this pattern could not be seen from particles path-lines either in neap and spring tide in August 2008. It could be caused by diurnal inequality of tides. The diurnal inequality means that the duration of flood and ebb tides are different. Furthermore, it can be seen that flood and ebb tide duration in spring condition relatively equal compared to the neap condition because the near-closed path-line was found in the spring tide. The drifting distance of particles when spring tide also further than neap condition, because current velocity is faster in spring than neap.

In addition to tides, the morphology of the bay also affected the particle movement, particularly at OAB. When flood tide, particles from OAB are difficult to enter the IAB because the narrow canal or sill obstruct their movement. The particles will be deflected back to OAB as shown in path-line of particle orange and red.

4. Conclusion

Verification results show that an average model error is 2.32%. Either on one neap and spring tidal cycle, surface particles dominantly move from inner bay to outer bay through the western side of the sill. It also can be seen from the surface particles path-lines that the drifting distance when spring is further than the neap tide.

5. References

[1] Wenno L and Anderson J J 1984 *Mar. Res. Indones.* 23 13-20
[2] Tarigan M S and Edward 2000 *Coastal and Beach of Indonesia IV* (Jakarta: LIPI)
[3] Wyrtki K 1961 *Naga Report Vol. 2 Scientific Results of Marine Investigation of the South China Sea and the Gulf of Thailand 1959-1961* (California: The University of California)
[4] Anderson JJ and Sapulete 1981 *Proc. 4th Int. Coral Reef Symp.* (Manila) p 360-374
[5] Corvianawatie 2014 *Water Mass Exchange Mechanism in Ambon Bay Based on Density Assimilation Model* (Bandung: Bandung Institute of Technology)
[6] Von Schwind J J 1980 *Geophysical Fluid Dynamics for Oceanographers* (Engelwood Cliffs, N J: Prentice-Hall)
[7] Kampf J 2010 *Advanced Ocean Modelling Using Open-Source Software* (Berlin: Springer)
[8] Hamzah M S and Wenno L F 1987 *Ambon I: Biology, Fisheries, Oceanography, and Geology* (Ambon: LIPI) p 91-101
[9] Stewart R H 2008 *Introduction to Physical Oceanography* (Texas: Texas A & M University)

IOP Conf. Series: Journal of Physics: Conf. Series **1127** (2019) 012041 doi:10.1088/1742-6596/1127/1/012041

Impact of surface data assimilation using an ensemble Kalman filter technique on the improvement of weather prediction

N J Trilaksono[1], M Taqiyya[2,3], N Dewani[4,5], I D G A Junnaedhi[1], E Riawan[1], and R Fitriani[4]

[1] Atmospheric Science Research Group, Faculty of Earth Sciences and Technology, Institut Teknologi Bandung, Jalan Ganesha 10, Bandung, ID
[2] Meteorology Program, Institut Teknologi Bandung, Jalan Ganesha 10, Bandung, ID
[3] International Graduate Program for Advanced Science, Tohoku University, 6-3 Aramaki Aza Aoba Sendai, JP
[4] Earth Sciences Program, Institut Teknologi Bandung, Jalan Ganesha 10, Bandung, ID
[5] UPC Renewables Indonesia, Sunset Road No. 29 C, Kuta Bali, ID

E-mail: jpatiani@meteo.itb.ac.id

Abstract. Initial condition data is known to be an important factor, which contributes to the accuracy of a weather prediction. One of an attempt to improve the initial condition is by applying data assimilation method. Ensemble Kalman Filter (EnKF) is a sophisticated data assimilation method that has shown great promise for atmospheric data assimilation in a way that the method uses flow-dependent error statistics estimated from relatively short period of prediction. Data assimilation of surface observations using an EnKF technique is evaluated for the potential improvement of a regional nonhydrostatic model. We use Weather Research and Forecasting - Advanced Research WRF (WRF-ARW) model to implement the EnKF data assimilation system.

The EnKF system combined with the WRF-ARW model on a 1-km horizontal grid spacing over Karimunjawa Island is cycled over a field course period from 14–18 August 2015. The analysis ensemble is generated every 6 h by assimilating surface observations from three different locations of automatic weather stations (AWSs). By assimilating the surface observation data, the models produce colder temperature over the sea near Karimunjawa Island compare to those without assimilation process. Thus, the assimilation data give a stronger and more defined sea breeze circulation because the circulation generated by temperature difference between the sea and land.

1. Introduction

Sea breeze is a coastal local wind that blows from sea to land, caused by the temperature difference between the sea and the adjacent land. The sea breeze has been recognized for centuries throughout the world and extensively studied using observational data and model simulation (e.g., [1][2]). In the tropics, which is characterized by a prominent feature of diurnal cycle of precipitation, the sea breeze is associated with the maximum peak of precipitation over land in the afternoon [3][4]. Although the important role of the sea breeze has been relatively well understood, there is still difficulty in forecasting the strength and correct phase of the phenomenon. Many of numerical studies of the diurnal cycle related to the sea breeze have reported shortcomings in reproducing the phase and amplitude of the phenomenon [5][6].

On the other hands, initial condition data for model simulation is one of the important aspect, which contribute to the accuracy of the model output. However, it also contributes to the source of error in the model simulation, whereas this problem has already recognized very long time ago by Bjerknes [7]. To provide better initial condition, data assimilation method can be employed to obtain the best estimate of state of the atmosphere [8]. Meanwhile, ensemble-based data assimilation system, such as the ensemble Kalman filter (EnKF) has shown to give reliable results for atmospheric data assimilation because the system use flow-dependent error statistics estimated from short-term ensemble forecasts [9].

This study describes the performance of an EnKF system over the sea breeze circulation over a small island. The primary goal of this study is to evaluate the use of automatic weather station data in improving daily weather prediction. This paper proceeds as follows. Section 2 describes the modeling and data assimilation system, while section 3 provides the results of the experiments. A summary and discussion are provided in section 4.

2. Experimental Design

In this study, ensemble analyses are generated every 6 h by cycling an EnKF from 06 UTC 15 to 00 UTC 16 August 2015. Figure 1 shows the computational domain, which consists of three nested domains. The outer domain has 192×130 grids with a grid spacing of 20 km, centered at 6.38°S, 110.15°E on a Mercator projection. While the two inner domains are nested with the grid spacing ratio of 3, thus they have 3 and 1 km grid spacings, respectively. All ensemble members are advanced in time using the Advanced Research Weather Research and Forecasting model (ARW-WRF, hereafter WRF; [10]) version 3.5.1 with 27 vertical levels. The WRF subgrid-scale parameterizations and their parameters follow the work by [11]. The National Centers for Environmental Prediction (NCEP) Global Forecast System (GFS) with the horizontal resolution of $0.25° \times 0.25°$ and the time interval of 3 h are used for both model initial and boundary data.

Ensemble initial and lateral boundary conditions are generated by drawing random perturbations from the NCEP error covariances contained in the WRF VAR system. Observation data from three Automatic Weather Stations (AWSs) sited in the Karimunjawa Island, — site 1: (5.885764°S, 110.4457°E), site 2: (5.838771°S, 110.421465°E), and site 3: (5.772914°S, 110.484846°E)—are assimilated using the Data Assimilation Research Testbed (DART; [12]), which is an implementation of the ensemble adjustment Kalman filter. Only three variables from each AWS are included in the assimilation process, namely u-wind and v-wind component and temperature.

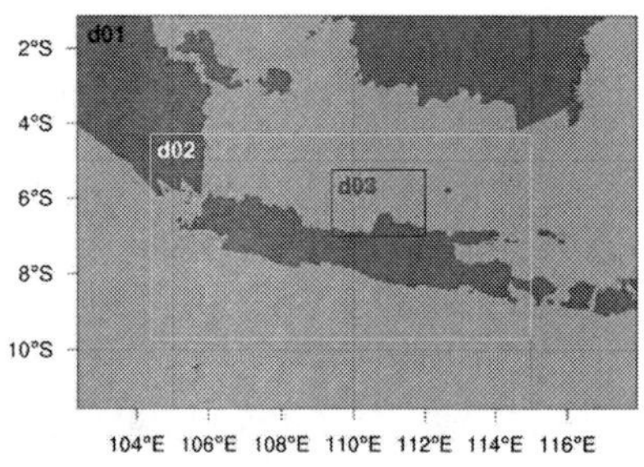

Figure 1. Computational domain. Symbols d01, d02 and d03 denote the domain number.

IOP Conf. Series: Journal of Physics: Conf. Series **1127** (2019) 012041 doi:10.1088/1742-6596/1127/1/012041

3. Results

3.1. Spatial comparison of cloud top temperature between satellite data and model output

A qualitative analysis is done by comparing the forecast without assimilation against the Himawari 8 satellite data to verify the numerical result. A comparison between the Himawari 8 satellite IR1 channel data and a derived variable, which mimic Multi-functional Transport Satellite 2 (MTSAT2) brightness temperature, from the model output shows a comparable scale of cloud feature. Generally, the model is able to reproduce the small scale feature of cloud particularly over the land area, regardless the periode of study that tend to be clear sky over southern part of Indonesia. However, we notice artificial features near the edge of computational domain, which associated with boundary effect problem.

As mention in the previous section, the initial and boundary condition are from GFS $0.25° \times 0.25°$ and the outer domain has 9 km grid spacing. However, the model output shows reproducibility of small scale features similar to the satellite data, which has about 5 km of horizontal resolution. This result implies that the model is usefull to represent the atmospheric condition to some extent.

3.2. Statistical evaluation of assimilation experiment

Statistical analyses are performed to evaluate the impact of data assimilation to the improvement of prediction, particularly the wind component. Figure 2 shows comparison of RMSE of u-wind and v-wind component for with and without data assimilation. Note that site 4 is another station sited at Jepara, which located in the mainland of Java Island. There is a significant improvement of RMSE for both u-wind and v-wind component compare to those without assimilation. The smaller RMSE for the assimilation result denote that the assimilation process reduce the error of the model compare to the observation. However, a higher RMSE for data assimilation experiment exists at site 2 for v-wind component.

Correlation analysis between wind from the model output and the observation also shows an increase of correlation coefficient, which implies an improvement of model output. In particular, the significant increase of the correlation coefficient are for v-wind component. Although, the correlation coefficient for u-wind component also increase, the correlation is in reverse relation at site 1–3.

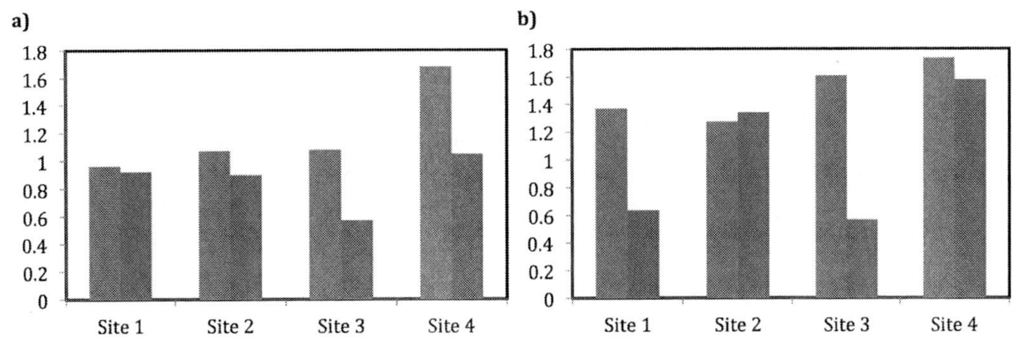

Figure 2. Comparison of RMSE between control run - observation and assimilation - observation for a) u-wind component and b) v-wind component. Blue color denotes control run - observation and red color denotes assimilation - observation.

IOP Conf. Series: Journal of Physics: Conf. Series **1127** (2019) 012041 doi:10.1088/1742-6596/1127/1/012041

3.3. Spatiotemporal distribution of surface temperature

The sea breeze circulation strongly relates with the temperature difference between the sea and adjacent land as many studies point out (see [1][2]). Therefore, it is important to examine the surface temperature parameter from the model output to determine the sea breeze existence. In Figure 3, we present comparison of spatial distribution of the surface temperature from the model output with and without assimilation. For the EnKF system, we only run five ensemble members due to limited of computational resources.

A relatively large difference of surface temperature appears near Karimunjawa Island between with and without assimilation. All of the ensemble member of EnKF system show a colder temperature in the sea surrounding the Karimunjawa Island compare to the result without assimilation. Although careful investigation are required, we still can expect that there will be stronger sea breeze circulation in the ensemble member of EnKF than in the control run (without assimilation) due to stronger temperature gradient between the sea and the land in Karimunjawa Island.

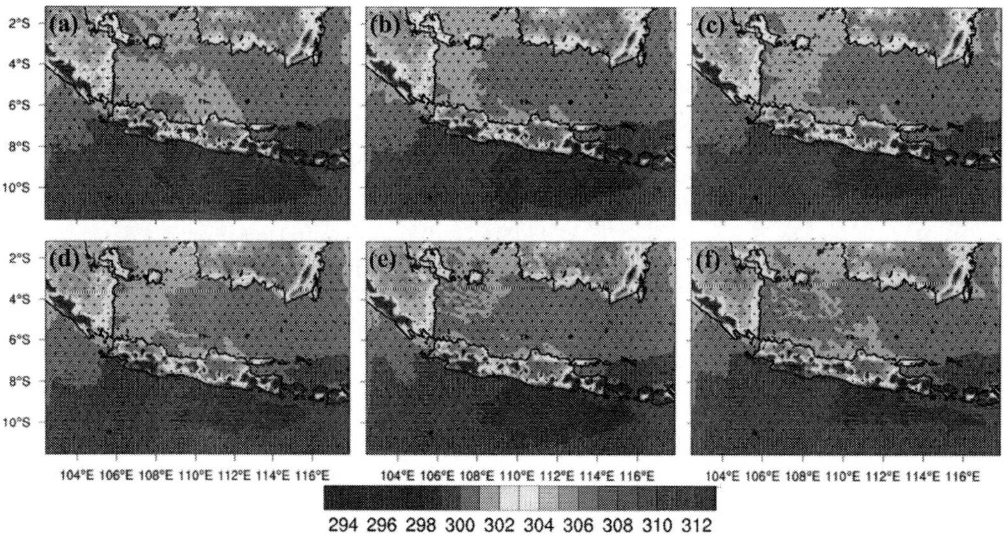

Figure 3. Comparison of surface temperature (shaded; in unit K) and horizontal wind (vector; in unit m s^{-1}) from model output a) without and b–f) with assimilation (member 1–5) for a particular time (15 August 2015 15:00 local time).

4. Summary and Conclusions

This study presents the impact of AWS data assimilation to the daily prediction that include local circulation over a small island. A comparison between the Himawari 8 satellite IR1 channel data and brightness temperature from the model output shows a good agreement of small scale features, particularly over the Java Island. Statistical analyses show that the data assimilation can improve the prediction as represented by decrease of RMSE values. EnKF data assimilation system produce colder surface temperature over the sea surrounding Karimunjawa Island, which can be expected produce stronger sea breeze circulation over the island.

Acknowledgments

We thank Glen Romine (NCAR) for his guidance to run WRF/DART system. This work is supported by funds provided by Institut Teknologi Bandung Research and Innovation Program 2016 under Grants FITB.PN-6-10-2016.

References

[1] Miller S T K, Keim B D, Talbot R W and Mao H 2003 Sea breeze: structure, forecasting, and impacts *Rev. Geophys.* **41** 1011
[2] Hadi T W, Horinouchi T, Tsuda T, Hashiguchi H and Fukao S 2002 *Mon. Wea. Rev* **130** 2153–66
[3] Sato T, Miura H, Satoh M, Takayabu Y N and Wang Y 2009 *J. Climate* **22** 4809–26
[4] Zhou L and Wang Y 2006 *J. Geoph. Res.* **111** D17104
[5] Teo C-K, Koh T-Y, Lo J C-F and Bhatt B C 2011 *J. Climate* **24** 4662–75
[6] Koo M-S and Hong S-Y 2010 *J. Geoph. Res.* **115** D05105
[7] Bjerknes V 1911 *Dynamic Meteorology and Hydrography. Part II. Kinematics* (New York: Carnegie Institute, Gibson Bros.)
[8] Kalnay E 2003 *Atmospheric Modeling, Data Assimilation and Predictability* (Cambridge: Cambridge Univ. Press) p 341
[9] Torn R D 2010 *Mon. Wea. Rev.* **138** 4375–92
[10] Skamarock W C, Klemp J B, Dudhia J, Gill D O, Barker D M, Duda M G, Huang X-Y, Wang W and Powers J G 2008 *A Description of the Advanced Research WRF Version 3* (Boulder, Colorado: National Center for Atmospheric Research) p 125
[11] Gustari I, Hadi T W, Hadi S and Renggono F 2012 *Jurnal Meteorologi dan Geofisika* **13** 119–30
[12] Anderson J L, Hoar T, Raeder K, Liu H, Collins N, Torn R and Arellano A 2009 *Bull. Amer. Meteor. Soc.* **90** 1283–96

IOP Conf. Series: Journal of Physics: Conf. Series **1127** (2019) 012042 doi:10.1088/1742-6596/1127/1/012042

Determination of The Type of Soil Using 2D Geoelectric Method And Laboratory Analysis for Landslide Area Cililin West Java

Ikah N P. Permanasari[1,2], M. Fadli Akbar[1], Gunawan Handayani[1]and Lilik Hendarajaya[1]

[1]Physics of Earth and Complex System Research Group, Faculty ofMathematics and Natural Science, Bandung Institute of Technology,Jl. Ganesha no. 10 Bandung, Indonesia, 40132.
[2]Physics Department,Institut Teknologi Sumatera,Jl. TerusanRyacudu Lampung , Indonesia

Abstract. Landslide occurrence at Mukapayung, Cililin West Java with material movement downward slope as far as 500m and hit residential areas of the village Nagrog cause eighteen people died and ten homes were destroyed and twenty-three heads of families evacuated. 2D Geoelectric measurements and index properties of soil analysis has been conducted to determine the type of soil slope composing. From the results of resistivity measurements obtained that the material making up the slope dominated by a mix of clay and sand, in which the soil types are prone to landslides if triggered by the presence of water.

1. Introduction

Landslide is one of natural hazard that affect humans and their livelihood especially in the mountainous area[1].A landslide is generally defined as the movement of rock, debris and soil down to slope driven by gravity and preconditioned by the landform. Landslides can be classified into different categories depending on the types of materials (earth, rocks and debris) involved and the types of movements (fall, topple, slide, spread and flow); for example, rock fall and earth flow.Steepness of slope, morphology, soil type and underlying geology are the most important factors for causing landslides[2,3].

The main aim of the present study is to test the capability of 2Dresistivity inversion of individual vertical electrical soundings(VES) to investigate the type of soil composing. The VES is common method to determined the geoelectricparameters[4,5,6,7]. Special attention is paid to explain the reasons behind the so-calledwaterless areaResistivity data are measured with the conventional Schlumberger array and processed and interpreted in both 2D manners. VES is based mainly on injecting electric current to the earth via two electrodes (current electrodes) and measuring the potential difference via two other electrodes (potential electrodes). This four electrode system has many types of arrays such as Schlumberger, Wenner and, Dipole–Dipole. The vertical resistivity sounding is based on the conceptof a homogeneous, isotropic, layered horizontal earth model. VES is effective in ground water exploration, particularly for water table detection. For the 1-D case, where the subsurface is restricted to a numberof horizontal layers, the linearfilter method iscommonly used to calculate the true resistivity from the measuredapparent resistivity[8].

In this paper we measure the resistivity of the soil Cililin Landslides area to determine the type of soil slope composing and have conducted laboratory measurement to see the properties index of the upper soil.

2. Theoretical Background

The resistivity of a substance or material is defined as the electrical resistance between the two surfaces facing each other of cubic unit material. Thus the resistivity can be referred to as the barrier type of material. In the case of the conductive cylinder with resistance R, length L, and the surface area of the material A, the resistivity values can be determined by the equation:

$$\rho = \frac{RA}{L} \tag{1}$$

The SI unit of resistivity ρ are *ohm-meter*(ohm. m). The resistivity is highly variable physical properties. Some minerals such as metals and graphite electric discharge via the flow of electrons. There are also minerals isolator which the electric current is carried through the rock via the ions in the water. Most rocks conduct electricity to the electrolytic process compared to electronic processes as usual[9].

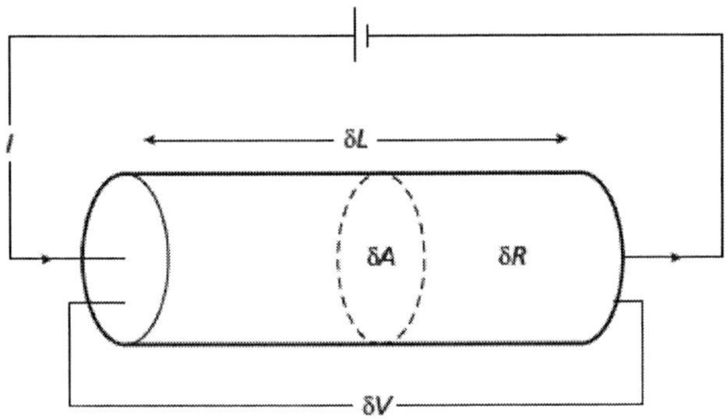

Figure1.The Parameters Used To Define Resistivity

Equation (1) refers to the conduction of electricity, but this equation can also be used to describe the effective resistivity of rock that the resistivity of rock and resistivity of water through the rock pores. Effective resistivity can be expressed in terms of resistivity and the volume of water in the rock pores, according to the empirical equation

$$\rho = a\phi^{-b} f^{-c} \rho_w \tag{2}$$

withϕ is porosity, f is the fraction of pores that contain water with a resistivity ρ_w. a, b and c are empirical constants. Water resistivity value may vary depending on the quantity and conductivity of the material that is passed thewater[4].

3. Result and Discussion

Mukapayung Landslide area is located in Cililin West Java Indonesia, about 35 km from the center of Bandung City (Fig. 2). The altitude of this Mukapayung, Cililin landslide area is around 974 meters about the mean sea level while the relief is dominated by the mountainous area. Type of ground surface is dominated by sandy soil moist and the presence of some water flow along the field measurements. Geoelectric measurements and sampling was conducted along 64 meters. The slope of the ground in the measuring area reached 45 degrees with the surface uneven ground because there are many piles of former land avalanches. Local conditions can be seen in Figure 3

IOP Conf. Series: Journal of Physics: Conf. Series **1127** (2019) 012042 doi:10.1088/1742-6596/1127/1/012042

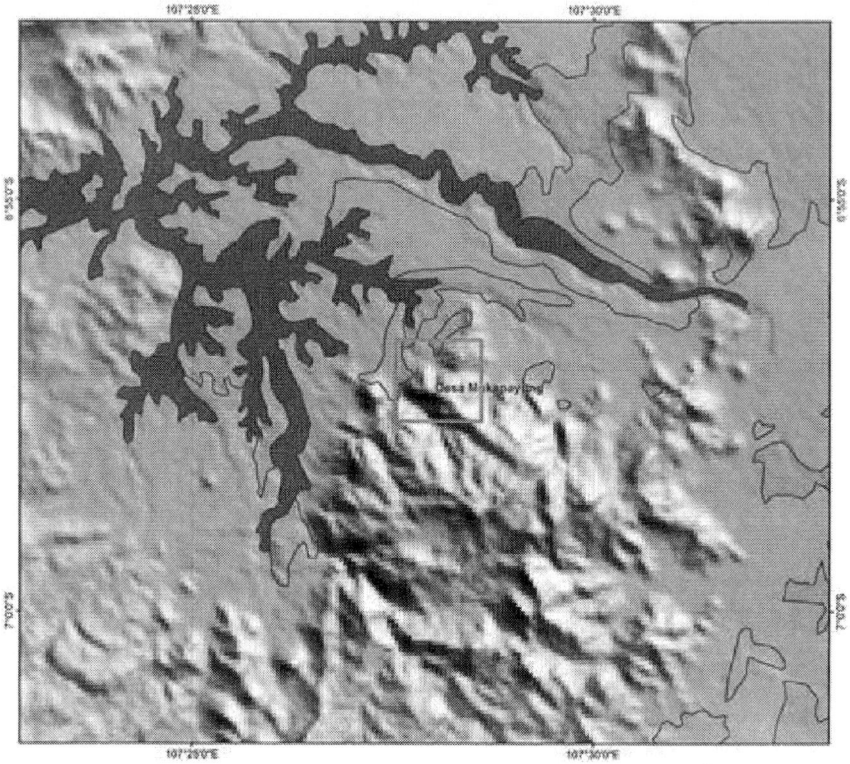

Figure 2.The location of The Mukapayung, CIlilin Landslide area

In general, the District Cililin morphology area is fairly steep hills with a slope between 15 ° - 30 °. But in some places several settlements located on a steep cliff with a slope of more than 40 °. Bedrock forming the slope by Geological Map Sheet Cianjur, West Java, consisting of lacustrine, while the upper part of weathering in the form of clay soil in the form of sand with a thickness of 4 to 6 meters.

Figure 3. Local area condition

And the topography of the area measurement are show in the figure 4 below :

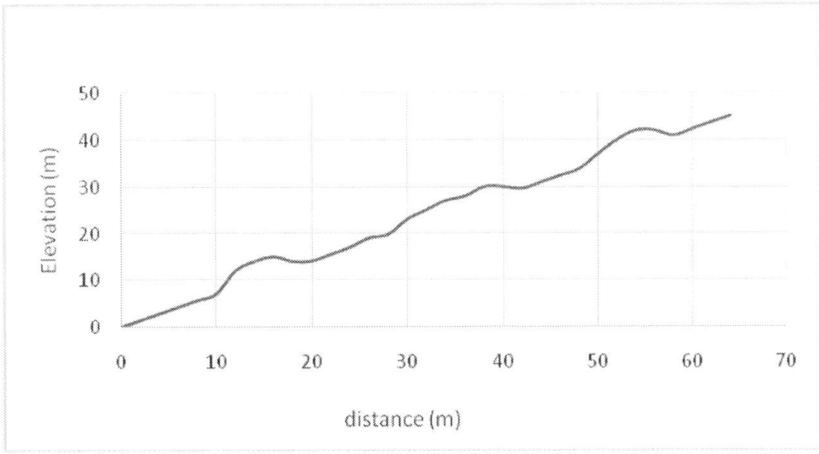

Figure 4.Topography of the area

Geoelectric data retrieval using a Wenner-Schlumbergerconfigurationwith 16 electrode (see Figure 5)

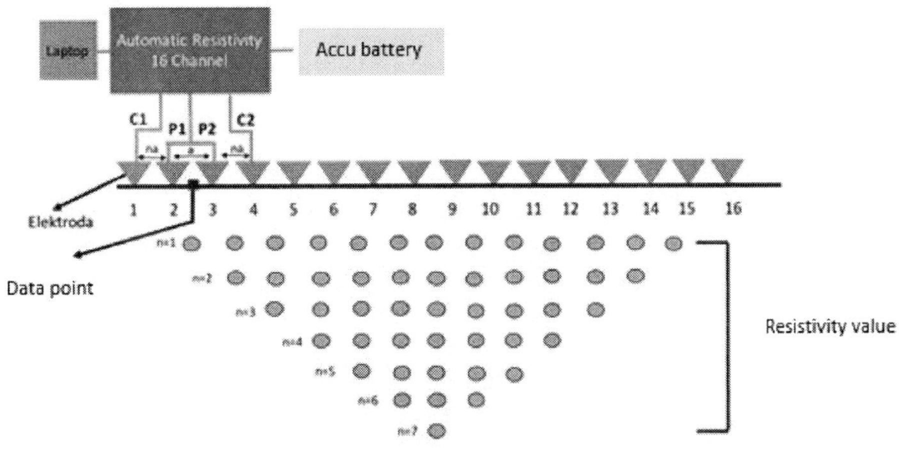

Figure 5. Geoelectric measurement with Wenner-Schlumberger configuration

From the measurement results is known that the area is dominated by the type of soil that has a resistivity values in the range 2.9-400 ohm.m. directly before the measurement can be seen that the ground is sand and clay-type soil. Resistivity values that small can be caused by a large flow of rain water infiltration on the surface of the soil so as to reduce the resistivity values of the measurements. This can be seen in the flow of water between the datum point 24-28 and 44-48.

IOP Conf. Series: Journal of Physics: Conf. Series **1127** (2019) 012042 doi:10.1088/1742-6596/1127/1/012042

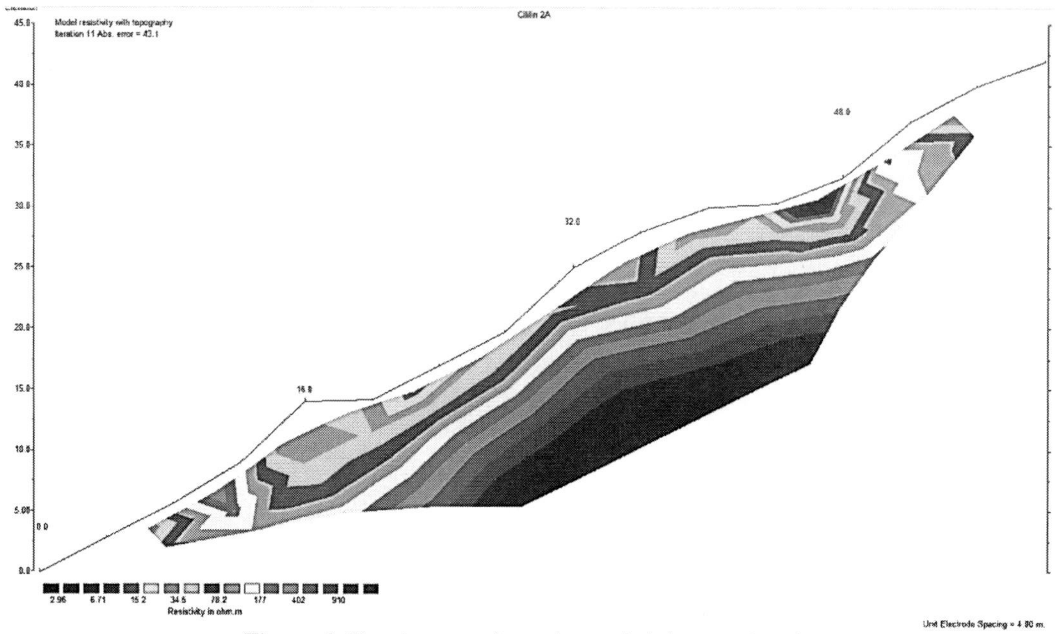

Figure 6. Sketches pseudosection resistivity geoelectric

Surface soil sample (disturbed) were took and measure the properties index of from the laboratory measurement of soil showed that the soil type material of the sloe is clay sand and the water content is 45,3% (see table 1)

Table 1. Index properties of soil

Measurement	Result
Water content	45.3 %
Wet density	1.63 (t/m3)
Dry density	1.12 (t/m3)
Specific gravity	2.60
Void ratio	1.31
Porosity	0.57
Saturated deg	89.15%
Soil type	Clay-sand

In clay soil the structure changes with the water contentdue to the swell-shrink dynamics of the material. important role played in the reactivations of landslides by the dry seasons and then rainy periods.Dry periods can cause a deeper cracking in the soils, which may facilitate water infiltration during rainy periods[10]. When water comes to the clay soil, in the long run soil will be saturated and induced the landslide.

4. Conclusion

Based on the result of the measurement ,real resistivity value range is at 2.96 to 910 ohm.m. Its surface is dominated apparent resistivity values 2.96 up to 177 ohm.m which is the sandy soil soaked with the flow of water in the ground two points of the field. The bottom layer of the soil is dominated by a mixture of sand and clay soil and there is a layer of high resistivity is> 910 ohm.m who suspected is a layer of sandstone boulders.This soil type is suspected of causing landslides during the certain

IOP Conf. Series: Journal of Physics: Conf. Series **1127** (2019) 012042 doi:10.1088/1742-6596/1127/1/012042

amount of water into the ground and saturate the soil.In line with the analysis from resistivity value, the index properties analysis shown that the upper soil type is clay sand that can induced the landslide when water gets into the soil

References

[1] HenkyNugraha, DhandhunWacono , Gilang Arya Dipayana et al 2014 *Procedia Environmental Sciences* 28 p 578-586.
[2] Sunguuo Wang, AlirezaMalehmir, MehrdadBastani 2016 *JournalTectonophysics* p 677-678.
[3] Tatiana Ivasuc, Ernest Olinic 2015 Journal Agriculture and Agricultural science Procedia 6 p 459-466
[4] E.V. Deev, N.N. Navedrova, I.D. Zol'nikov et, al 2012 *Journal Russian Geology and Geophysics* 53 p 92-107
[5] WadahMahmoodShakir Al-Khafaji, Hayder Abdul Zahra Al-Dabbagh 2016 *JournalOf Astronomy and Geophysics* (article in press)
[6] CassainoAnonioBortolozo, Marco AntonioCoutoJr, Jorge Luis Porsani, et al 2014 *Journal of Applied Geophysics* 111 p 33-46
[7] AkosGyulai, TamasOrmos, MihalyDobroka 2010 *Journal of Applied Geophysics* 72 p 232-241
[8] Mohamed A. Khalil, Fernando A. Montero Santos 2013 *JournalOf Applied Geophysisc*89 p 116-124
[9] Kearey, Philips., Brooks, Michael., Hill, Ian. 2002. *An Introduction to Geophysical Exploration 3rd Edition*. New York: John Wiley & Sons.
[10] C. Meisina 2006 Journal Natural Hazard and Earth System Sciences 6 p 825-838

Line Variations of Hα Emissions and Absorptions of Be Stars

Aprilia[1], R Muztaba, M Irfan, E Puspitaningrum, G E Ramadhania and H L Malasan

Department of Astronomy and Bosscha Observatory, Faculty of Mathematics and Natural Sciences, Institute Technology of Bandung, Jalan Ganesha 10, Bandung Indonesia

[1]aprilia@as.itb.ac.id

Abstract. We present the spectroscopic variations of 14 program Be stars observed within 2008 – 2009 at the Bosscha Observatory, Lembang, Indonesia. The Be stars show spectroscopic profiles as Be and Be-shell, and sometimes as B-normal. These profiles come from the disk-like envelope or rings surrounded the star's equatorial plane. Our program stars are: 5 as Be single-peaked, 2 as Be double-peak, and 7 as B-normal stars. We also measure the Ip/Ic ratio from the line profiles.

1. Introduction

B-emission (Be) stars are B-type stars characterized by Balmer emission (e.g. Hα lines) that originate in a circumstellar matter surrounding the stars (see, for example, Collins 1987, Owocki 2006). Be stars rotate rapidly (100 – 400 km/s, or more). Their rotation rates are typically 70 – 80 % of their critical rotational velocities (Owocki, 2006). Sepctral lines of Be stars show three main profiles (Doazan, 1982): Be single-peaked, Be double-peaked, and Be-shell line profiles. These profiles are caused by the evolution of envelope formation around the stars. Therefore, it is possible for the stars to have phase profile changing, from Be to Be-shell, or become B-normal, or vice versa, for some duration of times, typically within years. The mechanism that drives these kind of phenomena is still a topic of discussion nowadays (Porter & Rivinius, 2003). It also seems that one mechanism that fits to one stars, may not fit to other Be stars (Malasan, et al., 2014).

The emission line spectrum of Be stars show some variations, among others V/R and Ip/Ic. The V/R is time variation of relative intensities of violet (V) and red (R) components in double-peaked emission-line profiles, while Ip/Ic is time variation of emission-line intensities (Ip) relative to the adjacent continuum (Ic) (Kogure & Leung, 2007). These variations can quantify the strength of Hα line.

In this work, we present the spectroscopic observations of 14 program Be stars observed in 2008 – 2009. We also measure the Ip/Ic variation of the emission-line profiles. We have discussed the V/R variation of some of our Be stars and published in Puspitaningrum, et al. (2015).

2. Observations and Data Reduction

The 14 program Be stars have been observed spectroscopically at the Bosscha Observatory, Lembang, Indonesia, in 2008 – 2009, by using the 60cm Zeiss Double Refractor equipped with Bosscha Compact Spectrograph (BCS) and SBIG ST-8XME CCD camera. These observations are situated in the southern hemisphere, declination between +20° to –40°. Spectral coverage is limited from 6300 Å to 6800 Å. We

IOP Conf. Series: Journal of Physics: Conf. Series **1127** (2019) 012043 doi:10.1088/1742-6596/1127/1/012043

select the program stars with the range of declination (from +20° up to –40°) from the Bright Stars Catalogue (Hoffleit & Jascheck, 1991). The 14 selected stars are shown in table 1. The spectrograms were reduced by using the IRAF package from the NOAO (National Optical Astronomical Observatory), Tucson, Arizona, USA (standard procedure can be seen in http://iraf.noao.edu/).

3. Results and Discussions

Parameters related to the program stars are listed in table 1. Data in column 2 – 4 are taken from several references, while Ip/Ic values in column 5 are measured from the observed line profiles.

Table 1. Parameters of program stars. HR is the name of the stars in the Bright Stars Catalogue. V_{mag} is the stellar visual magnitude. Spectral type shows spectral and luminosity classification of the stars. *v sin i* states the stellar rotational velocity.

HR	V_{mag}	Spectral type	*v* sin *i* (km/s)	Ip/Ic
5953	2.32	B0.3IV	180	10
6141	4.79	B3V	232	5.18
6881	5.71	B5Ve	250	3.3
7342	4.58	Bpshe	133	5.31
7890	6.22	B7IIIne	282	8.95
7106	3.52	B7Ve	-	9.15
8402	4.70	B7IVe	227	10.49
4638	3.96	B3V	-	-
4898	4.04	B2IV	-	-
5528	4.31	B5IV	-	-
6700	4.73	B9V	140	-
6804	5.46	B2III	197	-
7166	5.52	B3V	-	-
7739	4.76	B3Ve	160	-

Figure 1 shows the observed Hα profiles with various shape of emission and absorption lines. The profiles shown in figure 1 consist of Be single-peaked for 5 stars, Be double-peaked for 2 stars and B-normal for 7 stars. Below is the brief description for the observed line profiles of the program stars:

- The stars HR 5953, HR 6141, HR 6881, HR 7342, and HR 7890 show single-peaked emission lines. These indicate that at the observational period, these Be stars are surrounding by their envelopes. The Hα lines are strong enough above the continuum level for the HR 5953 and HR 7890 (Ip/Ic is high enough). We need simultaneously observations in order to study whether the ratio Ip/Ic increases or decreases toward B-normal phase.
- The spectral line of HR 6881 indicates that the part of the envelope is going to change to B-normal phase, preceded by the Be double-peaked phase. This can be seen from the deep wing parts of the emission spectral line.
- The stars HR 7106 and HR 8402 show double strong emission peak. Both stars exhibit V/R variations: V/R = 0.7 for HR 7106 and V/R = 0.9 for HR 8402 (Puspitaningrum, et al., 2015). There are some interpretations about the V/R ratio, one of them indicates that the envelope is expanding if V/R < 1.
- The stars HR 4638, HR 4898, HR 5528, HR 6700, HR 6804, HR 7166, and HR 7739 are in B-normal phases. Among these stars, only HR 7739 is indicated as a Be star as stated in the Bright Stars Catalogue.

IOP Conf. Series: Journal of Physics: Conf. Series **1127** (2019) 012043 doi:10.1088/1742-6596/1127/1/012043

4. Conclusions

The program stars are observed in 2008 – 2009. The observed profiles exhibit various shapes of emission and absorption lines. Five stars are classified as Be single-peaked, two stars as Be double-peaked, and seven stars as B-normal stars. The Ip/Ic variations show that the stars have strong emission lines.

To know if the stars exhibit phase profile changes for some duration of times, we compare our data to data taken from BeSS (Be Star Spectra) database (see http://basebe.obspm.fr/basebe/). Most of the stars do not show phase change for these recent years. Only two stars show this changing: HR 6881, from Be single-peaked to Be double-peaked lines, and HR 7739, from B-normal to Be-shell.

To study the phase changing, the observations must be done simultaneously. From the changings, we can study the evolution of the stellar envelopes and then the mechanism of the envelope formation to understand the characteristic of the Be stars.

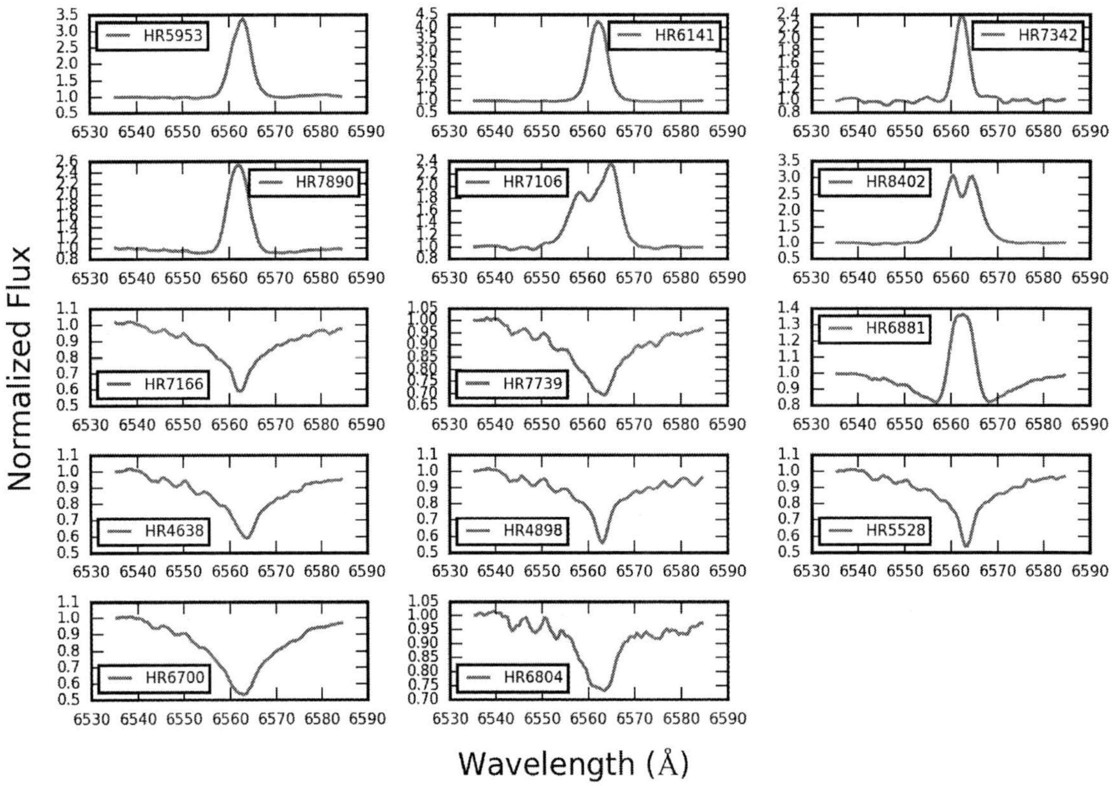

Figure 1. Hα line profiles of program stars

References

[1] Collins G W 1987 *Physics of Be Stars IAU Colloq.* vol 92, ed A Slettebak and T.P. Snow (Cambridge: Cambridge Univ. Press) p 3

[2] Doazan V 1982 *B Stars With and Without Emission Lines* (NASA SP-456)

[3] Hoffleit D and Jaschek C 1991 *Bright Star Catalogue* 5th *Revised Ed. (Preliminary Version)* (Yale University Observatory)

[4] Kogure T and Leung, K–C 2007 *The Astrophysics of Emission-Line Stars* (Berlin-Springer)

[5] Malasan H L, Irfan M, Aprilia and Dawanas D 2014 *The Tenth Pacific Rim Conference on Stellar Astrophysics* **482** 25

[6] Owocki S 2006 *ASP Conference Series* **355** 219

[7] Porter J M and Rivinius T 2003 *Pub. Astron. Soc. Pacific* **115** 1153
[8] Puspitaningrum E, Muztaba R, Ramadhania G E, Irfan M, Aprilia and Malasan H L 2015 *Prosiding Seminar Sains Antariksa* p174

Steady State Scenario in Asteroids Orbital Simulation with Yarkovsky Effect Inclusion

Judhistira Aria Utama[1,2*)], Taufiq Hidayat[3], Umar Fauzi[4], Ferry Mukharradi Simatupang[3]

[1]Astronomy Study Program, Institut Teknologi Bandung, Jl. Ganesha 10, Bandung
[2]Physics Education Department, Universitas Pendidikan Indonesia, Jl. Dr. Setiabudhi 229, Bandung
[3]Astronomy Research Division, Institut Teknologi Bandung, Jl. Ganesha 10, Bandung
[4]Geophysics & Complex System Research Division, Institut Teknologi Bandung, Jl. Ganesha 10, Bandung

*)Email: j.aria.utama@upi.edu

Abstract As one result in our orbital simulation of the 3372 samples of real NEAs (Near-Earth Asteroids) whose orbits were considered to be well-known, as many as 480 asteroids evolved out of the four major classes (Amor, Apollo, Aten and Atira) of NEAsat the end of the first 500,000 yearsperiodinto Mars-Crossers asteroids in majority. To test the steady statehypothesis for NEAs population, these evolved asteroids had been used as a sample for further simulation to obtain the entrance flux from the Intermediate Sources/IS (in this case the Intermediate Mars-Crossers/IMC only) to NEAs region. The entrance flux from IMC $\rightarrow$ NEAs region is 71 $\pm$ 17 objects per million year for H< 18. Asteroids of this amount as a contribution from IMC (along with contributions from other sources in the Main Asteroid Belt) will be delivered to near-Earth space to replenish this region due to the decay of NEAs population into the Sinks (collide with the Sun/planets or being thrown to the outer Solar System).After5 million years of simulation we found there were 1898 asteroids eliminated from further computational process with the majority endedtheir liveas Sun-grazer and ejected from the Solar System.The near-Earth space lost a total of 114 $\pm$ 17 objects per million year for H< 18.The median dynamical lifetime obtained directly from our NEAs sample decay is3.7x106 years.

Keywords: near-Earth asteroids, orbital evolution, Yarkovsky effect.

1. Introduction

Near-Earth space is defined as aregion that meets the inequality of perihelion distance q [= a(1 –e)] < 1.30 au(1 au is defined as Sun-Earth mean distance, which is equal 1.496x108 km) and aphelion distance Q [= a(1 + e)] > 0.98 au, which is inhabited by asteroid, comets and meteoroids or commonly known as Near-Earth Objects (NEOs). The asteroid is the largest population of NEOs, so we can be grouped this population as Near-Earth Asteroid (NEAs) which is our concern in this study. With the rise of space survey programs that scan the near-Earth space since the 1980s, NEAs is increasingly found in various sizes [1]. For NEAs with diameter D> 1 km (corresponding to an absolute magnitude H< 18) of Amor class (semimajor axes a> 1 au and 1.017 au<q< 1.3 au) the survey completeness has already reached 66%, whereas for Apollo class (a> 1.0 au and q< 1.02 au) and Aten class (a<1.0 au and Q> 0.98 au) the completeness are 33% and 45%,respectively [2]. Even the survey completeness of 100% has been achieved for NEAs with D> 7 km (corresponding to H<14) [3].The NEAs population is believed to be the source of projectiles that produce impact craters on terrestrial planets' (and also on the Moon's) surface.

The observations of craters formation on the surface of the Moon suggests that the rate of its formation nearly constant, which indicates thatNEAs populationis in a steady state within the last 3.7 – 3.8 billion years[4]. This means that the mechanism in the supply process of NEAs population is dominated by a continuous process rather than a sporadic. In the view of Classical Model (CM), supply process of NEAs populationis dominated by a catastrophic collision among asteroids in the Main Belt which often take place. Asteroids collide with a quite high speed, about5 kms^{-1}[5], are able to produce impact craters and fragmentation that would affect the orbit, the shape, internal structure and the state of the rotation of asteroids involved. Even the devastating collision is able to destroy the asteroids into a cluster of fragments that each fragment has orbital parameters of a, eccentricity e and inclination i are similar to each other forming anasteroid family. A high ejection speed and correct direction from the impact site can make fragments enter the zone of strong resonance generated by gravitational perturbations of the planets. While stay in the zone of strong resonance, fragment can evolve into highly elliptical orbit so that it intersects the orbit of planet Mars or even reach the near-Earth region.

There are gaps between the CM predictions with observations. Therefore, factor other than collisions and gravity, is necessary to be included in the study of the dynamics of NEAs, which is known as the thermal effect (Yarkovsky). Yarkovsky effect is a non-gravitational phenomena related to non-isotropic thermal emissions which cause the system to lose heat and momentum. This effect is believed to have an effective influence on the rotating objects with small diameter D (ranging from a few centimetres to several tens km) experiencing heating by solar radiation. Using the modified Swift integrator package [6], namely Swift_rmvs4y which has involved the Yarkovsky effect [7], as well as utilize the IS and NEAs steady state model exist, this workintend to demonstrate the validity of the NEAs steady state hypothesis. To the best of the authors' knowledge, direct proof through numerical simulation of steady-state NEAs population is not present yet in the literatures.

2. Method

With the help of Swift_rmvs4y integrator,the NEAs steady state hypothesis was tested. This means our method is numerical simulation. The numerical algorithm is an improvement of the one in [6] and including the thermal force (Yarkovsky effect). The main data were obtained from NASA JPL Small-Body Database Search Engine (http://ssd.jpl.nasa.gov/sbdb_query.cgi). Only four classes (Apollo, Amor, Aten and Atira) of asteroids with very well-known orbit (condition code = 0) were included. Using this filtering we obtained 3372 NEAs (as of 10thMarch 2016). Due to limitation of total number of bodies could be integrated, the computing process was executed by dividing the total samples into four batches (batch1 to batch3 consists of 1000 asteroids each and batch4 of 372 asteroids). The computations utilized Perseus cluster facilities and a PC of *Intel (R) Core (TM) i7-2600K CPU* (both are available at Astronomy Research Group of InstitutTeknologi Bandung)and other PC of *Intel (R) Core (TM) i3-3240T CPU* at Computational Physics Laboratory of Physics Education Department of UniversitasPendidikan Indonesia.

Computations were performed using a segmentation strategy, which meantwe usedmultiples of 500,000 years on the three facilities available.Time step was setto be 1/1000 years (~9 hours) and the results were logged every 1000 years. Thegravitational perturbations of all the planets except Pluto were taken into account in the simulations. The Moon was treated as separate body. The NEAs'orbital evolution werefollowed until theyeitherstruck the Sun or a planet, travelled outside 100 au from the

Sun, or the final integration time was reached. We called the first two conditions as Sinks. The collision among asteroids and disruption events were not considered in this work.

3. Results

After the first 500,000 years, we found as many as 480 NEAs evolved out of the four classes of their origin (Amor, Apollo, Aten andAtira) as shown in table 1. According to [2] one of the primaryIntermediate Sources/IS forNEAs population is Mars-crosserasteroidsbesides the ν_6, 3:1 resonance and the Outer Belt (OB) regions.Intermediate Mars-Crossers source region is adjacent to the Main Belt but has not reached orbit with q<1.3 au. All of these 480 asteroids were subsequently treated as sample of the Intermediate Mars-Crossers/IMC population to be followed their evolution up to 3x10^6 years to obtain the supply rate from the IMC to the near-Earth region. Oncethe IMC sample undergoes an episode as one of the four main classes of NEAs, the corresponding sample is immediately "eliminated". By elimination we mean its existence is not taken into account in the IMC sample anymore.

Table 1. Subpopulation of asteroids out of the four NEAs class at the end of the first 500,000 years of evolution.

Σ Particle	Orbit	Remark
384	1.3 au<q< 1.58 au	Deep Mars-Crossers (DMC)
34	1.58 au<q< 1.67 au	Shallow Mars-Crossers (SMC)
62	1.8 au<a< 25 au 1.6 au<q< 12.5 au 2 au<Q< 44 au	-

Figure 1 shows the number of remaining IMC asteroids as function of timeand the (absolute) fractional decay rate (r_sf) curve obtained by following procedure in [2]. The (absolute) r_sf curve constructed by fitting astraight line by the method of linear regression to ln[$N(t)$] inthe interval [0,T], with T varying from 0.6x10^6 to 3.5x10^{6}years at 1000 years steps.We decided to compute an average of (absolute) $r_sf$only the values between$T$ = 1.3x10^6 and T = 3.5x10^6 years(the time interval with stable value of r_sf), which gives r_sf = 0.2153 ± 0.0102 per million year.From the NEAs model [2] we can obtain the steady state number of IMC with $H<H_0$ in the IS region. In particular,N_{IMC}= 4000 ± 940 forH_0<18. This latter value multiplied by r_sf, yields entrance flux F from IMC to near-Earth space, that is 861 ±202 objects per million year for H< 18. Since IMC contributes only ~8% of the total IS contribution to near-Earth space, the entrance flux from IMC region to near-Earth space becomes F_{IMC} = 71 ± 17 objects per million year for H< 18. This value is consistent with the number of asteroids delivered from IMC region to near-Earth space (65 ± 15 objects per million year (H< 18) [2]) to keep NEAs population in steady state.

Figure 2 shows our remaining NEAs sample as the function of time (up to 5x10^6 years) and the corresponding (absolute) mean r_sf curve. Within the time interval of 2.5x10^6–5x10^6 years (the interval with stable value of r_sf) we obtained the absolute mean value of r_sf= 0.1188 ± 0.0016 per millionyear. According to the steady state model of NEAs, there are N_{NEAs} = 960 ± 120 for H_0< 18 [2]. So we can calculate the NEAs entranceflux to the Sinks, that is F_{NEAs} = 114±14 per million year for H< 18.The decay of NEAs population due to asteroids evolution into the Sinks will be replenished by IS population, one of which is our IMC sample.

From the decay plot of NEAs, we can determine the median dynamical lifetime of our sample directly, that is 3.7x10^6 years. This is the characteristic time at which the population is reduced to half itsinitial population. This new value is shorter than that obtained in [8], which is 1x10^7 years. This indicates the different role played by resonances in the region of 0.4 au<a< 2 au[9] against NEAs sample used in [8] and in this work. Our result shows that the resonances in this region have made the asteroids evolved into Sinks rapidly.

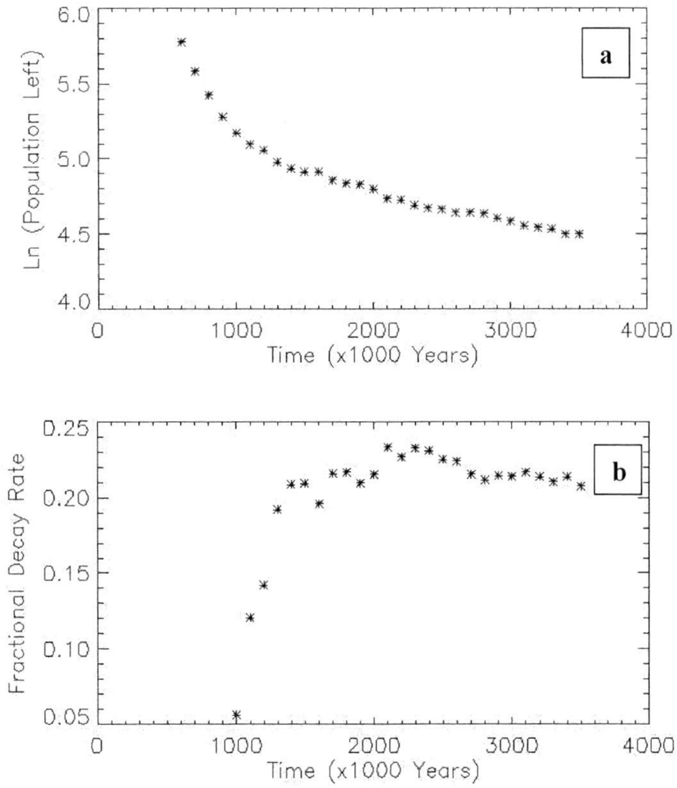

Figure 1. The decay rate from IMC region to near-Earth region: (a) natural logarithm of the population left in the IMC region as a function of time; (b) absolute fractional decay rate into near-Earth region.

We also plot the residence time of our NEAs sample during $5x10^6$ years of integration in figure 3. Starting from theinitial *epoch*, we have computed the time spent by all the asteroids in different (*a*, *e*, *i*) cells along the integration. For anasteroid in its initial location, we compute the time it spends in different cells of (*a*, *e*, *i*) over its whole evolution. We repeat this computation for all the asteroids and finally the cumulative time spent by all asteroids in each cell is normalized to thetotal time spent by all asteroids in all cells. Thisillustrates the steady-state distribution of objects in the a< 4.2 au region. These plots are constructed via a grid of (*a*, *e*, *i*) cells covering a< 4.2 au, e< 1.0 and i< 90° with delta value of each orbital element is 0.05 au, 0.02 and 2° respectively. These plots also tell us the mean residence time of our NEAs sample in different (*a*, *e*, *i*) cells in the inner Solar System during their future evolution.

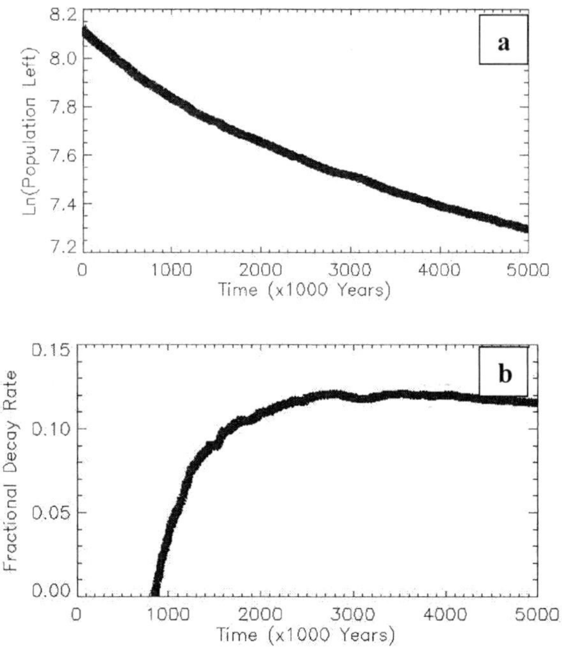

Figure 2. The decay rate from near-Earthspace to the Sinks: (a) natural logarithm of the NEAs population left as a function of time; (b) absolute fractional decay rate into the Sinks.

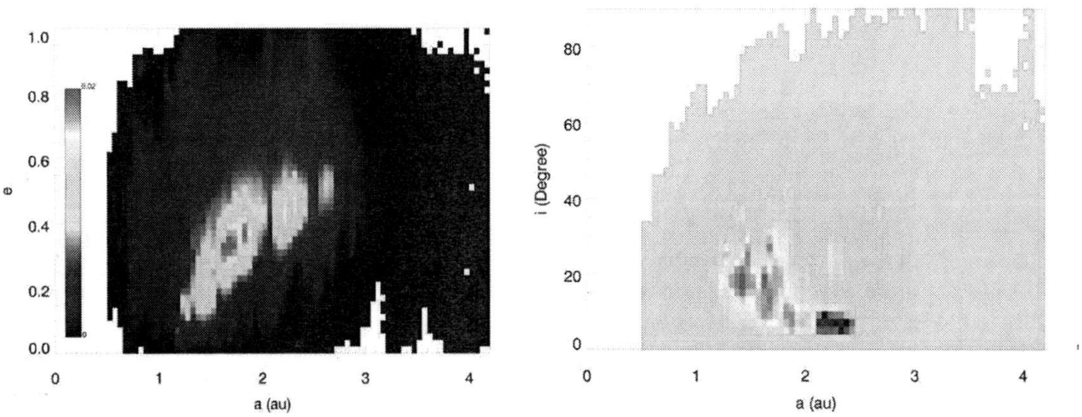

(a) (b)

Figure 3. Mean residence time of our NEAs sample in different cells of the (a,e) plane (a) and (a,i) plane (b) during $5\text{x}10^6$ years of evolution. The colour scale depicts the normalized average amount of time spent by asteroids in a particular cell. Red colour indicates where NEAs are most likely to spend their time (in contrast to black colour) and purely white regions are never visited.

4. Conclusion

Entrance flux of the IMC to the near-Earth space has been obtained from numerical simulation that includes Yarkovsky effect. The rate at which asteroids are delivered into near-Earth space is consistent with the model. Therefore the steady state hypothesis of NEAs population can be demonstrated as valid through our numerical simulation. The shorter median dynamical lifetime of NEAs population obtained in this work have indicated the different role played by resonances in the 0.4 au $<a<$ 2 au region which is affecting the dynamical lifetime of NEAs population.

Acknowledgments
JAU are grateful to the committee of this conference for financial support to attend this meeting. We also thank Astronomy Research Division of InstitutTeknologi Bandung, Mr. IkbalArifyanto and Computational Physics Laboratory of UniversitasPendidikan Indonesia for supporting on computational facilities.

References

[1] Bottke, W.F., Jedicke, R., Morbidelli, A., Petit, J.M., Gladman, B., 2000.Understanding the distribution of near-earth asteroids. *Science***288**, 2190–2194

[2] Bottke, W.F., Morbidelli, A., Jedicke, R., Petit, J.M., Levison, H.F., Michel, P., Metcalfe, T.S., 2002. Debiased orbital and absolute magnitude distribution of the near-Earth objects. *Icarus***156**, 399–433

[3] Rabinowitz, D.L., Helin, E., Lawrence, K., Pravdo, S., 2000. A reduced estimateof the number of kilometre-sized near-earth asteroids. *Nature***403**, 165–166

[4] Strom, R.G., Renu, M., Xiao, Z.Y., Takashi, I., Fumi, Y., Lilian, R.O., 2015. The inner solar system cratering record and the evolution of impactor populations.*RA&A***15**, 147–165

[5] Bottke, W.F., Vokrouhlický, D., Rubincam, D.P., Nesvorný, D., 2006. The Yarkovskyand YORP effects: implications for asteroid dynamics. *Annu. Rev. Earth. Planet. Sci.***34**, 157–191

[6] Levison, H.F., Duncan, M.J., 1994. The long-term dynamical behavior of short-period comets. *Icarus***108**, 18–36

[7] Dermawan, B., Hidayat, T., Utama, J.A., 2013. Pengembangan integrator swift_rmvs4 dengan melibatkan efek termal. Proceedings of the Indonesia Astronomical Society Seminar.

[8] Gladman, B., Michel, P.,Froeschlé, Ch., 2000. The near-Earth object population.*Icarus***146**, 176–189

[9] Michel, P., Froeschlé, Ch., 1997.The location of secular resonances for semimajor axes smaller than 2 AU.*Icarus***128**, 230–240

IOP Conf. Series: Journal of Physics: Conf. Series **1127** (2019) 012045 doi:10.1088/1742-6596/1127/1/012045

Bosscha Robotic Telescope (BRT) - a 0.35 meter telescope on Bosscha Observatory

M. Yusuf[1], P. Mahasena[1,2], A.T.P. Jatmiko[1], D. Mandey[1], E.I. Akbar[1,2], A. Setiawan[1], M. Sulaeman[1]

[1] Bosscha Observatory, FMIPA, Institute of Technology Bandung, Indonesia
[2] Astronomy Research Division, FMIPA, Institute of Technology Bandung, Indonesia

E-mail: yusuf@as.itb.ac.id

Abstract. We introduce the second generation of the Bosscha Robotic Telescope (BRT), a remotely and automatically operated 0.35-meter telescope. This paper will describe the BRT technical properties, system, and the data from early observations. We have been working on the robotic automation of the telescope system in order to be ready for the national observatory along with international telescope networks. The use of the telescope includes monitoring observation of variable stars, exoplanet, near earth asteroid; and observational for education.

1. Introduction

Robotic telescope is a telescope that can operate without human interaction during the night time observation. Recent advances in a robotic telescope made it possible to do an observation in a very remote location that offer very dark sky and optimum condition in order to obtain high quality data. Bosscha Robotic Telescope (BRT) is included as one of the preparation steps for the national observatory that will be located at a very remote area in Timor Island (East Nusa Tenggara), Indonesia.

BRT (figure 1) is a 0.35-meter robotic telescope located at the Bosscha Observatory on Lembang. This telescope is the second generation of our robotic telescope with more capable mounting system, bigger telescope and a better camera as advancements from the previous generation [1]. The first generation of BRT has relocated to Nusa Cendana University, East Nusa Tenggara. This system is planned to be applied to a 0.5 meter telescope in Timor.

2. System

2.1. Optical Tube Assembly

BRT is a telescope system with f/7.2 Corrected Dall-Kirkham (CDK) design from PlaneWave Instruments[1] with a 0.35-meter diameter primary mirror. It has an ellipsoidal primary mirror with spherical secondary mirror, and a combination of two additional corrector lens to produce a 52mm diameter circle of coma-free flat field with no off-axis astigmatism. Table 1 and figure 2 show the detail specification and the spot diagram of the telescope.

The telescope has a motorized focuser capable of 18 kg lifting within a 30 mm (30000 microns) range with a 1 micron step. It is also equipped with a dew heater, capable to maintain the

[1] http://planewave.com/products-page/telescopes/14-inch-cdk-optical-tube-assembly/

IOP Conf. Series: Journal of Physics: Conf. Series **1127** (2019) 012045 doi:10.1088/1742-6596/1127/1/012045

temperature of the telescope at a desire value to prevent a dew build up. The telescope is equipped with a self-made rotator and spacer in order to get a fine adjustment of CCD camera position angle.Figure 3 shows our custom rotator and spacer made at our workshop.

Table 1. Optical tube assembly specification.

Optical Design	Corrected Dall-Kirkham (CDK)
Aperture	356mm
Focal Length	2563mm
Focal ratio	F/7.2
Central Obstruction	23.5% by surface area; 48.5% by diameter
Optical Performance	3.1 micron RMS at 13mm off-axis
	6.0 micron RMS at 35mm off-axis
Optimal Field of View	70mm Image Circle

Figure 1. BRT Telescope.

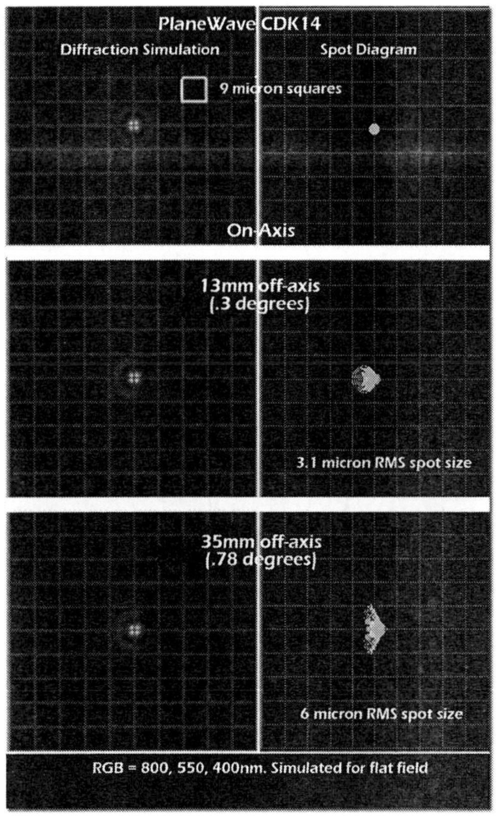

Figure 2. Telescope Spot Diagram.

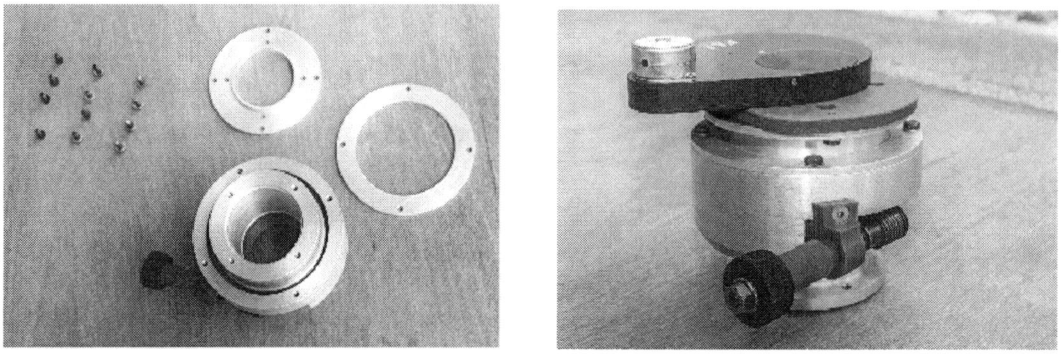

Figure 3. Custom rotator and spacer.

2.2. Mount

The telescope mount is an Astrophysics 1600 GTO[2] with a German-Equatorial design. This design requires a pier-flip when tracking a target across the observer's meridian from east to west, that requires about 1 minute delay in data acquisition. The mount has an RMS pointing accuracy of $\sigma = 4$ arcmin around the mean pointing position, and < 0.3 arc seconds periodic error tracking. The error most probably due to flexure in the telescope and a small errors in the polar alignment. The pointing will be automatically corrected by the software and we eliminate the tracking error with an external guiding scope. The mount also has a relatively fast slew with a 5 degrees/s speed.

2.3. Science Camera

We use FLI Proline 11002 Monochrome camera features a Kodak Interline KAI-1102M sensor. The camera is equipped with an integral five-position filter wheel with a 2-inch diameter *BVRI* filter. The camera is capable to maintain -55°C below the ambient temperature as we usually operate the camera at -30°C. The CCD specification are listed in table 2.

Table 2. CCD camera specification.

Detector sensor	4008 × 2672
Pixel sise	9 µm
CCD field of view	48.38 arcmin × 32.26 arcmin
Full well depth	60000 electrons
Dark current	< 0.005 electrons/pixel/sec at -30°
Typical system noise	9 electrons at 1 MHz; 16 electrons at 12 MHz

2.4. Control System

Not like the previous generation, to cut the development time and resources, we decided to use the ready-made software ACP (Astronomers Control Program) to control the instruments and equipments. ACP is a product of DC-3 Dreams[3] that takes advantages of ASCOM standards environment to interface a range of astronomy equipments.

[2] http://www.astro-physics.com/index.htm?products/mounts/1600gto/1600gto
[3] http://acp.dc3.com

Observations have been carried out either by writing a plan then submit it to a scheduler, or by a direct command from console in a real time. The scheduler accepted inputs as a list of observing request and set of constrains to perform the observation automatically. The scheduler also capable of interrupting the run to allow quick response of transient events (e.g. GRB and supernova follow-ups) and will respond to changes in observing condition. It is also qualified to dispatch requests that can be perform in the current condition, and will retry failed observation automatically. [2].

The focusing is also done automatically by searching the best Full Width at Half Maximum (FWHM) of a selected 7-8 magnitude star. The typical focusing process is about 3 minutes, including the slewing time. We have a predefined focus offset to skip focusing through each filter. The focusing is performed in the beginning of each night or if the statistical value of FWHM changed by 50%.

The flat field, dark, and bias images are taken regularly at dusk and dawn time. The flats are taken using a sky flats where the telescope picks a position in the sky on the solar circle near the zenith, offset in the anti-solar direction by 15 degrees. This should be "close enough" for most uses, including precision photometry [3].

3. Science Programs and Current Usage

3.1. Exoplanet

We used BRT to observe the transiting exoplanet WASP-74b, one of the brightest system accessible to the southern hemisphere telescopes. It is a 0.95 M_{Jup} planet with a moderately bloated radius of 1.5 R_{Jup} in a 2-d orbit around a slightly evolved F9 star [4]. We were able to observe delta magnitude of $dV = 0.010 \pm 0.001$mag. Figure 4 shows the transit light curve of WASP-74b on June 27, 2016.

3.2. Variable Stars

This telescope is able to make observation of variable stars monitoring with two or three target per night depending on time resolution. In our monitoring project we choose a variable stars that follows: short variability period, relatively bright, and have a modest delta magnitude [5].

3.3. Astrometry of Near Earth Asteroid

We are also able to observe and do astrometry on Near Earth Asteroid (NEA) up to 18.4 magnitude (figure 5). The telescope is able to track a fast moving object by modifying the tracking speed of both axis, RA and DEC. Table 3 shows the example of astrometry observation of NEA with 3-sigma uncertainty from minor planet center[4].

3.4. Education

BRT also functions as a valuable teaching and educational resource for undergraduate student at astronomy research division of Institute of Technology Bandung. The system control has made it possible to operate telescope from anywhere so it opens wide opportunities for collaborative observation with other institution. Figure 6 is an example of astrophotography from BRT. With the advancement of operating system that has been developed, we hope to encourage future citizen scientists and amateur astronomers, groups which historically have analyzed a lot of astronomical data and made numerous discoveries.

[4] http://www.minorplanetcenter.net/

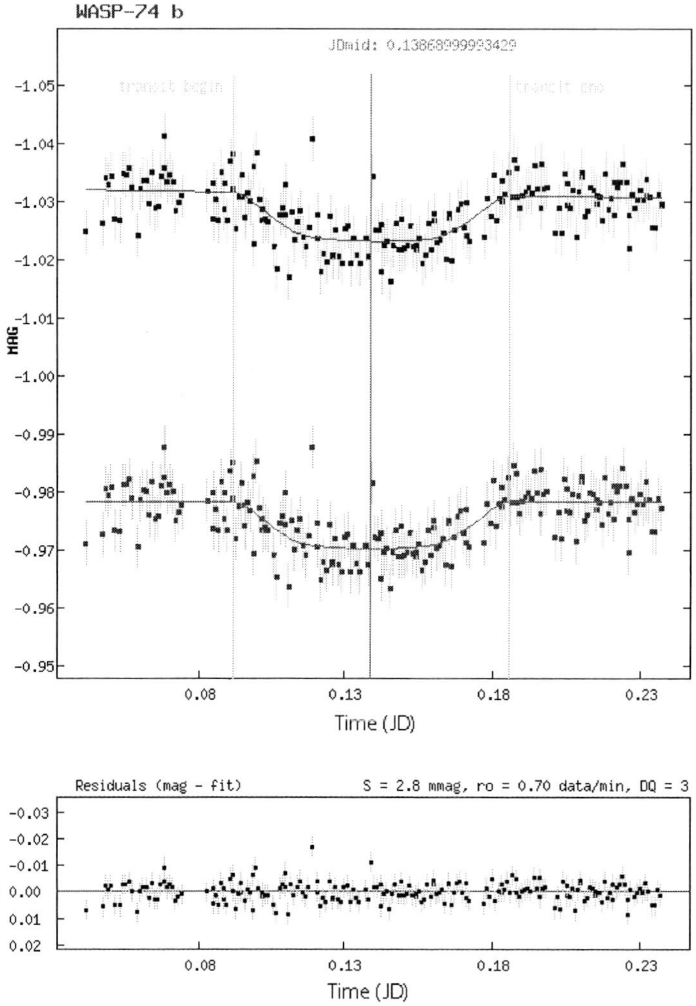

Figure 4. Transit light curve of WASP-74b.

4. Summary

In this paper, we presented current hardware and software of BRT. BRT is a fully operated 0.35-meter telescope installed at Bosscha Observatory, that can be controlled remotely from anywhere. The use of the telescope includes monitoring observation of a variable stars, exoplanet, near earth asteroid; and observational for education. We are planning to do a long term monitoring of variable stars and near earth asteroid.

Table 3. Example observation of NEA.

Object	Mag (V)	Time (UT)	RA predicted (d m s)	DEC predicted (d m s)	Sky Motion ("/min)	RA observed (d m s)	DEC observed (d m s)
1998 GL10	17.8	2015 03 24 20:15:44	11 51 16.9	-05 51 13	1.33	11 51 16.9	-05 51 12
2015 MU59	18.4	2015 08 29 19:59:45	23 54 01.7	-07 34 16	1.23	23 24 01.7	-07 34 16
2016 KB1	18.4	2016 06 26 21:48:33	00 54 28.5	-22 17 32	7.17	00 54 28.8	-22 17 28

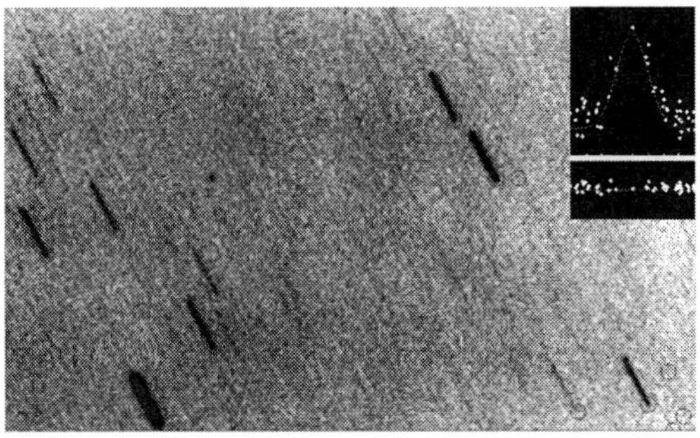

Figure 5. Near Earth Asteroid 2015 MU59 (purple circle) with 18.4 magnitude and 1.23 arcsec/min motion. Reference stars that were used for the astrometric data reduction are circled blue.

Figure 6. A *BVR* composite color image of M 20 taken with BRT.

References

[1] Yusuf M., Putra M., Mandey D. et al., 2013, *Seminar HAI 2013 - 90 Tahun Observatorium Bosscha.*

[2] Denny, Robert B., 2004 Dispatch Scheduling of Automated Telescope, *Society for Astronomical Sciences Annual Symposium*, **Vol. 23**.

[3] Chromey F., Hasselbacher D., 1996The Flat Sky: Calibration and Background Uniformity in Wide-Field Astronomical Images, *PASP 108*: **944-949**.

[4] Hellier C., Anderson D., Collier C. A., Delrez L. et al., 2015 Three WASP-South transiting exoplanets: WASP-74b, WASP-83b & WASP-89b, *Astron. J.*, **150**, 18.

[5] Jatmiko A.T.P., 2016 Light Curve Analysis of Eclipsing Binary System ASAS 172533-1221.4, *The 6th International Conference of Mathematics and Natural Sciences*, **in prep**.

Extracting diffuse interstellar bands from cool star spectra: tracing Galactic ISM structures, kinematics, and properties

L Puspitarini[1], **R Lallement**[2], **H-C Chen**[3], **A Monreal-Ibero**[2]
[1]Bosscha Observatory and FMIPA Institut Teknologi Bandung Indonesia
[2]GEPI, Observatoire de Paris, CNRS UMR8111, Universit Paris Diderot, Place Jules Janssen, 92190 Meudon, France
[3]Institute of Astronomy, National Central University, Chungli, Taiwan

E-mail: lucky.puspitarini@gmail.com

Abstract. Diffuse interstellar bands (DIBs) are broad and weak absorption features that appear in spectra recorded towards stars that lie behind interstellar (IS) clouds. They are known as a long-standing challenge in astronomical spectroscopy because of their unknown carriers. However, they are also a potential IS tracers, in particular to probe distant clouds because they are not easily saturated. We devised automated DIB-fitting methods appropriate for cool star spectra and multiple IS components. The data were fitted with a combination of a synthetic stellar spectrum, a synthetic telluric transmission, and empirical DIB profiles. The initial number of DIB components and their radial velocity were guided by HI 21 cm emission spectrum, or, when available in the spectral range, IS neutral sodium (NaI) absorption lines. For NaI, radial velocities of IS NaI lines and DIBs were maintained linked during a global simultaneous fit. Using the tools, we have analyzed Gaia-ESO Survey (GES) spectra of stars that probe between 2-10 kpc long line-of-sight (LOS) in few different regions of the Milky Way (fields). For all fields, the DIB strength and target extinction are well correlated. We show that, toward distance-distributed target stars, DIBs can be used to locate dense clouds in the Galaxy (i.e., Galactic arm) and probe a LOS kinematical structure. In addition, we have devised the DIB-DIB global fit to study the relation between DIBs and can be a potential tool to trace interstellar environmental conditions.

1. Introduction

A longstanding challenge in astronomical spectroscopy is to uncover the carriers of the diffuse interstellar bands (DIBs). They are broad absorption features observed in stellar spectra. The first observational record of DIBs was made by [6] and systematic studies began with the work of [10] who detected several DIBs in stellar spectra. Today, around 500 DIBs have been identified in the UV, visible, and infrared part of the electromagnetic spectrum (see [13, 9] and references therein). These DIBs behave like interstellar (IS) lines with regard to occurrence, strength, and Doppler shift. Instead of being narrow and sharp, they are somewhat widened and irregular. They are described as being diffuse as their widths are greater than the absorption features of well-known atoms or molecules. Using a high resolution spectrograph, there is an evidence that the widths are due to unresolved blending of individual rotational/vibrational lines as could arise

IOP Conf. Series: Journal of Physics: Conf. Series **1127** (2019) 012046 doi:10.1088/1742-6596/1127/1/012046

in the electronic spectrum of gas phase molecules or molecular ions [13, 4]. Therefore, candidate carriers might be sought among a large number of possible carbon-based organic molecules.

Although we do not know yet the carriers of DIBs, they can still be a promising tool to trace and to study interstellar medium (ISM) in the Galaxy, and also in nearby galaxies (e.g., see [3]). First of all, DIBs are numerous. Moreover, they are not easily saturated even for stars located at great distance or behind dense clouds, which is a strong advantage for Galactic-scale ISM mapping [8]. Furthermore, they may indicate the physical conditions in the ISM owing to their differential behaviors, i.e., some DIBs are sensitive to the stellar radiation field, and some are not. DIBs are correlated with extinction with a considerable dispersion and differences for each band.

In this article, we will show methods on extracting DIBs from cool (late-type) star spectra. We will also show that DIBs can be used to locate dense clouds in the Galaxy (i.e., Galactic arm), to probe a line-of-sight (LOS) kinematical structure and to trace IS environmental conditions.

2. Data

Data from Gaia-ESO Spectroscopic Survey (GES; [5]) and some analysis have been presented in [11]. In this article, we only review 2 GES fields, namely CoRoT anti-center and CoRoT center (see table 1), due to their interesting kinematical structures. The fields are almost diametrically opposite on the sky (see figure 1). All target stars were observed with the FLAMES multi-object spectrograph (GIRAFFE, $R \simeq 17000$ and UVES, $R \simeq 47000$) with different observation settings.

Table 1. The selected GES Fields (these data are part of [11]).

Field	GIR. targ.	UVES targ.	l_c (°)	b_c (°)	D_{min} (kpc)	D_{max} (kpc)	$A0_{min}$ mag	$A0_{max}$ mag	Ang. size (°)	Setting(s)	Studied DIBs
COROT-AC	57	7	212.9	-2.0	0.1	8.6	0.0	2.5	0.4	UVES5800,HR15N,HR21	6283,6614,8620
COROT C	105	5	37.5	-7.0	0.1	16.1	0.0	2.1	0.3	UVES5800,HR15N,HR21	6283,6614

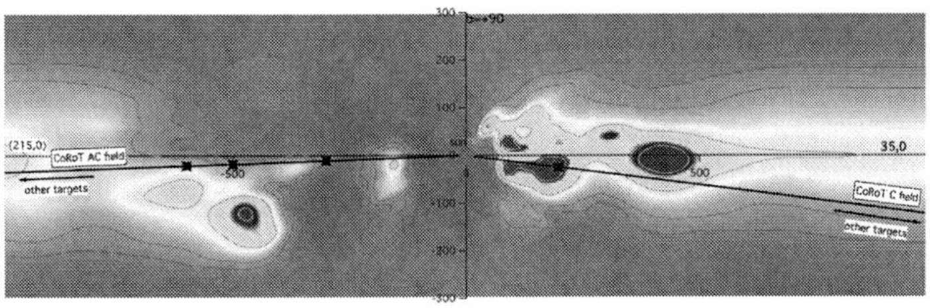

Figure 1. CoRoT fields (black line) and the target stars (black marks) in a slice of the 3D ISM map [8]. Color-excess is shown in rainbow color. Purple shows higher reddening region. Red shows lower reddening region. The CoRoT anti-center direction (left) encounters gradually denser clouds, whereas the CoRoT center field (right) encounters nearby dense region (probably associated with Aquila rift).

3. Method

We have devised several fitting methods to extract DIBs from cool star spectra. The principle of the fitting methods is essentially the same as in [2]. The spectrum is fitted with a combination of a synthetic stellar spectrum, a synthetic telluric transmission, and empirical DIB profile(s). The main difference compared to [2] is that we allow for multicomponent DIBs and subsequently we extract kinematic information. See [11] for the complete description.

For multicomponent DIB fitting, it is necessary to start with initial parameters that are close to the actual solution to avoid secondary minima and to converge more rapidly toward the final solution. In the absence of any IS absorption line (as for GIRAFFE spectra), the initial guesses for the number of required velocity components and DIB radial velocities come from a simplified decomposition of the HI emission spectrum [7] taken from the spectral HI cube in the direction of the target star. Whereas for UVES spectra, the initial guesses come from IS NaI lines.

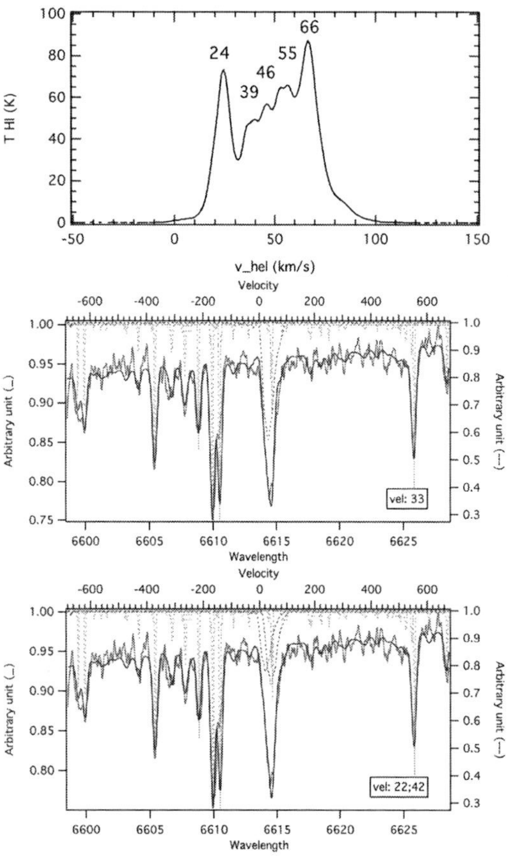

Figure 2. Illustration of the multi-component 6614 Å DIB fitting. See text. Figure is taken from [11].

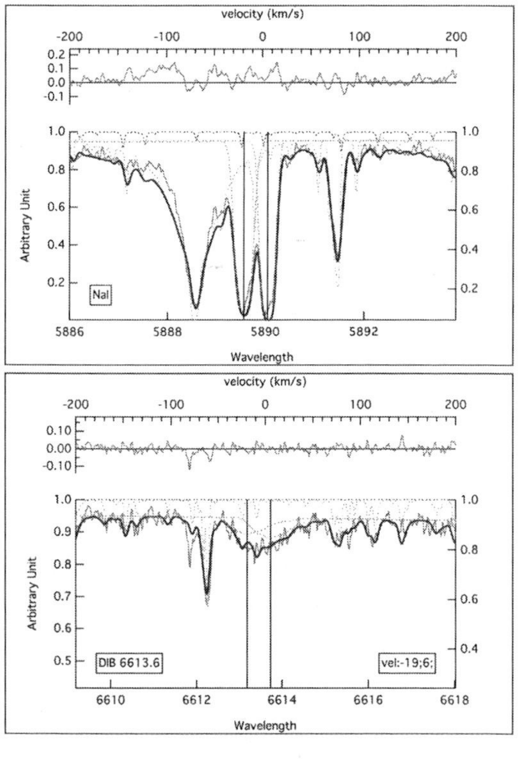

Figure 3. Illustration of the NaI doublet-6614 Å DIB global multi-component analysis. The radial velocities of the NaI and DIBs components are kept linked (see the black vertical line). Figure is taken from [11].

3.1. Hierarchical DIB fitting

When no IS lines are available in the spectral range, independent HI emission spectrum can be used. Since the HI emission spectrum represents the totality of the IS clouds, both in front of and beyond the target star, it should be used carefully. The HI emission spectrum was therefore used to construct a table of velocity guesses (v_{rHI}) and provide upper and lower limits to the velocity range. Then, the DIBs were fitted using the hierarchical sequence of the velocity prior values. Figure 2 shows an illustration of the multi-component 6614 Å DIB fitting. The red line shows the stellar spectrum. The dotted lines are the models: stellar (orange), telluric (green), empirical DIBs (grey). The thick blue line is the optimal model adjustment. The initial guesses for the DIB velocity centroids are 24, 40, 50 km/s and based on the HI spectrum in the same direction (see top plot). The middle panel is an example of a preliminary adjustment with a unique DIB component. The DIB velocity is found to be $\sim$ 33 km/s. The large difference from the initial guess (24 km/s) demonstrates the need for the introduction of a second cloud component. The lower panel is an example of a subsequent adjustment with two DIB components. The two fitted velocities are now close to the first two HI emission peaks.

3.2. Global analysis of lines and DIBs

IS NaI lines are not only used as sources of the first guesses of the cloud parameters, but also enter in the global analysis of lines and DIBs, which means that they are simultaneously measured together with the DIB components. Their radial velocities are linked so that they remain identical throughout the adjustment, component by component. Figure 3 shows an illustration of the NaI doublet-6614 Å DIB global multi-component analysis. The NaI region is shown at top panel and the DIB region at lower panel. The red line shows the stellar spectrum (lower plot) and the fitting residuals (upper plot). The dotted lines are the models: stellar (orange), telluric (green), and interstellar components (blue) respectively. The thick blue line is the fitting result.

3.3. Global analysis of DIBs

In addition to the methods that have been presented in [11], we have devised DIB-DIB global analysis. The global analysis is applied between two different diffuse bands in which their radial velocities are linked. This method can be used to derive the ratio variability between two DIBs. It can be a strong advantage in the detection itself too: as an example, it helps constraining the radial velocity interval for a very wide DIB from its association with a much narrower one, or it helps detecting a weak DIB from the constraint on its radial velocity deduced from a stronger one. See an example of the fitting in figure 4.

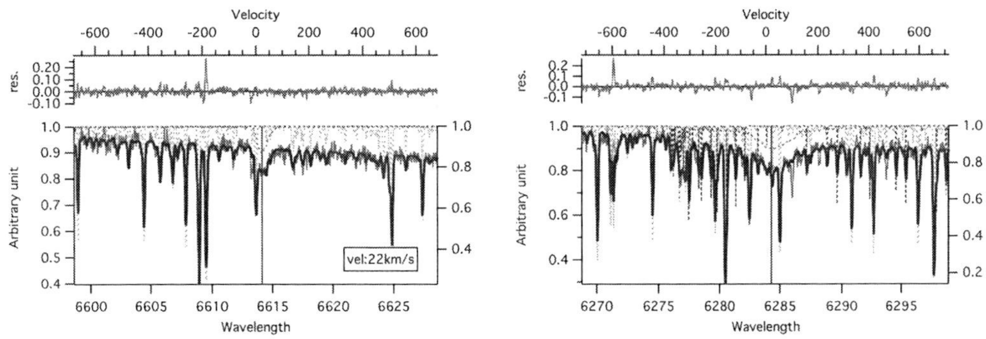

Figure 4. Illustration of the Global analysis: the 6614 Å DIB and 6283 Å DIB for star 06432980-0050546 in CoRoT anti-center field. Vertical black lines show the radial velocity.

IOP Conf. Series: Journal of Physics: Conf. Series **1127** (2019) 012046 doi:10.1088/1742-6596/1127/1/012046

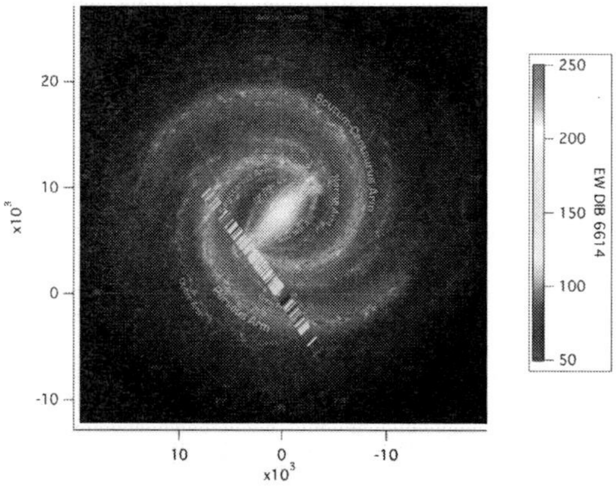

Figure 5. Distribution of target stars projected in Galactic plane and superimposed with Galaxy sketch by Robert Hurt of the Spitzer Science Center. Colors show 6614 DIB strength (mÅ). Unit in pc.

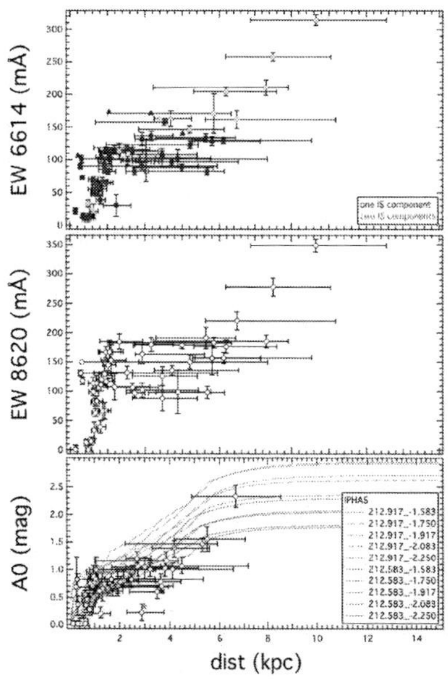

Figure 6. DIBs can be used to locate Galactic arm. Abrupt increases corresponds to the crossings of dense clouds and plateaus to interclouds or interarms. Shown in the plots are results for target stars in CoRoT anti-center field. Figure is taken from [11].

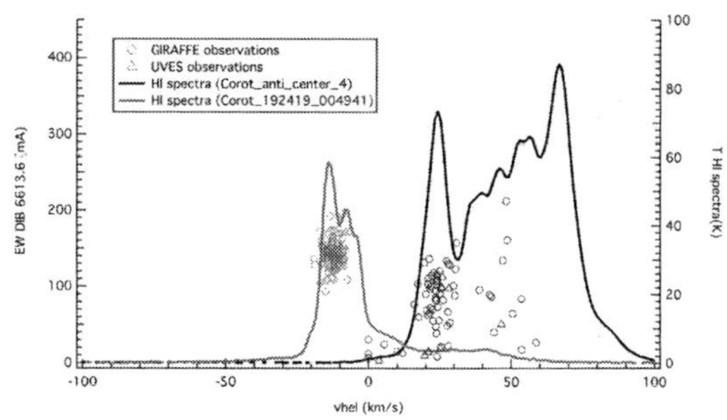

Figure 7. DIBs can be used to trace LOS kinematical structure. The figure shows comparison between the DIB radial velocities and the HI 21 cm emission spectra. Black (red) markers and lines are DIB fitting results and HI spectra from CoRoT anti-center field (CoRoT center field). For CoRoT anti-center field, a significant number of spectra requires two velocity components, which very likely correspond to the Local and Perseus arms. Figure is taken from [11]

4. Results and perspectives

Figure 5 shows the two directions of the selected fields in the Galactic Plane and measurement results of 6614 Å DIB.

For the CoRoT anti-center field, the target stars are located about the Galactic plane and are widely distributed in distances (from 0 to 7 kpc from the Sun). This allows us to probe not only the local arm, but also external Galactic arm (i.e., Perseus arm). As the distance of the target star increases, the LOS intersects a greater part of the ISM and therefore the EW of the DIB is expected to increase, with abrupt increases corresponding to crossings of dense clouds and plateaus to interclouds or interarms (see figure 6). From upper to lower panels are EW 6614 Å DIB vs. estimated distance from Bayesian method [1], EW 8620 Å DIB vs. distance, and A_0 vs. distance. Circles show GIRAFFE observations. Triangles show UVES observations. Colors correspond to the number of IS components used to fit the IS line or band. All nearby targets (D $\leq$ 1kpc) have only single IS component. Distant targets have more than one velocity component, in agreement with the crossing of at least one external arm. Comparison with IPHAS data [12] is also shown in the figure. Here is a demonstration that DIBs can be used to locate dense clouds in the Galaxy (i.e., Galactic arm).

DIBs can also be used to trace LOS kinematical structure (see figure 7). Here, we compare the fitted DIB radial velocities and the HI 21 cm emission spectra. In addition, we have UVES data in which DIBs are linked to IS NaI lines. Thus, DIBs, HI emission spectrum, and IS NaI lines radial velocities are actually in agreement.

In addition, we have applied the 6614 and 6283 Å DIBs global analysis to the GES/UVES targets in the two CoRoT fields. There were 12 target stars that are tested: 7 targets in the anti-center direction and 5 in the center direction. Interestingly, we found that the two directions differ in terms of equivalent width (EW) ratios. With $\frac{EW\ 6614}{EW\ 6283} \simeq 0.1 - 0.2$ for the CoRoT anti-center and $\frac{EW\ 6614}{EW\ 6283} \simeq 0.3$ for the CoRoT center. These differences may be explained by the specific behavior found in several studies. Environmental conditions (especially radiation field) may destroy or favor the DIB carrier(s). The 6283 Å DIB in particular has been shown to be one of the DIBs sensitive to the stellar radiation field. This a preliminary test that DIBs can be used to trace interstellar environmental conditions.

Altogether, these results show how DIBs can be used to reconstruct the large-scale distribution of the ISM in the Galaxy, and can be especially useful for distant clouds because in this case they are strong enough, but not saturated. These results pave the way to three-dimensional mapping of the Galactic ISM based on DIB absorption measurements from current or future stellar spectroscopic surveys. Like all three-dimensional maps, future DIB-based maps are expected to gain in accuracy in a considerable way when Gaia parallax measurements will be available soon in the next data release.

References

[1] Babusiaux C *et al.* 2014 *Astronomy and Astrophysics* **563** A15
[2] Chen H -C *et al.* 2013 *Astronomy and Astrophysics* **550** A62
[3] Cox N L J and Cordiner M A 2008 *IAU Symposium* **251** 237
[4] Galazutdinov G A, LoCurto G, and Krelowski J 2008 *Astrophysical Journal* **682** 1076
[5] Gilmore G *et al.* 2012 *The Messenger* **147** 25
[6] Heger M L 1922 *Lick Observatory Bulletin* **10** 141
[7] Kalberla P M W *et al.* 2005 *Astronomy and Astrophysics* **440** 775
[8] Lallement R *et al.* 2014 *Astronomy and Astrophysics* **561** A91
[9] McCall B J and Griffin R E 2013 *Proc. R. Soc. A* **469** 2151
[10] Merrill P W 1934 *PASP* **46** 206
[11] Puspitarini L *et al.* 2015 *Astronomy and Astrophysics* **573** AA35
[12] Sale S E *et al.* 2014 *arXiv:1406.0009*
[13] Sarre P J 2006 *Journal of Molecular Spectroscopy* **238** 1

IOP Conf. Series: Journal of Physics: Conf. Series **1127** (2019) 012047 doi:10.1088/1742-6596/1127/1/012047

Geoid determination with Hotine's integral based on terrestrial gravity data in Semarang city

L. M. Sabri, Leni S. Heliani T. A. Sunantyo, Nurrohmat Widjajanti

Program Studi Doktor Ilmu Teknik Geomatika, Departemen Teknik Geodesi.
Program Pascasarjana Fakultas Teknik Universitas Gadjah Mada, Jl. Grafika No. 2
Yogyakarta, Indonesia - 55281

laodesabri@gmail.com

Abstract. GNSS surveys can accurately determine the geodetic position, but the elevation values do not have physical meaning for the purposes of engineering work. Physical height can be determined through the gravimetric geoid model representing the relationship between the geometric and physical shapes of the earth. Theoretically, the gravimetric geoid model is a solution of Boundary Value Problems (BVP) to obtain the boundary value in the form of the geoid surface. Before the era of GNSS, a BVP was calculated using Stokes or Molodensky approaches. Another approach that was quite difficult to be practiced in that era was Hotine approach. This approach requires gravity disturbance data on surface of the earth or on the geoid itself. This paper presents an accurate gravimetric geoid modelling based on terrestrial gravity data in Semarang using Hotine's approach. The main gravity data was measured in March 2016 using Scintrex CG-5 gravimeter which has an accuracy of 5 microgal. It covered an area about 1 arc degree squares. Additional gravity data for improving the accuracy of the geoid models was measured by some government and private agencies using analogue gravimeters. It covered Jawa island. The global geopotential model was GECO with maximum degree 2190. The results showed that the computation of Hotine's integral for modelling the geoid of Semarang city had an accuracy of ± 0.044 m. In the region of land subsidence, the geoid model is quite prospective to be implemented as the Vertical Reference System (VRS). Measurement of elevation of a distant point using geoid can be completed in a matter of hours, while using spirit levelling completed within a matter of days.

1. Introduction

In geodesy, accurate elevation can be measured by two methods, namely spirit leveling and geodetic leveling. Measurement of elevation with spirit leveling to produce physical height value is affected by the gravity field equipotential of reference and measurement points [1]. Determining height with a spirit leveling to a certain point located far away from the reference point requires a series of measurements of height difference which is lengthy and time consuming. Measurement a section of spirit leveling within 2 km requires about 1 day of measurement.

Another technique which is accurate and productive for height determination is the measurement using the Global Navigation Satellite System or GNSS. GNSS is the term of satellite-based positioning system that combines satellite of the United States called the Global Positioning System or GPS satellite system, GLONASS belongs to Russia, the satellite system Galileo belongs to the EU, the satellite system COMPASS of China, and other new satellite systems. GNSS survey does not require visibility among points as required in the survey with sprit level. Settings of equipment in GNSS were accomplished only on point of interest or Bench Mark (BM). Those conditions significantly minimize equipment setting errors, so that the remaining error in the survey is just the intrinsic bias of signals from GNSS satellites to receivers [2]. Although GNSS survey can accurately determine the geodetic position, but the resulting height values do not yet have a physical sense as desired in engineering work. Physical height can be calculated through a reduction of the geoid undulation that represents the relationship between the geometric shape of the earth and the physical shape of the earth.

Gravimetric geoid modeling is basically solving Boundary Value Problems or BVP to obtain the boundary value in the form of the geoid surface. Before GNSS era, BVP was solved by Stokes approach and the approach Molodensky [3]. Stokes approach, as used in methods SHGeo, requires gravity data, and the rock density below measured locations [4]. Accuracy of Mexico geoid using SHGeo or Stokes-Helmert Geoid was ± 0.599 m [5], while the accuracy of geoid in Sudan and Tanzania modeled by KTH method were ± 0.290 m [6] and ± 0.278 m [7].

One approach that is quite difficult to be applied before GNSS era is a GPS BVP due to lack of geodetic height data referred to ellipsoid [1]. Recently, coordinate of gravity stations can be measured accurately using relative GNSS positioning. Hotine approach applied to solve GPS BVP requires gravity disturbance data on earth's surface. Testing in Western Australia showed that the geoid modeled from the 60 gravity disturbance data had a similar shape to that modeled geoid from Free-Air gravity anomaly data [8]. The standard deviation Western Australia geoid computed by Stokes approach was ± 0.165 m, while the standard deviation of Hotine's approach was of ± 0.267 m. Application of Hotine's approach for geoid modeling in China using GNSS data and gravity measured in 4870 locations and tested at 65 points introduced an accuracy of ± 0.024 m [9].

In the region of land subsidence, a geoid model with an accuracy of ± 0.024 m is quite prospective to be used as the vertical reference system or VRS. Generally, VRS can be divided in leveling based VRS and geoid based VRS. Elevation measurement of long baseline occupying GNSS receivers and geoid can be completed in a matter of hours, while sprit level measurement of a distant section needs many days of measurement.

Another advantage of the geoid based VRS is a realization that is simpler than the leveling based VRS. Geoid based VRS does not require many physical BM. In a modern measurement system, the role of BM can be replaced by one or several Continuously Operating Reference Station or CORS that can be used as a measurement reference. Deformation of CORS points in the active region can be monitored, because these stations record data continuously. GNSS data can be calculated relative to the International GNSS Service (IGS) stations. Users can utilize the data from the CORS for geodetic position measurement any time with a static or kinematic method, in real time or post processing. Precise orthometric precision depends on quality of GNSS measurement and quality of geoid as reduction model. This research aims to produce precise geoid based on terrestrial gravity data measured in small area.

2. Geoid determination using Hotine's integral

Ellipsoid is a mathematical model to fit the shape of the geoid. Theoretically, gravity potential of geoid is equal to normal gravity potential on ellipsoid. Irregular shapes of the earth and density variation introduce discrepancies between the quantities in normal earth model and actual earth. The difference between the actual gravity potential on the earth's surface (W) and the potential of normal gravity (U) at the earth's surface is called the potential disturbance or potential disruption (T) defined by equation (1) as follow

$$T = W - U \ . \tag{1}$$

The difference between the actual gravity on the earth's surface (g_P) and normal gravity on earth's surface (γ_P) is called the gravity disturbance or disruption of gravity (δg) defined by equation (2) as follow

$$\delta g = g_P - \gamma_P \ . \tag{2}$$

The difference between the actual gravity on the geoid (g_Q) and normal gravity on the ellipsoid (γ) called the gravity anomaly or gravity anomalies (Δg) defined by equation (3) as follow

$$\Delta g = g_Q - \gamma \ . \qquad (3)$$

Measured gravity data are affected by three kind of wave, namely short, medium, and long wave. Short wave was contributed by topographical effect, while long wave was contributed by global gravity of the earth. Medium wave itself is gravity data contributed by density contrast below earth surface. Process of elimination and preservation of short and long wave is known as the Remove-Compute-Restore or RCR. First step of RCR procedure is to calculate the residual gravity disturbance (δg_{res}) by removing the effects of global gravity disturbance (δg_1) from actual gravity observations (δg_{obs}) using equation (4) as follow

$$\delta g_{res} = \delta g_{obs} - \delta g_1 \qquad (4)$$

Global gravity disturbance (δg_1) computation is based on degree (n) and order (m) of spherical harmonic coefficient, cosines coefficient ($\Delta\overline{C}_{n.m}$), sinus coefficient ($\Delta\overline{S}_{n.m}$) as follow

$$\delta g_1(\theta,\lambda) = \frac{GM}{R^2}\sum_{n=2}^{N}(n+1)\sum_{m=0}^{n} P_{n.m}(\cos\varphi)\left[\Delta\overline{C}_{n.m}.\cos m\lambda + \Delta\overline{S}_{n.m}.\sin m\lambda\right]. \qquad (5)$$

In equation (5), G denotes Newton gravitational constant, M denotes earth's mass, R denotes earth's radius, $P_{n,m}$ denotes legendre function, while φ and λ denote latitude and longitude of point of interest.

Residual gravity disturbance ($\delta g(\varphi',\lambda')$) are then interpolated using krigging method to ensure dense distribution of contribution points. Spherical distance (ψ) between computation point in geodetic coordinate (φ , λ) and contribution point in geodetic coordinate (φ' , λ') can be calculated using equation (6) as follow

$$\cos\psi = \sin\varphi.\sin\varphi' + \cos\varphi.\cos\varphi'.\cos(\lambda'-\lambda) \qquad (6)$$

Residual geoid undulation (N_{res}) can be calculated using Brun's formula as seen in equation (7)

$$N_{res}(\varphi,\lambda) = \frac{R}{4.\pi.\gamma}\int_{\lambda'=0}^{2\pi}\int_{\varphi'=-\pi/2}^{\pi/2}\delta g(\varphi',\lambda').H(\psi).\cos\varphi'.d\varphi'.d\lambda' \qquad (7)$$

where Hotine function (H) is declared in equation (8)

$$H(\psi) = \frac{1}{\sin\frac{\psi}{2}} - \ln\left(1 + \frac{1}{\sin\frac{\psi}{2}}\right) \qquad (8)$$

Restoration of Global Geopotential Model effect on local geoid (N_1) to obtain definitive undulation (N) can be calculated using equation (9)

$$N = N_{res}(\varphi,\lambda) + N_1 \qquad (9)$$

where

$$N_1(\varphi,\lambda) = R.\sum_{n=2}^{N}\sum_{m=0}^{l}\overline{P}_{nm}.(\sin\varphi).\left(\Delta\overline{C}_{nm}.\cos m\lambda + \Delta\overline{S}_{nm}.\sin m\lambda\right) \qquad (10)$$

3. Method of research

Geoid precision analysis occupies comparison of gravimetric and geometric geoid models. Gravimetric geoid modeling requires terrestrial gravity data, global geopotential model, and geodetic coordinates of gravity point. Geometric geoid is obtained by calculating the difference between geodetic height and orthometric height. Geodetic height was measured using relative static method of GNSS, while orthometric height was measured by Wild NAK2 spirit level.

Global geopotential model used in this research is data GECO created from the combination of data EGM2008 with data from the GOCE satellite. GECO has a harmonic coefficients to degree 2190 representing an area of 5 arc minute spatial resolution. Gravity data measured using Gravimeter Scintrex CG-5 with accuracy of 5 microgal. First geoid was calculated using gravity data measured at 62 locations with a density of about 2 km, as shown in Figure 1. Second geoid was calculated using gravity data measured at 186 points which was concentrated in the city of Semarang with additional measurements at the points until it reached an area of 1 arc degree squares, as shown in Figure 2.

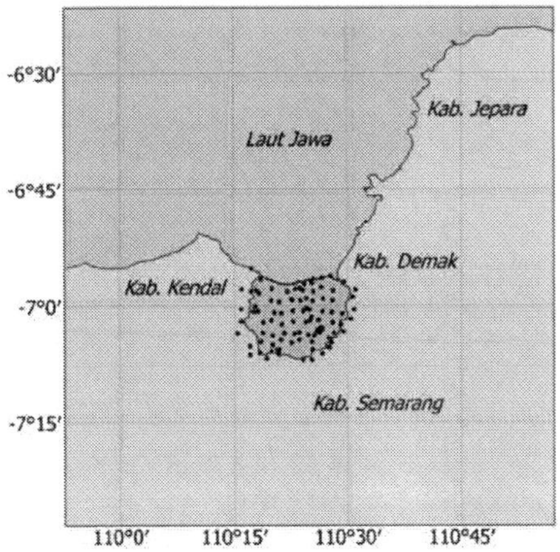

Figure 1. Gravity measurement location for 15 arc minute geoid modelling

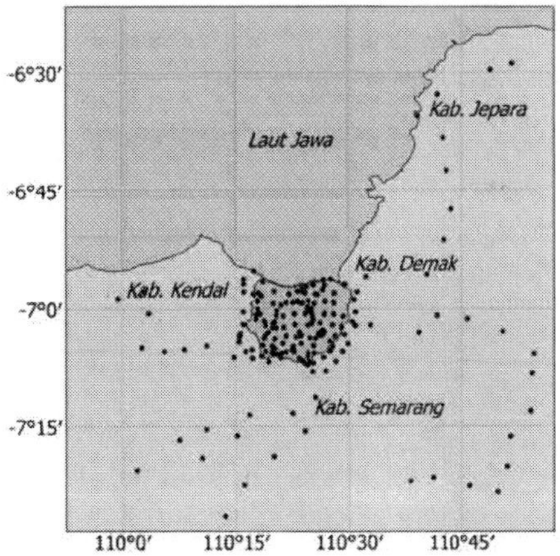

Figure 2. Gravity measurement location for 1 arc degree geoid modelling

Geodetic coordinates of gravity measurement locations was measured accurately using Topcon Hiper II GNSS receivers that recorded the GPS and GLONASS satellite signals. Measurements were made with rapid static method that was connected to GNSS CORS station. Reduction baseline performed using software Topcon Tools 8. Gravsoft 2.9.7 was occupied for gravimetric geoid calculation, especially GEOEGM and STOKES modules. GEOEGM was used to calculate the gravity disturbance and gravimetric geoid undulation of global geopotential model. For this research purpose, stokes function in STOKES module was modified to Hotine function.

4. Results

According to GNSS measurement, geodetic elevations of gravimetric stations were between 26.741 m to 356.04 m above the WGS84 reference ellipsoid. To obtain high accuracy positioning, the satellite observations carried out for 1 to 4 hours to ensure the value of the phase integer ambiguity was resolved. Solving phase ambiguity in Topcon Tools 8.5 produced geodetic coordinates with horizontal accuracy from ±0.004 m to ±0.087 m, while vertical accuracy from ±0.006 m to ±0.058 m.

Measurement of sprit level was divided into three loops of measurement, namely: northern loop, southern loop, and UNDIP loop. Path length of sprit level in the northern loop was 13.805 km with disclosure about +0.004 m. Based on SNI 19-6988-2004, the leveling net was fit the LAA classification. Length of southern loop is 51.520 km with misclosure +0.096 m which fit to LD classification.

Range of geoid undulation based on GECO data only was from 24.6 m to 25.4 m, as shown in Figure 3. Geoid undulation rose dramatically towards the south or towards Mount Ungaran. Drastic increase in geoid undulations was found in the northeast or the direction of Mount Muria in Jepara. Visually, the geoid undulation in Semarang and surrounding areas were modeled from GECO had a similar shape to the geoid obtained from direct measurements by Scintrex CG-5, as shown in Figure 4.

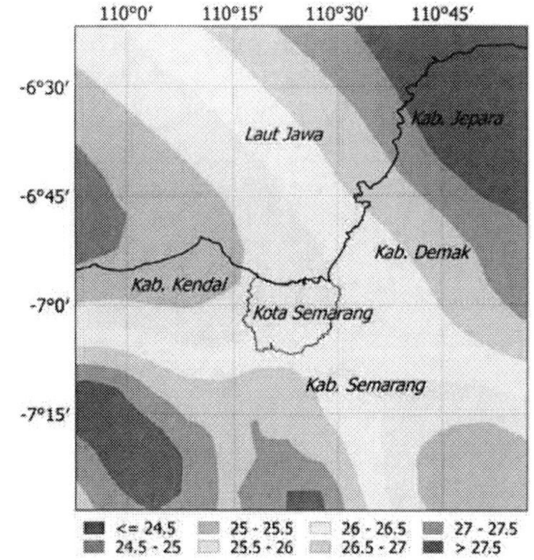

Figure 3. Geoid undulation from GECO only

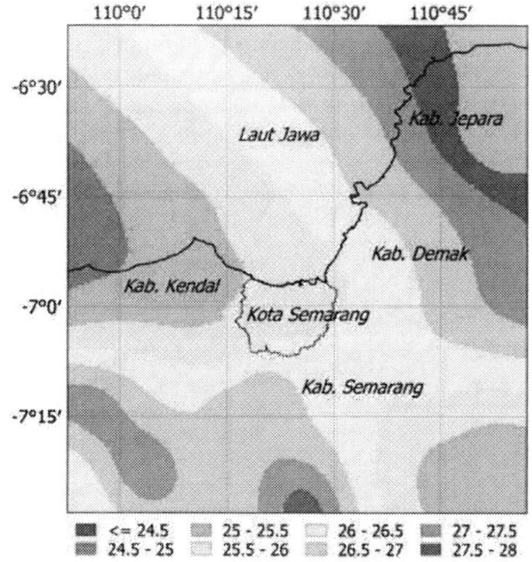

Figure 4. Geoid undulation from combination of GECO and terrestrial gravity data

Accuracy of gravimetric geoid was validated using 30 points of geometric geoid which was measured using GNSS and spirit leveling. Validation point was distributed along southern loop. Accuracy of geoid modeled from combination of GECO and 62 gravity data was ± 0.045 m, while geoid modeled using 186 gravity data was ± 0.044 m.

5. Conclusion

The conclusions of this research are:
- Accuracy of modeling the geoid with Hotine approach is suitable for engineering purposes
- Accuracy of geoid can be improved by enlarging gravity data measurement.

Acknowledgement

Thanks to Prof. Rene Forsberg of DTU who has allowed the use of Software Gravsoft. Thanks to Badan Informasi Geospasial (Geospatial Information Agency of Indonesia) for providing the gravimeter.

References

[1] Hofmann-Wellenhof, Bernhard, Helmut Moritz. (2005). Physical Geodesy. Springer
[2] Leick, Alfred. (2004). GPS Satellite Surveying. John Wiley & Sons.
[3] Sneeuw, Nico. (2006). Lecture Notes : Physical Geodesy. Ianstitute of Geodesy Universitat Stuttgart.
[4] Vanicek, Petr, Robert Kingdon, Marcelo Santos. (2012). Geoid versus quasigeoid: a case of physics versus geometry. Contributions to Geophysics and Geodesy Vol. 42/1, 2012 (101–

117)
[5] Hernandez-Navarro, Antonio. (2004). The submetric Geoid of Mexico. FIG Working Week 2004. Athens, Greece, May 22-27, 2004
[6] Abdalla, Ahmed. (2009). Determination of a gravimetric geoid model of Sudan using the KTH Method. Tesis. Royal Institute of Technology (KTH), Stockhlom, Swedia.
[7] Ulotu, Prosper E. (2009). Geoid Model of Tanzania from Sparse and Varying Gravity Data Density by the KTH Method. Disertasi. Royal Institute of Technology (KTH), Stockhlom, Swedia.
[8] Kirby, J. F. (2003). On the combination of gravity anomalies and gravity disturbances for geoid determination in Western Australia. Journal of Geodesy (2003) 77: 433-439
[9] Shu, Chan-Fang, Fei Li, Ming-Feng Li, Jie Zhang. (2011). Determination of GPS/Gravity quasi-geoid using the Bjerhammar method. Chinese Journal of Geophysics. Vol 54 (10), 2503-2509.
[10] Badan Standarisasi Nasional. (2004). SNI 19-6988-2004: Jaring kontrol vertikal dengan sipat datar.
[11] Sanso, Fernando, Michael G. Sideris. (2013). Geoid determination: Theory and methods. Springer

Light curve analyses of eclipsing binary system ASAS 172533–1221.4

Agus T P Jatmiko[1], M Yusuf[1], and M Putra[1,2]

[1]Bosscha Observatory, Institut Teknologi Bandung, Jl. Peneropongan Bintang, Lembang, Bandung, West Java, Indonesia

[2]Astronomy Study Program, Institut Teknologi Bandung, Jl. Ganesha No. 10 Bandung, West Java, Indonesia

E-mail: agustrionopj@alumni.itb.ac.id

Abstract. Using the data taken from our 0.36 m f/7.2 robotic telescope, we performed a very first light curve (LC) analyses of eclipsing binary ASAS 172533–1221.4, one of target stars which is part of program stars in our variable star survey project. The LC of this star was constructed by using LEMON, a semi–automatic photometric pipeline written in Python. We refined a Time of Minima (ToM, T_0) and variability period of this system, P and updated its ephemerides as $HJD(min\ I) = 2457200.255578 + 0.678861 * \phi$. The LC modeling of the system was conducted with the PHOEBE (PHysics Of Eclipsing BinariEs) software built on top of the widely used WD program . The assorted LC modeling solutions are shown as follows: mass ratio $q = 0.811 \pm 0.009$, inclination $i = 70.62 \pm 0.01°$, temperature of primary and secondary component $T_1 = 5559.23 \pm 83.51\ K$ and $T_2 = 3871.64 \pm 43.66\ K$, respectively, and modified Kopal potentials which are a function of primary's and secondary's radii $\Omega_1 = 3.436 \pm 0.018$ and $\Omega_2 = \Omega_{cr} = 2.980$, respectively. It is concluded that ASAS 1725533–1221.4 is found to be near–contact system with almost similar size between primary and secondary components, with its secondary component is already filling its Roche lobe.

1. Introduction

The All Sky Automated Survey (ASAS, [1]) project started in 1996 with the observations in the Southern Hemisphere aimed at classification and investigation about 10^7 stars that show photometric variability in their light curves. The observations were carried out with the CCD cameras having 4K and 2K resolutions in the V and I filters at the stations located in Chile and Hawaii. The main scope of the survey was to form the ASAS Catalog of Variable Stars (ACVS).

Among those stars in ACVS, there are many of them which are still do not have a published physical and geometric parameters yet. We are trying to fill the gap by observing unpublished variable stars by using our new robotic telescope (Bosscha Robotic Telescope, BRT). BRT has 0.36 m f/7.2 telescope equipped with 4K × 4K CCD camera [2]. Our telescope system has a bigger aperture and camera resolution, in comparison with ASAS', so aiming for better signal–to–noise ratio (S/N) and temporal resolution should not be a problem. This work also acts as benchmark model for our next works to develop the next generation robotic telescopes, which is part of Indonesia National Observatory Project in Amfoang, East Nusa Tenggara [3]. This new observatory is planned to do its first light in 2019.

2. Observation

The observations were conducted from June to August 2015. Target stars was chosen with several criteria as follows: (1) They have a relatively short variability period (from several hours to couple of days) so hopefully we could cover one orbital period during a single night or several nights of observations; (2) They are relatively bright ($m_{\text{vis}} < 14$) in order to achieve good S/N within allotted time; (3) They have a modest (but not too small) delta magnitude ($\Delta m > 0.15$) so we could easily recognize variability in target stars. With this criteria in mind, we managed to observe four target stars only, due to unstable weather during our observing time, by using differential photometry technique. BRT gives us a Field of View (FoV) of $\sim 42' \times 42'$ so it is sufficient to ensure that several comparison and check stars do fit into one frame along with target star. The target stars and their observation time are listed in table 1 below and the properties of the systems given by ACVS are shown in table 2.

In this work, we present a LC modeling and analyses of ASAS 172533–1221.4 which will be explained in details in the following section. We focused only on ASAS 172533–1221.4 because we do not have a good coverage of light curves for other listed stars due to weather problem. Analyzing those data will give us an unreliable conclusion about the systems.

Table 1. BRT Target Stars (June–August 2015).

Observation Time	Target Stars
June 25, 2015	ASAS 172533–1221.4, ASAS 185542–0123.1
June 26, 2015	ASAS 172533–1221.4, ASAS 185542–0123.1
August 16, 2015	ASAS 193943–1116.6, ASAS 185542–0123.1
August 24, 2015	ASAS 005328+2536.4
August 28, 2015	ASAS 172533–1221.4, ASAS 185542–0123.1
August 29, 2015	ASAS 172533–1221.4, ASAS 185542–0123.1
August 31, 2015	ASAS 172533–1221.4, ASAS 185542–0123.1

Table 2. Properties of the systems given by ACVS. Magnitudes are given in V filter. α and δ are right ascension and declination of the systems.

System	α_{2000}(h m s)	δ_{2000} (d m s)	P (days)	Magnitude
ASAS 172533–1221.4	17 25 33	–12 21 24	0.678839	9.97
ASAS 185542–0123.1	18 55 42	–01 23 06	1.161599	10.15
ASAS 193943–1116.6	19 39 43	–11 16 36	1.925659	11.02
ASAS 005328+2536.4	00 53 28	+25 36 24	0.345571	10.80

3. LC Construction and Period Analysis

We applied a standard data reduction technique, i.e. bias, dark and flat correction in order to obtain a noise–free data. The LC construction for ASAS 172533–1221.4 was done by using `LEMON`, a semi–automatic photometric pipeline written in Python [4]. This code does plate solving automatically to the data and updates their file header. It also could boost S/N by combining multiple files into one 'mosaic' file. This step, even though it is optional, could make determination of target star's magnitude easier, especially when dealing with data which have low S/N. Next process is to obtain an instrumental magnitude for target star. We modify `LEMON` code to automatically include 200 brightest stars in the image frame as a comparison and check star, aside from target star itself. By default, `LEMON` will include all of stars which it could find in a whole frame. In our case this code could detect more than 5000 stars in a single frame. By modifying the core program, we cut down the need of extensive computing power and time because `LEMON` only needs to process less stars compared to the unaltered code. Another workaround to achieve the same goal is to provide a list–file which contains coordinates of several stars which are being used as a comparison and check stars. We consider this method was more tedious because we had to select manually the stars which we use for that purposes. Last but not least, the next step is generating the LC itself.

ACVS provides an estimation of variability period and T_0 of target star. We tried to fine–tune these parameters by developing a Python code and apply it to the ACVS data. Our code for period determination is based on Phase Dispersion Minimization (`PDM`) techniques [5] which is easy and efficient to compute. The ACVS data contain grade C and D points that are mentioned as "not measured" and "useless" in the header of the data files, therefore, we excluded these points from our solutions. For T_0 determination, we took the estimated value taken from ACVS as initial value and iterate it within `PHOEBE`. As result, we update the ephemerides of ASAS 172533–1221.4 as

$$HJD(min\ I) = 2457200.255578 + 0.678861 * \phi$$

where HJD is Heliocentric Julian Day, and ϕ is normalized epoch of observation (or phase).

4. LC Modeling and Physical Parameters Determination

We were using `PHOEBE` [6] which is based on `WD` program [7] for data fitting to obtain physical properties of the star. Complete solutions need spectroscopic data to be elaborated with photometric data. This, however, is not covered in this work, therefore, physical properties of the star are obtained solely based on LC modeling and are provided 'as is'. The most important parameter for LC data fitting is mass ratio, q, which can be determined accurately only through high resolution spectroscopic radial velocity measurements. In this case, we adopted value from ACVS and consider it fixed until other parameters reach a certain degree of convergency, then we iterate it as well. We also held period fixed during the whole iteration process.

Basically, we set mass ratio q, inclination i, temperature of primary and secondary components T_1 and T_2, modified Kopal potentials for primary and secondary components, Ω_1, Ω_2, and luminosity of primary component L_1 as free parameters. We took the initial values from ACVS for these aforementioned parameters and iterate them within `PHOEBE`. In addition, since the system was hypothesized to be semi–detached binary—according to a quick look into the LC profile—the analyses were performed in the suitable mode of the `PHOEBE` software. Therefore, Ω_2 was set to the critical potential value of the first Lagrangian point, Ω_{cr} . We also incorporate the gravitational darkening coefficients g_1, g_2 which were derived from the values given by [8]. The results of physical properties determination are tabulated in table 3. The LC yielded by solutions is presented in figure 1. We also present the geometric configuration of ASAS 172533–1221.4 at $\phi = 0.25$ in figure 2. Reader should take a note that all uncertainties presented here are a formal one, it means that . In order to determine whether the solutions are a global minimum or not, one should apply a more sophisticated algorithm, such as Bayesian statistics, e.g. Markov Chain Monte Carlo (MCMC), in particular. The next generation of `PHOEBE`, i.e. `PHOEBE2` will adopt this algorithm along with 'classical' algorithm which are already applied in the current `PHOEBE` code [9].

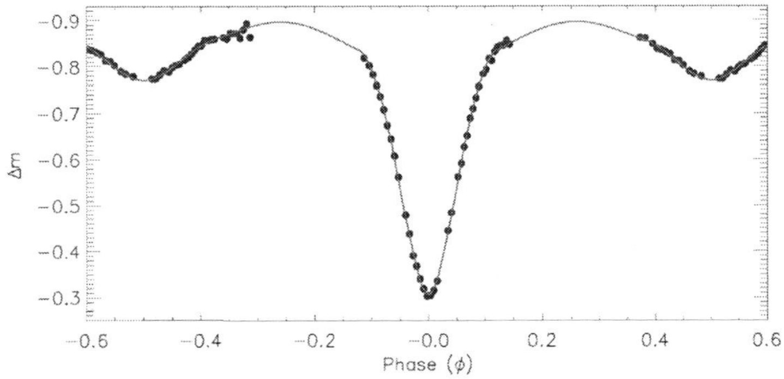

Figure 1. The theoretical LC compared to observation data for ASAS 172533–1221.4. The dots are the observed data points, and the thick line is best fits of a LC model to data points.

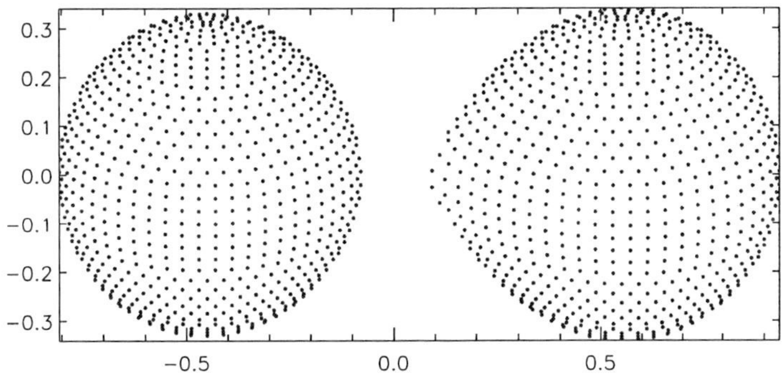

Figure 2. The geometric configurations of ASAS 172533–1221.4 at $\phi = 0.25$.

Table 3. LC Solutions of ASAS 172533–1221.4.

Parameter	Result from this work
q	0.811 ± 0.009
i (°)	70.62 ± 0.01
T_1 (K)	5559.23 ± 83.51
T_2 (K)	3871.64 ± 43.66
Ω_1	3.436 ± 0.018
Ω_2	$\Omega_{cr} = 2.980$
$r_1(=\frac{1}{R_1})$	0.291 ± 0.005
$r_2(=\frac{1}{R_2})$	0.277 ± 0.005
T_0	2457200.255578
P (days)	0.678861

IOP Conf. Series: Journal of Physics: Conf. Series **1127** (2019) 012048 doi:10.1088/1742-6596/1127/1/012048

5. Discussion and Concluding Remarks

We present the LC solutions of ASAS 172533–1221.4, one of binary system selected from ASAS database as part of variable stars observation campaign in Bosscha Observatory. The filling factor of the primary component ($f = \frac{r_1}{r_L}$) of ASAS 172533–1221.4 was found to be 81% using the volume radius of the Roche lobe given by [10]:

$$r_L = \frac{0.49q^{\frac{2}{3}}}{0.6q^{\frac{2}{3}} + \ln(1 + q^{\frac{1}{3}})}$$

Therefore, the component is quite close to the Roche lobe which confirm our hypothesis. It is concluded that ASAS 172533–1221.4 is a near–contact (or semi–detached) system. Both components have almost a similar sizes with its secondary component is already filling its Roche Lobe. We also got a slightly higher q (i.e. 0.811 from this work, compared with $q = 0.707$ from ACVS). This indicates that the secondary component are more massive than previously thought. The more accurate q value should be obtained from spectroscopy observation which is not available in our data and should be put as a constant parameter in LC modeling and physical parameters determination process.

More data are always desirable to complete the gap, especially between $\phi \approx 0.2$ to $\phi \approx 0.6$ in our data. These data points will improve fitting accuracy and they will ensure that we get more reliable results. Spectroscopic data, i.e. radial velocity measurements, surely will improve the LC solutions, especially on determining q value which is crucial to LC modeling in order to get absolute parameters of the system. Implementation of Bayesian statistics into data is highly recommended in order to ensure that the LC solutions are already at a global minimum.

References

[1] Pojmanski, G. 1997 *AcA* **47** 467
[2] Yusuf, M. 2016 6th *ICMNS* (in prep., ABS–279)
[3] Hidayat,T. et al. 2014 *Experimental Astron.* **37** 85–108
[4] Terrón, V. and Fernández, M. 2011 *Highlights of Spanish Astrophysics Conf. VI* 755–761
[5] Stellingwerf, R.F. 1978 *ApJ* **224** 953–960
[6] Prša, A. and Zwitter, T. 2005 *ApJ* **628** 426–438
[7] Wilson, R. E. & Devinney, E. J. 1971 *ApJ* **166** 605–619
[8] van Hamme, W. 1993 *AJ* **106** 2096
[9] Prša, A. et al. 2016 `http://arxiv.org/abs/1609.08135`
[10] Eggleton, P.P. 1983 *ApJ* **268** 368

CCD observation of daylight crescent moon at Bosscha observatory

P. Mahasena(1,2), **M. Yusuf**(2), **M. Irfan**(2), **E.I. Akbar**(1,2), **A.T.P. Jatmiko**(2), **D. Mandey**(1,2), **A. Setiawan**(2), **M. Sulaeman**(2), **T. Hidayat**(1,2), **D. Herdiwijaya**(1,2), **M. Raharto**(1,2), **B. Dermawan**(1,2)

[1]Astronomy Research Division, FMIPA, Institute of Technology Bandung, Indonesia
[2]Bosscha Observatory, FMIPA, Institute of Technology Bandung, Indonesia

E-mail: mahasena@as.itb.ac.id

Abstract. Bosscha Observatory, Institut Teknologi Bandung, has been experimenting to observe crescent moon using small telescope equipped with digital cameras since 2007. The initial motivation was to provide evidence, *i.e.* digital image, of the first crescent moon visible after new moon (hilal) for moslem community in determining the first day of Ramadan and Shawwal. Initially, observation was done after sunset, following common believe that crescent moon can be detected only after glare from the Sun disappear. In early 2000's it was shown that crescent moon can be detected using CCD camera before sunset (in daylight) with special setup. We catched up the idea in 2013. In this paper we summarize what we have learned so far. Now, detection of crescent moon with elongation $e \geq 4^\circ$ is not a problem. Detailed report will be written somewhere else.

1. Introduction

There are several islamic organizations in Indonesia who define their own (lunar) calendars, with different sets of criteria. At least, for determining the beginning of three months (Ramadan, Shawwal, and Dhul-Hijjah) some organizations require sighting of the new moon crescent to determine its first day. On the other hand, some organizations define their calendar only by (modern) calculation of relative positions of the Moon from the Sun. This occasionally results in uncomfortable situation in the country when there were two versions of the 1st of Shawwal, for example [1]. Apart from the difference in the definition for the new month, weather and inaccuracies in the sighting report contribute to the situation. Incorrect positive reports (mostly statement under oath; without hard evidence) from unaided-eye observers in "impossible" cases (*e.g.* because the Sun-Moon angular distance were too small) caused efforts to reconcile the two different definitions more difficult. To improve positive sighting report, by providing digital images, was our first motivation to do experiment on observing crescent moon using digital camera [2].

Soon after we embarked with our first experiment back in 2007 we found that it was not a trivial task. Using small telescope ($D = 66$ mm, $F/5.9$) equipped with pocket digital camera, we followed the tradition to observe crescent moon after sunset (new moon) or before sunrise (old moon). For the first five years, 2007 – 2011, we worked in video mode streaming our after-sunset observation to Internet. Mostly we failed to detect the crescent when moon age was ≤ 18 hours,

Table 1. List of our first successful observations of crescent moon in 2012 and 2013. Observations number 1 – 5 used DSLR camera after sunset, number 6 – 8 used CCD camera (+ I filter) in daylight. Details including (contrast-enhanced) photos of these observations can be found at `http://bosscha.itb.ac.id/hilal/m/data`.

No.	Place	Date	Time	Moon age	Δ Alt	Δ Az	Elongation
1	Subang	22-05-2012	18:08 LT	$+37^h\,14^m$	$+14°\,51'$	$+03°\,20'$	$15°\,14'$
2	Subang	20-06-2012	18:06 LT	$+19^h\,27^m$	$+08°\,42'$	$-02°\,05'$	$08°\,57'$
3	Bangkalan	20-07-2012	17:46 LT	$+30^h\,13^m$	$+13°\,33'$	$-06°\,19'$	$14°\,57'$
4	Kupang	18-08-2012	17:55 LT	$+18^h\,03^m$	$+07°\,26'$	$-06°\,36'$	$09°\,57'$
5	Makassar	18-08-2012	18:19 LT	$+18^h\,24^m$	$+07°\,01'$	$-07°\,23'$	$10°\,11'$
6	Lembang	15-09-2012	14:54 LT	$-17^h\,00^m$	$-11°\,38'$	$-03°\,17'$	$11°\,56'$
7	Lembang	16-10-2012	09:46 LT	$+13^h\,23^m$	$-08°\,13'$	$+08°\,31'$	$09°\,18'$
8	Lembang	13-03-2013	07:53 LT	$+29^h\,58^m$	$-14°\,19'$	$-07°\,47'$	$16°\,02'$

or when moon illumination was $\leq 1\%$. The hardest challenges were (i) often not clear sky above west horizon, (ii) low contrast between the crescent and the background sky, (iii) polar alignment and pointing accuracy of our portable mounting, (iv) wind in the telescope area, and (v) low effective resolution of our low-cost video system.

We gave up working in video mode and then experimenting in picture mode using DSLR camera, with better results. At about the same time we tried a technique used by Martin Elsasser [3] and Thierry Legault [4] to photograph crescent moon using CCD camera and Johnson/Bessel I filter (to increase contrast; referred to as I filter hereafter), in daylight. We list in Table 1 our first positive results using DSLR and CCD camera (+ I filter), which are comparable regarding their elongation. The smallest elongation was $\sim 9°$.

In the followings we will show why I filter is essential for contrast enhancement, describe our current instrument setup, briefly describe our observation procedure and software algorithm, and show some latest results.

2. Wavelength dependence of contrast

To see in which wavelength crescent moon is most contrast with its sky background, in visual range, we took some CCD spectral images with slit orientation shown in Figure 1 (inset). The spectra of waning crescent were taken using Shelyak Lhires III spectrograph attached to an 8-inches reflecting telescope when the sky background was relatively blue.

Using spectroscopy data reduction packages within *IRAF* we extract spectra from two segments of the slit indicated by "Sky + Moon" and "Sky." We then plot the ratio between the two spectra as a function of wavelength in Figure 1. We can see that contrast between the crescent and its sky background is higher at the red-end of the spectra, above 740 nm. Since response profile of I filter also peaks around 740 nm, daylight crescent moon will be easiest to be recorded using CCD + I filter, in visual range. Note that quantum efficiency of visual CCD usually has decreased to about 40% at 740 nm.

3. Instruments setup

In order to increase crescent image contrast, apart from utilising I filter, we use a homemade yet very effective baffle to reduce stray light. The baffle is a cylinder, extension of the original telescope baffle of same diameter, made from black cardboard. The cylinder is about 1 meter long, and at every 4 cm we put black cardboard rings (inside). The effectiveness of this baffle had been judged by pointing the telescope to different angular distances from the Sun while taking

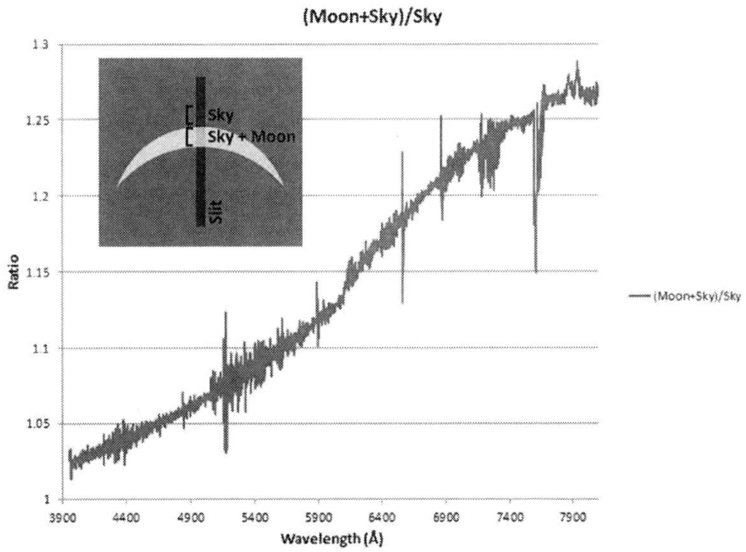

Figure 1. Ratio of spectra extracted from slit segments indicated by "Sky + Moon" and "Sky" (in the inset) as a function of wavelength. Inset: slit orientation with respect to crescent arc. Shelyak Lhires III spectrograph had been used to obtain the spectra.

images of the sky. In moderately blue sky we do not see significant sky brightness increase even if the telescope is pointing to only $\sim 3°$ from the Sun.

We use William Optics 66 mm ($F/5.9$) telescope and Celestron Skyris 445M (or ImagingSource DMK 41AU02.AS) CCD camera. With this configuration full moon is about 100% (or 80%) of the short axis of the CCD and the sampling is $\sim 2''$/pixel (or $\sim 2.4''$/pixel). The camera can take images up to 30 fps (or 15 fps) and its variable shutter speed (0.0001 s – 30 s) is very convenient for taking bright sky images near the Sun where saturation usually is a problem.

In order to have good pointing to and tracking of the Moon we use fixed mounting AstroPhysics 1600 GTO with pointing accuracy of $\sim 3''$. This mounting can track the Moon both in the right ascension and declination direction. With this mounting we are confident that the crescent would stay on the same pixels for few minutes.

4. Observation procedure and software algorithm

Before taking images of the sky in the crescent direction, it is very important to create "flat image" from an adjacent direction (estimation: far enough to exclude the crescent). Experience shows that pattern of flat image changes if we move too far from crescent direction, which makes subsequent image reduction less effective. The pattern changes with elongation as well, so we need to renew our flat image from time to time.

With the above instruments setup, we point the telescope to the direction where the crescent is supposed to be. We then create a stacked image (call it S) from N individual frames, each of which is processed with the following steps: (i) divide the frame with current best flat image, (ii) do histogram equalization to the resulted image to increase contrast, and (iii) put the resulted image into the stack. Contrast of the final stacked image is further enhanced using Contrast Limited Adaptive Histogram Equalization (CLAHE) method proposed in [5] and summarized in [6]. The resulted image is the image S.

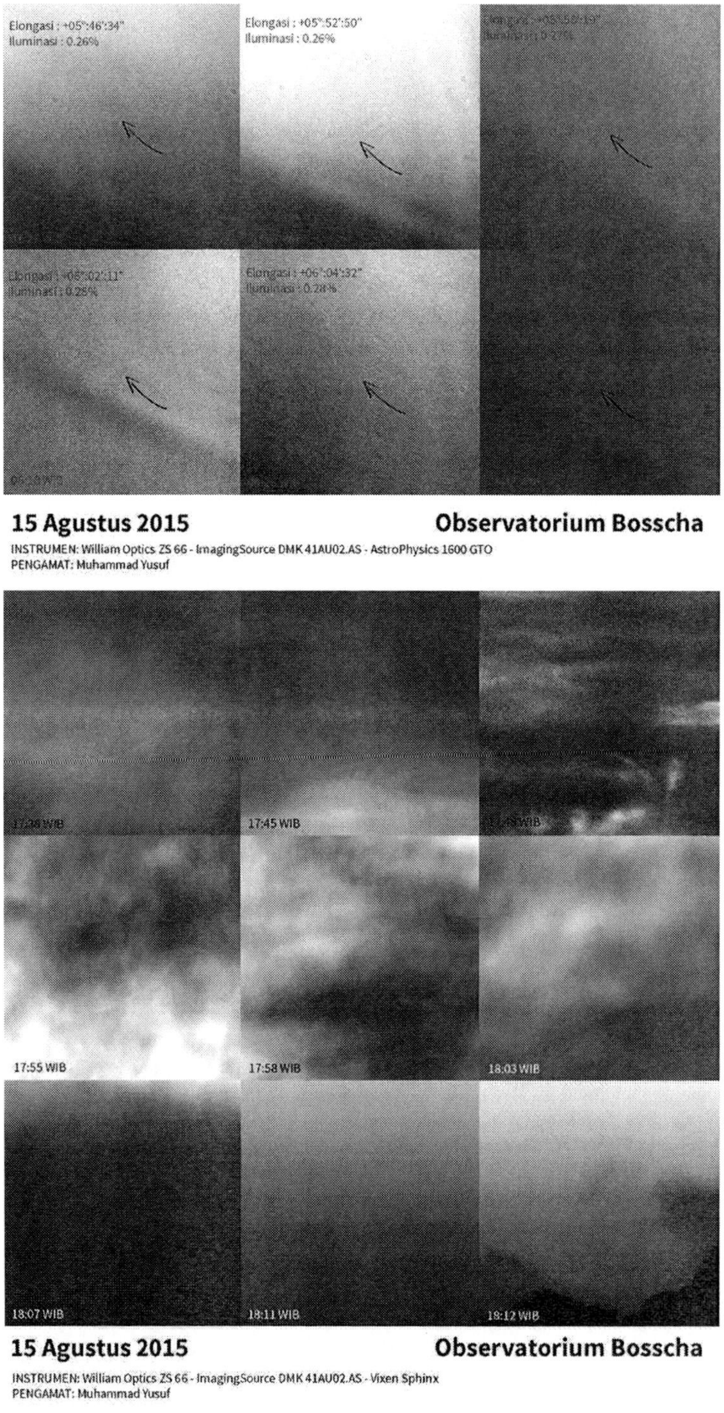

Figure 2. Upper panels: before sunset. Lower panels: after sunset.

The repetition of image acquisition and processing are done by a software that we have developed. For example, if $N = 50$ then we will get one image S every 2 seconds, because the camera can work at maximum 30 fps. We can thus see a near-real-time video observation of crescent moon. At this point, it is easy to broadcast/stream the observation to Internet (to `youtube.com`, for example).

5. Results

In 15 August 2015 we succeeded in tracking crescent moon from Bosscha Observatory. The tracking started at around 07:30 LT when illumination was 0.2% and elongation was $05^\circ\,46'$, and ended at around 18:10 LT when the moon set at the horizon with illumination of 0.7% and elongation of $09^\circ\,44'$. We were surprised that our system works even if there are considerable amount of cloud. We show images of before- and after-sunset crescent moon in Figure 2. Now we are confident that, provided the sky is moderately blue, crescent moon with elongation $e \geq 4^\circ$ will be detectable using our system.

Acknowledgments

We sincerely thank staff at Bosscha Observatory ITB for their technical and administrative support which make this experiment possible, and the conference LOC for their hospitality.

References

[1] Djamaluddin T 2000 Crescent visibility in Indonesia *Warta LAPAN* **2** 137
[2] Hidayat T *et al.* 2010 Developing information system on lunar crescent observations *ITB J. Sci.* **42A** 67
[3] Elsasser M 2011 *Private communication*
[4] Legault T 2011 *Private communication*
[5] Pizer S M *et al.* 1987 Adaptive Histogram Equalization and its Variations *Comput. Vision Graph* **39** 355
[6] Zuiderveld K 1994 *Contrast Limited Adaptive Histogram Equalization* (Graphics Gems IV) ed P S Heckbert (Cambridge MA: Academic Press) chapter VIII.5 pp 474-485

A Model of Eclipsing Binary System V525 Sgr From CCD Photometry

Gabriela K. Haans, Hakim L. Malasan

Astronomy Study Program, Faculty of Mathematics and Natural Sciences, Institut Teknologi Bandung, Bandung 40132, Indonesia

,

Bosscha Observatory, Faculty of Mathematics and Natural Sciences, Institut Teknologi Bandung, Lembang 40391, Indonesia

E-mail: gabrielahaans@gmail.com

Abstract. New *BVRI* CCD photometry of eclipsing binary system V525 Sgr is presented. It is found that the light curves of V525 Sgr show a typical Eb-type light variation with an O'Connell effect. Photometric solutions and model of V525 Sgr were derived by qualitative fitting method using a Wilson-Devinney program. It showed that the mass ratio, q, of V525 Sgr is 0.55 and it is one of the near-contact binary in which the primary fills its Roche lobe. As the result of light curve fitting. We propose that the O'Connell effect may be attributed to the gas impacts of mass transferring in the system which is represented as a hot spot on the secondary star.

1. Introduction

V525 Sgr was discovered as an eclipsing variable star in 1934 by O'Connell (1935) and was recognized having a difference in its maxima also by O'Connell (1949). Later this difference in magnitude of its maxima is known as O'Connell effect. Cileé & Lindsay (1953) classified this system as EB-type and Payne-Gaposchkin (1949) gave A2 as its spectral type. Since the previous observations, there is no detailed study about this system.

2. Observation

The observations of V525 Sgr in the B, V, R, and I bands were carried out on seven nights from July to September 2013 with the 0.2 m ($f/10$) Schmidt Cassegrain telescope installed at Gunma Astronomical Observatory-Institut Teknologi Bandung Remote Telescope System (GAO-ITB RTS) house in Bosscha Observatory, Lembang, Indonesia. The telescope is equipped with a CCD camera SBIG ST9-XE. The field of view of the photometric system is 17.26 arcmin $\times$ 17.26 arcmin. The differential photometry method has been applied for the observations. CD-30°16666 and CD-30°16667 were chosen as the comparison and the check star, respectively. The coordinates and magnitudes of the variable star, the comparison star, and the check star are listed in Table 1.

Table 1. Coordinates and magnitudes of V525 Sgr, the comparison, and the check stars.

Stars	RA (J2000.0)	Dec. (J2000.0)	V
V525 Sgr	$19^h\ 07^m\ 13^s.530$	-30° 09′ 37″.70	8.28-8.93
The comparison	$19^h\ 07^m\ 19^s.489$	-30° 11′ 12″.89	9.87
The check	$19^h\ 07^m\ 19^s.915$	-30° 13′ 34″.75	9.86

A standard reduction using calibration images (bias, darks, and flat) has been performed for every night of observations. The instrumental magnitudes were measured by applying aperture photometry method. IRAF[1] has been performed for processing the observed images either for photometry reduction using `APPHOT` task or for standard reduction using `CCDRED` task. The CCD light curves in B, V, R, and I bands with respect to the linear ephemeris given by Kreiner (2004):

$$HJD = 2452500.648 + 0.7051217 \times E \tag{1}$$

are shown in Fig. 1.

3. Photometric Solutions

The light curves displayed in Fig. 1 are asymmetric in B, V, R, and I bands and show typical EB-type variations. This asymmetric in light curves of eclipsing binaries is known as O'Connell effect. The photometric solutions and model of V525 Sgr were obtained by qualitative fitting method using a Wilson-Devinney program (W-D program, Wilson & Devinney 1971). Relative flux in each light curve in Fig. 1 is a normalization of the magnitude difference between V525 Sgr and the comparison star. The absolute parameters of V525 Sgr are not available due to the lack of spectroscopic observations for this object.

The light curves of V525 Sgr have shown a pattern of O'Connell effect where the maximum I (after the primary minimum) is brighter than the maximum II (after the secondary minimum) and also the differences between the two maxima is bigger in B band than other bands which is redder. These features are in agree with previous observations by O'Connell (1949) and Cileé & Lindsay (1953). The consistent of appearance of this O'Connell effect may be caused by a hot spot due to the impact of mass transferring of gas stream in the system from the primary to secondary star. In order to obtain the photometric parameters, the temperature of primary star was adopted about 9250 K which corresponds to its spectral type. The other adopted parameters are the limb darkening coefficients for B, V, R, and I bands as given in van Hamme (1993), the gravity darkening coefficients as given in Lucy (1967), and the period of the system as given in Kreiner (2004). We also assumed that this system has a circular orbit and there is no reflection effect and no differential rotation. The adjustable parameters were the orbital inclination i, the mean temperature of secondary star T_2, the mass ratio q, and the spot parameters. The spot parameters are consist of coordinate (co-latitude and longitude), radius, and temperature factor of the hot spot. The light curves are fitted with a hot spot located on secondary star which in agreement with the mass transferring that may occurred in this system. Its photometric solutions are listed in Table 2.

The calculation for determining the configuration have been carried out for each value of q. The final configuration of the system is a semi-detached system or near-contact system where the primary filling its Roche lobe and the secondary is very close to its Roche lobe. The model of its configuration is shown in Fig. 2.

[1] Image Reduction and Analysis Facility (IRAF) is distributed by the National Optical Astronomy Observatory (NOAO), which is operated by the Association of Universities for Research in Astronomy, Inc.

IOP Conf. Series: Journal of Physics: Conf. Series **1127** (2019) 012050 doi:10.1088/1742-6596/1127/1/012050

Table 2. Photometric solutions for V525 Sgr.

Parameters	Photometric Elements
g_1	1.000
g_2	0.320
X_{1B}	0.543
X_{2B}	0.860
X_{1V}	0.458
X_{2V}	0.668
X_{1R}	0.380
X_{2R}	0.516
X_{1I}	0.304
X_{2I}	0.413
T_1	9250 K
T_2	6200 K
P	0.7051217 d
q (M_2/M_1)	0.550
Ω_{in}	3.959
Ω_{out}	3.540
i	74°
$L_1/(L_1+L_2)$ (B)	0.901
$L_1/(L_1+L_2)$ (V)	0.877
$L_1/(L_1+L_2)$ (R)	0.862
$L_1/(L_1+L_2)$ (I)	0.828
r_1 (pole)	0.417
r_1 (side)	0.444
r_1 (back)	0.478
r_2 (pole)	0.301
r_2 (side)	0.314
r_2 (back)	0.343
Co-latitude$_{spot}$	90°
Longitude$_{spot}$	73°.7
Radius$_{spot}$	19°
Temperature factor$_{spot}$	1.75

4. Discussion and Summary

O'Connell (1949) found that the light curves of V525 Sgr were asymmetric from the photographic observations where the maximum after the primary minimum is brighter than the maximum after the secondary minimum. The asymmetries of the B, V, R, and I bands are all less than 0.1 mag as shown in Fig. 1. The O'Connell effect in eclipsing binary systems has been the most challenging topic in binary star studies. There are some theories that have been suggested to explain the nature of O'Connell effect such as the starspots theory, the gas stream impacts theory, the circumstellar matter theory or the asymmetric circumfluence due to coriolis forces theory (Wilsey & Beaky, 2009). But there is no theory that has been successfully applied to an excessively number of eclipsing binary systems. The starspots theory could not explain the consistent of appearance in the maxima of V525 Sgr, but the gas stream impacts theory may explain this phenomenon. The last two theories that have been mentioned do not have the attention as much as the first two theories. Therefore, the hot spot due to the gas stream impact is included in the analysis.

The photometric solution using W-D program indicates that V525 Sgr is a semi-detached or near-contact eclipsing binary system with a mass ratio of $q = 0.55$. Due to the lack of spectroscopic observations, the absolute parameters of this system have not derived. The orbital inclination, i, is obtained 74°. Further CCD photometric and spectroscopic observations are needed to gain information about the absolute parameters and to confirm if the nature of the O'Connell effect in this system is due to the gas stream impacts theory.

Acknowledgements
The authors are very grateful to Ms. J. Suherli for her kind assistance during the observations in Bosscha Observatory and all the constructive comments and suggestions for a better improvement of this study.

References
[1] Cilleé G C and Lindsay E M 1953 *Mon. Not. R. Astron. Soc.* **113** 516
[2] Kreiner J M 2004 *Acta Astron.* **54** 207
[3] Lucy L B 1967 *Zeitschr. Astrophys.* **65** 89
[4] O'Connell D 1935 *Pub. of the Riverview College Obs.* **1** 9
[5] O'Connell D 1949 *Pub. of the Riverview College Obs.* **2** 78
[6] Payne-Gaposchkin C 1949 *Harvard Ann.* **115** 138
[7] van Hamme W 1993 *Astron. J.* **106** 2096
[8] Wilsey N J and Beaky M M 2009 *Soc. Astron. Sci. Ann. Symp.* **28** 107
[9] Wilson R E and Devinney E J 1971 *Astrophys. J.* **166** 605

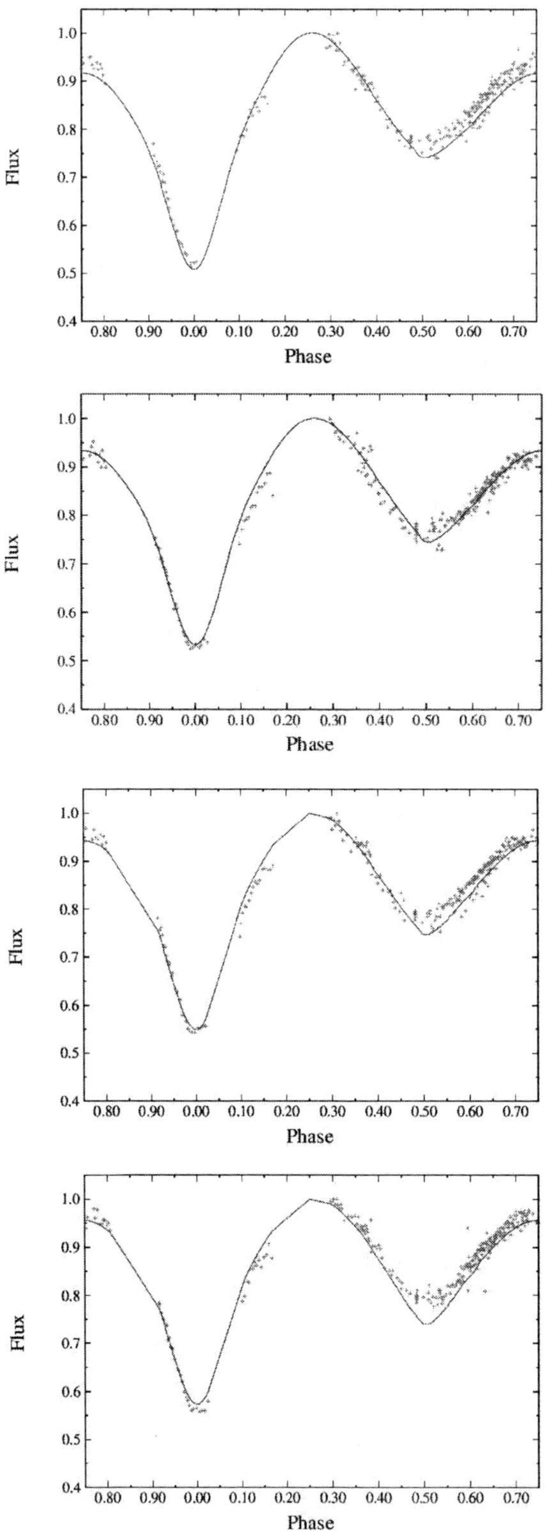

Figure 1. The light curve of V525 Sgr in the B, V, R, and I bands. The plus symbols and solid lines represent the observational and theoretical light curves, respectively.

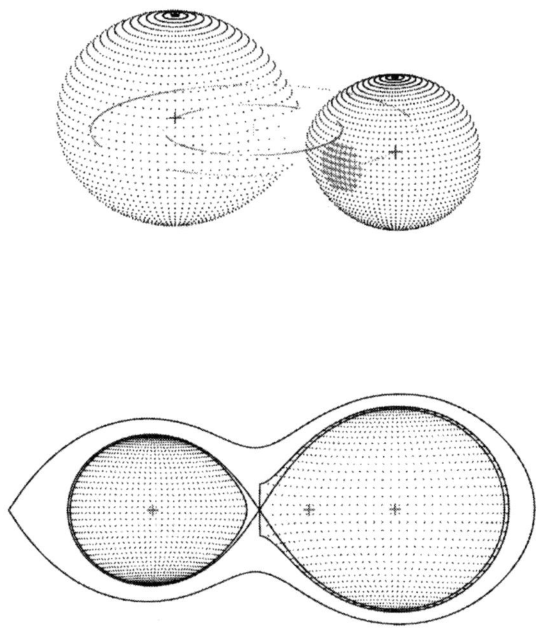

Figure 2. The final configuration of V525 Sgr in phase 0.17 and its Roche lobe model. The red spot in the final configuration represents the location of the hot spot on the surface of secondary star.

IOP Conf. Series: Journal of Physics: Conf. Series **1127** (2019) 012051 doi:10.1088/1742-6596/1127/1/012051

Exoplanet Survey Observations of NGC 6494 and NGC 6633 at Bosscha Observatory

Denny Mandey (1,2), Muhammad Yusuf (1), and Mahasena Putra (1,2)

1. Bosscha Observatory, Faculty of Mathematics and Natural Science, Institut Teknologi bandung, Indonesia.

2. Astronomy Study Program, Faculty of Mathematics and Natural Science, Institut Teknologi bandung, Indonesia.

Mandey.de@gmail.com

Abstract. We present the preliminary result of exoplanet survey observation of NGC 6494 and NGC 6633 at Bosscha Observatory. This research is part of a long-term program of exoplanet survey observations in the open star cluster near the celestial equator. Transit method is used to detect planets orbiting stars in the open star clusters. We are using the 11 inch diameter telescope and targeting transit depth up to 0.01 magnitudes. So far, we still do not detect transiting exoplanets. However, we detect several stars showing variability in its flux resembling the characteristic of variable star. Follow up observation would be needed to identify the type of variable stars.

Keyword: Exoplanet, transit, open star cluster.

1. Introduction

1.1. Progress on Exoplanet Discoveries

Exoplanet research is one of the fastest growing field of research in astronomy. Exoplanet exploration began since the first confirmed discovery of exoplanet 51 pegasi b in 1995 by the Geneve Astronomers Michel Mayor and Didier Queloz [4]. Since then, exoplanets become a popular topic in astronomy observation and theoretical. Using variety of detection methods and equipment both ground-based and space-based, until now it has been confirmed the existence of 3540 exoplanets in 2655 planetary systems. The annual progress on exoplanet discoveries can be seen at Figure 1.

The use of Kepler spacecraft in last 5 year has pushed the discovery of exoplanet rapidly. Kepler spacecraft using the transit method in the search for exoplanets. Although Kepler benefit from its position outside the Earth's atmosphere but it shows that the transit method has a great chance in finding exoplanets compared to the use of other methods. Transit method allows small-sized telescopes to detect the presence of exoplanets. Small telescopes generally have a wide field of view so that it can scan the vast skies in a relatively short time. This is the reason of exoplanet survey observations using a small telescope in large quantities. Where the main task is to search the stars of exoplanet candidates as much as possible.

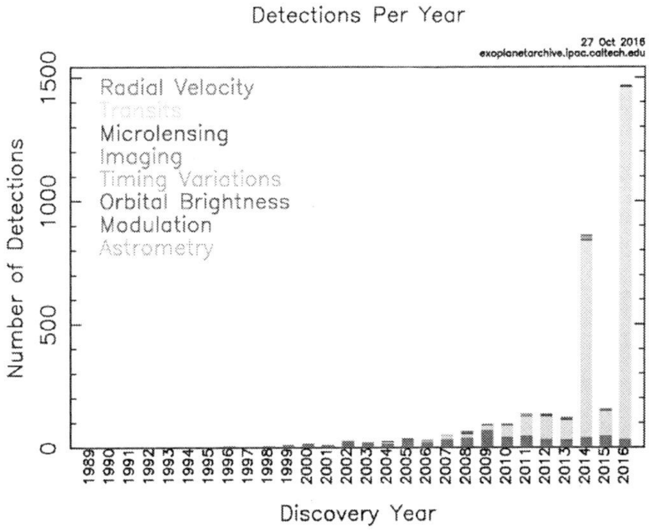

Figure 1. Progress of exoplanet discoveries and the detection methods used.

1.2. Research Background

Currently, out of 3535 confirmed exoplanets only 7 exoplanets discovered orbiting stars that are members of open cluster [2] [5] [8]. In his paper, Meibom [5] showed that the frequency of finding exoplanets in star clusters as large as the outside of a star cluster. Theories of star formation suggest that the star was born in the form of clusters. Open star cluster is able to sustain its group of stars for hundreds of millions to several billion years. Whereas the planet, especially the giant gas planet, can form within 5 million years since the parent star is formed [1] [3] [7]. This indicates that we should have discovered exoplanets in the open star cluster more than what we find today.

Exoplanet survey observations on open star cluster is a long-term research program conducted at Bosscha observatory. The purpose of this research is to find the star of exoplanet candidates in the open star cluster system using transit method. We observe several open cluster located near the celestial equator. The selected star cluster will be observed continously. The stars that show periodic changes in the flux will be observed further to identify whether changes occur as a result of flux variability star or planetary transits. For this paper, we only report result from two open cluster NGC 6494 and NGC 6633. Currentlly, only these two open cluster where its observation data available for periodicity searching.

2. Observations

2.1. Instruments

In the process of collecting observational data for this reasearch, we use one of robotic capable telescope at Bosscha Observatory, the Bosscha Robotic Telescope Experiment 2 (BRTX2). We use these telescopes because it has a wide field of view that is sufficient for the survey observation of the open star cluster. During image acquisitions, filters are not used in order to get sufficient flux in a short time. This research aims to find exoplanet candidates so that use of the filter is not required. Technical specifications of the telescope can be seen in the Figure 2.

Bosscha Robotic Telescope Experiment 2 (BRTX2)

Main Telescope	
Optical type	: Reflector Schmidt-Cassegrain
Diameter	: 279 mm
Focal length	: 1760 (with focal reducer f0.63)
Detector	: CCD Camera SBIG ST-8XME
Filters	: Narrow band (Bessel BVRI) Wide band (RGBL)
Field of view	: 26,8' x 17,9'
Guide Telescope	
Optical type	: Refractor ED Apochromat
Diameter	: 102 mm
Focal length	: 820 mm
Detector	: CCD Camera SBIG ST-402
Filters	: Wide band (RGBC)
Field of view	: 29' x 19,3'

Figure 2, Technical specification of the Bosscha Robotic Telescope Experiment 2.

2.2. Object of Observations

The objects that we present in this paper is open star cluster NGC 6494 and NGC 6633. Both star clusters located near the celestial equator and relatively bright cluster with brightness 5-6 in visual magnitude. More parameters of the clusters can be seen at Table 1. Figure 3 shows samples of the cluster images taken in our observation sessions.

Table 1. The Cluster Parameters

	NGC 6494	NGC 6633
RA (J2000)	$17^h\ 57^m\ 04^s$	$18^h\ 27^m\ 15^s$
Dec (J2000)	-18° 59' 06"	+06° 30' 30"
Log (age)	8.477	8.63
Diameter	29 arc min	20 arc min
Distance	628 parsec	376 parsec
Member	1342 stars	872 stars
Magnitude (V)	6	5

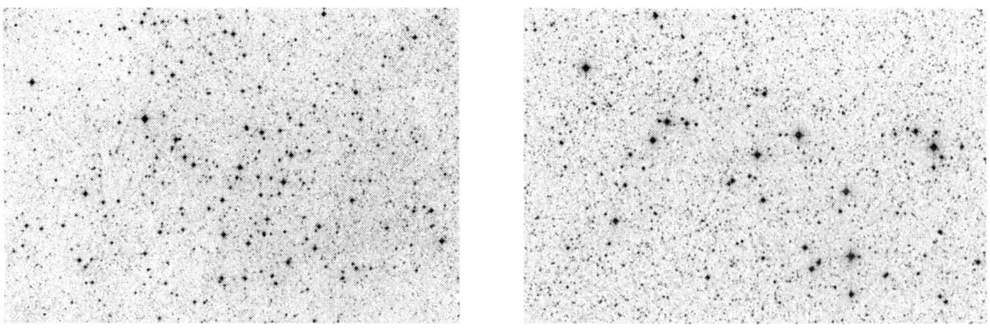

Figure 3. Sample of inverted images of NGC 6494 (left panel) and NGC 6633 (right panel).

IOP Conf. Series: Journal of Physics: Conf. Series **1127** (2019) 012051 doi:10.1088/1742-6596/1127/1/012051

3. Result

These are the preliminary results of exoplanet survey observation program on open star cluster conducted at Bosscha Observatory. As mentioned at previous section, we only have observation data from NGC 6494 and NGC 6633 that available for periodicity searching. We use data from six night observations of NGC 6494 and three nights observations of NGC 6633. LEMON program were used to measure fluctuation of star brightness [9]. This program use differential photometry method. For now, we measure every star detected on the direction of the cluster. Vstar program were used to search the periodicity in the lightcurve and produce the phasefold lightcurves. We aim to find hot Jupiter type exoplanet. This type of exoplanet usually have short orbital periode and produce relatively deep transit depth on its lightcurve. Using 28 cm diameter telescope we targeting transit event to the depth (delta magnitude) of 0.01 magnitude.

Currently, reduction and analysis process of star images is still done manually. We still analizing less than 100 lightcurves out of nearly 2000 lightcurves generated by the LEMON program. So far, we have not found any fluctuation of star brightness that indicates the transit event of exoplanet. However, we found some changes on the star brightness that showed the characteristic of pulsating variable star and eclipsing binary star. This variable star candidates need further observation focused on each of the stars to determine the type of the variable star. Follow-up observations will be made on a separate observation program. Figure 4 shows the phaseplot lightcurve of the variable star candidates.

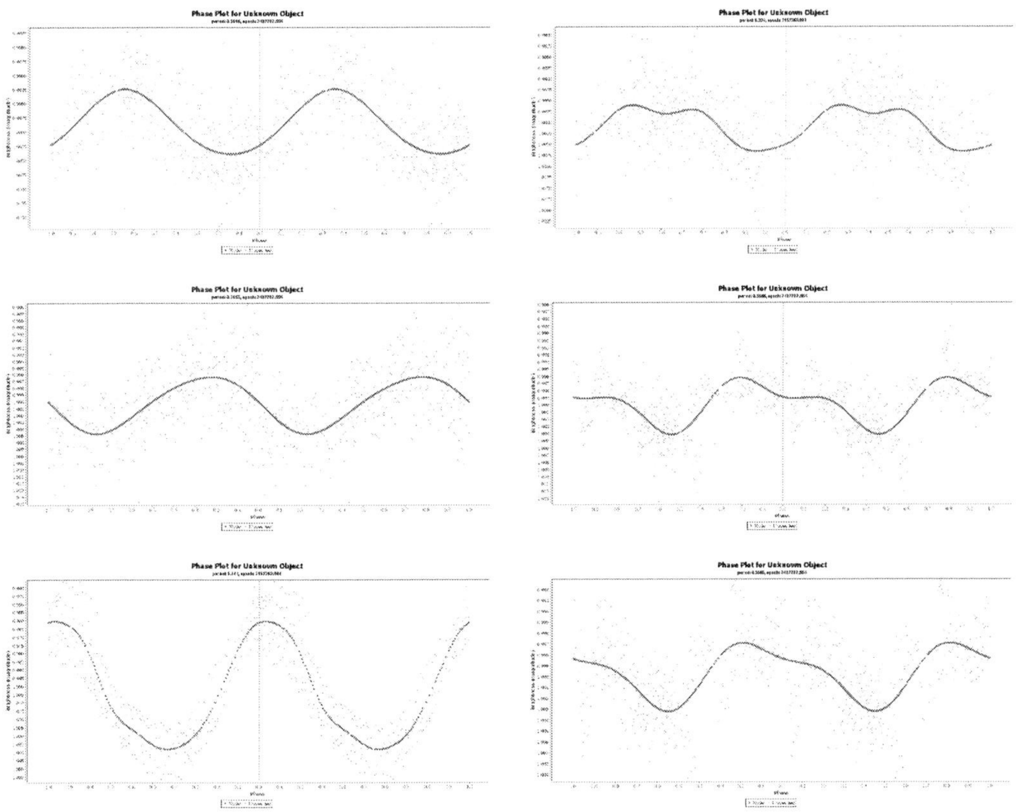

Figure 4. Sample images of lightcurves of pulsating stars (left panel) and eclipsing binaries (right panel) found in the observed open clusters.

IOP Conf. Series: Journal of Physics: Conf. Series **1127** (2019) 012051 doi:10.1088/1742-6596/1127/1/012051

References

[1] Boss A P 2001 *Astrophysical Journal* **563** 376

[2] Brucalassi A, Pasquini L, Saglia R, Ruiz M T, Bonifacio P, Bedin L R, Biazzo K, Melo C, Lovis C and Randich S 2014 *Astronomy & Astrophysics* **561L** 9B

[3] Haisch K E Jr, Lada E A and Lada C J 2001 *Astrophysical Journal* **553** 153

[4] Mayor M and Queloz D 1995 *Nature* **378** 355

[5] Meibom S, Torres G, Fressin F, Latham D W, Rowe J F, Ciardi D R, Bryson S T, Rogers L A, Henze C E, Janes K et al 2013 *Nature* **499**

[6] Pollacco D L, Skillen I, Collier Cameron A, Christian D J, Hellier C, Irwin J, Lister T A, Street R A, West R G, Anderson D R et al 2006 *The Publications of the Astronomical Society of the Pacific* **118** 1407

[7] Pollack J B, Hubickyj O, Bodenheimer P, Lissauer J J, Podolak M and Greenzweig Y 1996 *Icarus* **124** 62

[8] Quinn S N, White R J, Latham D W, Buchhave L A, Cantrell J R, Dahm S E, Furesz G, Szentgyorgyi A H, Geary J C, Torres G et al 2012 *Astrophysics Journal* **756** 33

[9] Terron V dan Fernandez M 2011 *Highlights of Spanish Astrophysics VI Proceedings* **755**

IOP Conf. Series: Journal of Physics: Conf. Series **1127** (2019) 012052 doi:10.1088/1742-6596/1127/1/012052

Solar coronal holes and their geo-effectiveness

Santi Sulistiani[1,2] **and Dhani Herdiwijaya**[1]

[1] Department of Astronomy, Bandung Institute of Technology (ITB), Jl. Ganesha 10, Bandung 40132, Indonesia
[2] Space Science Center, National Institute of Aeronautics and Space (LAPAN), Jl. Dr. Djundjunan No. 133, Bandung 40173, Indonesia

E-mail: santi.sulistiani@lapan.go.id

Abstract. Solar corona as the outermost part of the Sun is constantly emanating solar wind consisting of charged particles. Coronal holes are rigidly rotating phenomena in the solar corona that are magnetically open to interplanetary space, which allow the solar wind particles to escape from the Sun. The study of coronal holes and their associated high speed solar wind streams is important because they are the sources of space weather disturbances during minimum solar activity. We present some results on the relationship between large solar coronal holes close to the central meridian and solar wind stream with the aim to analyse their geo-effectiveness. The analysis were carried out by considering geomagnetic and solar wind parameters such as total geomagnetic field, B_z component, proton density and plasma speed during Earth-side large solar coronal holes appearance. Using the data during declining phase solar activity we found that a solar coronal hole is geo-effective during and after its transit on the central meridian but not all of them can be the proxy of high speed solar wind.

1. Introduction

Solar corona is the outermost part of the Sun with density of ~ 10^5 m^{-3} and temperature of millions of Celsius. This region of hot gas is constantly emanating solar wind, consisting charged particles (protons, electrons, and helium ions). Coronal holes are regions of low-density plasma on the solar corona that appear dark in EUV [1] or X-ray [2] images of the Sun. They have open magnetic field structures which allow solar wind particles to escape easily into interplanetary space [3].

Space weather often associated with transient events on the Sun such as flares and coronal mass ejections (CMEs) that can affect the Earth's magnetosphere. Despite the importance of these transient events, coronal holes, especially during the minimum solar activity, also play important role in space weather because they are the source region of the high speed solar wind streams (HSSs) [4]. Combination of coronal holes and their associated HSSs with solar rotation are the dominant contributors to space weather disturbances at times of quiet solar activity because they are sources of recurrent geomagnetic storms [5]. The aim of this paper is to study the relationship between large solar coronal holes and solar wind streams in order to analyse their geo-effectiveness, i.e. their effects on the Earth's environment.

2. Data analysis and methods

Our study is based on the following data sets: The interplanetary magnetic field (B and B_z) and solar wind parameters (proton density and plasma speed) accessible at OMNIWeb Plus (http://omniweb.gsfc.nasa.gov/) prepared by Goddard Space Flight Center/Space Physics Data Facility; Solar Dynamics Observatory (SDO) Atmospheric Imaging Assembly (AIA [6]) 193 Å quick-look images accessible at SDO (http://sdo.gsfc.nasa.gov/). We focus on large coronal holes in the declining phase of the Solar Cycle 24, when the coronal mass ejection (CME) and solar flare activity were very low.

3. Results

We have selected four coronal holes between February-May 2015 including equatorial, polar and trans-equatorial ones. The results are as follows.

3.1. *Equatorial coronal hole on 6-15 February 2015*

This coronal hole was transiting the central meridian around 11 February 2015 (Figure 1). Figure 2 shows the total magnetic field strength, z-component, proton density and plasma speed during its appearance on the SDO/AIA 193 Å images. The total magnetic field decreased significantly on 14 February 2015 around 3 UT and increased at 5 UT. Another fluctuation occurred on 15 February 2015 with minimum magnetic field of 4 nT around 4 UT coincided with proton density peak of 22.1 particles/cm^3 while the B_z was negative (southward). Later, the total magnetic field reached its maximum of 10.9 nT around 13 UT followed by proton density enhancement up to 16.6 particles/cm^3 a day later around 13 UT. This fluctuating total magnetic field and peaking proton density are suggested to be originated from the large coronal hole as it rotated to the western hemisphere.

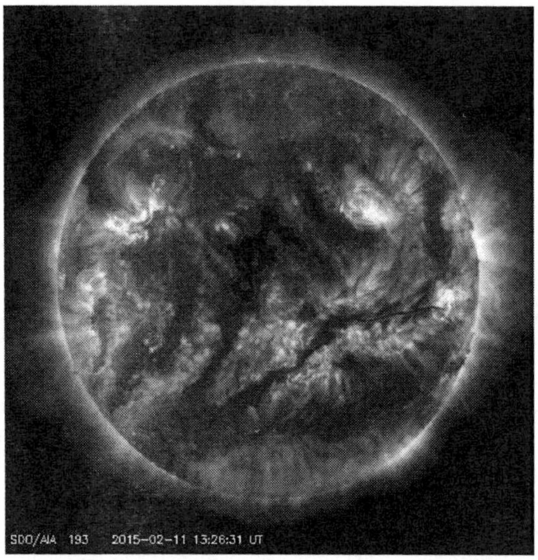

Figure 1. SDO/AIA 193 Å image showing an equatorial coronal hole (blue line) transiting the central meridian on 11 February 2015.

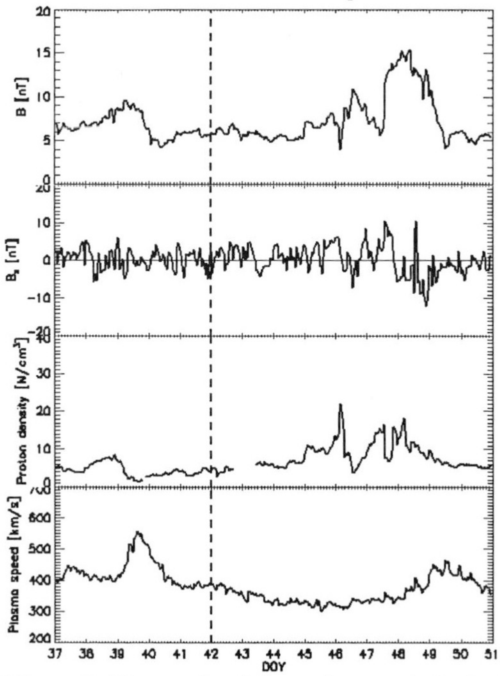

Figure 2. Time series (1-hour data resolution) of B (nT), B_z (nT), solar wind proton density (N/cm^3) and solar wind plasma speed (km/s), from top to bottom, respectively, during the appearance of equatorial coronal hole (6-19 February) on SDO/AIA 193Å images. The time is expressed in the day-of-year form, DOY, for 2015. The vertical dashed-line denote the approximate time of coronal hole transit.

IOP Conf. Series: Journal of Physics: Conf. Series **1127** (2019) 012052 doi:10.1088/1742-6596/1127/1/012052

Another enhancement on total magnetic field took place at 11 UT on 16 February 2015 to a maximum at 17 February 2015 at around 9 UT. This was also originated from the same coronal hole. As shown in figure 2, B_z was mostly negative during the maximum of total magnetic field coincided with proton density enhancement and followed by solar wind acceleration. The plasma speed increased from 400-460 km/s on 18 February 2015.

3.2. Polar coronal hole on 23 February-3 March 2015

The coronal hole on the Sun's southern pole is a very large one (Figure 3). Figure 4 shows some parameters during the appearance of this coronal hole, similar to figure 2. The strong magnetic field, density and solar wind speed at the beginning of its appearance were affected by another coronal hole on the west limb.

Solar wind stream from the polar coronal hole caused proton density enhancement started on 27 February 2015 and peaked on 28 February 2015 at around 11 UT with maximum of 24.6 particles/cm^3. This enhancement was followed by increasing total magnetic field and plasma speed starting from 28 February 2015. The total magnetic field peaked at 12.6 nT on 1 March 2015, while solar wind speed was fluctuating between 560-640 km/s and B_z was mostly negative and oscillating between -8 to 3 nT.

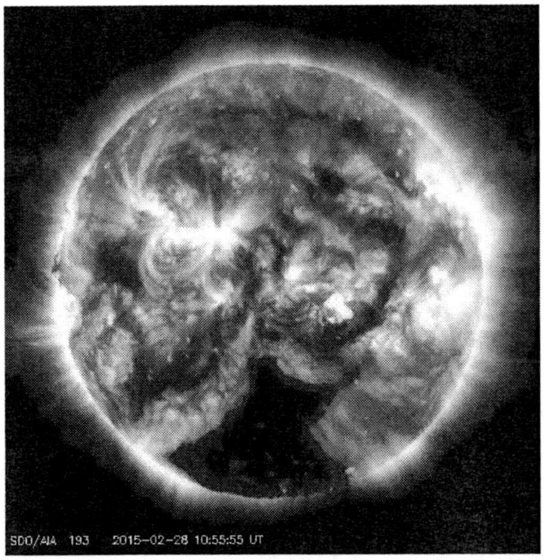

Figure 3. SDO/AIA 193 Å image showing a large polar coronal hole (blue line) transiting the central meridian on 28 February 2015.

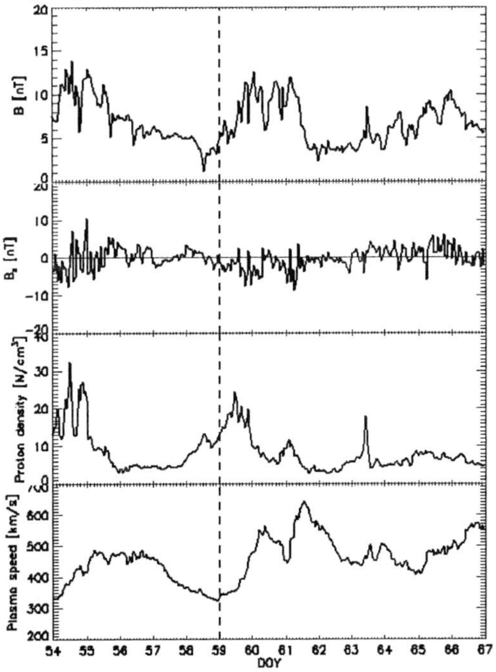

Figure 4. Similar to figure 2, during the appearance of polar coronal hole (23 February-7 March 2015) on SDO/AIA 193Å images.

3.3. Trans-equatorial coronal hole on 26 March-6 April 2015

This large coronal hole stretched from the solar north-pole to equator (Figure 5). During its appearance, there were no strong events and the Dst index was relatively stable. However, there was plasma speed enhancement on 3-5 April 2015 (Figure 6). This enhancement to about 550 km/s was originated from our trans-equatorial coronal hole. It is interesting that the proton density was rapidly declining when the solar wind speed was maximum on 3 April 2015. A rapid decrement also shown in the total magnetic field while B_z was slightly oscillating between -1 and 3 nT. The relatively quiet total magnetic field and proton density were continued to occur until the proton speed decreased to about

320 km/s on 9 April 2015. After that, all considered parameters were highly fluctuated as the effect of CME on 6 April 2015.

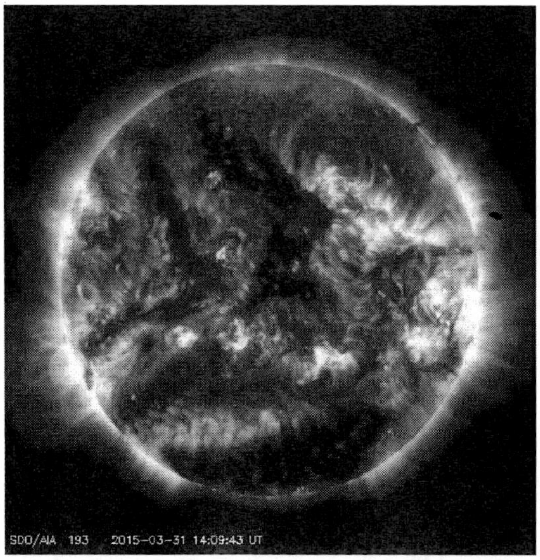

Figure 5. SDO/AIA 193 Å image showing a trans-equatorial coronal hole (blue line) transiting the central meridian on 31 March 2015.

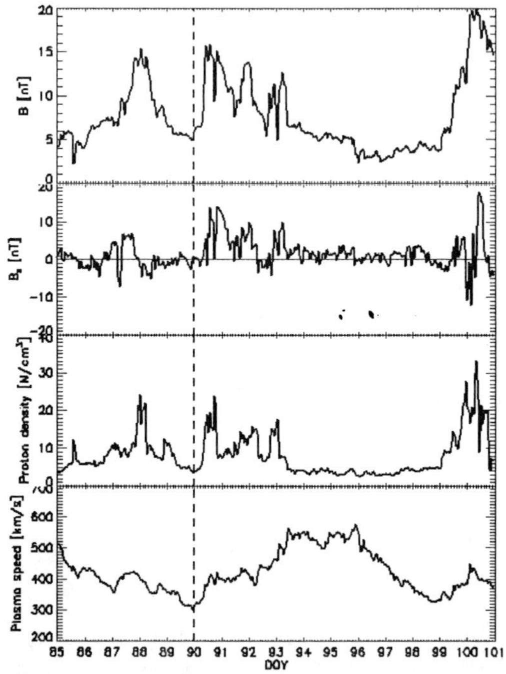

Figure 6. Similar to figure 2, during the appearance of trans-equatorial coronal hole (26 March-10 April 2015) on SDO/AIA 193Å images.

3.4. *Equatorial coronal hole on 19-30 May 2015*

There were no large events occurred during this equatorial coronal hole appearance. However, there was an enhancement in the proton density on 23 and 26 May 2015 (Figure 8). Proton density enhanced from around 5 to around 12 particles/cm^3 only within 4 hours on 23 May 2015. The Solar Influences Data Analysis Center (SIDC-http://sidc.oma.be/) reported it as a sector boundary crossing (SBC) of the interplanetary magnetic field, i.e. when the interplanetary magnetic field reversed from positive to negative. Another SBC, interplanetary magnetic field reversal from negative to positive, occurred on 26 May 2015 with stronger enhancement of proton density up to 21 particles/cm^3. The solar wind plasma speed was considerably low, i.e. around 300 km/s, during both SBCs.

IOP Conf. Series: Journal of Physics: Conf. Series **1127** (2019) 012052 doi:10.1088/1742-6596/1127/1/012052

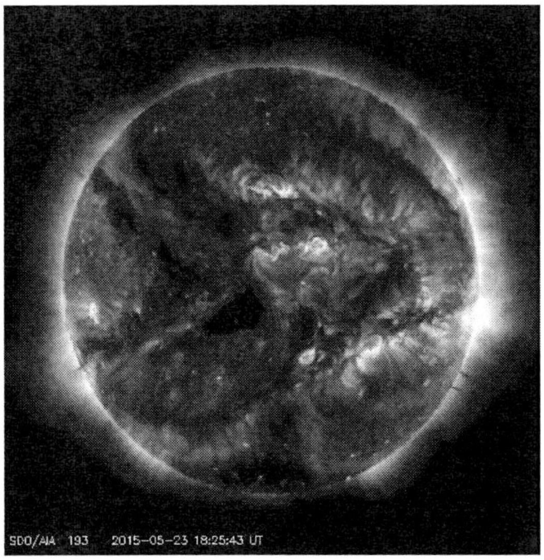

Figure 7. SDO/AIA 193 Å image showing an equatorial coronal hole (blue line) near the central meridian on 23 May 2015.

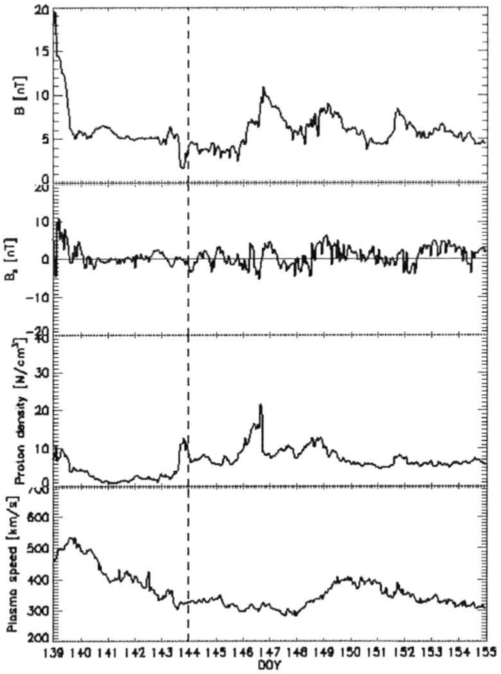

Figure 8. Similar to figure 2, during the appearance of equatorial coronal hole (19 May-3 June 2015) on SDO/AIA 193Å images.

4. Discussion and conclusion

Based on four coronal holes considered in this study, we found that a solar coronal hole can be geo-effective, i.e. producing high-speed solar wind stream affecting the Earth's magnetic field, during and after its transit on the central meridian when the z-component (B_z) is negative (southward). The effect of the coronal hole can be seen as significant changes in plasma speed, proton density, and total magnetic field.

However, the equatorial coronal hole on 19-30 May 2015 produced a relatively low plasma speed during its appearance. This suggested that not all of the large coronal holes can be the proxy of high speed solar wind. We suppose that this is because we did not consider the polarity of the coronal holes, as a recent study found that coronal holes with opposite polarity (southern negative, and northern positive) has the smallest geo-effectiveness [7].

Further analysis on the coronal holes evolution is necessary in order to study their possible impact to Earth's magnetic field. We need to collect more data with various positions, i.e. equatorial, polar and trans-equatorial, to get a comprehensive conclusion on the relationship between solar coronal hole evolution and its potential to disturb the Earth's magnetic field.

References

[1] Newkirk J G 1967 Structure of the solar corona *Ann. Rev. Astron. Astrophys.* **5** 213
Del Zanna G and Bromage B J I 1999 The elephant's trunk: spectroscopic diagnostics applied to SOHO/CDS observations of the August 1996 equatorial coronal hole *J. Geophys. Res.* **104** 9753
[2] Altschuler M D, Trotter D E and Orrall FQ 1972 Coronal holes *Solar Phys.* **26** 354
[3] Tsurutani B T, Gonzalez W D, Gonzalez A L C, Tang F, Arballo J K and Okada M 1995 Interplanetary origin of geomagnetic activity in the declining phase of the solar cycle *J. Geophys. Res.* **100** 21717

Gosling J T and Pizzo V J 1999 Formation and evolution of corotating interaction regions and their three dimensional structure *Space Sci. Rev.* **89** 21

Cranmer S R 2009 Coronal holes *Living Rev. Solar Phys.* **6** 3

[4] Krieger A S, Timothy A F and Roelof E C 1973 A coronal hole and its identification as the source of a high velocity solar wind stream *Solar Phys.* **29** 505

[5] Rotter T, Veronig A M, Temmer M and Vršnak B 2015 Real-time solar wind prediction based on SDO/AIA coronal hole data *Solar Phys.* **290** 1355

[6] Lemen J R *et al.* 2012 The Atmospheric Imaging Assembly (AIA) on the Solar Dynamics Observatory (SDO) *Solar Phys.* **275** 17

[7] Abunina M, Abunin A, Belov A, Gaidash S, Tassev Y, Velinov P I Y, Mateev L and Tonev P 2014 Properties of magnetic fields in coronal holes and geoeffective disturbances in Solar Cycle 24 *C. R. Acad. Bulg. Sci.* **67** No 5 699

IOP Conf. Series: Journal of Physics: Conf. Series **1127** (2019) 012053 doi:10.1088/1742-6596/1127/1/012053

The updated statistics of binary star clusters in the Large Magellanic Cloud

Rhorom Priyatikanto[1,2], Mochamad Ikbal Arifyanto[1], Hesti R.T. Wulandari[1] and Suryadi Siregar[1]

[1] Department of Astronomy, Faculty of Mathematics and Natural Sciences, Institut Teknologi Bandung, Indonesia

[2] Space Science Center, National Institute of Aeronautics and Space (LAPAN), Indonesia

E-mail: rhorom.priyatikanto@lapan.go.id

Abstract. Large Magellanic Cloud (LMC) has been known to host a significant fraction of binary star clusters whose dynamical and physical nature become interesting issue in accordance with the history of the galaxy. To understand the characteristics and the history of the binary clusters, it is important to establish a well-sampled statistics of this population. Among ~ 3100 star clusters (excluding associations), there are 634 clusters considered as binary/multiple cluster candidates. In this paper, the empirical statistics of the population, including spatial, size, separation, and age distribution are constructed using updated data. The current cluster compilation includes the ages of 88% group components, which is three times more complete compared to the previous study. From the established statistics, we found an evidence that cloud fissions become the most probable mechanism of binary star cluster formation. Size and separation distribution of the clusters in binary/multiple systems show the importance of dynamical interaction within the systems.

1. Introduction

Large Magellanic Cloud (LMC) is recognized as the host of significant fraction of binary and multiple star clusters [1, 2, 3] whose dynamical and physical nature are interesting. The expected lifetime of this system is $\lesssim 100$ Myr since internal and external dynamics push the system to merger or separation ending [4, 5, 6]. Several formation mechanism have been proposed to explain the existing systems one of which is primordial binary cluster formation triggered by oblique collision of two molecular clouds [7, 8]. The latest collision between two Magellanic Clouds ~ 150 Myr ago increased the velocity dispersion within the galaxies and the probability of cloud-cloud collisions that produce binary and multiple star clusters. To confirm this scenario, well-sampled statistics of binary and multiple star clusters is needed.

Years ago, Dieball et al. [9] published a statistical study of binary and multiple clusters in the LMC as a continuing endeavour to understand the characteristics of this population. The statistics was based on the extended catalog containing 4089 clusters [10]. The study showed that more than half of detected pairs can not be considered as chance pairs. The binaries tend to be young ($\lesssim 300$ Myr) and have components with similar size and age. However, the study was unable to satisfactorily establish the cluster age distribution because of the lack of data.

In this study, we update the statistics of binary/multiple star cluster population in the LMC by exploiting the compiled star cluster properties from literature. The main objectives are to

IOP Conf. Series: Journal of Physics: Conf. Series **1127** (2019) 012053 doi:10.1088/1742-6596/1127/1/012053

establish the empirical distribution of binary/multiple cluster candidates in order to understand their characteristic and to investigate the history of this population in relation to the global history of the LMC. The data and method used in this study are presented in section 2 while the results are presented and discussed in section 3. The obtained conclusions are provided in section 4.

2. Data and Method

The main catalog used in this study comes from Bica et al. [11]. The catalog contains 9305 extended objects in Large Magellanic Cloud, Small Magellanic Cloud, and the interconnecting bridge. Among those objects, there are 3086 star clusters located in the LMC (excluding associations). For every object, Bica et al. [11] provide positions (α, δ), dimension (major and minor axis), and classification (e.g. C for clusters). The clusters' equatorial coordinates were transformed into $2D$ cartesian system using Gnomon projection [12] centered on $(\alpha_0, \delta_0) = (80°, -69°45')$:

$$x = -\frac{\cos\delta \sin(\alpha - \alpha_0)}{\sin\delta \sin\delta_0 + \cos\delta \cos\delta_0 \cos(\alpha - \alpha_0)}, \tag{1}$$

$$y = \frac{\sin\delta \sin\delta_0 - \cos\delta \sin\delta_0 \cos(\alpha - \alpha_0)}{\sin\delta \sin\delta_0 + \cos\delta \cos\delta_0 \cos(\alpha - \alpha_0)}. \tag{2}$$

From their positions in $2D$ space, cluster pairs were identified as two clusters with projected separation $s < 20$ pc ($< 1.37'$ assuming distance modulus of 18.49 [13]). Using this criterion, 281 binary and multiple systems were identified: 10 overlapped systems, 217 doubles, 39 triples, 9 quadruples, and 5 multiple systems with more than four components. This group list actually consists of 634 individual clusters.

The ages of the clusters were compiled from literature [14, 15, 16, 17, 18, 19, 20, 21, 22]. Various approaches were used in those literature, including SWB classification of young clusters, color-magnitude diagram fiitting, and spectral energy distribution modeling. If any cluster has more than one published ages, then the arithmatic mean of those ages were adopted. However, the compiled cluster ages only cover 65% of the identified pairs/groups. To increase the completeness, a number of conspicious clusters were re-analysed photometrically such that their ages could be estimated. The photometric data for these analysis came from the Optical Gravitational Lensing Experiment (OGLE, [23]) and the Magellanic Clouds Photometric Survey (MCPS, [24]) catalogue. PARSEC isochrones [25] were used to model the observed color-magnitude diagram of the clusters. This process increased the number of clusters with known age upto 88% (557) of the sample.

3. Results and Discussion

From the compiled list of clusters, both individual or the one that belongs to multiple systems, several empirical distribution were constructed as the platform of the analyses. Spatial distribution of the binary/multiple candidates is displayed in Figure 1, suplemented the age and average separation of the system. Group candidates tend to be located at the denser region of the LMC, confirming the results stated in [1] and [9]. Closer examination reveals no spatial clustering of the systems with specific projected separation. Both systems with small and large separation can be found in dense region or in the outer part of the LMC. It is noteworthy that systems with small separation and located outside the dense region have a higher probability to be considered as physical or real pairs.

On the other side, there is noticeable spatial clustering of age. Group candidates with $\log t < 7.5$ can be found in 30 Doradus, constellation III, and the outer bar region (see Figure 1 for designations). The first two regions are considered as the most active star forming region

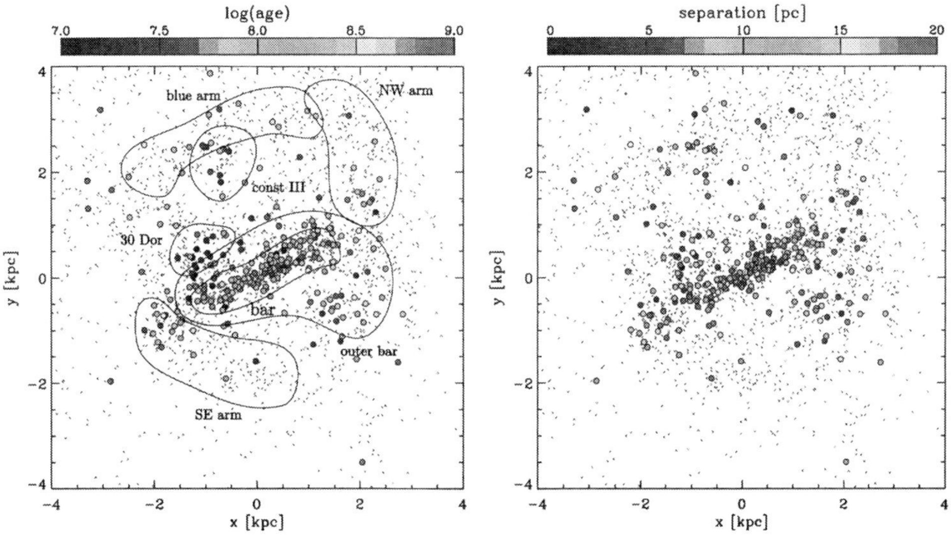

Figure 1. Spatial distribution of the detected groups (circles) among the cataloged clusters in the LMC. Color code (available in the on-line version) in the left panel represents system age ($\log t$), while the one in the right represents projected separation (s).

[26] induced by the latest SMC-LMC collision about 200 Myr ago [27]. Most of the intermediate age systems ($7.5 \leq \log t < 8.5$) can be found in the bar region, northwest arm, southeast arm, and blue arm. Almost all the systems in the SE arm have $8.0 \leq \log t < 8.5$, in agreement with the increasing star formation history of this region at that time window [26]. The older groups ($\log t \geq 8.5$) are more concentrated in the bar region. The existence of these groups can not be explained by primordial formation. They are either statistical pairs or the products of tidal encounter.

The history of star cluster system in the LMC can also be examined from the age distribution (dN/dt) in Figure 2. The age distribution of single clusters was constructed using 1780 clusters with known age (58% of total sample), while the distribution of group components was constructed using 557 clusters (88% of total sample). The inverse ratio between the two can be considered as a proxy of binary fraction in the LMC over time. This fraction needs to be corrected from the selection bias (completeness) and observational bias (decontaminating chance pairs), but this correction is beyond the scope of this current study.

In general, the number of clusters decays with time as the clusters lose mass and dissolve to the field of stars. Abrupt increase of star formation rate at any epoch can be traced from the bumps at $t \approx 150$ Myr and $t \approx 1500$ Myr, in agreement with the finding of Baumgardt et al. [28]. The age distribution of group components is slightly different, particularly at the leftmost and the rightmost part. The excess of binary/multiple cluster candidates with very young ($\log t \lesssim 7$) or higher binary fraction during the last 10 Myr supports the primordial formation and also the hierarchical cluster formation [29]. The binary fraction seems to be stable around 25% at 30 − 300 Myr, but there is pronounced increase of binary fraction at $t \approx 90$ Myr while the LMC-SMC collision assumed to occur at $t \approx 150$ Myr. This increase is a probable evidence of binary cluster formation triggered by cloud-cloud collision. The binary fraction fluctuates at the right-end of the plot as the number of old clusters decreases and the Poisson error grows.

Size distribution (dN/dD) is presented in Figure 3. For the clusters with diameter of $D \geq 10$ pc, the distribution can be approximated using power law $dN/dD \propto D^{-\eta}$. Size distribution

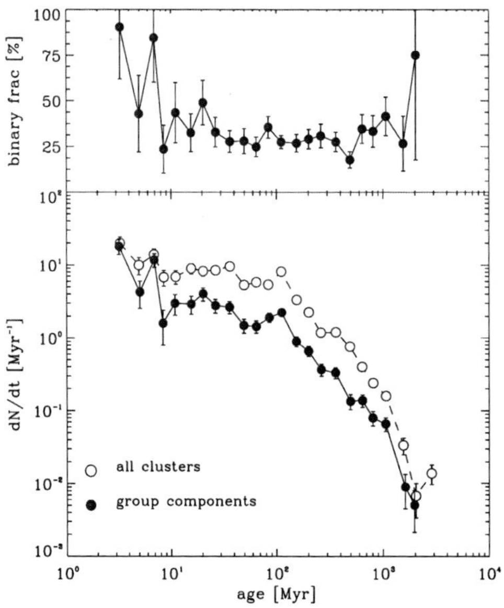

Figure 2. Bottom panel: age distribution of group components (filled circle) compared to the distribution of all clusters (with known age). Top panel: the binary fraction over time.

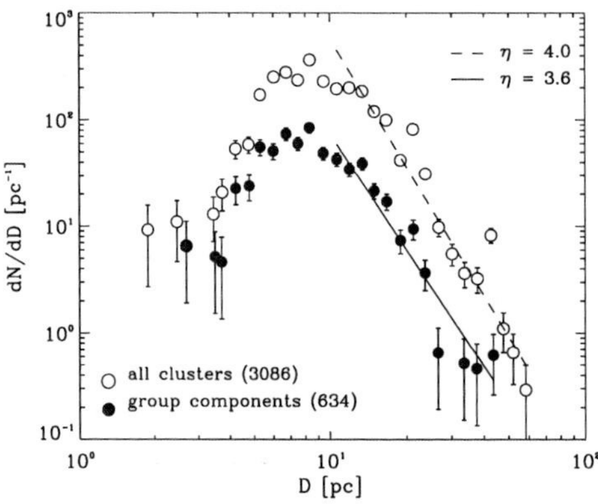

Figure 3. Size (diameter) distribution of group components (filled circle) compared to the distribution of all clusters. Power law $dN/dD \propto D^{-\eta}$ were fitted for $D \geq 10$ pc.

of individual clusters is well-fitted using $\eta = 4.0$, while the distributions of group components is fitted using $\eta = 3.6$. Shallower gradient for group components is caused by the excess of clusters with large diameter which is in accordance with the dynamical paradigm that binary and multiple clusters suffer mutual tide that heats and inflates the clusters [6].

The other interesting distribution is constructed from the projected separation of cluster pairs (dN/ds, see Figure 4). A trend of increasing dN/ds is obvious as the number of chance pairs increases, e.g. $(dN/ds)_{chance} \propto s$. Dieball et al. [9] found multi-modal distribution of separation that peak at 6 and 15 pc, while this study found peaks at 6 and 12 pc. To emphasize the

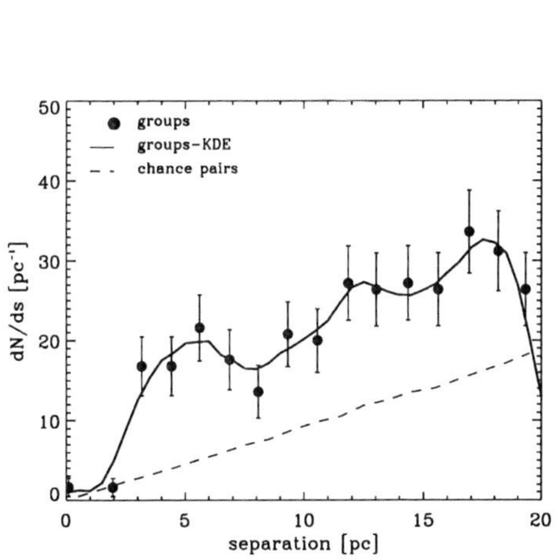

Figure 4. Distribution of the cluster pairs projected separation (filled circle) compared to the expected statistical/chance pairs (dashed line).

bimodal nature of dN/ds, the empirical distribution was also calculated using Kernel Density Estimation (KDE). Dieball et al. argued that the dip of dN/ds around 10 pc is caused by observational bias. However, three-dimensional separation distribution of cluster pairs in Milky Way also shows bimodality [30] though the observational bias is minimized. Dynamical evolution of binary star clusters can be involved in shaping bimodal distribution of separation. The dip between two peaks is the region where the inward gravitational force that leads to merger balances the outward force that disrupts the systems [5, 6].

4. Conclusions

In this study, several empirical distribution of binary/multiple cluster candidates in the LMC have been constructed using the updated cluster data. From those empirical distributions, some conclusions can be drawn:

(i) Binary/multiple cluster candidates tend to be young and there is convincing evidence that the binary fraction increases as the rise of star formation rate. The last LMC-SMC collision at $t \approx 150$ Myr may be related to the increase of binary fraction at $t \approx 90$ Myr. This finding supports the theory of binary clusters formation induced by cloud-cloud collisions.

(ii) There is an excess of the number of group components with large diameter. Dynamical interation between clusters in binary or groups may heat and inflate the clusters.

(iii) Distribution of cluster pairs separation has multiple peaks. Considering the similar result obtained for Milky Way binary cluster candidates, the observed bimodal distribution is thought as the product of dynamical evolution of the system, not merely observational bias.

Acknowledgments

This study was supported by the Ministry of Research Technology and Higher Education throuh Hibah Riset Disentralisasi Ristekdikti ().

References

[1] Bhatia R K and Hadzidimitriou D 1988 *Mon. Not. Royal Astron. Soc.* **230** 215–221
[2] Bhatia R K, Read M A, Hatzidimitriou D and Tritton S 1991 *Astron. & Astroph. Supp.* **87** 335
[3] Pietrzyński G and Udalski A 2000 *Acta Astron.* **50** 355

[4] Bhatia R K 1990 *Pub. Astron. Soc. Japan* **42** 757–767
[5] de la Fuente Marcos R and de la Fuente Marcos C 2010 *Astroph. J.* **719** 104–118
[6] Priyatikanto R, Kouwenhoven M B N, Arifyanto M I, Wulandari H R T and Siregar S 2016 *Mon. Not. Royal Astron. Soc.* **457** 1339
[7] Fujimoto M and Kumai Y 1997 *Astron. J.* **113** 249
[8] Bekki K, A B M, Forbes D A and Couch W J 2004 *Astroph. J.* **602** 730
[9] Dieball A, Müller H and Grebel E K 2002 *Astron. & Astroph.* **391** 547564
[10] Bica E, Schmitt H, Dutra C and Oliveira H 1999 *Astron. J.* **117** 238
[11] Bica E, Bonatto C, Dutra C and Santos, Jr J 2008 *Mon. Not. Royal Astron. Soc.* **389** 678–690
[12] Coxeter H 1969 *Introduction to geometry, 2nd ed.* (New York: Wiley)
[13] de Grijs R, Wicker J and Bono G 2014 *Astron. J.* **147** 122–135
[14] Bica E, Claria J J, Dottori H, C S J F and Piatti A E 1996 *Astroph. J. Supp.* **102** 57
[15] Pietrzyński G and Udalski A 2000 *Acta Astron.* **50** 337
[16] Walborn N R, Maíz-Apellániz J and Barbá R H 2002 *Astron. J.* **124** 1601
[17] Gouliermis D, Kontizas M, Kontizas E and Korakitis R 2003 *Astron. & Astroph.* **405** 111
[18] Glatt K, Grebel E K and Koch A 2010 *Astron. & Astroph.* **517** A50
[19] Popescu B, Hanson M M and Elmegreen B G 2012 *Astroph. J.* **751** 122
[20] Asa'd R S and Hanson M M 2012 *Mon. Not. Royal Astron. Soc.* **419** 2116
[21] Piatti A E, Guandalini R, Ivanov V D, Rubele S, Cioni M R L, de Grijs R, For B Q, Clementini G, Ripepi V, Anders P and Oliveira J M 2014 *Astron. & Astroph.* **570** A74
[22] Palma T, Gramajo L V, Clariá J J, Lares M, Geisler D and Ahumada A V 2016 *Astron. & Astroph.* **586** A41
[23] Udalski A, Kubiak M and Szymański M 1997 *Acta Astron.* **47** 319
[24] Zaritsky D, Harris J, Thompson I B and Grebel E K 2004 *Astron. J.* **128** 1606
[25] Bressan A, Marigo P, Girardi L, Salasnich B, Dal Cero C, Rubele S and Nanni A 2012 *Mon. Not. Royal Astron. Soc.* **427** 127
[26] Harris J and Zaritsky D 2009 *Astron. J.* **138** 1243–1260
[27] Bekki K and Chiba M 2007 *Pub. Astron. Soc. Australia* **24** 21
[28] Baumgardt H, Parmentier G, Anders P and Grebel E 2013 *Mon. Not. Royal Astron. Soc.* **430** 676–685
[29] Elmegreen B G and Efremov Y N 1996 *Astroph. J.* **466** 802
[30] de la Fuente Marcos R and de la Fuente Marcos C 2009 *Astron. & Astroph.* **500** L13–L16

IOP Conf. Series: Journal of Physics: Conf. Series **1127** (2019) 012054 doi:10.1088/1742-6596/1127/1/012054

Comparative Petrographic Analysis of Volcanic Rocks in Five Different Areas Using "R"

J Ramdani[1], **F Septiana**[1], **D E Irawan**[2*], **A Darul**[1]

[1]Institut Teknologi dan Sains Bandung, Kompleks Industri Deltamas, Bekasi
[2]Institut Teknologi Bandung, Bandung, 40132
[3]Author to whom any correspondence should be addressed, dasaptaerwin@gmail.com

dasaptaerwin@outlook.co.id

Abstract. Petrographical analysis is a visual method to describe the mineral properties of a rock sample, related to its composition and classification. The mineral composition is mostly determined using a comparator, thus it is merely a quantitative method. This paper uses multivariable statistics as tools to support the visual classification. We took 89 thin sections from five locations, based on secondary data, gathered from five final projects from Department of Geology ITB: Mt. Lamongan-Probolinggo (LAM), Mt. Banyuresmi-Bogor (BAN), Mt. Kromong-Palimanan (KRM), Mt. Sangkur-Garut (SAN), and Mt. Wangi-Purworejo (WAN). The locations is Quaternary volcanic (LAM, KRM, SAN, and WAN) and Tertiary (BAN), consists of 85 igneous rock samples, three pyroclastics, and one sedimentary rock sample. The mineral percentage data was then analyzed using Principal Component Analysis (PCA) and Cluster Analysis (CA) with R. Our data frame size was 89 rows x 24 columns. Our preliminary results show a separation between Quartenary and Tertiary samples on PCA and CA plots. We identified three clusters: cluster 1 BAN; cluster 2 LAM, KRM, SAN, WAN, and BAN, and cluster 3 LAM, KRM, SAN, WAN, and BAN. Based on our PCA plot, cluster 1 shows a strong influence of: chlorite, calcite, and quartz; cluster 2: piroxene, plagioklas, olivine, glass and cluster 3: olivine, piroxene, opaque, glass, and feldspar. Following the results we see that there are samples from BAN detached from the rest. We believe this is due to an anomalous volcanic process on that spot. On the other hand we see the similarity between samples from different locations. Based on those analyses, we perceive that multivariable statistics can support visual petrographic analysis by reading the data structure, selecting stronger variables and extracting clusters.

Key words: petrography, multivariable statistics, volcanic rock classifications

1. Background

Petrographical analysis is one of the most used method in geological mapping for rock classification. This method uses visual strength to identify the microscopic structure and texture of minerals. The interesting bit that we would like to introduce is whether a quantitative analysis, based on multivariable statistics principles, can support the visual analysis by reading mineral percentage data set, reducing the number of variables, creating a new set of variables, and extracting clusters.

We used Principal Component Analysis (PCA) and Cluster Analysis (CA) technique in this case. The methods have been widely used in the field of hydrogeology [1, 2], hot water classification [3], and ecology [4], to capture hidden patterns in the data structure [5].

2. Materials and methods

The analysis in this paper began with the preparation of thin section of rock sample[6] and petrographic analysis under a microscope[7]. Then the mineral identified in previous activity was tabulated before went to the statistical analysis. Our toy data set was secondary data set extracted from five final project reports published by The Department of Geology ITB library repository. The thin section samples were taken from five locations: Mount Lamongan-Probolinggo (LAM)[8], Banyuresmi-Bogor (BAN)[9], Gunung Kromong-Palimanan (KRM)[10], Mount Sangkur-Garut (SAN)[11], and Mount Wangi-Purworejo (WAN)[12] (see Figure 1). Based on the time scale, the five locations are Quartenary volcanic system (LAM, KRM, SAN, and WAN) and Tertiary volcanic system (BAN). The total samples were 89 consists of 85 (igneous rocks) from basalt to andesite, three samples (pyroclastics) tuffaceous sands and tuff crystal, and one sedimentary deposit. We stored curated data as separate open access data and hosted online[13].

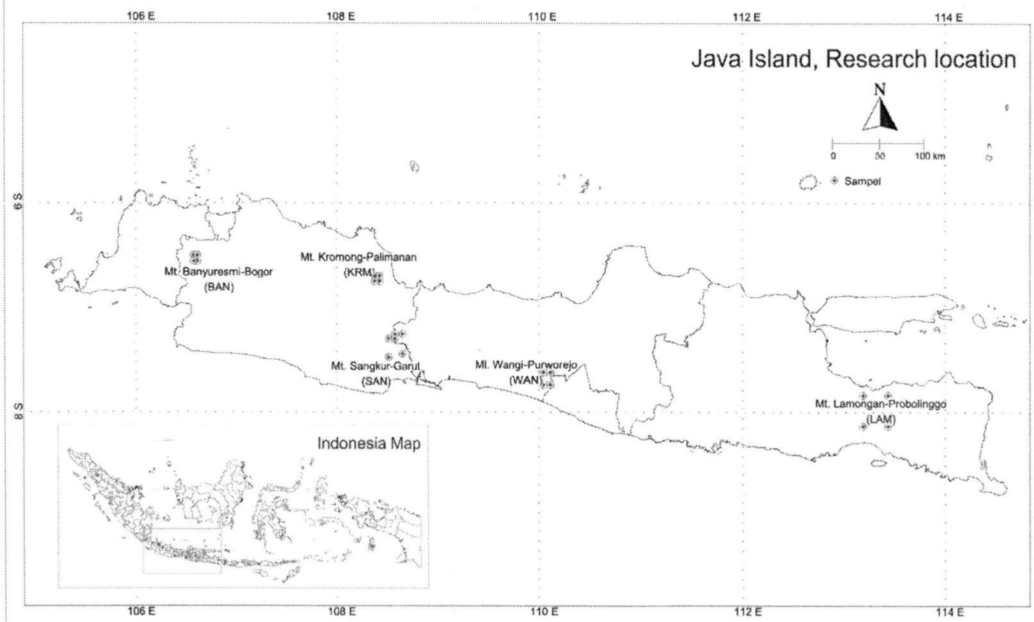

Figure 1. Study are five locations with active volcanism

Basically, we would like to know the classification of the volcanic rocks in those five locations based on mineral composition derived from previous petrographical analysis under statistical equations. Our particular interest is how we can draw connections between samples from different locations and how we markup anomalous samples using PCA and CA[14]. The PCA will reduce the number of variables and creates a set of new variables (or grouping of old variables), while CA will make a classification of samples based on the data structure (Figure 2).

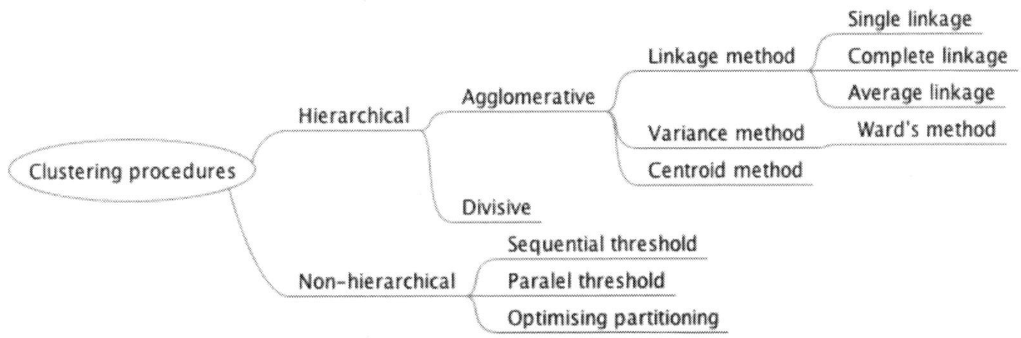

Figure 2 Various cluster The concept of PCA and CA[15]

We used R, an free and open source statistical software [14] and R Studio IDE[15]. The dataset was parsed as csv (comma separated value) and then analyzed using the following packages: ggplot2[16], dplyr[17], factomineR[18], factoExtra[19], cluster 20], ggcorrplot[21], and ape[22].

3. Results and discussions

Our preliminary results show a separation between Quartenary and Tertiary samples on PCA and CA plots and also volcanic rocks from sedimentary rocks. Figure 3 shows clustering of variables in four quadrants:

- Quadrant 1 (upper left), 2 (upper right), and 4 (lower left) show strong variables from volcanic rocks,
- While Quadrant 3 (lower right) shows strong variables for sedimentary rocks.

We identified three groups with various strong variables (see Figure 5 and Figure 6):

- Group 1: LAM, KRM, SAN, WAN, and BAN with a strong influence of: piroxene, plagioklas, olivine, glass, clay, iron oxide;
- Group 2: LAM, KRM, SAN, WAN, and BAN with strong influence of: piroxene, plagioklas, olivine, glass;
- Group 3: WAN and BAN with strong control of clay, quartz, and *foraminifera.*

The following results we see that there are samples from BAN detached from the rest. We believe this is due to an anomalous volcanic process on that spot. On the other hand we see the similarity between samples from different locations. Based on those analyses, we perceive that multivariable statistics can support visual petrographic analysis by reading the data structure, selecting stronger variables and extracting clusters.

Figure 4 shows three PCA groups:

- Group 1: LAM, KRM, SAN, WAN, and BAN with a strong influence of: piroxene, plagioklas, olivine, glass, clay, iron oxide;
- Group 2: LAM, KRM, SAN, WAN, and BAN with strong influence of: piroxene, plagioklas, olivine, glass;
- Group 3: WAN and BAN with strong control of clay, quartz, and foraminifera.

As we see in Figure 4, there are many points assembled at the center of the plot indicating a proportional influence of the variables, and some are plotted further to the side to indicate stronger influence from certain variables than the other. This plot clearly could show us the role of variables building the system spatially.

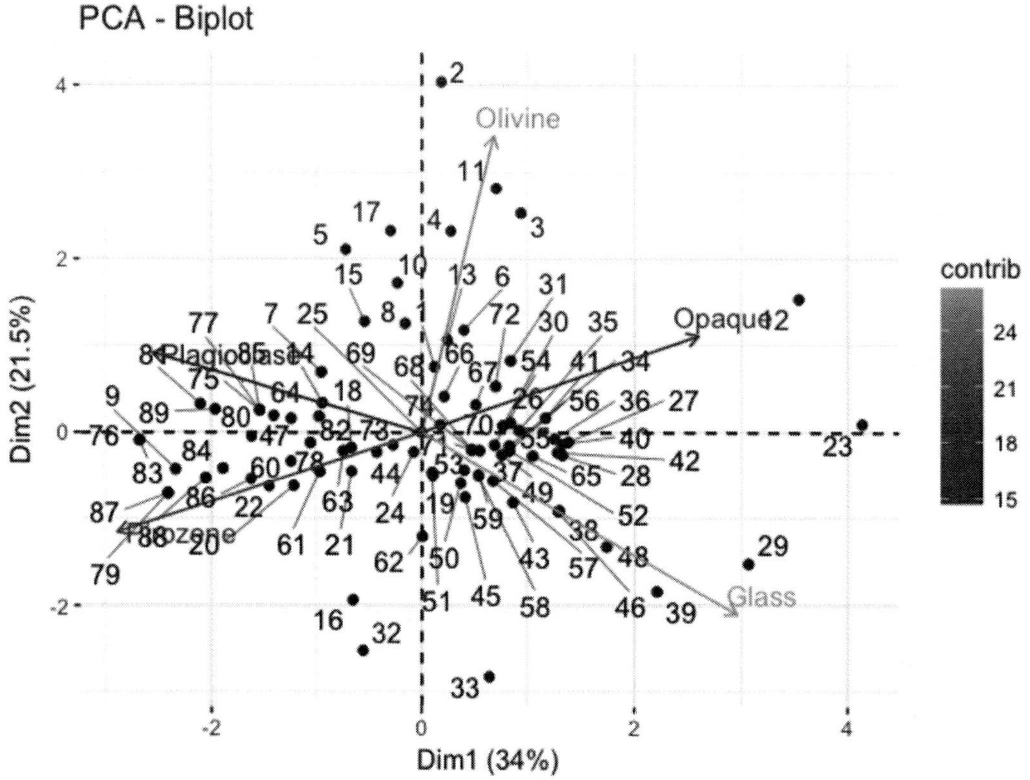

Figure 3 Variables factor map-PCA

Another interesting point is the occurrence of foraminifera (as organic substances) in the cluster where most volcanic rock samples are located. This should be an indication of interaction between volcanic rock with sedimentary rock, possibly under the volcanic rock, possibly in erosion process during volcanic eruption and other unknown reasons. First impression from us, the clustering is able to capture a minimum value of foraminifera content and point its spatial location in the plot.

Consistently, Figure 5 shows three clusters:

- Cluster 1: LAM, KRM, SAN, WAN, and BAN dominated by volcanic-based minerals: olivine, pyroxene (piroksen), opaque, plagioclase, and glass,
- Cluster 2: mostly WAN samples with clay mineral, foraminifera, and a small fraction of opaque and quartz.
- Cluster 3: mostly BAN samples with plagioclase, quartz, dan calcite, as cluster marking.

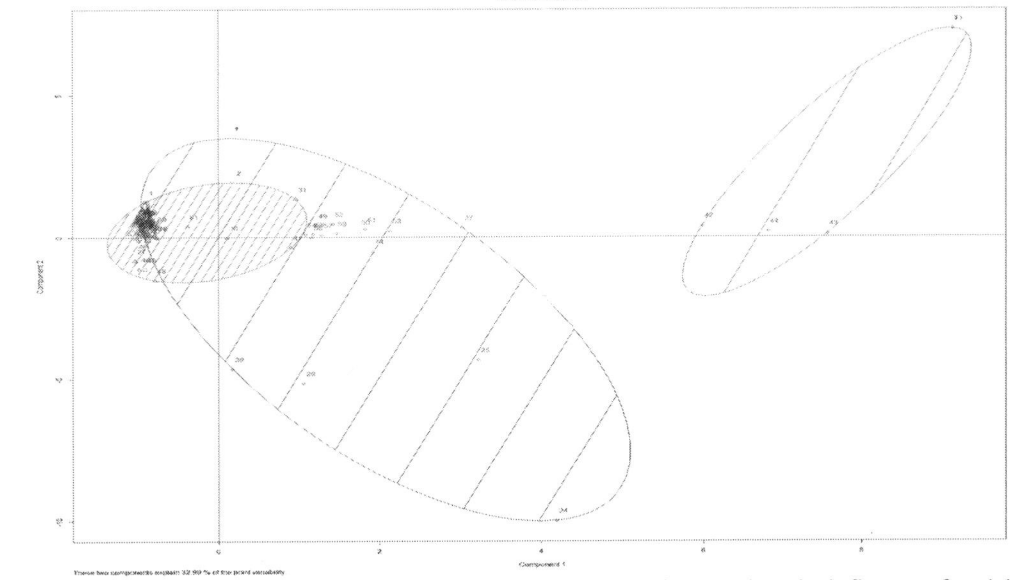

Figure 4 Clusplot PCA to show the spatial distribution of samples based on the influence of variables in each quadrant.

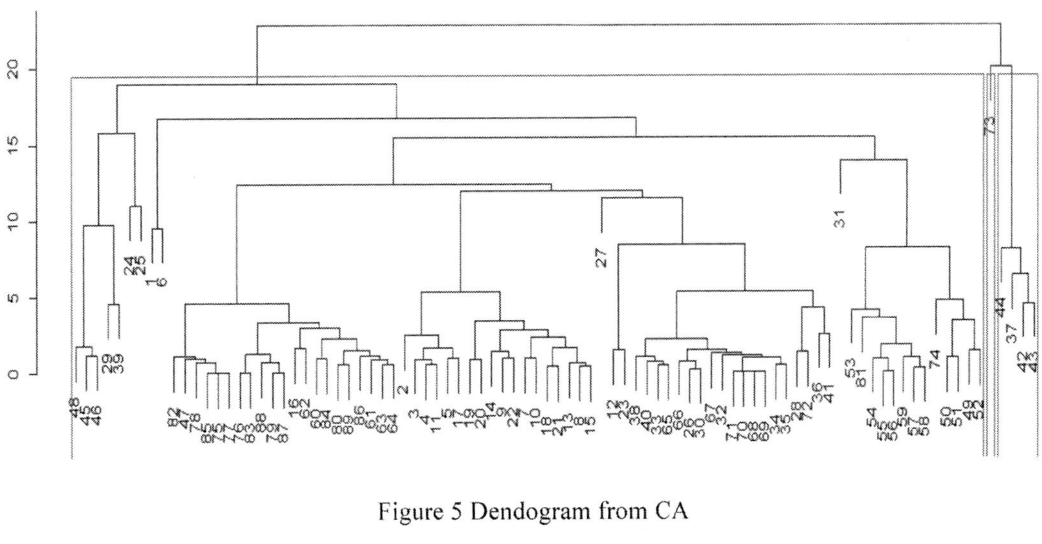

Figure 5 Dendogram from CA

4. Remarks

According to our experiment, the cluster results are consistent with our visual observations. This could be an initial signal that this method can be used for massive petrograpical sample analysis. We suggest more toy data set with various volcanic forming history should be added to test the clustering system in this paper. Another question we have is whether this method could also be applied for other rock type, limestones or sand and clay. This method benefits to a lot of petrologists to make their rock classification easier and time effective. More detail description is needed to add more values to our model.

The method could detect strong and weak roles of the variables in the model. This could help geoscientists to evaluate and specifically classify large amount of data effectively and efficiently. But we need to watch for computer memory usage in relation to the number of data. Memory size of 8 GB is recommended for such purpose. We also believe the usage of open source software are the future in geology. R and Python are the two main programming languages in this context as they provide a straight forward programming algorithm with thousands of free and usable add on package to make coding as a pleasant activity.

In this paper, we also need to point out the importance of open access practice in research. To support this movement and setting an example, aside to only put the manuscript online, we also provide related data and R code freely available under CC-BY (Creative Commons Attribution) 4.0 license in INARxiv preprint server, a community-driven server for Indonesian academia.

REFERENCES

[1] Irawan DE, Puradimaja DJ, Notosiswoyo S, and Sumintadireja P 2009 Hydrogeochemistry of Volcanic Hydrogeology Based on Cluster Analysis of Mount Ciremai, West Java, Indonesia. *Journal of Hydrology* 376: 221–234, url: https://doi.org/10.1016/j.jhydrol.2009.07.033.

[2] King AC, Raiber M, and Cox ME 2014 Multivariate statistical analysis of hydrochemical data to assess alluvial aquifer–stream connectivity during drought and flood: Cressbrook Creek, Southeast Queensland, Australia, *Hydrogeology Journal*, 22, 481–500, doi:10.1007/s10040-013-1057-1, url: http://link.springer.com/10.1007/s10040-013-1057-1.

[3] Sumintadireja P, Irawan DE, Rezky Y, Gio PU, and Agustin A 2016 Classifying hot water chemistry: Application of multivariate statistics, *ITB International Geothermal Workshop*. Zenodo. url: http://doi.org/10.5281/zenodo.45630.

[4] Wiharto M 2013 Analisis Klaster Menggunakan Bahasa Pemrograman R Untuk Kajian Ekologi, url: http://ojs.unm.ac.id/index.php/bionature/article/view/1450, accessed October 20, 2016.

[5] Hatvani IG, Magyar N, Zessner M, Kovács J, and Blaschke AP 2014 The Water Framework Directive: Can more information be extracted from groundwater data? A case study of Seewinkel, Burgenland, Eastern Austria, *Hydrogeology Journal*, 22, 779–794, doi:10.1007/s10040-013-1093-x, url: http://dx.doi.org/10.1007/s10040-013-1093-x.

[6] Subagyo dan Tukidjo 2003 Perparasi sayatan tipis dan poles contoh batuan dari Jumbang I, Kalan - Kalimantan Barat dan Ketapang, Madura - Jawa Timur, *Kumpulan laporan hasil penelitian BATAN tahun 2003*, ISBN 978-979-99141-2-5, url: http://digilib.batan.go.id/e-prosiding/File%20Prosiding/Geologi/Laporan_Pen._2004-2006_PPGN_berkas_A/artikel/subagyo_es_472.pdf.

[7] Open University 2017 An introduction to minerals and rocks under the microscope, Chapter Minerals and the microscope, url: http://www.open.edu/openlearn/science-maths-technology/science/introduction-minerals-and-rocks-under-the-microscope/content-section-2.3.2.

[8] Jhonny 2006 Petrografi Gunung Api Lamongan Kabupaten Probolinggo-Lumajang Jawa Timur, *Skripsi S1 Prodi Teknik Geologi*, Institut Teknologi Bandung.

[9] Tanghamap DY 2009 Geologi Dan Petrografi Batuan Vulkanik Daerah Banyuresmi Dan Sekitarnya Kecamatan Cigudeg Kabupaten Bogor Jawa Barat, *Skripsi S1 Prodi Teknik Geologi*, Institut Teknologi Bandung.

[10] Hadinata J 2009 Geologi dan Alterasi Permukaan Gunung Kromong Dan Sekitarnya Provinsi Jawa Barat, *Skripsi S1 Prodi Teknik Geologi*, Institut Teknologi Bandung.
[11] Wisnu A 1989 *Geologi Daerah Banjar Dan Sekitarnya Dan Petrografi Satuan Vulkanik Gunung Sangkur Kabupaten Ciamis*, Skripsi S1 Prodi Teknik Geologi, Institut Teknologi Bandung.
[12] Wahyuningsih R 1992 *Geologi Dan Petrogenesa Batuan Vulkanik Daerah Gunung Wangi dan Sekitarnya Purworejo Jawa Tengah*, Skripsi S1 Prodi Teknik Geologi, Institut Teknologi Bandung.
[13] Ramdani J, Septiana F, Irawan DE, and Darul A 2017 *Comparative Petrographic Analysis of Volcanic Rocks in Five Different Areas Using R*. Skripsi S1 Teknik Pertambangan, ITSB, Open Science Framework. url: https://osf.io/sj5m4.
[14] Wilkinson DJ 2014 Multivariate Data Analysis using R: a course notes, Tech. rep., url: https://www.staff.ncl.ac.uk/d.j.wilkinson/teaching/mas8381/notes14.pdf .
[15] Gio PU and Irawan DE 2015 *Belajar Statistika dengan R*, USU Press, ISBN: 979 458 806 7, url: http://www.olahdatamedan.com/.
[16] R Core Team 2016 R: A Language and Environment for Statistical Computing, R Foundation for Statistical Computing, Vienna, Austria, url: https://www.R-project.org/.
[17] R Studio Team 2015 RStudio: Integrated Development for R. RStudio, Inc., Boston, MA, url: http://www.rstudio.com/.
[18] Wickham H 2009 ggplot2: Elegant Graphics for Data Analysis, Springer-Verlag New York, url: http://ggplot2.org.
[19] Wickham H and Francois R 2016 dplyr: A Grammar of Data Manipulation, url: https://CRAN.R-project.org/package=dplyr, R package version 0.5.0.
[20] Lê S, Josse J, and Husson F 2008 FactoMineR: A Package for Multivariate Analysis, Journal of Statistical Soft-ware, 25, 1–18, url: http://dx.doi.org/10.18637/jss.v025.i01.
[21] Kassambara A and Mundt F 2016 factoextra: Extract and Visualize the Results of Multivariate Data Analyses, url: https://CRAN.R-project.org/package=factoextra, R package version 1.0.3, 2016.
[22] Maechler M, Rousseeuw P, Struyf A, Hubert M, and Hornik K 2016 cluster: Cluster Analysis Basics and Extensions, url: https://cran.r-project.org/web/packages/cluster/index.html, R package version 2.0.4.
[23] Kassambara A 2016 ggcorrplot: Visualization of a Correlation Matrix using 'ggplot2', url: https://CRAN.R-project.org/package=ggcorrplot, R package version 0.1.1, 2016.
[24] Paradis E, Claude J, and Strimmer K 2004 APE: analyses of phylogenetics and evolution in R language, *Bioinformatics*, 20, 289–290, url: http://bioinformatics.oxfordjournals.org/content/20/2/289.abstract.

Light Curve Analysis for The Transit of Exoplanet WASP-74b Observed at Bosscha Observatory

R. W. Wibowo[1,2], **M. Yusuf**[1], **T. Hidayat**[2], **P. Mahasena**[1,2], **D. Mandey**[1,2], **B. Dermawan**[2]

[1]Bosscha Observatory, Institut Teknologi Bandung, Jalan Peneropong Bintang, Lembang, Bandung Barat 40391, Indonesia

[2]Astronomy Study Program, Institut Teknologi Bandung, Jalan Ganesha 10, Bandung 40132, Indonesia

E-mail: `ridlo.w.wibowo@gmail.com`

Abstract. WASP-74b is an exoplanet in one of the brighter systems accessible to Southern telescope. We report the analysis of light curve from the transit event of WASP-74b observed at Bosscha Observatory. This report is focused on deriving physical parameters from a single transit lightcurve of WASP-74b at around JD 2457597.14 using Markov Chain Monte Carlo sampler (MCMC) and simple trends model. Physical parameters of the system are successfully derived, in agreement with the discovery paper which uses data from longer monitoring and more thorough observation.

1. Introduction

There are several methods to detect exoplanet. The simplest and most widely used is transit method. Transit happens when an exoplanet moves in front of a host star. During the transit, the brightness of the host star will decrease. Although this dimming is very small, it is feasible to measure it using small telescope and inexpensive CCD. Finding new exoplanet with transit method will be one of the main research at the new national observatory that is planned to be built at Timau, East Nusa Tenggara (see Hidayat, T. *et al.*, 2012).

WASP-74 is a slightly evolved F9 star with mass of $1.48 \pm 0.12\ M_\odot$ and radius $1.64 \pm 0.05\ R_\odot$. It is a star of V=9.7 located at RA = $20^h18^m09.^s32$ and declination=$-01°04'32''.6$ (J2000). First detection of exoplanet WASP-74b was carried out using WASP-South survey telescope. It is a 0.95 $M_{\rm Jup}$ planet with a radius of 1.5 $R_{\rm Jup}$ orbiting the host star with period around 2 days. More detail on this system can be found on the discovery paper (Hellier *et al.*, 2015).

In this paper, we present the light curve from the transit event of WASP-74b observed at Bosscha Observatory. It is important to point out that this paper will focus on the analysis of a single light curve to derive physical parameters using Bayesian approach (MCMC method).

2. Data and Analysis

2.1. Observation

WASP-74b was observed at Bosscha observatory by M. Yusuf using Planewave CDK telescope (14", f/7.2), with CCD detector FLI Proline 11002 in R filter. The observations were carried out for two nights (July $27^{\rm th}$ and $29^{\rm th}$ 2016). Due to bad weather conditions however, reliable data used in this analysis is only from observation on July 27th, 2016. Differential photometry was carried out with total length of observation about 4.7 hours and a cadance of $\sim$ 1.4 minutes.

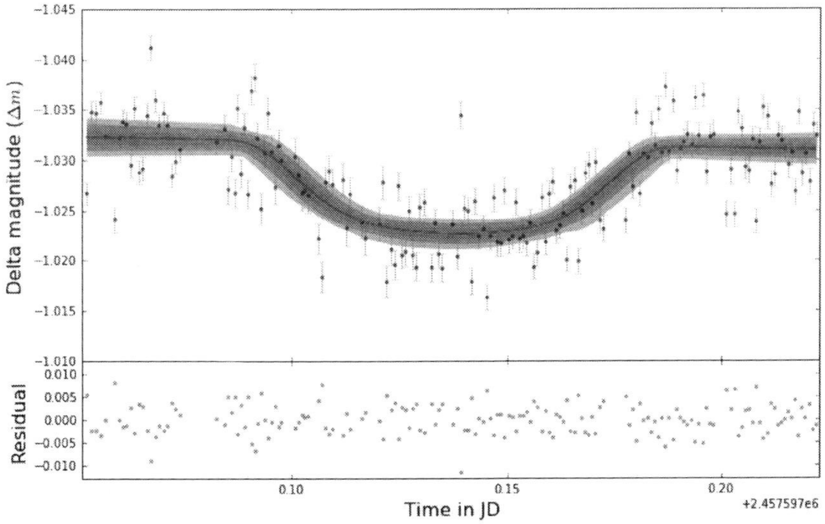

Figure 1. A single transit light curve of WASP-74b (*top*) and its residual from the best-fit model (*bottom*). Red solid line represents the best-fit model. From the darkest green region to the lighter ones show median-centred at 68%, 95%, and 99.7% central posterior intervals (correspond to approximately 1, 2, and 3σ intervals respectively if the distribution is normal).

2.2. Light Curve Analysis

Figure 1 presents the light curve extracted from our observation. We subjectively remove some bad data points with large errorbar which we suspect due to thin cloud during observation. The removed points can be seen as gap around JD 2457597.08. This light curve may still contain contamination from some correlated (red) noise, e.g. from instruments, astrophysical source, and airmass. This red noise can be removed if we have a good model of the noise source. In many cases, the modelling is difficult to be performed. We follow the method used in Exoplanet Transit Database (ETD, `http://var2.astro.cz/ETD/`) to simplify the red noise model for a single transit lightcurve. We add 'quadratic' trend to the model of the light curve,

$$\Delta m(t_i) = A - 2.5 \log F(\theta) + B(t_i - \bar{t}) + C(t_i - \bar{t})^2 \tag{1}$$

We use transit model $F(\theta)$–with θ as parameter vector–from Mandel & Agol (2002). Detail of the transit model can be seen in this paper. We use the implementation of this model in `PyTransit` (Parviainen, 2015). Variable A in Equation 1 represent the zero-point shift of the magnitudes, while B and C describe systematic trends in the data. Linear and quadratic terms are computed with respect to the mean time of observations $\bar{t} = \sum_i t_i/N$ to supress numeric errors.

Instead of using Lavenberg-Marquardt non-linear least squares for fitting procedure as in ETD, we use MCMC sampling to obtain marginal posterior density estimates for the model parameters. The joint model has 11 free parameters in total, all listed in Table 1, along with their priors. The limb darkening uses parameterisation as suggested in Kipping (2013). This parameterization allows us to use uniform priors from 0 to 1 to cover the whole physically sensible limb darkening coefficients (u, v)-space. We assume the data contains normally distributed white noise only. Therefore, we employ simple normal likelihood without taken error of the data into account to reduce computing time.

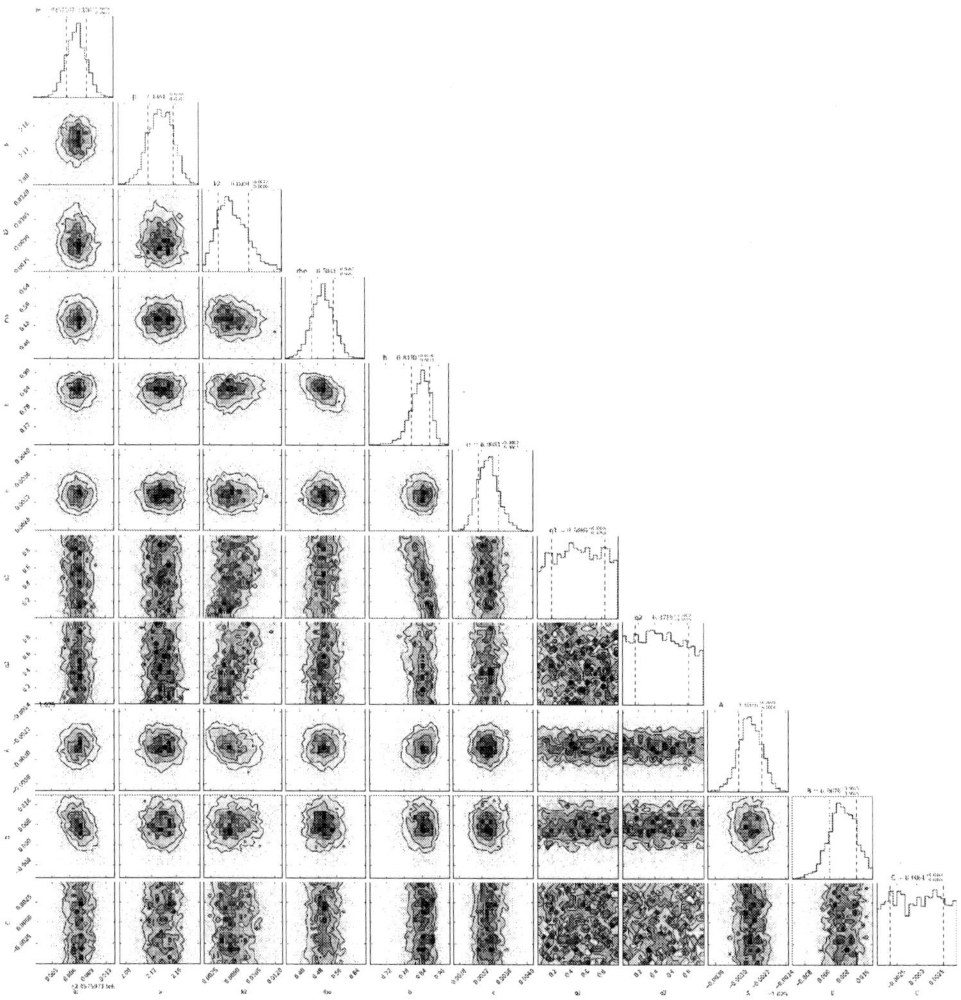

Figure 2. Marginal distributions and covariances for 11 parameters in unconstrained model (free all the parameters). Histogram from the left to the right are for each parameter: transit center, period, area ratio, star density, impact parameter, white noise, limb darkening parameter 1, 2, detrending constant A, B, and C.

We sample the posterior distribution with `emcee` package (Foreman-Mackey *et al.*, 2013), a Python implementation of the Affine Invariant MCMC sampler (Goodman & Weare, 2010). To reduce burn-in time, we initialize the `emcee` sampler (walkers) using the parameter vector population found with PyDE, an implementation of the differential evolution global optimization algorithm (Storn & Price, 1997). We complete the analysis using NumPy, matplotlib (Hunter, 2007), and `corner.py` (Foreman-Mackey, 2016).

The posterior sampling is carried out by running the `emcee` sampler with 100 simultaneous chains (walkers) for 10000 markov chain steps. After burn-in period of 2000 steps and a thinning

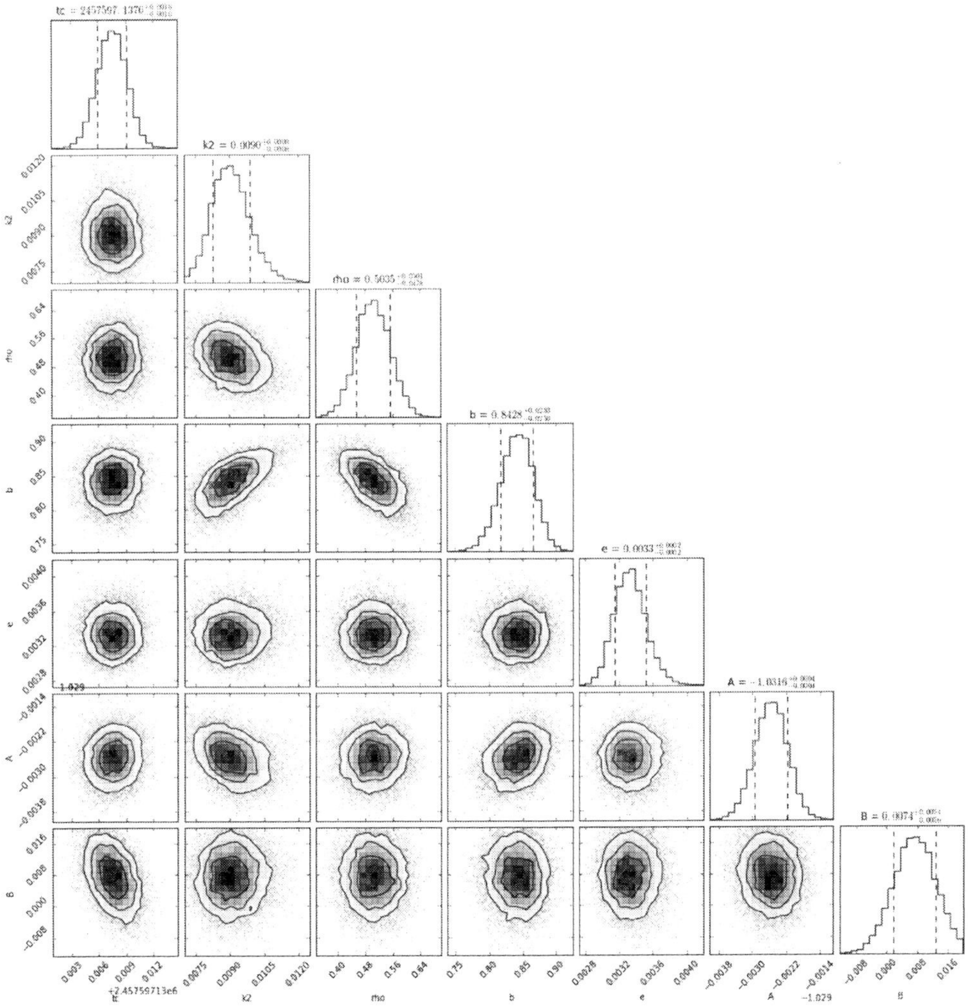

Figure 3. Marginal distributions and covariances for 7 parameters in constrained model (period, limb darkening constants, and detrending constant C are fixed). Correlation between impact parameter b and area ratio k^2, also anticorrelation between b and star density $\rho_\star$ more clearly seen in this constrained model. We can also see anticorrelation between detrending constant B and transit center t_c.

factor of 281 based on the chain autocorrelation length estimates, we obtain 2900 independent posterior samples. Afterwards, we constrain our model with a fixed value for 4 parameters: period (P), limb darkening coefficients (u, v), and quadratic detrending term C. Thinning factor in constrained model is 84, resulting 9600 independent posterior samples.

We fix the period in constrained model, since the data is for a single transit event only. In many cases, when more than one transits are observed, the period can be obtained by folding and binning the light curve. Limb darkening constants (u, v) are fixed because their value are not so sensitive to the fitting result of our data (indicated by nearly uniform posterior as seen in Figure 2). We also remove quadratic term (C) in constrained model, since adding it as free

IOP Conf. Series: Journal of Physics: Conf. Series 1127 (2019) 012055 doi:10.1088/1742-6596/1127/1/012055

parameter does not improve the fit significantly.

3. Result and Discussion

We list the model parameters for both unconstrained and constrained model in Table 1. The values correspond to the posterior medians (values) and the 16th and 84th percentiles (errors), closely correspond to the posterior mean $\pm$ 1σ estimates (if the posteriors are close to normal). We present the marginal distributions for all parameters in Figure 2 and Figure 3 which correspond to unconstrained and constrained model respectively. We also plot the data and best-fit of unconstrained model in Figure 1.

Table 1. Parameters, priors, and the fitting result. Transit center t_c and period P are in days, density of the star $\rho_\star$ is in gr/cm^3. We used normal $\mathcal{N}$, Jeffreys $\mathcal{J}$, and uniform $\mathcal{U}$ distribution as priors. The uniform and Jeffreys distribution is defined by its lower and upper bounds, while the normal distribution is defined by its mean and standard deviation. Fixed value for several parameters in constrained model are also given.

Parameters	Prior	Free all params	Constrained
Transit center (t_c)	$\mathcal{N}(2457597.13, 0.03)$	$2457597.1376^{+0.0016}_{-0.0017}$	$2457597.1376^{+0.0016}_{-0.0016}$
Period (P)	$\mathcal{N}(2.13775, 0.02)$	$2.1381^{+0.0193}_{-0.0202}$	2.13775(fixed)
Area ratio (k^2)	$\mathcal{J}(0.007, 0.0125)$	$0.0091^{+0.0012}_{-0.0009}$	$0.0090^{+0.0009}_{-0.0008}$
Stellar density ($\rho_\star$)	$\mathcal{N}(0.5, 0.05)$	$0.5044^{+0.0487}_{-0.0497}$	$0.5035^{+0.0501}_{-0.0478}$
Impact parameter (b)	$\mathcal{U}(0, 0.99)$	$0.8410^{+0.0286}_{-0.0371}$	$0.8428^{+0.0233}_{-0.0250}$
White noise std (σ)	$\mathcal{U}(0.002, 0.02)$	$0.0033^{+0.0002}_{-0.0002}$	$0.0033^{+0.0002}_{-0.0002}$
Limb darkening par. (q_1)	$\mathcal{U}(0.0, 1.0)$	$0.5086^{+0.3268}_{-0.3283}$	0.49(fixed)
Limb darkening par. (q_2)	$\mathcal{U}(0.0, 1.0)$	$0.4719^{+0.3409}_{-0.3120}$	0.36(fixed)
Detrending constant A	$\mathcal{U}(-1.53, -0.53)$	$-1.0316^{+0.0004}_{-0.0004}$	$-1.0316^{+0.0004}_{-0.0004}$
Detrending constant B	$\mathcal{U}(-0.02, 0.02)$	$0.0076^{+0.0055}_{-0.0056}$	$0.0074^{+0.0054}_{-0.0056}$
Detrending constant C	$\mathcal{U}(-0.005, 0.005)$	$0.0001^{+0.0032}_{-0.0035}$	0.0 (fixed)
Derived parameters			
Radius ratio ($k = R_p/R_\star$)		$0.0953^{+0.0061}_{-0.0048}$	$0.0951^{+0.0045}_{-0.0041}$
Scaled semimajor axis ($a/R_\star$)		$4.9564^{+0.1584}_{-0.1684}$	$4.9546^{+0.1593}_{-0.1622}$
$0.0953^{+0.0061}_{-0.0048}$ Inclination (i)		$80.24^{+0.59}_{-0.57}$	$80.20^{+0.53}_{-0.54}$
Limb darkening coef. (u)		$0.5876^{+0.5408}_{-0.4036}$	0.5(fixed)
Limb darkening coef. (v)		$0.0328^{+0.4309}_{-0.4280}$	0.2(fixed)

The computation cost of this fitting method largely depends on the number of walkers and MC steps. It is however rather cheap for a single transit with a simple model like presented in this work. It only takes several minutes in standard laptop for a single run. We compare our result with the discovery paper in Table 2, we can see that the sigma of our result are larger.

True physical parameters of the planet can not be derived by using transit data only. Therefore, we present the result in the form of ratios between parameters. It is possible to get true physical parameters by combining our result with other independent measurement such as radial-velocity (RV) or spectral observation of the host star (and isochrone fitting). For example, if we know the mass of the host star is $1.48 \pm 0.12\ M_\odot$ (Hellier *et al.*, 2015), the radius of the planet derived from our data is $1.53 \pm 0.12\ R_{\rm Jup}$ (still within 1σ from their result,

Table 2. Some of the paramaters can be compared with the discovery paper Hellier *et al.*, (2015). We must aware that the model used in here are different, so the direct comparison like this should be treated wisely.

Parameters	Our result	Hellier *et al.*, (2015)
Period (days)	$2.1381^{+0.0193}_{-0.0202}$	2.137750 ± 0.000001
Area ratio	$0.0091^{+0.0012}_{-0.0009}$	0.00961 ± 0.00014
Stellar density (gr/cm^3)	$0.5044^{+0.0487}_{-0.0497}$	0.477 ± 0.025
Impact parameter	$0.8410^{+0.0286}_{-0.0371}$	0.860 ± 0.006

$1.56 \pm 0.06\ R_{\rm Jup}$). We do not have high resolution spectrograph to do RV observation yet, but with our current instruments we may be able to rule out the grazing eclipsing binary by using transit color signature (see Tingley *et al.*, 2014).

4. Summary

We derive some physical parameters of WASP-74b from the transit light curve data observed at Bosscha Observatory. We follow the model used in ETD, but change the fitting procedure to Bayesian approach (MCMC sampling). With low computation cost, this method can be used to fit a single-noisy light curve data like presented in this paper or amateur data collected by ETD. In the future analysis, we can add our own spectral observation and isochrone fitting to derive mass of the host star independently.

Acknowledgments

The author acknowledges funding from Faculty of Mathematics and Natural Science ITB and Leids Kerkhoven-Bosscha Fonds.

References

[1] David M Kipping 2013 *MNRAS* **435** 2152
[2] Foreman-Mackey D, Hogg D W, Lang D and Goodman J 2013 *PASP* **123** 306
[3] Foreman-Mackey D 2016 *The Journal of Open Source Software* **24** 1
[4] Goodman J, & Weare J 2010 *Communications in Applied Mathematics and Computational Science* **5** 65
[5] Hellier C *et al.* 2015 *ApJ* **150** 18
[6] Hidayat T, Mahasena P, Dermawan B, Hadi T W, Premadi P W and Herdiwijaya D 2012 *MNRAS* **427** 1903
[7] Hunter J D 2007 *Comput. Sci. Eng.* **9** 90
[8] Mandel K & Agol E 2002 *ApJ* **580** L171
[9] Parviainen H 2015 *MNRAS* **450** 3233
[10] Storn R, & Price K 1997 *Journal of Global Optimization* **11** 341
[11] Tingley B, Parviainen H, Gandolfi D, Deeg H J, Palle E, Montañés Rodriguez P, Murgas F, Alonso R, Bruntt H, and Fridlund M 2014 *A&A* **567** A14

The Hα line-profile variations for B-emission stars

W C Setiadi[1] and Aprilia[1]

[1]Department of Astronomy, Faculty of Mathematics and Natural Sciences, Institute Technology of Bandung, Jalan Ganesha 10, Bandung Indonesia

wendicahya@gmail.com

Abstract. Spectroscopic variations of Hα line profiles from 9 B-emission (Be) stars in the southern sky, taken from the 5th revised of Bright Star Catalogue year 1991, are analyzed. These variations, emission and/or absorption lines, are due to the presence of the circumstellar envelopes surrounding the stars and also describe the phase changing/evolution phases of the stellar spectra if we have data for some years. Therefore, we compare the spectra to data from BeSS database and references for the recent years. Five stars show double-peaked emission line profile of Hα. One star shows Be phase with a single peaked Hα emission line, and three stars show evolution phases of the Hα line profile. By studying this evolution of the line variations, we can fit the suitable envelope models for each star taken from references.

1. Introduction

Be star is a non-supergiant B star whose spectrum has, or had at some time, one or more Balmer lines in emission [6]. This definition is considered too common, because there are also observed other stars that have emission lines like Herbig Ae/Be stars and Algol systems. Therefore, in this discussion we use the 'Classical Be-emission Star' to refer only to a rapidly-rotating (100-400 km/s), B-type star of luminosity class III-V [1]. Be stars show emission on its early Balmer lines (Hα and Hβ) and sometimes on its neutral helium [4]. It is estimated that 20% of all B-type stars are Be stars, and the highest fraction of Be stars appears around spectral type B1e-B2e ([1],[5]).

Be stars have different line profiles, i.e. Be (Be single-peak, Be double-peak) and Be-shell, and sometimes as B normal (without emission line). Initially, these were interpreted in terms of different angles of inclination on the line of sight on the rotation axis of the stars. However, from observations, it is found that some Be stars show phases changing as: B Be Be-Shell B. Therefore this changing is preferred to call as Be phenomenon.

2. Methods

The data is taken from Bright Star Catalogue Rev 5th [2], selected only for stars in the southern hemisphere and of luminosity class V and IV. We use Be Stars Spectra (BeSS) database for the spectral lines. All the spectra are processed using Python syntax and IRAF package released by National Optical Astronomical Observatories (NOAO). A standard procedure for using IRAF can be seen in http://iraf.noao.edu/. The stellar parameter for the selection data are shown in table 1.

Table 1. Program stars

No.	HR	HD	Star Name	Ra (⁰)	Dec (⁰)	Spectra	Vmag
1	4621	105435	δ Cen	182.09	-50.7225	B2IVne	2.6
2	6510	158427	α Ara	262.96	-49.8761	B2Vne	2.95
3	472	10144	α Eri	24.4287	-57.2367	B3Vpe	0.46
4	2356	45725	β Mon	97.2042	-7.0328	B3Ve	4.6
5	2364	45871	IY CMa	97.1633	-32.3711	B4Vnpe	5.74
6	3745	81753	-	141.687	-28.7875	B6Ve	6.1
7	3971	87543	-	150.893	-61.8839	B7IVne	6.14
8	8386	209014	η PsA	330.209	-28.4536	B8V	5.42
9	9098	225132	2 Cet	0.935	-17.3361	B9.5Vn	4.55

3. Results

Figure 1 – 9 show the Hα line profiles of the selected stars. Stars that show evolution phase are HD 10144, from B normal to Be-shell (from 2010 – 2015, figure 1), HD 45725, from B normal to Be (2002, 2007 – 2015, figure 2), and HD 81753 from Be-shell to Be (2009 and 2014, figure 3). Stars that show Be phase with double peak Hα emission lines for certain period are HD 209014 (figure 4), HD 45871 (figure 5), HD 225132 (figure 6), HD 158427 (figure 7), and HD 105435 (figure 8). Star that shows Be phase with single peak Hα emission line is HD 87543 (figure 9).

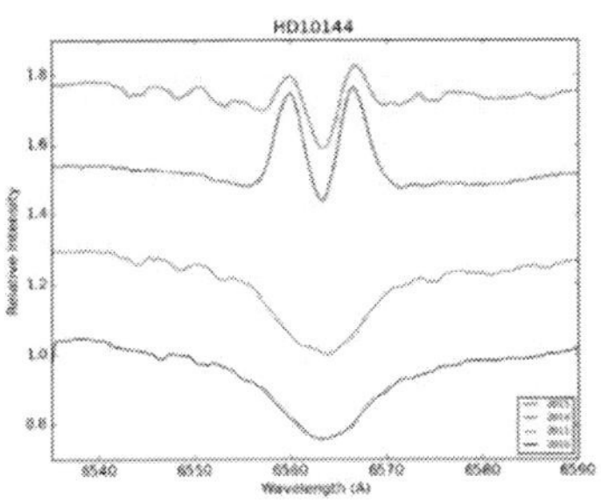

Figure 1. Spectra of HD 10144 from 2010-2011 and 2014-2015

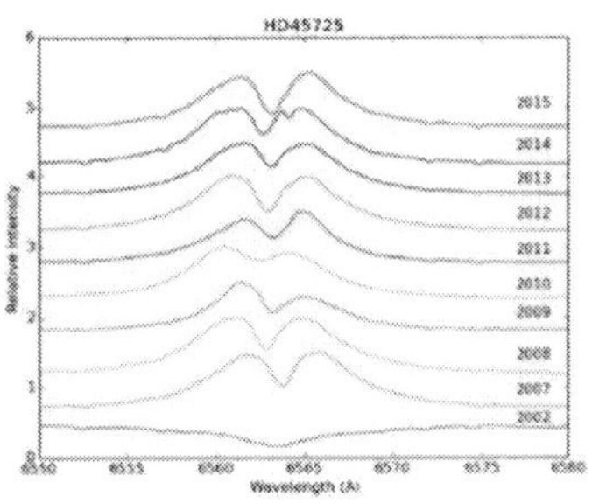

Figure 2. Spectra of HD 45725 from 2002 and 2007-2015

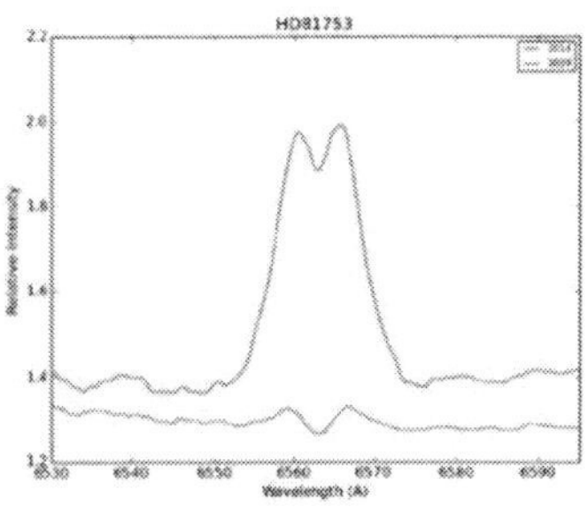

Figure 3. Spectra of HD 81753 from 2009 and 2014

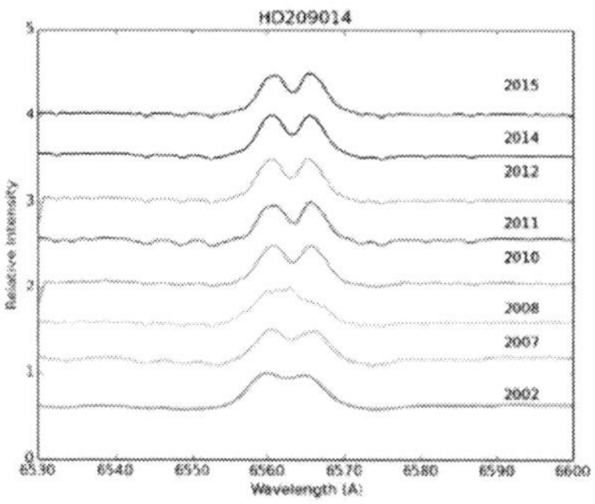

Figure 4. Spectra of HD 209014
from 2002 and 2007-2015

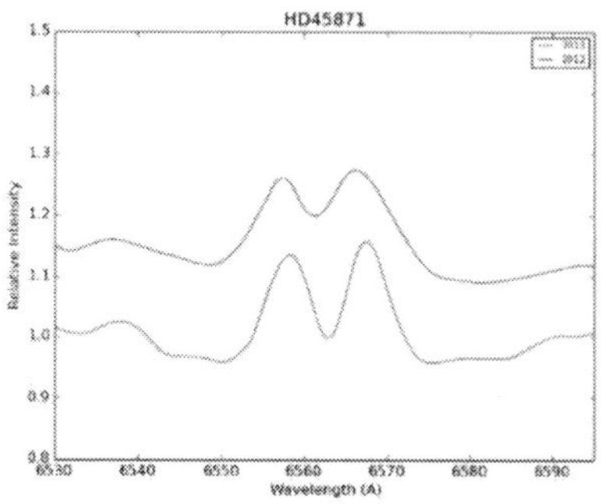

Figure 5. Spectra of HD 45871
from 2012-2013

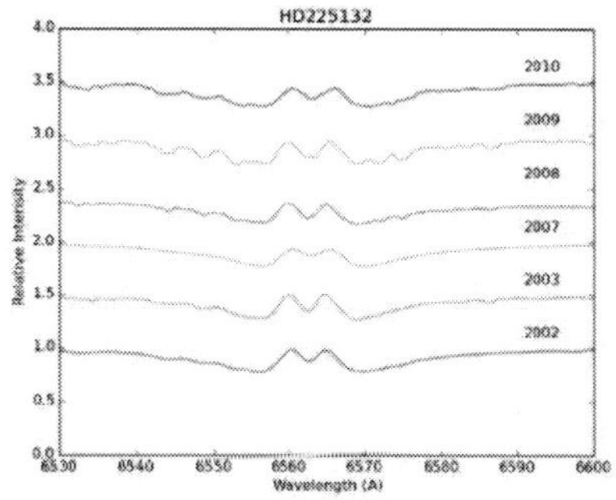

Figure 6. Spectra of HD 225132
from 2002-2003 and 2007-2010

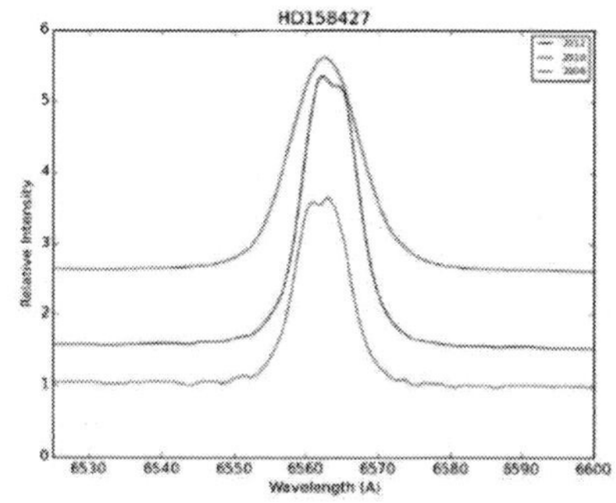

Figure 7. Spectra of HD 158427
in 2008, 2010, and 1012

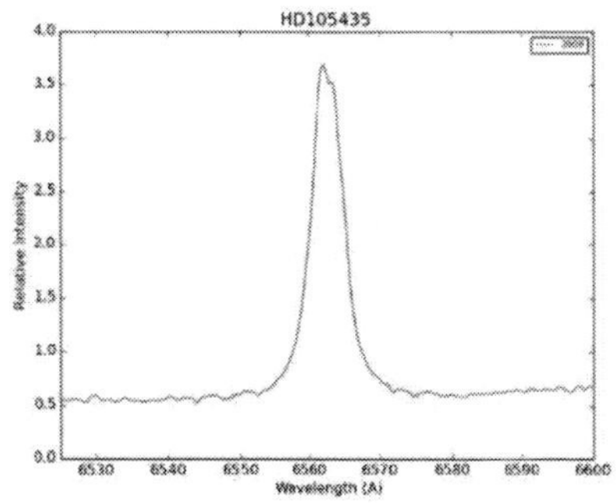

Figure 8. Spectra of HD 105435
in 2008

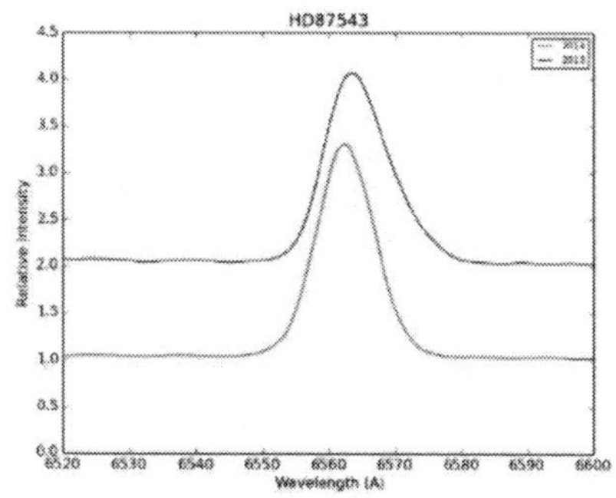

Figure 9. Spectra of HD 87543
from 2013-2014

4. Conclusion

From the nine selected stars, one star exhibits single-peak emission line profile (HD 87543), five stars show double-peak emission line profiles (HD 209014, HD 45871, HD 158427, HD 225132, HD 105435), and the other three stars show evolution phases of the Hα emission lines (HD 45725, HD 10144, HD 81753). The evolution phase of HD 45725 are from B to Be double-peak line. For HD 81753, the changes are from Be-shell to Be double-peak line. HD 10144 changes from B to Be-shell. HD 10144 is a member of binary star system, and the Be phenomenon is probably caused by the mass transfer to its companion [3]. The profile of Hα absorption and emission lines are due to the presence of circumstellar envelopes surrounding the stars. For better understanding the Be phenomenon, for further research, it is important to study the suitable envelope models that fit the stars.

References

[1] Doazan V 1982 *B Stars With and Without Emission Lines* (NASA SP-456)
[2] Hoffleit D and Jaschek C 1991 *Bright Star Catalogue* 5th *Revised Ed. (Preliminary Version)* (Yale University Observatory)
[3] Kervella P, Thévenin F, Ségransan D, Berthomieu G, Lopez B, Morel P and Provost J 2003 *Astronomy & Astrophysics* **404** 1087
[4] Malasan H L, Irfan M, Aprilia and Dawanas D 2014 *The Tenth Pacific Rim Conference on Stellar Astrophysics* **482** 25
[5] Porter J M and Rivinius T 2003 *Pub. Astron. Soc. Pacific* **115** 1153
[6] Slettebak A 1988 *Pub. Astron Soc. Pacific* **100** 770

IOP Conf. Series: Journal of Physics: Conf. Series 1127 (2019) 012057 doi:10.1088/1742-6596/1127/1/012057

Numerical Simulation of Milky Way Galaxy and Sagittarius Dwarf Galaxy Collision

Ramadhani Putri Ayu[1] & M. Ikbal Arifyanto[2]

[1]Department of Computational Science, Faculty of Mathematics and Natural Science, Institut Teknologi Bandung, Indonesia

[2]Astronomy Research Division, Institut Teknologi Bandung, Indonesia

E-mail: ramadhani.putri92@gmail.com

September 2016

Abstract. The Milky Way has its own system of dwarf satellite galaxies, which have much lower mass. Interaction between the Milky Way and its satellites can impact the structure. Sagittarius Dwarf Galaxy is one of the nearest and brightest Milky Way's satellite galaxies. But it has been torn apart because of potential of the Galaxy. In this paper, we investigate a few models of Milky Way collision with its satellite. Sagittarius Dwarf Galaxy will be constructed as spherical and disk dwarf galaxy. We also included the stars, gaseous, and dark matter as the building blocks of the dwarf galaxy in calculation of this tidal interaction. The calculation was performed using N-body and Smoothed Particle Hydrodynamic program, GADGET-2 code.

1. Introduction

First model of galactic formation called ELS model was proposed by Eggen et al. [1]. In this model, rotating spherical gas cloud collapsed and formed a galaxy. The rotating gas cloud then formed either disk or elliptical galaxy depend on rotational velocity. This model could not explain the various ages of cluster on halo galaxy. Another model of galactic formation proposed that galaxy was formed by aggregation of collapsed cloud fragments, called protogalaxy. This interaction could form a bigger galaxy.

Galactic formation models have led us to think about our galaxy, Milky Way. λ Cold Dark Matter (λCDM) theory said that Milky Way must had tens or hundreds of dwarf galaxy forming an observed halo galaxy. In fact, the observed satellite galaxy of Milky Way were not that much. It was probably due to the domination of dark matter in our halo. The evidence of that interaction was the existence of the satellite galaxies and observed star streams around our Galaxy. Sagittarius Dwarf Galaxy (SDG) was one of the big satellite galaxy of Milky Way.

This paper is constructed as follows. In Section 2, we describe our simulation, including the initial condition and method. In Section 3 we show the result of the

IOP Conf. Series: Journal of Physics: Conf. Series **1127** (2019) 012057 doi:10.1088/1742-6596/1127/1/012057

Table 1. Potential Parameter of Milky Way Galaxy.

Parameter	Value
Bulge	
Mass, M_b	$9.489 \times 10^9 M_\odot$
Scale Length, b_b	0.23 kpc
Disk	
Mass, M_d	$6.626 \times 10^{10} M_\odot$
Scale length, a_d, b_d	4.22 kpc, 0.292 kpc
Halo	
Virial Mass (M_{200})	$3.299 \times 10^{12} M_\odot$
Characteristic radius(r_s)	45.02 kpc

simulation fitted with the observational data, and discuss the result. Finally, in Section 4 we summarize our work.

2. Computational Model

Milky Way model consists of bulge, disk, and halo with Hernquist, Miyamoto-Nagai, and NFW (Nafarro, Frenk and White) profile potential respectively. Milky Way potential is considered fixed, with total mass is $\sim 10^{12} M_\odot$. Stars, gaseous, and dark matter was also included in the calculation of this model.

Similar with Milky Way, SDG models consist of bulge and halo with Hernquist and NFW potential respectively. We used two model of SDG. First SDG model have spherical form with additional Plummer potential. Second SDG model have disk form with additional Miyamoto-Nagai potential. Detail composition of Milky Way and SDG mass were shown in the table 1 and 2.

Table 2. Potential Parameter of Sagittarius Dwarf Galaxy.

	Parameter	Value
	Spherical	
Model	Mass	$4 \times 10^6 M_\odot$
I	Scale length	0.45 kpc
	Particles	21000
	Disk	
Model	Mass, M_d	$4 \times 10^6 M_\odot$
II	Scale length, a_d	0.23 kpc
	Particles	21000
	Mass, M_b	$5.3 \times 10^5 M_\odot$
Bulge	Scale length, b_b	0.23 kpc
	Particles	5000
	Virial Mass (M_{200})	$8.03 \times 10^7 M_\odot$
Halo	Characteristic Radius (r_s)	4 kpc
	Particles	4774

IOP Conf. Series: Journal of Physics: Conf. Series **1127** (2019) 012057 doi:10.1088/1742-6596/1127/1/012057

N-body/SPH simulation was performed using cosmological simulation code, `GADGET-2`[2], with initial condition of particle distribution was generate using open source `DICE`[3] code. We used `GADGET-2` to simulate 2×10^5 particles in Newtonian space and ignores the cosmological space expansion. Initial position of SDG was derived by simulate point mass particle backward from current position of SDG ($X_{gal}, Y_{gal}, Z_{gal} = (19.41, 2.69, -6.9)$; $v_x, v_y, v_z = (248.8, -44.1, 172.9)$) to 5 Gyr ago. This simulation produced an orbit of SDG with eccentricity, apogalactic, and perigalactic are 0.5, 43.9, and 14.4 respectively. N-body/SPH simulation started from apogalactic position of the orbit.

3. Result and Discussion

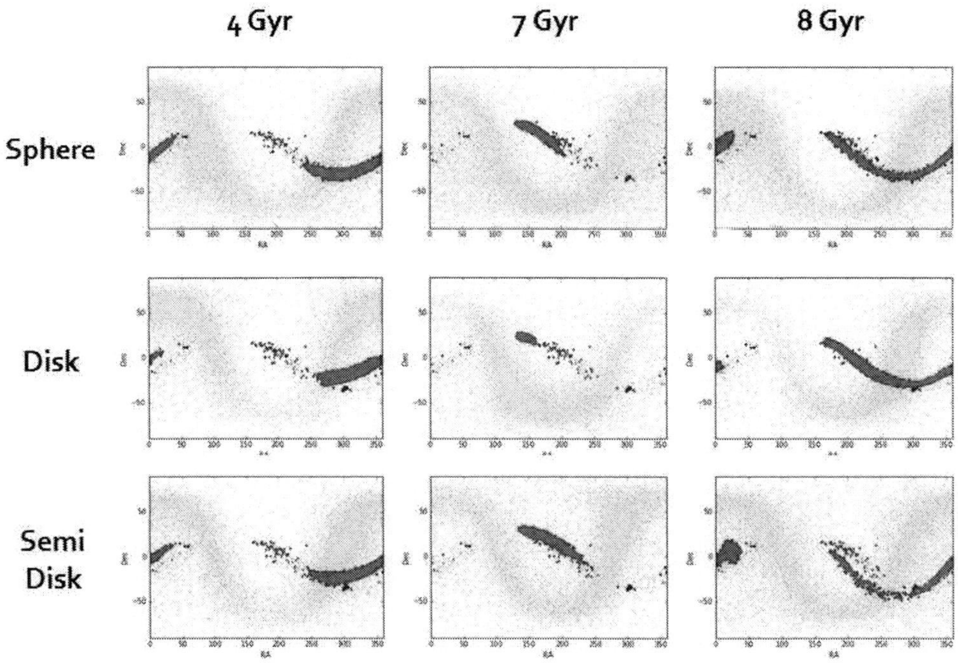

Figure 1. Result of N-body simulation of Milky Way and SDG interaction in equatorial coordinate projection after 4, 7, and 8 Gyr. Top, middle, and bottom figure show the simulation result of Milky Way and Model I (spherical SDG), Model II (disk SDG), semi-disk SDG respectively. Cyan, red, yellow, blue, and green colors respectively show particle of Milky Way's plane from simulation, simulation result for particle member of SDG, observed SDG by Law et al. [4], leading and trailing tail of SDG by Majewski et al. [5].

Simulation result of Milky Way and SDG interaction was shown in figure 1. As comparison, we also performed a simulation of interaction Milky Way and mixed spherical and disk SDG, or semi-disk SDG (see bottom part of figure 1). Observational

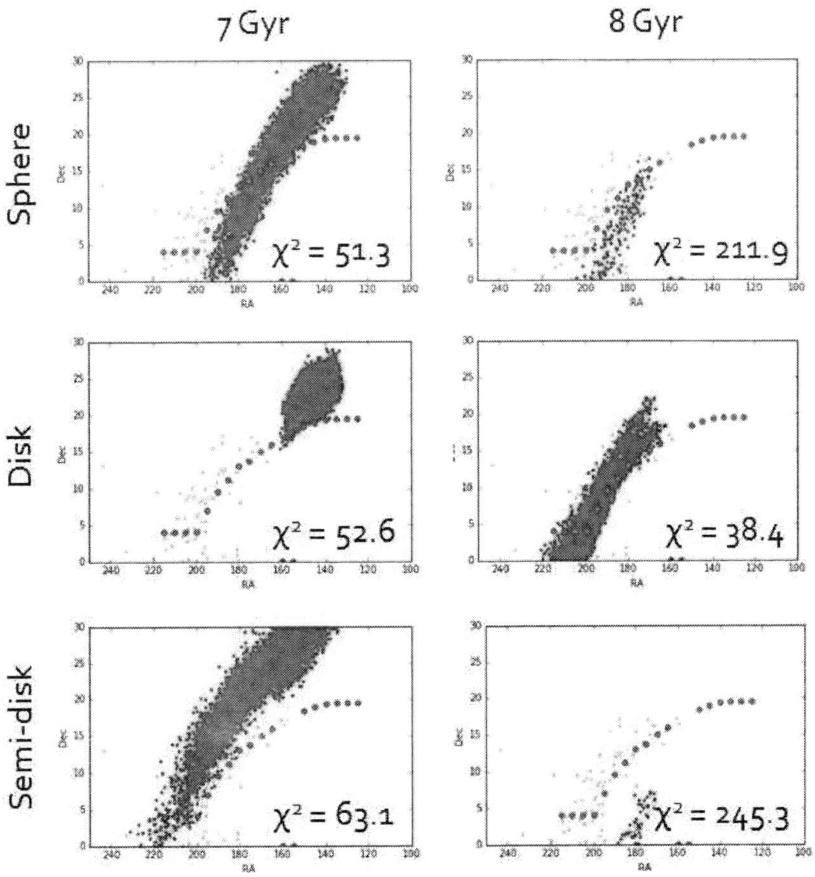

Figure 2. Data fitting of simulation result with leading stellar stream by Belokurov et al. [6]. Red color shows simulation result of SDG. "x" symbols show mean of simulation data in range of RA every 5°. Blue color show observed leading tail by Law et al. [4]. Mean data of observed SDG by Belokurov et al. [6] show in "o" symbol.

data of Law et al. [4] and Majewski et al. [5] was used to confirm the result (yellow, blue, and green colors in figure 1).

In simulation result, star stream was formed after SDG has orbited Milky Way for ~4 Gyr. This result agreed with the reported age of SDG by Bellazzini et al. [7]. They reported that SDG star stream population was started to be formed while progenitor SDG was colliding with Milky Way in 3-4 Gyr. However, the best fit simulation result with observational data was obtained after 7 Gyr interaction (see center and right part of figure 1). Then, the simulation star stream was fully followed the feature of observed stream along with equatorial coordinate, while SDG was dominated by M giant star population with age ≥ 5 Gyr (5.5-9.5 Gyr), and average 8 Gyr (Bellazzini et al. [7]).

Another report (Ibata et al. [8]; Law & Majewski et al. [9]; and Pennarubia et al. [10]) said that SDG was torn apart and formed stellar stream after interacted for 3 Gyr with progenitor total mass of $\sim 10^8 M_\odot$. The mass fraction was dominated by baryon particles (star and gas). Meanwhile, Gibbons et al. reported that SDG stellar stream was formed -as observed now- after interacting for 4 Gyr with progenitor SDG total mass of $10^{10} M_\odot$ and 90% of the mass fraction dominated by dark halo. In this simulation, the mass fraction of progenitor SDG was too large so the stream was formed much longer (more than 4 Gyr) then expected.

$$\chi^2_X = \sum_i \frac{(X_{obs_i} - X_{simulation_i})^2}{X_{simulation_i}} \tag{1}$$

$$\chi^2_{sum} = \chi^2_{RA} + \chi^2_{dec} + \chi^2_{distance} \tag{2}$$

Stellar stream feature in the observed area by Belokurov et al. [6] presented in the simulation result after 7 Gyr. Trend of the stream was similar with observational data. The fitting data was calculated using equation (1). From the equation (1), X is the mean of position (RA, declination, or galactocentric distance). Observational data (X_{obs}) derived from Belokurov et al. [6]. Total of χ^2 calculated by summing up χ^2 of RA, dec, and distance (equation 2). The best fit of the simulation result was derived after 8 Gyr of the interaction Milky Way and disk SDG, with $\chi^2 = 38.4$.

4. Summary

We have reported the simulation of Milky Way and Sagittarius Dwarf Galaxy interaction using cosmological simulation code, `GADGET-2`. Initial conditions of particle distribution in the galaxy was generated using open source code `DICE`. We used a fix model of Milky Way that composed of bulge, disk, and halo (included dark halo). For the Sagittarius Dwarf Galaxy (SDG), we considered two models, spherical and disky SDG. Initial position of SDG was derived by simulating the SDG orbit backward in 5 Gyr from current position. The simulation was performed using $\sim 2 \times 10^5$ particles until 8 Gyr forward. The tidal stream was formed after SDG orbiting the Milky Way 4, 7, and 8 Gyr.

From 2 (+1) models, tidal stream of disky SDG after orbiting Milky Way 8 Gyr fitted with observed star stream with $\chi^2 = 38.4$. Nevertheless, the tidal stream was formed after certain times. It might be happened due to the overestimated mass fraction and length scale of the SDG bulge. Varying initial position and mass fraction of SDG are necessary for next studies to get better result.

Acknowledgments

RPA gratefully acknowledge the financial support from DIKTI fresh graduate scholarship and FMIPA ITB for this research.

References

[1] Eggen, O.J., Lynden-Bell, D., and Sandage, A.R. 1962 *Astrophysical Journal*, **136**, 748766.

[2] Springel, V. 2005 *Monthly Notices of the Royal Astronomical Society*, **364.4**, 1105-1134.
[3] Perret, V., Renaud, F., Epinat, B., Amram, P., Bournaud, F., Contini, T., Teyssier, R. and Lambert, J.C. 2014 *Astronomy and Astrophysics*, **562**, A1.
[4] Law, D.R., Johnston, K.V., and Majewski, S.R. 2004 *arXiv preprint astro-ph/0407566.*
[5] Majewski, S.R., Skrutskie, M.F., Weinberg, M.D., and Ostheimer, J.C. 2003 *Astrophysical Journal*, **599**, 1082.
[6] Belokurov, V., et al. 2006 *Astrophysical Journal*, **642**, L137.
[7] Bellazzini, M., Correnti, M., Ferraro, F.R., Monaco, L. and Montegriffo, P. 2006 *Astronomy and Astrophysics*, **4462**, L1-L4.
[8] Ibata, R.A., Wyse, R.F.G., Gilmore, G., Irwin, M.J, and Suntzeff, N.B. 1997 *Astronomical Journal*, **113.2**, 634.
[9] Law, D.R., and Majewski, S.R. 2010 *Astronomical Journal*, **714**. 229.
[10] Penarrubia, J., Belokurov, V., Evans, N.W., Martnez-Delgado, D., Gilmore, G., Irwin, M., Niederste-Ostholt, M., and Zucker, D.B. 2010 *Monthly Notices of the Royal Astronomical Society*, **408**, L26L30.
[11] Gibbons, S.L.J., Belokurov, V. and Evans, N.W., 2016 *arXiv preprint arXiv:1607.00803.*

IOP Conf. Series: Journal of Physics: Conf. Series **1127** (2019) 012058 doi:10.1088/1742-6596/1127/1/012058

The 3-rainbow index of amalgamation of some graphs with diameter 2

Zata Yumni Awanis[1] **and A.N.M. Salman**[2]

[1,2]Combinatorial Mathematics Research Group, Faculty of Mathematics and Natural Sciences, Institut Teknologi Bandung, Jalan Ganesa 10, Bandung 40132, Indonesia

E-mail: [1]zata.yumni@s.itb.ac.id, [2]msalman@math.itb.ac.id

Abstract. Let $G = (V, E)$ be a nontrivial, connected, and edge-colored graph with n vertices, and let k be an integer with $2 \leq k \leq n$. A tree T in G is a rainbow tree, if no two edges of T receive the same color. A k-rainbow coloring of G is an edge coloring of G having property that for every set S of k vertices of G, there exists a rainbow tree T such that $S \subseteq V(T)$. The minimum number of colors needed in a k-rainbow coloring of G is the k-rainbow index of G, denoted by $rx_k(G)$. The distance $d(x, y)$ of two vertices x and y in G is the length of a shortest $x - y$ path in G. The greatest distance between any two vertices in G is the diameter of G, denoted by $diam(G)$. Let $\{G_1, G_2, \ldots, G_t\}$ be a finite collection of graphs and each graph G_i have a fixed vertex v_{o_i} called a terminal. The amalgamation of $G_1, G_2, \ldots, G_t$, denoted by $Amal(G_1, v_{o1}, G_2, v_{o2} \ldots, G_t, v_{ot})$, is a graph obtained by taking all the G_i's and identifying their terminals. In case $G_i \cong G$ and $v_{o_i} = u$, the amalgamation of $G_1, G_2, \ldots, G_t$ is denoted by $Amal(G, t, u)$. In this paper, we determine the 3-rainbow index of amalgamation of some graphs $Amal(G, t, u)$ with $diam(G) = 2$.

1. Introduction

All graphs considered in this paper are simple, finite, and connected. An edge-colored graph G is *rainbow connected*, if every two vertices x and y of G is connected by a path whose edges have distinct colors. Such a path is called a *rainbow path*. An edge coloring of G that results in a rainbow connected graph is a *rainbow coloring* of G. The *rainbow connection number* $rc(G)$ of G, introduced by Chartrand et al. [2], is the minimum number of colors needed in a rainbow coloring of G. Some results about the rainbow connection number of graphs can be found in [6, 7, 8, 9, 10].

In [3], they generalized the concept of rainbow connection number. A tree T in G is a *rainbow tree*, if no two edges of T receive the same color. Let k be an integer with $2 \leq k \leq n$. An edge coloring of G is called a *k-rainbow coloring*, if for every set S of k vertices of G, there exists a rainbow tree in G containing the vertices of S. Such a rainbow tree is called a *rainbow S-tree* [4]. The *k-rainbow index* $rx_k(G)$ of G is the minimum number of colors needed in a k-rainbow coloring of G. Thus $rx_2(G)$ is the rainbow connection number $rc(G)$ of G. For every connected graph G of order n, it is easy to check that $rx_2(G) \leq rx_3(G) \leq \cdots \leq rx_n(G)$.

The *distance* $d(x, y)$ of two vertices x and y in G is the length of a shortest $x - y$ path in G. The greatest distance between any two vertices in G is the *diameter* of G, denoted by $diam(G)$ [5]. The *Steiner distance* $d(S)$ of a set S of vertices in G is the minimum size of a tree in G containing S. Such a tree is called a *Steiner S-tree* or simply a *Steiner tree*. The *k-Steiner diameter* of G, denoted by $sdiam_k(G)$, is the maximum Steiner distance of S among all sets of S with k vertices in G. Thus if

$k = 2$ and $S = \{x, y\}$, then $d(S) = d(x, y)$ and $sdiam_2(G) = diam(G)$. In [3], they provided a simple upper bound and a lower bound for $rx_k(G)$ as follows.

Lemma 1. *Let G be a nontrivial connected graph of order $n \geq 3$. For each integer k with $3 \leq k \leq n-1$, $rx_k(G) \leq n - 1$ whereas $rx_n(G) = n - 1$.*

Lemma 2. *For every connected graph G of order $n \geq 3$ and each integer k with $3 \leq k \leq n$, $k - 1 \leq sdiam_k(G) \leq rx_k(G)$.*

In this paper, we determine the lower and upper bounds for 3-rainbow index of $Amal(W_n, 2, u)$ for $n \geq 4$ and $Amal(F_n, 2, u)$ for $n \geq 3$.

2. Main Results

The following definition of amalgamation of graph is taken from [1]. For $t \in \mathbb{N}$ and $i \in \{1, 2, \ldots t\}$, let G_i be a simple connected graph and $|V(G_i)| = k_i \geq 2$ for some $k_i \in \mathbb{N}$. For $t \geq 2$, let $\{G_1, G_2, \ldots, G_t\}$ be a finite collection of graphs and each graph G_i has a fixed vertex v_{o_i} called a *terminal*. The *Amalgamation* of $G_1, G_2, \ldots, G_t$, denoted by $Amal(G_1, v_{o_1}, G_2, v_{o_2} \ldots, G_t, v_{o_t})$, is a graph obtained by taking all the G_i's and identifying their terminals. In case $G_i \cong G$ and $v_{o_i} = u$, the amalgamation of $G_1, G_2, \ldots, G_t$ is denoted by $Amal(G, t, u)$.

Let n be a positive integer. Let W_n be a wheel on $n + 1$ vertices, F_n be a fan on $n + 1$ vertices, and u be a vertex in W_n or F_n where the degree of u is $d(u) = n$. The vertex u is called the *center vertex*. It is easy to check that $diam(W_n) = 2$ for $n \geq 4$ and $diam(F_n) = 2$ for $n \geq 3$.

Let $G \cong Amal(G_1, v_{o_1}, G_2, v_{o_2})$, and we denote the identified vertex as u. In this paper, we consider the lower and upper bounds for 3-rainbow index of $G \cong Amal(G_1, v_{o_1}, G_2, v_{o_2})$ where for each $i \in \{1, 2\}$, G_i is a wheel W_n or a fan F_n on $n + 1$ vertices and v_{o_i} is the center vertex. For simplifying, we define $[a, b] = \{x \in \mathbb{Z} | a \leq x \leq b\}$.

First, we provide the lower and upper bounds for 3-rainbow index of $G \cong Amal(W_n, 2, u)$ for $n \geq 4$.

Theorem 1. *Let n be an integer with $n \geq 4$, W_n be a wheel on $n+1$ vertices, and $G \cong Amal(W_n, 2, u)$ where u is the center vertex of W_n. Then $rx_3(G) = 4$ for $n \in [4, 8]$ and $rx_3(G) \leq 5$ for $n \geq 9$.*

Proof. Let $V(G) = \{u, v_i, w_i | i \in [1, n]\}$ and $E(G) = \{uv_i, uw_i, v_iv_{i+1}, w_iw_{i+1} | i \in [1, n]\}$ where $v_{n+1} = v_1$ and $w_{n+1} = w_1$. It is easy to check that $sdiam_3(G) = 3$. We divide the proof into two cases based on n as follows.

Case 1. $n \in [4, 8]$

To show that $rx_3(G) \leq 4$, we define a 3-rainbow 4-coloring $c : E(G) \longrightarrow [1, 4]$ as follows.

$$c(uv_i) = c(w_iw_{i+1}) = \begin{cases} 1, & \text{for } i \in [1, n] \text{ and } i \equiv 1 \text{ mod } 2; \\ 2, & \text{for } i \in [1, n] \text{ and } i \equiv 0 \text{ mod } 2; \end{cases}$$

$$c(uw_i) = c(v_iv_{i+1}) = \begin{cases} 3, & \text{for } i \in [1, n] \text{ and } i \equiv 1 \text{ mod } 2; \\ 4, & \text{for } i \in [1, n] \text{ and } i \equiv 0 \text{ mod } 2. \end{cases}$$

We now show that c is a 3-rainbow 4-coloring of G. Let S be a set of three vertices of G. We consider seven subcases.

Subcase 1.1. $S = \{u, v_i, v_j\}$, where $i < j$.
Subcase 1.2. $S = \{u, v_i, w_j\}$.
Subcase 1.3. $S = \{u, w_i, w_j\}$, where $i < j$.
Subcase 1.4. $S = \{v_i, v_j, v_k\}$, where $i \neq j, i \neq k, j \neq k$.
Subcase 1.5. $S = \{v_i, v_j, w_k\}$, where $i \neq j$.
Subcase 1.6. $S = \{w_i, w_j, v_k\}$, where $i \neq j$.
Subcase 1.7. $S = \{w_i, w_j, w_k\}$, where $i \neq j, i \neq k, j \neq k$.

Table 1. A rainbow S-tree for Case 1 of Theorem 2.1

x	y	z	n	Condition ($i,j,k \in [1,n]$)	A rainbow S-tree
u	v_i	v_j	-	i and j have same parity, $i<j$	$\{uv_i, uv_{j-1}, v_{j-1}v_j\}$
			-	i and j have different parity, $i<j$	$\{uv_i, uv_j\}$
u	v_i	w_j	-	-	$\{uv_i, uw_j\}$
u	w_i	w_j		the proof is similar with case $x=u, y=v_i, z=v_j$	
v_i	v_j	v_k	4	$i<j<k$	$\{v_iv_j, v_jv_k\}$
			5	i,j,k have same parity, $i<j<k$	$\{uv_i, uv_{j+1}, v_{j+1}v_j, v_{j+1}v_k\}$
				$i=1, j=5$, k is even	$\{uv_i, v_iv_j, uv_k\}$
				i,j are odd, k is even, $i<j$ i,j are even, k is odd, $i<j$	$\{uv_i, v_iv_{i+1}, v_{i+1}v_j, uv_k\}$
			6	i,j,k have same parity, $i<j<k$	$\{uv_i, uv_{j+1}, v_{j+1}v_j, v_{j+1}v_k\}$
				i,j are odd, k is even, $i<j$ i,j are even, k is odd, $i<j$	$\{uv_i, v_iv_{i+1}, v_{i+1}v_j, uv_k\}$
			7	$i=7, j=1, k=3$ i,j,k have same parity, $i<j<k$	$\{uv_i, uv_{j+1}, v_{j+1}v_j, v_{j+1}v_k\}$
				$i=1, j=7$, k is even	$\{uv_i, v_iv_j, uv_k\}$
				i,j are odd, k is even, $i<j$ i,j are even, k is odd, $i<j$	$\{uv_i, v_iv_{i+1}, v_{i+1}v_j, uv_k\}$
			8	i,j,k have same parity, $i<j<k$	$\{uv_i, uv_{j+1}, v_{j+1}v_j, v_{j+1}v_k\}$
				i,j are odd, k is even, $j-i=2$ i,j are even, k is odd, $j-i=2$	$\{uv_i, v_iv_{i+1}, v_{i+1}v_j, uv_k\}$
				i,j are odd, k is even, $\lvert j-i\rvert=4, \lvert k-i\rvert=1$ i,j are even, k is odd, $\lvert j-i\rvert=4, \lvert k-i\rvert=1$	$\{uv_k, v_kv_i, uv_j\}$
				i,j are odd, k is even, $\lvert j-i\rvert=4, \lvert k-j\rvert=1$ i,j are even, k is odd, $\lvert j-i\rvert=4, \lvert k-j\rvert=1$	$\{uv_i, uv_k, v_kv_j\}$
w_i	w_j	w_k		the proof is similar with case $x=v_i, y=v_j, z=v_k$	
v_i	v_j	w_k	-	i,j,k are odd, $i<j$ i,j,k are even, $i<j$	$\{uv_i, uv_{j-1}, v_{j-1}v_j, uw_k\}$
			-	i,j are odd, k is even, $i>j$ i,j are even, k is odd, $i<j$	$\{uv_i, uv_{j+1}, v_{j+1}v_j, uw_k\}$
			-	i,j have different parity, $i<j$	$\{uv_i, uv_j, uw_k\}$
w_i	w_j	v_k		the proof is similar with case $x=v_i, y=v_j, z=w_k$	

We can check that for each subcase, there exists a rainbow S-tree in G as shown in Table 1. So, c is a 3-rainbow 4-coloring of G. Thus $rx_3(G) \leq 4$.

Next we show that $rx_3(G) \geq 4$ by using a contradiction. Suppose that $rx_3(G) \leq 3$, then there exists a 3-rainbow 3-coloring $c^* : E(G) \longrightarrow [1,3]$. For $n=4$, consider $S=\{v_i, v_j, w_k\}$ or $S=\{v_i, w_j, w_k\}$ with $d(v_i, v_j)=2$ or $d(w_j, w_k)=2$. Note that the only S-tree of size 3 is $T=\{uv_i, uv_j, uw_k\}$ or $T=\{uv_i, uw_j, uw_k\}$, respectively. Firstly, consider $S_1=\{v_1, v_3, w_1\}$, then $T_1=\{uv_1, uv_3, uw_1\}$. Without loss of generality, write $c^*(uv_1)=1, c^*(uv_3)=2$, and $c^*(uw_1)=3$. Secondly, consider $S_2=\{v_1, v_3, w_3\}$, then $T_2=\{uv_1, uv_3, uw_3\}$. Since $c^*(uv_1)=1$ and $c^*(uv_3)=2$, we have $c^*(uw_3)=3$. Next, consider $S_3=\{v_1, w_1, w_3\}$, then $T_3=\{uv_1, uw_1, uw_3\}$, but $c^*(uw_1)=c^*(uw_3)=3$, it follows that there is no rainbow S_3-tree, which is a contradiction. So, $rx_3(G) \geq 4$. Therefore, $rx_3(G)=4$ for $n \in [4,8]$.

Case 2. $n \geq 9$

To show that $rx_3(G) \leq 5$, we define a 3-rainbow 5-coloring $c : E(G) \longrightarrow [1,5]$ as follows.

$$c(uv_i) = \begin{cases} 1, & \text{for } i \in [1,n] \text{ and } i \equiv 1 \bmod 3; \\ 2, & \text{for } i \in [1,n] \text{ and } i \equiv 2 \bmod 3; \\ 3, & \text{for } i \in [1,n] \text{ and } i \equiv 0 \bmod 3; \end{cases}$$

$$c(uw_i) = \begin{cases} 5, & \text{for } i \in [1,n] \text{ and } i \equiv 1 \bmod 3; \\ 4, & \text{for } i \in [1,n] \text{ and } i \equiv 2 \bmod 3; \\ 3, & \text{for } i \in [1,n] \text{ and } i \equiv 0 \bmod 3; \end{cases}$$

$$c(v_iv_{i+1}) = \begin{cases} 4, & \text{for } i \in [1,n] \text{ and } i \equiv 1 \bmod 2; \\ 5, & \text{for } i \in [1,n] \text{ and } i \equiv 0 \bmod 2; \end{cases}$$

$$c(w_iw_{i+1}) = \begin{cases} 1, & \text{for } i \in [1,n] \text{ and } i \equiv 1 \bmod 2; \\ 2, & \text{for } i \in [1,n] \text{ and } i \equiv 0 \bmod 2. \end{cases}$$

We now show that c is a 3-rainbow 5-coloring of G. Let S be a set of three vertices of G. Then, we consider seven subcases like in Case 1. The main idea of obtaining a rainbow S-tree is similar with Case 1. So, it is easy to check that c is a 3-rainbow 5-coloring of G. Thus, $rx_3(G) \leq 5$ for $n \geq 9$. This completes the proof. □

For illustration, a 3-rainbow 4-coloring on $G \cong Amal(W_8, 2, u)$ and a 3-rainbow 5-coloring on $G \cong Amal(W_{12}, 2, u)$ are given in Figure 1.

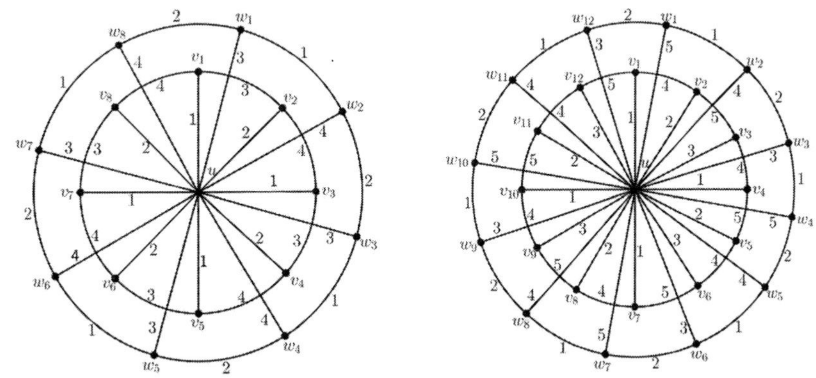

(a) (b)

Figure 1. (a) A 3-rainbow 4-coloring on $G \cong Amal(W_8, 2, u)$ and (b) a 3-rainbow 5-coloring on $G \cong Amal(W_{12}, 2, u)$

With the aid of Theorem 1, we get the following corollary.

Corollary 1. *Let n_1 and n_2 be two positive integers with $n_1 \geq 4$ and $n_2 \geq 4$, and let $G \cong Amal(W_{n_1}, W_{n_2}, u)$ where u is the center vertex of W_n. Then $rx_3(G) = 4$ for $n_1, n_2 \in [4,8]$ and $rx_3(G) \leq 5$ for $n_1 \in [4,8]$ and $n_2 \geq 9$ or $n_1, n_2 \geq 9$.*

Proof. Let $V(G) = \{u, v_i, w_j | i \in [1, n_1], j \in [1, n_2]\}$ and $E(G) = \{uv_i, uw_j, v_iv_{i+1}, w_jw_{j+1} | i \in [1, n_1], j \in [1, n_2]\}$ where $v_{n_1+1} = v_1$ and $w_{n_2+1} = w_1$. This results follow directly from Case 1 of Theorem 1 for $n_1, n_2 \in [4,8]$, and from Case 2 of Theorem 1 for $n_1 \in [4,8]$ and $n_2 \geq 9$ or $n_1, n_2 \geq 9$. □

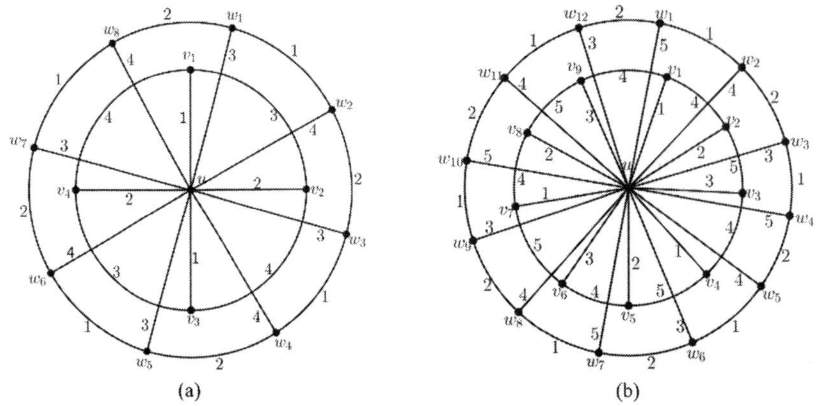

Figure 2. (a) A 3-rainbow 4-coloring on $G \cong Amal(W_4, W_8, u)$ and (b) a 3-rainbow 5-coloring on $G \cong Amal(W_9, W_{12}, u)$

For illustration, a 3-rainbow 4-coloring on $G \cong Amal(W_4, W_8, u)$ and a 3-rainbow 5-coloring on $G \cong Amal(W_9, W_{12}, u)$ are given in Figure 2.

Next, we provide the lower and upper bounds for 3-rainbow index of $G \cong Amal(F_n, 2, u)$ for $n \geq 3$.

Theorem 2. *Let n be an integer with $n \geq 3$, F_n be a fan on $n+1$ vertices, and $G \simeq Amal(F_n, 2, u)$ where u is the center vertex of F_n. Then $rx_3(G) = 4$ for $n \in [3, 6]$ and $rx_3(G) \leq 5$ for $n \geq 7$.*

Proof. Let $V(G) = \{u, v_i, w_i | i \in [1, n]\}$ and $E(G) = \{uv_i, uw_i, v_iv_{i+1}, w_iw_{i+1} | i \in [1, n-1]\}$. It is easy to check that $sdiam_3(G) = 3$. We divide the proof into two cases based on n as follows.
Case 1. $n \in [3, 6]$
To show that $rx_3(G) \leq 4$, we define a 3-rainbow 4-coloring $c : E(G) \longrightarrow [1, 4]$ as follows.

$$c(uv_i) = c(w_jw_{j+1}) = \begin{cases} 1, & \text{if } i \in [1,n], j \in [1,n-1], \text{ and } i,j \text{ are odd;} \\ 2, & \text{if } i \in [1,n], j \in [1,n-1], \text{ and } i,j \text{ are even;} \end{cases}$$

$$c(uw_i) = c(v_jv_{j+1}) = \begin{cases} 3, & \text{if } i \in [1,n], j \in [1,n-1], \text{ and } i,j \text{ are odd;} \\ 4, & \text{if } i \in [1,n], j \in [1,n-1], \text{ and } i,j \text{ are even.} \end{cases}$$

We now show that c is a 3-rainbow 4-coloring of G. Let S be a set of three vertices of G. We can check that there exists a rainbow S-tree in G. So, c is a 3-rainbow 4-coloring of G. Thus $rx_3(G) \leq 4$.

Next, we show that $rx_3(G) \geq 4$ by using a contradiction. Suppose that $rx_3(G) \leq 3$, then there exists a 3-rainbow 3-coloring $c^* : E(G) \longrightarrow [1, 3]$. By using the same argument as in the proof of lower bound in Theorem 1 for $n \in [4, 8]$, we obtain a contradiction. So, $rx_3(G) \geq 4$. Therefore, $rx_3(G) = 4$ for $n \in [3, 6]$.
Case 2. $n \geq 7$
To show that $rx_3(G) \leq 5$, we define a 3-rainbow 5-coloring $c : E(G) \longrightarrow [1, 5]$ as follows.

$$c(uv_i) = \begin{cases} 1, & \text{for } i \in [1,n] \text{ and } i \equiv 1 \bmod 3; \\ 2, & \text{for } i \in [1,n] \text{ and } i \equiv 2 \bmod 3; \\ 3, & \text{for } i \in [1,n] \text{ and } i \equiv 0 \bmod 3; \end{cases}$$

$$c(uw_i) = \begin{cases} 5, & \text{for } i \in [1,n] \text{ and } i \equiv 1 \bmod 3; \\ 4, & \text{for } i \in [1,n] \text{ and } i \equiv 2 \bmod 3; \\ 3, & \text{for } i \in [1,n] \text{ and } i \equiv 0 \bmod 3; \end{cases}$$

$$c(v_iv_{i+1}) = \begin{cases} 4, & \text{for } i \in [1, n-1] \text{ and } i \equiv 1 \bmod 2; \\ 5, & \text{for } i \in [1, n-1] \text{ and } i \equiv 0 \bmod 2; \end{cases}$$

$$c(w_iw_{i+1}) = \begin{cases} 1, & \text{for } i \in [1, n-1] \text{ and } i \equiv 1 \bmod 2; \\ 2, & \text{for } i \in [1, n-1] \text{ and } i \equiv 0 \bmod 2. \end{cases}$$

We now show that c is a 3-rainbow 5-coloring of G. Let S be a set of three vertices of G. We can check that there exists a rainbow S-tree in G. So, c is a 3-rainbow 5-coloring of G. Thus, $rx_3(G) \leq 5$ for $n \geq 7$. This completes the proof. □

For illustration, a 3-rainbow 4-coloring on $G \cong Amal(F_6, 2, u)$ and a 3-rainbow 5-coloring on $G \cong Amal(F_8, 2, u)$ are given in Figure 3.

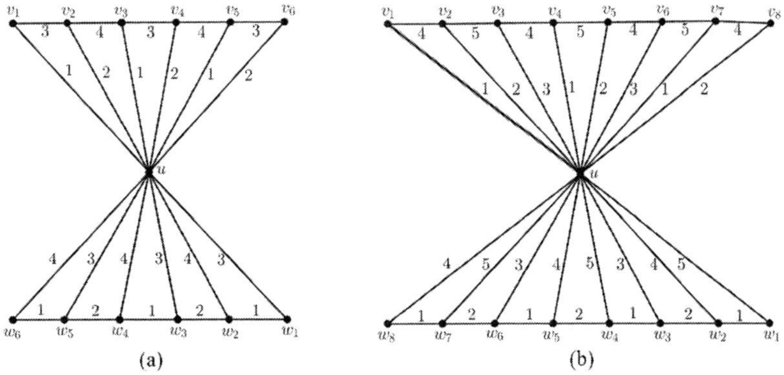

Figure 3. (a) A 3-rainbow 4-coloring on $G \cong Amal(F_6, 2, u)$ and (b) a 3-rainbow 5-coloring on $G \cong Amal(F_8, 2, u)$

This results suggest two open problems:
(1) whether the upper bounds presented are tight and;
(2) lower bound for 3-rainbow index of $G \cong Amal(G_1, v_{o_1}, G_2, v_{o_2})$, where for each $i \in \{1, 2\}$,

- G_i is a wheel W_n on $n+1$ vertices and v_{o_i} is the center vertex of W_n for $n \geq 9$;
- $G_i \cong W_{n_i}$ and v_{o_i} is the center vertex of W_{n_i} for $n_1 \in [4, 8]$ and $n_2 \geq 9$ or $n_1, n_2 \geq 9$;
- G_i is a fan F_n on $n+1$ vertices and v_{o_i} is the center vertex of F_n for $n \geq 7$.

3. References

[1] Carlson K 2006 Generalized books and C_m-snakes are prime graphs *Ars Combin.* **80** 215-21
[2] Chartrand G, Johns G L, McKeon K A and Zhang P 2008 Rainbow connection in graphs *Math. Bohem.* **133** 85-98
[3] Chartrand G, Okamoto F and Zhang P 2010 Rainbow trees in graphs and generalized connectivity *Networks* **55** 360-7
[4] Chen L, Li X, Yang K and Zhao Y 2013 The 3-rainbow index of a graph *arXiv:1307.0079V3 [math.CO]*
[5] Diestel R 2000 *Graph Theory* 2nd ed (New York: Springer-Verlag)
[6] Fitriani D and Salman A.N.M. 2016 Rainbow connection number of amalgamation of some graphs *AKCE International Journal of Graphs and Combinatorics* **13(1)** 90-9
[7] Kumala I S and Salman A.N.M. 2015 The rainbow connection number of a flower (C_m, K_n) graph and a flower (C_3, F_n) graph *Procedia Computer Science* **74** 168-72
[8] Nabila S and Salman A.N.M. 2015 The rainbow connection number of origami graphs and pizza graphs *Procedia Computer Science* **74** 162-7
[9] Resty D and Salman A.N.M. 2015 The rainbow connection number of an n-crossed prism graph and its corona product with a trivial graph *Procedia Computer Science* **74** 143-50
[10] Simamora D N S and Salman A.N.M. 2015 The rainbow (vertex) connection number of pencil graphs *Procedia Computer Science* **74** 138-42

(H_1, H_2)-supermagic labelings for some shackles of connected graphs H_1 and H_2

Yeva Fadhilah Ashari[1] and A.N.M. Salman[2]

[1,2] Combinatorial Mathematics Research Group, Faculty of Mathematics and Natural Sciences, Institut Teknologi Bandung, Jalan Ganesa 10, Bandung 40132, Indonesia

E-mail: [1]yevafadhilahashari@students.itb.ac.id, [2]msalman@math.itb.ac.id

Abstract. Let H_1 and H_2 be two graphs. A simple graph $G = (V(G), E(G))$ admits an (H_1, H_2)-covering, if every edge in $E(G)$ belongs to at least one subgraph of G isomorphic to H_1 or H_2. The graph G is called (H_1, H_2)-magic, if there are two fixed positive integers k_1 and k_2, and a bijective function $f : V(G) \cup E(G) \rightarrow \{1, 2, ..., |V(G)| + |E(G)|\}$ such that $\sum_{v\in V(H')} f(v) + \sum_{e\in E(H')} f(e) = k_1$ and $\sum_{v\in V(H'')} f(v) + \sum_{e\in E(H'')} f(e) = k_2$, for every subgraph $H^{'} = (V(H^{'}), E(H^{'}))$ of G isomorphic to H_1 and for every subgraph $H^{''} = (V(H^{''}), E(H^{''}))$ of G isomorphic to H_2. Moreover, it is said to be super (H_1, H_2)-magic, if $f(V(G)) = \{1, 2, ..., |V(G)|\}$.

This paper aims to study an (H_1, H_2)-supermagic labelings for some shackles of connected graphs H_1 and H_2 such as cycle, flower, and prism graph. A shackle of $G_1, G_2, G_3, \ldots, G_k$ denoted by $shack(G_1, G_2, G_3, \ldots, G_k)$ is a graph constructed from nontrivial connected and ordered graphs such that for every i and $j \in [1, k]$ with $|i - j| > 2$, G_i and G_j have no common vertex, and for every $i \in [1, k-1]$, G_i and G_{i+1} share exactly one common vertex, called linkage vertex, where the $k - 1$ linkage vertices are all distinct. In case G_i is isomorphic to H_1 for odd i and G_i is isomorphic to H_2 for even i, we denote such shackle by $shack(H_1, H_2, k)$. We give a sufficient condition for some $shack(H_1, H_2, k)$ being (H_1, H_2)-supermagic for even k.

Keywords: (H_1, H_2)-covering, (H_1, H_2)-supermagic, shackle

1. Introduction

All graphs considered in this paper are finite, simple, and connected. Let $G = (V(G), E(G))$ be a graph. A *labeling* of a graph G is a map that carries all of the elements of G to the positive or non-negative numbers. A graph G admits an H*-covering*, if every edge in $E(G)$ belongs to at least one subgraph of G isomorphic to H. Suppose that G admits an H-covering. A bijective function $f : V(G) \cup E(G) \rightarrow \{1, 2, ..., |V(G)| + |E(G)|\}$ is an H*-magic labeling* of G, if there exists a positive integer k, such that for any subgraph $H^{'} = (V(H^{'}), E(H^{'}))$ isomorphic to H, we have $\sum_{v\in V(H')} f(v) + \sum_{e\in E(H')} f(e) = k$ (constant). Moreover, the labeling f is an H*-supermagic labeling* of G, if $f(V(G)) = \{1, 2, ..., |V(G)| + |E(G)|\}$. If G admits an H-supermagic labeling then we say that G is H*-supermagic*.

The concept of *H-(super)magic* labeling was first introduced by Gutiérrez and Llado [2] in 2005. Such a labeling is an extension of an edge magic total labeling defined by Kotzig and Rosa [3]. In [2], Gutiérrez and Llado proved that a complete bipartite graph $K_{m,n}$ is a star $K_{1,h}$-supermagic and a path P_n and a cyce C_n are P_h-supermagic for $2 \leq h < n$. Llado and Moragas [4] constructed a cycle-supermagic labelings for wheels, prisms, books, subdivided wheels, and windmills. Maryati *et al.* [6, 7]

and Salman *et al.* [16] investigated that some families of trees are path-supermagic. In 2010, Maryati *et al.* [9] showed that shackles and amalgamations constructed from copies of a connected graph H is H-supermagic. Some other results on H-supermagic labelings of another classes of graphs can be found [1, 5, 8, 9, 10, 11, 12, 14, 15].

Recently, we introduced an (H_1, H_2)-*(super)magic labeling* of a graph in [13]. Let H_1 and H_2 be two graphs. A graph G admits an (H_1, H_2)-*covering*, if every edge in $E(G)$ belongs to at least one subgraph of G isomorphic to H_1 or H_2. Suppose that G admits an (H_1, H_2)-covering. A bijective function $f : V(G) \cup E(G) \rightarrow \{1, 2, ...|V(G)| + |E(G)|\}$ is an (H_1, H_2)-*magic labeling* of G, if there exist two positive integer k_1, and k_2, such that for any subgraph $H' = (V(H'), E(H'))$ isomorphic to H_1, we have $\sum_{v\in V(H')} f(v) + \sum_{e\in E(H')} f(e) = k_1$ and for any subgraph $H'' = (V(H''), E(H''))$ isomorphic to H_2, we have $\sum_{v\in V(H'')} f(v) + \sum_{e\in E(H'')} f(e) = k_2$. Additionally, the labeling f is called an (H_1, H_2)-*supermagic labeling* of G, if $f(V(G)) = \{1, 2, ..., |V(G)| + |E(G)|\}$. If G admits an (H_1, H_2)-supermagic labeling then we say that G is (H_1, H_2)-*supermagic graph*.

Maryati *et al.* [7] defined that a *shackle* of $G_1, G_2, \ldots, G_k$, denoted by $shack(G_1, G_2, G_3, \ldots, G_k)$, is a graph constructed from nontrivial connected and ordered graphs such that for every i and $j \in [1, k]$ with $|i - j| > 2$, G_i and G_j have no common vertex, and for every $i \in [1, k-1]$, G_i and G_{i+1} share exactly one common vertex, called *linkage vertex*, where the $k - 1$ linkage vertices are all distinct. In case G_i are isomorphic to H_1 for odd i and G_i are isomorphic to H_2 for even i, we denote such shackle by $shack(H_1, H_2, k)$. In this paper we prove that for even k, graph $shack(C_m, C_n, k)$ is (C_n, C_m)-supermagic for any positive integers m and n at least 3 and $shack(F_m, Y_n, k)$ is (F_m, Y_n)-supermagic for any positive integers m and for any odd integer n with m and n at least 3, where C_n, F_m and Y_n are a cycle on n vertices, a flower with $2m + 1$ vertices and a prism with $2n$ vertices, respectively. We set $\{a|x \leq a \leq y\}$ as $[x, y]$ and $\sum_{x\in X} x$ as $\sum X$.

2. Technique of k-balanced multiset and $(k_1, k_2, l_1, l_2, \theta_1, \theta_2)$-balanced multiset

In order to prove our main result, in this section we develop a technique of k-balanced multiset that is introduced by Maryati *et al.* [9]. We also introduce a new technique of partitioning a multiset called $(k_1, k_2, l_1, l_2, \theta_1, \theta_2)$-balanced multiset. A multiset is a set that allows the existence of the same elements in it. Let $k \in \mathbb{N}$ and Y be a multiset that contains some positive integers. If α appears r times in X and s times in Y where r and s are two non-negative integers, then α appears $r + s$ times in the $X \uplus Y$. A multiset Y is said to be a k-balanced if there exist k subsets of Y, i.e. $Y_1, Y_2, ..., Y_k$ such that for every $i \in [1, k]$, $|Y_i| = \frac{|Y|}{k}$, $\sum Y_i = \frac{\sum Y}{k} \in \mathbf{N}$, and $\biguplus_{i=1}^{k} Y_i = Y$. Then, Y_i is called a k-balanced subset of Y if those hold for every $i \in [1, k]$.

The following three lemmas have been proved by Maryati *et al.* [9, 10].

Lemma 1. [9] *Let* $y \in \mathbb{N}$ *and* $k \geq 2$ *be an integer, then a multiset* $Y = [y, 3k + y] \uplus [y + 1, k + y - 1]$ *is* k*-balanced.*

Lemma 2. [9] *Let* $y \in \mathbb{N}$ *and* $k \geq 2$ *be an integer. If* k *is even and a multiset* $Y = [y, 2k + y] \uplus [y, k + y - 1] \setminus \{\frac{k}{2} + y\}$ *then* Y *is* k*-balanced.*

Lemma 3. [10] *Let* $k, x,$ *and* y *be integers such that* $1 \leq x \leq y$ *and* $k > 1$*. If* $X = [x, y]$ *and* $|X|$ *is a multiple of* $2k$ *then* X *is* k*-balanced.*

Then, we generalize Lemma 1 and Lemma 2 into two following lemmas.

Lemma 4. *Let* y *and* k *be positive integers with* $k \geq 2$ *and* z *be a non negative integer. Then the multiset* $Y = [y, k + y] \uplus [k + y + z + 1, 3k + y + z] \uplus [y + 1, k + y - 1]$ *is* k*-balanced.*

Proof. For every $i \in [1,k]$, define $Y_i = \{a_i, b_i, c_i, d_i\}$ where

$$\begin{aligned} a_i &= y+i-1; \\ b_i &= y+i; \\ c_i &= 2k+y+z+1-i; \\ d_i &= 3k+y+z+1-i. \end{aligned}$$

Next, let

$$\begin{aligned} A &= \{a_i | i \in [1,k]\} = [y, k+y-1]; \\ B &= \{b_i | i \in [1,k]\} = [y+1, k+y]; \\ C &= \{c_i | i \in [1,k]\} = [k+y+z+1, 2k+y+z]; \\ D &= \{d_i | i \in [1,k]\} = [2k+y+z+1, 3k+y+z]. \end{aligned}$$

It is obvious that $A \uplus B \uplus C \uplus D = Y$ and $\biguplus_{i=1}^{k} Y_i = Y$. For every $i \in [1,k]$, we have that $|Y_i| = 4$ and $\sum Y_i = 5k + 4y + 2z + 1$. Then, we conclude that Y is k-balanced. □

Lemma 5. *Let y and k be positive integers with $k \geq 2$ and z be a non negative integer. If k is even and the multiset $Y = [y, k+y] \uplus [k+y+z+1, 2k+y+z] \uplus [y, k+y-1] \setminus \{\frac{k}{2}+y\}$, then Y is k-balanced.*

Proof. For every $i \in [1,k]$, define $Y_i = \{a_i, b_i, c_i\}$ where

$$\begin{aligned} a_i &= \lceil \frac{i}{2} \rceil + y - 1; \\ b_i &= \frac{k}{2} - \lceil \frac{i}{2} \rceil + y + 1; \\ c_i &= 2k+y+z+1-i. \end{aligned}$$

Next, let

$$\begin{aligned} A &= \{a_i | i \in [1,k]\} = [y, \frac{k}{2}+y-1] \uplus [y, \frac{k}{2}+y-1]; \\ B &= \{b_i | i \in [1,k]\} = [\frac{k}{2}+y, k+y] \uplus [\frac{k}{2}+y+1, k+y-1]; \\ C &= \{c_i | i \in [1,k]\} = [k+y+z+1, 2k+y+z]. \end{aligned}$$

It can be verified that $A \uplus B \uplus C = Y$ and $\biguplus_{i=1}^{k} Y_i = Y$. For every $i \in [1,k]$, we obtain that $|Y_i| = 3$ and $\sum Y_i = 2k + 3y + \frac{k}{2} + z$. Hence, Y is k-balanced. □

Let $k_1, k_2, l_1, l_2, \theta_1$ and θ_2 be positive integers. We define that a multiset X is $(k_1, k_2, l_1, l_2, \theta_1, \theta_2)$-*balanced*, if there exist $k_1 + k_2$ subsets of X, i.e. $A_1, A_2, ..., A_{k_1}, B_1, B_2, ..., B_{k_2}$ such that $|A_i| = l_1$, $|B_j| = l_2$, $\sum A_i = \theta_1$, $\sum B_j = \theta_2$ and $\biguplus_{i=1}^{k_1} A_i \uplus \biguplus_{j=1}^{k_2} B_j = X$ for every $i \in [1, k_1]$ and $j \in [1, k_2]$. We use the proof of Lemma 3 to derive the following next lemma.

Lemma 6. *Let h_1, h_2, k, n_1, n_2 and z be positive integers. If h_1, h_2 and k are even and $X = [z+1, z+(h_1+h_2)k]$ then X is $(\frac{k}{2}, \frac{k}{2}, h_1, h_2, \theta_1, \theta_2)$-balanced where $\theta_1 = \frac{h_1}{2}(1 + h_1\frac{k}{2}) + (h_1 - n_1)n_2\frac{k}{2} + h_1 z$, $|B_i| = h_2$ and $\theta_2 = \frac{h_2}{2}(1 + h_2\frac{k}{2}) + (h_2 - n_2)h_1\frac{k}{2} + n_1 n_2\frac{k}{2} + h_2 z$.*

Proof. Let $n_i \leq h_i$ for $i \in [1,2]$. Then, for each $i \in [1, \frac{k}{2}]$ define $A_i = \{a^i_j | j \in [1, h_1]\}$ and $B_i = \{b^i_j | j \in [1, h_2]\}$ where

$$a^i_j = \begin{cases} \frac{k}{2}(j-1)+i+z, & \text{for odd } j \text{ and } j \in [1, n_1]; \\ \frac{k}{2}j - i + z + 1, & \text{for even } j \text{ and } j \in [1, n_1]; \\ \frac{k}{2}(j-1)+i+z+\frac{n_2 k}{2}, & \text{for odd } j \text{ and } j \in [n_1+1, h_1]; \\ \frac{k}{2}j - i + z + 1 + \frac{n_2 k}{2}, & \text{for even } j \text{ and } j \in [n_1+1, h_1]; \end{cases}$$

and

$$b^i_j = \begin{cases} \frac{k}{2}(j-1)+i+z+\frac{n_1 k}{2}, & \text{for odd } j \text{ and } j \in [1, n_2]; \\ \frac{k}{2}j - i + z + 1 + \frac{n_1 k}{2}, & \text{for even } j \text{ and } j \in [1, n_2]; \\ \frac{k}{2}(j-1)+i+z+\frac{h_1 k}{2}, & \text{for odd } j \text{ and } j \in [n_2+1, h_2]; \\ \frac{k}{2}j - i + z + 1 + \frac{h_1 k}{2}, & \text{for even } j \text{ and } j \in [n_2+1, h_2]. \end{cases}$$

It is easy to check that for every $i \in [1, \frac{k}{2}]$ satisfies $\biguplus_{i=1}^{\frac{k}{2}} A_i \uplus \biguplus_{i=1}^{\frac{k}{2}} B_i = X$, $|A_i| = h_1$, $\sum A_i = \frac{h_1}{2}(1+h_1\frac{k}{2})+(h_1-n_1)n_2\frac{k}{2}+h_1 z$, $|B_i| = h_2$ and $\sum B_i = \frac{h_2}{2}(1+h_2\frac{k}{2})+(h_2-n_2)h_1\frac{k}{2}+n_1 n_2\frac{k}{2}+h_2 z$. Therefore X is $(\frac{k}{2}, \frac{k}{2}, h_1, h_2, \theta_1, \theta_2)$-balanced with $\theta_1 = \sum A_i$ and $\theta_2 = \sum B_i$. □

3. Main Results

In this section, we prove that a $shack(C_m, C_n, k)$ is (C_m, C_n)-supermagic and a $shack(F_m, Y_n, k)$ is (F_m, Y_n)-supermagic. We use Lemma 4, Lemma 5 and Lemma 6 to prove the following results.

3.1. Graph $shack(C_m, C_n, k)$

Let n be a positive integer with $n \geq 3$. *A cycle graph*, denoted by C_n, is a graph on n vertices with the set of vertices $V(G) = \{v_i | 1 \leq i \leq n\}$ and the set of edges $E(G) = \{v_i v_{i+1} | 1 \leq i \leq n-1\} \cup \{v_n v_1\}$. We construct a $shack(C_m, C_n, k)$.

Theorem 1. *Let $k \geq 2$ be an even integer, m and n be any two positive integers at least 3. Then graph $shack(C_m, C_n, k)$ is (C_m, C_n)-supermagic.*

Proof. Let G be a $shack(C_m, C_n, k)$ with $|V(G)| = \frac{k}{2}(m+n-2)+1$ including $k-1$ linkage vertices and $|E(G)| = \frac{k}{2}(m+n)$. Therefore, the set of labels used to label all vertices and edges of G is $Z = [1, (m+n-1)k+1]$. It is easy to verify that G contains exactly $\frac{k}{2}$ subgraphs isomorphic to C_n and $\frac{k}{2}$ subgraphs isomorphic to C_m. Now, partition Z into two sets such that $Z = P \cup Q$ where $P = [1, k+1] \cup [(m+n-3)k+2, (m+n-1)k+1] \uplus [2,k]$ and $Q = [k+2, (m+n-3)k+1]$.

Apply Lemma 4 to partition P into k subsets P_i where $i \in [1,k]$, with $y = 1$ and $z = k(m+n-4)$. We obtain that P is k-balanced with $|P_i| = 4$ and $\sum P_i = k(2m+2n-3)+5$ for every $i \in [1,k]$. Next, we set $n_1 = m-2, n_2 = n-2, h_1 = 2m-2$, and $h_2 = 2n-2$. Since h_1, h_2 and k are even, by using Lemma 6, Q is $(\frac{k}{2}, \frac{k}{2}, h_1, h_2, \sum A_i, \sum B_i)$-balanced. Let A_i and B_i be a balanced subset of Q for every $i \in [1, \frac{k}{2}]$, as defined in the proof of Lemma 6. We have $\sum A_i = \frac{1}{2}(m-2)(k(2m+n-2)+6)$ and $\sum B_i = \frac{1}{2}(n-2)(k(3m+2n-6)+6)$ for $i \in [1, \frac{k}{2}]$.

Next, we define a total labeling f on G as follows. Consider G_i and G_{i+1} share exactly one vertex, linkage vertex, named u_i for $i \in [1, k-1]$. Label the linkage vertex u_i with $i+1 \in P_i$ for $i \in [1, k-1]$. Now, let $R_i = P_{2i-1} \setminus [2,k] \uplus A_i$ for $i \in [1, \frac{k}{2}]$ and $R_i = P_{2(i-\frac{k}{2})} \setminus [2,k] \uplus B_{i-\frac{k}{2}}$ for $i \in [\frac{k}{2}+1, k]$. Use the elements of R_i for $i \in [1, \frac{k}{2}]$ to label the remaining vertices and edges of every G_j for odd j with $j \in [1,k]$. Next, use the elements of R_i for $i \in [\frac{k}{2}+1, k]$ to label the remaining vertices and edges of G_j for even j with $j \in [1,k]$. Use the smaller label for vertices so every labels of vertices of G is smaller than the labels of edges of G.

Hence, the sum of vertex and edge labels of every subgraph C_m and every subgraph C_n of G are constant, namely $f(C_m) = \sum P_i + \sum A_i = \frac{1}{2}(2km^2 + kmn - 2km + 2kn - 2k + 6m - 2)$, and $f(C_n) = \sum P_i + \sum B_i = \frac{1}{2}(3kmn - 2km + 2kn^2 - 6kn + 6k + 6n - 2)$. Therefore, G is (C_n, C_m)-supermagic. □

For illustration, we can see a (C_4, C_7)-supermagic labeling on graph $shack(C_4, C_7, 6)$ in Figure 1.

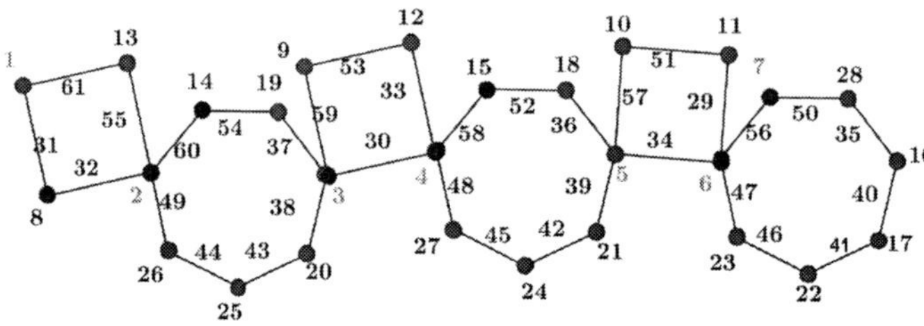

Figure 1. A (C_4, C_7)-supermagic labeling on graph $shack(C_4, C_7, 3)$

3.2. Graph $shack(F_m, Y_n, k)$

Let m and n be a positive integers at least 3. *A flower graph* on $2m + 1$ vertices, denoted by F_m, is the graph obtained from a wheel, W_m, on $m + 1$ vertices by attaching a pendant edge on each vertex of the m-cycle, then joining each pendant vertex to the central vertex of W_m. *A prism graph* on $2n$ vertices, denoted by Y_n, is the Cartesian product of a cycle C_n and a path with 2 vertices P_2 whose vertex set is $V(C_n) \times V(P_2)$ and whose edge set is the set of all pairs $(u_1, v_1)(u_2, v_2)$ such that either $u_1u_2 \in E(C_m)$ and $v_1 = v_2$ or $v_1v_2 \in E(P_2)$ and $u_1 = u_2$.

Theorem 2. *Let $k \geq 2$ be an even integer, m be any positive integer at least 3, and n be an odd positive integers at least 3. Then a $shack(F_m, Y_n, k)$ is (F_m, Y_n)-supermagic.*

Proof. Let G be a $shack(F_m, Y_n, k)$ with $|V(G)| = k(m + n - \frac{1}{2}) + 1$ including $k - 1$ linkage vertices and $|E(G)| = k(2m + \frac{3}{2}n)$. Therefore, the set of labels used to label all vertices and edges of G is $Z = [1, \frac{1}{2}(k(6m + 5n - 1) + 2)]$. It is easy to verify that G contains exactly $\frac{k}{2}$ subgraphs isomorphic to F_m and $\frac{k}{2}$ subgraphs isomorphic to Y_n. Now, partition Z into two sets such that $Z = P \cup Q$ where $P = [1, k + 1] \cup [\frac{1}{2}(k(6m + 5n - 3) + 4), \frac{1}{2}(k(6m + 5n - 1) + 2)] \uplus [1, k] \setminus \{\frac{k}{2} + 1\}$ and $Q = [k + 2, \frac{1}{2}(k(6m + 5n - 3) + 2)]$.

Apply Lemma 5 to partition P into k subsets P_i where $i \in [1, k]$, with $y = 1$ and $z = \frac{1}{2}k(6m+5n-5)$. We obtain that P is k-balanced with $|P_i| = 3$ and $\sum P_i = 3km + \frac{5}{2}kn + 3$ for every $i \in [1, k]$. Next, we set $n_1 = 2m - 1, n_2 = 2(n - 1), h_1 = 6m - 2$ and $h_2 = 5n - 3$. Since h_1, h_2 and k are even, by using Lemma 6, Q is $(\frac{k}{2}, \frac{k}{2}, h_1, h_2, \sum A_i, \sum B_i)$-balanced. Let A_i and B_i be a balanced subset of Q for every $i \in [1, \frac{k}{2}]$, as defined in the proof of Lemma 6. We have $\sum A_i = 9km^2 + m(4k(n - 1) + 9) - kn - 3$ and $\sum B_i = \frac{1}{4}(k(25n^2 + 4m(11n - 5) - 26n + 5) + 6(5n - 3))$ for $i \in [1, \frac{k}{2}]$.

Next, we define a total labeling f on G as follows. Consider G_i and G_{i+1} share exactly one vertex, linkage vertex, named u_i for $i \in [1, k - 1]$. Label the linkage vertex u_i with $S_i \cap S_{i+1}$ where $S_i = P_i \cap [1, k] \setminus \{\frac{k}{2} + 1\}$ for $i \in [1, k - 1]$. Now, let $R_i = P_{2i-1} \setminus [1, k] \setminus \{\frac{k}{2} + 1\} \cup A_i$ for $i \in [1, \frac{k}{2}]$ and $R_i = P_{2(i-\frac{k}{2})} \setminus [1, k] \setminus \{\frac{k}{2} + 1\} \cup B_{i-\frac{k}{2}}$ for $i \in [\frac{k}{2} + 1, k]$. Use the elements of R_i for $i \in [1, \frac{k}{2}]$ to label the remaining vertices and edges of G_j for odd j with $j \in [1, k]$. Use the elements of R_i for

$i \in [\frac{k}{2}+1, k]$ to label the remaining vertices and edges of every G_j for even j with $j \in [1, k]$. Use the smaller label for vertices so every labels of vertices of G is smaller than the labels of edges of G.

Hence, the sum of vertex and edge labels of every subgraph F_m and every subgraph Y_n of G are constant, namely $f(F_m) = \sum P_i + \sum A_i = \frac{1}{2}(18km^2 + 8kmn - 2km + 3kn + 18m)$, and $f(Y_n) = \sum P_i + \sum B_i = \frac{1}{4}(44kmn - 8km + 25kn^2 - 16kn + 5k + 30n - 6)$. Therefore, G is (F_m, Y_n)-supermagic. □

For illustration, we can see a (F_5, Y_7)-supermagic labeling on graph $shack(F_5, Y_7, 4)$ in Figure 2.

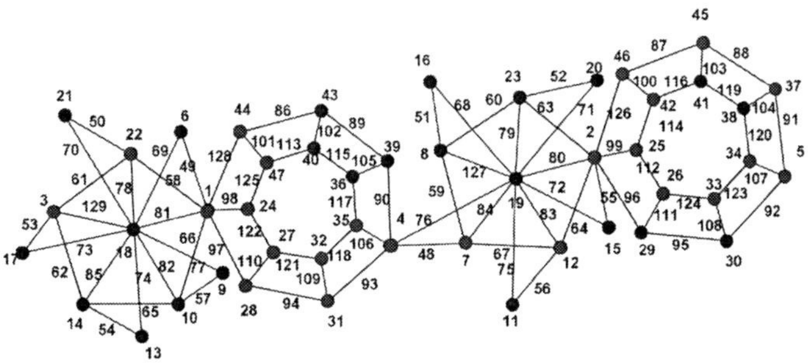

Figure 2. An (F_5, Y_7)-supermagic labeling on graph $shack(F_5, Y_7, 4)$

Based on the results, we propose the following open problem.

Open Problem. Let k be a positive integer and H_1 and H_2 be two connected graphs. Determine a sufficient condition such that a $shack(H_1, H_2, k)$ is (H_1, H_2)-supermagic graph.

References

[1] Gallian J A 2015 A dynamic survey of graph labeling, *Electronic Journal Combinatorics* **17** ♯DS6.
[2] Gutiérrez A and Llado A 2005 Magic Coverings *J. Combin. Math. Combin. Comput.* **55** 43–56
[3] Kotzig A and Rosa 1970 A, Magic Valuations of Finite Graphs, *Canad. Math. Bull.* **13** 451–461
[4] Llado A and Moragas J 2007 Cycle-Magic Graphs *Discrete Math* **307** 2925–2933
[5] Marbun H T and Salman A N M 2013 Wheel-Supermagic Labelings for A Wheel k-Multilevel Corona with A Cycle *AKCE International Journal of Graphs and Combinatorics* **10** 1–9.
[6] Maryati T K, Baskoro E T and Salman A N M 2008 P_h-super(magic) Labelings of Some Trees *J. Combin. Math. Comput.* **65** 197–204
[7] Maryati T K, Baskoro E T, Salman A N M and Irawati 2015 On The Path-(Super)Magicness of A Cycle With Some Pendants *Util. Math.* **96** 319–330
[8] Maryati T K, Salman A N M and Baskoro E T 2008 The Supermagicness of A Disjoint Union of Isomorphic Connected Graphs *Proceed. 4th IMT-GT Internat. Conf. Math. Stat. Appl.* **3** 1–5
[9] Maryati T K, Salman A N M, Baskoro E T, Ryan J and Miller M 2010 On H-Supermagic Labelings for Certain Shackles and Amalgamations of a Connected Graph *Utilitas Mathematica* **83** 333–342
[10] Maryati T K, Salman A N M and Baskoro E T 2013 Supermagic Coverings of The Disjoint Union of Graphs and Amalgamations *Discrete Mathematics* **313** 397–405
[11] Ngurah A A G, Salman A N M and Sudarsana I W 2010 On Supermagic Coverings of Fans and Ladders *SUT J. Math* **46** 67–78
[12] Ngurah A A G, Salman A N M and Susilowati L 2010 H-Supermagic Labelings of Graphs *Discrete Math* **310** 1293–1300
[13] Salman A N M and Ashari Y F 2016 (H_1, H_2)-(Super)magic Labelings of A Graph ***Submitted***
[14] Salman A N M and Maryati T K 2010 On Graph-(Super)Magic Labelings of A Path-Amalgamation of Isomorphic Graphs *The Proceedings of The* 6th *IMT-GT International Conference on Mathematics, Statistics and Its Applications* 228–233
[15] Salman A N M and Purnomo A D, Some Cycle-(Super)Magic Labelings of Some Complete Bipartite Graphs 2010 *East-West Journal of Mathematics* 283–291
[16] Salman A N M, Ngurah A A G and Izzati N 2010 On (Super) Edge-Magic Total Labellings of A Subdivision of a Star S_n, *Utilitas Mathematica* **81** 275–284

The 3-rainbow index of some graphs that constructed by joining a graph with a trivial graph

Dinda Kartika[1], **A.N.M. Salman**

Combinatorial Mathematics Research Group, Faculty of Mathematics and Natural Sciences
Institut Teknologi Bandung, Jalan Ganesa 10 Bandung, Indonesia

E-mail: [1]dkartikanda@gmail.com

Abstract. Let $G = (V, E)$ be a nontrivial, finite, and connected graph. A tree T in an edge-colored graph is said to be rainbow tree, if no two edges on the tree have the same color. An edge-coloring of G is called 3-rainbow coloring, if for any three vertices in G there exists a rainbow tree connecting them. The 3-rainbow index of G, denoted by $rx_3(G)$, is the minimum number of colors that are needed in a 3-rainbow coloring of G. The join of G_1 and G_2, denoted by $G = G_1 + G_2$, is the graph with vertex set $V(G_1) \cup V(G_2)$ and edge set $E(G_1) \cup E(G_2) \cup \{uv | u \in V(G_1), v \in V(G_2)\}$. In this paper, we determine $rx_3(G)$ where G is the join of a graph with a trivial graph.

1. Introduction

All graphs considered in this paper are simple, finite, and connected. The concept of rainbow connection in graphs was first introduced by Chartrand et al. [1] in 2008. Let G be a graph with vertex set $V(G)$ and edge set $E(G)$. Define a coloring $c : E(G) \to \{1, 2, \cdots, k\}$, for some $k \in \mathbb{N}$, such that adjacent edges can get the same color. A path P in G is called a *rainbow path*, if no two edges in the path have the same color. Graph G is said to be *rainbow connected*, if for every two vertices x and y in G there exists a rainbow path. An edge coloring on G that makes G rainbow connected is called a *rainbow coloring*. A function c is called a *rainbow k-coloring*, if it uses k colors. The *rainbow connection number of* G, denoted by $rc(G)$, is the smallest positive integer k such that G has a rainbow k-coloring. Some results on the rainbow connections can be found in [4, 5, 6, 7, 8].

In 2010, Chartrand et al. [2] generalized the concept of rainbow path to rainbow tree. A tree T in G is called a *rainbow tree*, if no two edges in the tree have the same color. For $S \subseteq V(G)$, a *rainbow S-tree* is a rainbow tree that connects the vertices of S. Let k be a fixed integer with $2 \leq k \leq n$. An edge coloring on G is called a *k-rainbow coloring*, if for every k-subset S of $V(G)$ there exists a rainbow S-tree. The *k-rainbow index of* G, denoted by $rx_k(G)$, is the smallest positive integer k such that G has a k-rainbow coloring. Clearly, when $k = 2$, $rx_2(G)$ is the rainbow connection number $rc(G)$ of G. For every connected graph G of order n, it is easy to check that $rx_2(G) \leq rx_3(G) \leq \cdots \leq rx_n(G)$.

The *diameter* $diam(G)$ of G is the greatest distance between two vertices of G. The *Steiner distance* $d(S)$ of a set S of vertices in G is the minimum size of a tree in G containing S. Such a tree is called a *Steiner S-tree* or simply a *Steiner tree*. The *k-Steiner diameter* of G is the maximum $Steiner$ distance of S among all sets S with k vertices in G, denoted by $sdiam_k(G)$. Clearly, if $k = 2$ and $S = \{x, y\}$, then $d(S) = d(x, y)$ and the $2 - Steiner$ diameter $sdiam_2(G) = diam(G)$. The $k - Steiner$ diameter then provides a lower bound for the k-rainbow index of G.

IOP Conf. Series: Journal of Physics: Conf. Series **1127** (2019) 012060 doi:10.1088/1742-6596/1127/1/012060

Lemma 1. *[2]. For every connected graph G of order $n \geq 3$ and each integer k with $3 \leq k \leq n$,*

$$rx_k(G) \geq sdiam_k(G) \geq k-1.$$

There are some results about k-rainbow index of a graph. In 2010, Chartrand et al. [2] determined the k-rainbow index of trees and cycles. Besides that, in 2015 Chen et al. [3] determined the 3-rainbow index of complete bipartite graphs and wheels.

The join of G_1 and G_2, denoted by $G = G_1 + G_2$, is the graph with vertex set $V(G_1) \cup V(G_2)$ and edge set $E(G_1) \cup E(G_2) \cup \{uv | u \in V(G_1), v \in V(G_2)\}$ [6]. A graph with single vertex is trivial, denoted by K_1.

In this paper, we determine the 3-rainbow index of some graphs that constructed by joining a graph with a trivial graph. Let a, b, p and q be positive integers. For simplifying, we define $[a, b] = \{x \in \mathbb{N} | a \leq x \leq b\}$ and

$$p \ mod^* \ q = \begin{cases} p \ mod \ q \ \ if \ p \neq kq \ \ for \ \ any \ \ k \in \mathbb{N}; \\ q \ \ if \ \ p = kq \ \ for \ \ some \ \ k \in \mathbb{N}. \end{cases}$$

2. Main Results

2.1. The 3-Rainbow Index of $P_n + K_1$

Definition 1. *Let n be a positive integer with $n \geq 3$. A $P_n + K_1$ graph or known as a fan graph on $n+1$ vertices, denoted by F_n, is a graph which is obtained by joining a path on n vertices and a trivial graph such that $V(F_n) = \{v\} \cup \{v_i | i \in [1, n]\}$ and $E(F_n) = \{vv_i | i \in [1, n]\} \cup \{v_i v_{i+1} | i \in [1, n-1]\}$.*

Theorem 1. *Let n be a postive integer with $n \geq 3$ and F_n be a fan graph on $n+1$ vertices. Then the 3-rainbow index of F_n is*

$$rx_3(F_n) = \begin{cases} 2 \ \ for \ n = 3; \\ 3 \ \ for \ [4, 6]; \\ 4 \ \ for \ [7, 16]; \\ 5 \ \ for \ n \geq 17. \end{cases}$$

Proof. We divide the proof into four cases.

Case 1. $n = 3$

First, we show that $rx_3(F_n) \geq 2$. Since $sdiam_3\ (F_n)$ is 2, by using Lemma 1 we obtain $rx_3(F_n) \geq 2$. Next we show that $rx_3(F_n) \leq 2$. Define a 2-coloring, $c : E(F_n) \rightarrow [1, 2]$ as follows:

$$c(e) = \begin{cases} 1, \ \ if \ e \in \{vv_2, vv_3, v_1v_2\}; \\ 2, \ \ if \ e \in \{vv_1, v_2v_3\}. \end{cases}$$

We can check that for every three vertices x, y, and z in $V(F_n)$, there exists a rainbow 3-tree as shown in Table 1.

Table 1. A rainbow 3-tree on F_n for $n = 3$

x	y	z	Condition (i, j, k in [1,n])	Rainbow tree
v	v_i	v_j	$\lvert i-j \rvert = 1$	vvi, v_iv_j
			$\lvert i-j \rvert = 2$	vvi, vv_j
v_i	v_j	v_k		v_iv_j, v_jv_k

IOP Conf. Series: Journal of Physics: Conf. Series **1127** (2019) 012060 doi:10.1088/1742-6596/1127/1/012060

Case 2. $4 \leq n \leq 6$
First, we show that $rx_3(F_n) \geq 3$. Since $sdiam_3\ (F_n)$ is 3, by using Lemma 1 we obtain $rx_3(F_n) \geq 3$. Next we show that $rx_3(F_n) \leq 3$. For $n \geq 6$, define a 3-coloring $c : E(F_n) \to [1, 3]$ as follows:

$$c(e) = \begin{cases} 1, & if\ e \in \{vv_1, vv_2, v_2v_3\}; \\ 2, & if\ e \in \{vv_3, vv_4, v_1v_2, v_5v_6\}; \\ 3, & if\ e \in \{vv_5, vv_6, v_3v_4\}. \end{cases}$$

Therefore, we have $rx_3(F_6) = 3$.
When $n = 5$, we provide a rainbow 3-coloring of F_5 obtained from the rainbow 3-coloring of F_6 by removing vertex v_6 and when $n = 4$, we provide a rainbow 3-cloring of F_4 obtained from the rainbow 3-coloring of F_6 by removing v_6 and v_5.
We can check that for every three vertices x, y, and z in $V(F_n)$, there exists a rainbow 3-tree as shown in Table 2.

Table 2. A rainbow 3-tree on F_n for $4 \leq n \leq 6$

x	y	z	Condition (i, j, k in [1,n])	Rainbow tree
v	v_i	v_j	$\|i-j\| = 1$	vvi, v_iv_j
			$\|i-j\| \geq 2$	vvi, vv_j
v_i	v_j	v_k	$\|i-j\| = 1$ and $\|j-k\| = 1, i < j < k$	v_iv_j, v_jv_k
			$\|i-j\| = 1$ and $\|j-k\| = 2, i < j < k$	$v_iv_j, v_jv_{j+1}, v_{j+}v_k$
			$\|i-j\| = 1$ and $\|j-k\| > 2, i < j < k$	v_iv_j, vv_j, vv_k
			$\|i-j\| \geq 2$ and $\|j-k\| \geq 2, i < j < k$	vv_i, vv_j, vv_k

Case 3. $7 \leq n \leq 16$
First, we show that $rx_3(F_n) \geq 4$.
Assume, to the contrary, that $rx_3(F_n) \leq 3$. Then there is a 3-rainbow coloring c^* for F_n. Without loss of generality, assume that $c^*(vv_1) = 1$. Consider v_1, v_3, and v_5. Since $d(v_1, v_3, v_5) = 3$ and the only (v_1, v_3, v_5)-tree of size 3 for v_1, v_3, and v_5 is $vv_1 \cup vv_3 \cup vv_5$, it follows that $c^*(vv_3) = 2$ and $c^*(vv_5) = 3$. Then, consider v_1, v_3, and v_7. Since $d(v_1, v_3, v_7) = 3$ and the only (v_1, v_3, v_7)-tree of size 3 for v_1, v_3 and v_7 is $vv_1 \cup vv_3 \cup vv_7$, it follows that $c^*(vv_7) = 3$. Then, consider v_3, v_5 and v_7. Since $d(v_3, v_5, v_7) = 3$ and the only (v_3, v_5, v_7)-tree of size 3 for v_3, v_5, and v_7 is $vv_3 \cup vv_5 \cup vv_7$, there is no rainbow (v_3, v_5, v_7)-tree. We get a contradiction. Thus, $rx_3(F_n) \geq 4$.
Next we show that $rx_3(F_n) \leq 4$. For $n \geq 16$, define a 4-coloring, $c : E(F_n) \to [1, 4]$ as follows:

$$c(e) = \begin{cases} 1, & if\ e \in \{vv_1, vv_3, vv_6, vv_8, v_4v_5, v_9v_{10}, v_{11}v_{12}, v_{14}, v_{15}\}; \\ 2, & if\ e \in \{vv_9, vv_{11}, vv_{14}, vv_{16}, v_1v_2, v_3v_4, v_6v_7, v_{12}, v_{13}\}; \\ 3, & if\ e \in \{vv_2, vv_4, vv_{13}, vv_{15}, v_5v_6, v_7v_8, v_{10}v_{11}\}; \\ 4, & if\ e \in \{vv_5, vv_7, vv_{10}, vv_{12}, v_2v_3, v_8v_9, v_{13}v_{14}, v_{15}v_{16}\}. \end{cases}$$

Therefore, we have $rx_3(F_{16}) = 4$.
When $7 \leq n \leq 15$, we provide a 4-rainbow coloring of F_n obtained from the 4-coloring of F_{16} similarly with **Case 2.**
We can check that for every three vertices x, y, and z in $V(F_n)$, there exists a rainbow 3-tree as shown in Figure 2.
Case 4. $n \geq 17$
First, we show that $rx_3(F_n) \geq 5$. Assume, to the contrary, that $rx_3(F_n) \leq 4$. Then there is a 4-rainbow coloring c^* for F_n. Since $deg(v) = 17 \geq 4 \times 4$, there exists $A \subseteq V(P_n)$ such that $|A| = 5$ and all edges in $\{uv|u \in A\}$ are colored the same, say 1.

Subcase 4.1
Suppose that $A = \{v_1, v_2, v_3, v_4, v_5\}$. Consider, v_1, v_4, and v_i, where $7 \leq i \leq n$. There is no rainbow (v_1, v_4, v_i)-tree. Contradiction.
Subcase 4.2
Suppose that $A = \{v_{n-4}, v_{n-3}, v_{n-2}, v_{n-1}, v_n\}$. Consider, v_n, v_{n-1}, and v_j, where $1 \leq j \leq n-6$. There is no rainbow (v_n, v_{n-1}, v_j)-tree. Contradiction.
Subcase 4.3
For $2 \leq i \leq n-5$. Let $A = \{v_i, v_{i+1}, v_{i+2}, v_{i+3}, v_{i+4}\}$, $B = \{v_1, \cdots, v_{i-1}\}$, and $C = \{v_{i+5}, \cdots, v_n\}$. Without loss generality, $|B| \leq |C|$. Then, consider $v_i, v_{i+2} \in A$ and $v_k \in C$ and $k \geq i+7$. Since, $d(v_i, v_{i+2}, v_k) = 3$ and the (v_i, v_{i+2}, v_k)-tree with size less than or equal to 4 is $vv_{i-1} \cup v_{i-1}v_i \cup vv_{i+2} \cup vv_k$, without loss of generality it follows that $c^*(vv_{i-1}) = 2$, $c^*(v_{i-1}, v_i) = 3$ and $c^*(vv_k) = 4$. In the similar way, color all edges vv_k where $v_k \in C$ with 4. We get a contradiction.
Subcase 4.4
Supppose that $A = \{v_i, v_{i+1}, v_{i+2}, v_{i+3}, v_{i+5}\}$, where $1 \leq i \leq n$. Consider v, v_{i+2}, and v_{i+5}. There is no rainbow (v_i, v_{i+2}, v_{i+5})-tree. Contradiction.
Thus, $rx_3(F_n) \geq 5$.
Next we show that $rx_3(F_n) \leq 5$. Define a 5-coloring, $c : E(F_n) \to [1,5]$ as follows:

$$c(e) = \begin{cases} i \ mod^* \ 5, & if \ e = vv_i \ and \ i \in [1,n]; \\ (i+3) \ mod^* \ 5, & if \ e = v_iv_{i+1} \ and \ i \in [1, n-1]. \end{cases}$$

We can check that for every three vertices x, y, and z in $V(F_n)$, there exists a rainbow 3-tree as shown in Table 3. □

Table 3. A rainbow 3-tree on F_n for $n \geq 17$

x	y	z	Condition (i, j, k in [1,n])	Rainbow tree
v	v_i	v_j	$i \ mod^* \ 5 = j \ mod^* \ 5$	$vvi, vv_{j+1}, v_jv_{j+1}$
			$i \ mod^* \ 5 \neq j \ mod^* \ 5$	vvi, vv_j
v_i	v_j	v_k	$i \ mod^* \ 5 = j \ mod^* \ 5 = k \ mod^* \ 5$	$vvi, vv_{j-1}, v_{j-1}v_j, vv_{k+1}, v_kv_{k+1}$
			$i \ mod^* \ 5 = j \ mod^* \ 5 \neq k \ mod^* \ 5$	$vvi, vv_{j-1}, v_{j-1}v_j, vv_k$
			$i \ mod^* \ 5 \neq j \ mod^* \ 5 \neq k \ mod^* \ 5$	vvi, vv_j, vv_k

For illustration, a rainbow 2-coloring of F_3 and a rainbow 3-coloring of F_6 are given in Figure 1. (a) and Figure 1. (b), respectively. A rainbow 4-coloring of F_{16} and a rainbow 5-coloring of F_{17} are given in Figure 2 and Figure 3, respectively.

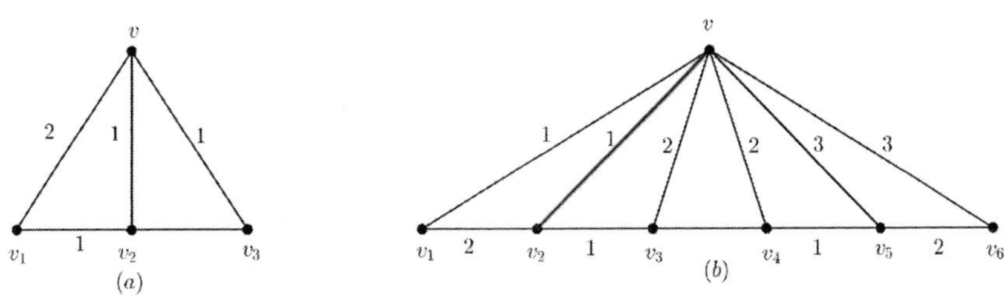

Figure 1. (a) A 2-rainbow coloring of F_3, (b) A 3-rainbow coloring of F_6

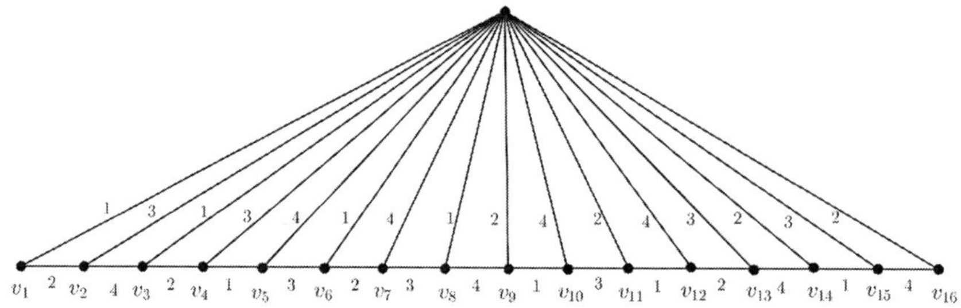

Figure 2. A 4-rainbow coloring of F_{16}

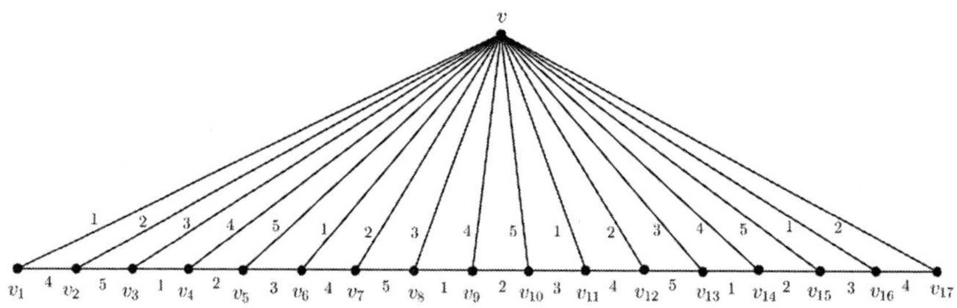

Figure 3. A 5-rainbow coloring of F_{17}

2.2. The 3-Rainbow Index of $K_{n,n} + K_1$

Definition 2. *Let n be a positive integer with $n \geq 3$. A $K_{n,n} + K_1$ graph on $2n + 1$ vertices, denoted by Kb_n, is a graph which is obtained by joining a complete bipartite graph on $2n$ vertices and a trivial graph such that $V(Kb_n) = \{u\} \cup \{v_i | i \in [1, n]\} \cup \{w_i | i \in [1, n]\}$ and $E(Kb_n) = \{uv_i | i \in [1, n]\} \cup \{uw_i | i \in [1, n]\} \cup \{v_i w_i | i \in [1, n]\}$.*

In this subsection we determine the 3-rainbow index of Kb_n as follows.

Theorem 2. *Let n be a postive integer with $n \geq 3$ and Kb_n be a graph which is obtained by joining a complete bipartite graph on $2n$ vertices and a trivial graph. Then the 3-rainbow index of Kb_n is*

$$rx_3(Kb_n) = 3.$$

Proof. First, we show that $rx_3(Kb_n) \geq 3$. Since $sdiam_3(Kb_n)$ is 3, by using Lemma 1 we obtain $rx_3(Kb_n) \geq 3$. Next we show that $rx_3(Kb_n) \leq 3$. Define a 3-coloring, $c : E(Kb_n) \to [1, 3]$ as follows:

$$\begin{aligned} c(v_i w_j) &= \begin{cases} 1, & for\ 1 \leq i = j \leq n; \\ 2, & for\ 1 \leq j < i \leq n; \\ 3, & for\ 1 \leq i < j \leq n; \end{cases} \\ c(uv_i) &= 3, \quad for\ i \in [1, n]; \\ c(uw_i) &= 2, \quad for\ i \in [1, n]. \end{aligned}$$

□

We can check that for every three vertices $x, y,$ and z in $V(Kb_n)$, there exists a rainbow 3-tree as shown in Table 4. For illustration, a rainbow 3-coloring of Kb_5 is given in Figure 4.

Table 4. A rainbow 3-tree on Kb_n for $n \geq 3$

x	y	z	Condition (i, j, k in [1,n])	Rainbow tree
u	v_i	v_j	$i < j$	v_iw_j, v_jw_j, uw_j
u	w_i	w_j	$i < j$	w_iv_j, v_jw_j, uv_j
u	v_i	w_j		w_i, u, w_j
v_i	v_j	v_k	$i < j < k$	w_jv_i, w_jv_j, w_jv_k
w_i	w_j	w_k	$i < j < k$	v_jwi, v_jw_j, v_jw_k
			$i = k$	v_iw_i, v_jw_i
v_i	v_j	w_k	$j = k$	v_iw_j, v_jw_j
			$i, j \neq k$	v_iw_k, v_jw_k
			$i = k$	v_iw_i, v_iw_j
w_i	w_j	v_k	$j = k$	v_jw_i, v_jw_j
			$i, j \neq k$	v_kw_i, v_kw_j

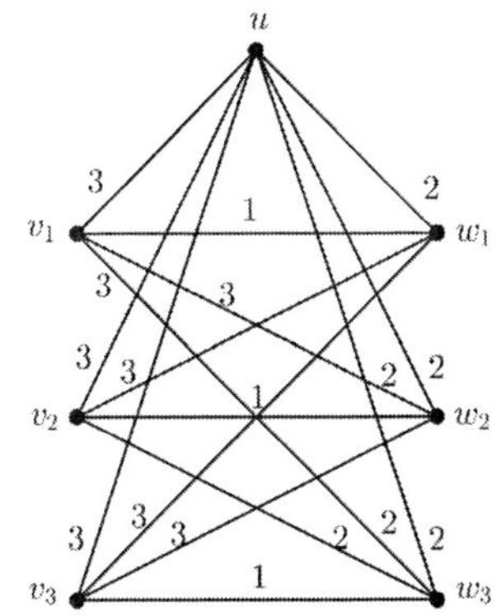

Figure 4. A 3-rainbow coloring of Kb_5

3. References

[1] Chartrand G, Johns G L, McKeon K A and Zhang P 2008 Rainbow connection in graphs *Mathematica Bohemica* **133** 85–98

[2] Chartrand G, Okamoto F and Zhang P 2010 Rainbow trees in graphs and generalized connectivity *Networks* **55** 360–7

[3] Chen L, Li X L, Yang K and Zhao Y 2015 The 3-rainbow index of a graphs *Graph Theory* **35** 81–94

[4] Fitriani D and A.N.M. Salman 2016 Rainbow connection number of amalgamation of some graphs *AKCE International Journal of Graphs and Combinatorics* **13** 90–9

[5] Kumala I S and A.N.M. Salman 2015 The rainbow connection number of a flower $(C_m; K_n)$ graph and a flower $(C_m; F_n)$ graph *Procedia Computer Sciences* **74** 168–172

[6] Li X, Shi Y and Sun Y 2013 Rainbow connections of graphs: a survey *Graphs and Combinatorics* **29** 1–38

[7] Nabila S and A.N.M. Salman 2015 The rainbow connection number of origami graphs and pizza graphs *Procedia Computer Sciences* **74** 162-7

[8] Resty D and A.N.M. Salman 2015 The rainbow connection number of an n-crossed prism graph and its corona product with a trivial graph *Procedia Computer Sciences* **74** 143–150

IOP Conf. Series: Journal of Physics: Conf. Series **1127** (2019) 012061 doi:10.1088/1742-6596/1127/1/012061

Boundedness of Modified Fractional Integral Operators on Morrey Spaces

Sofihara Al Hazmy
Bandung Institute of Technology, Ganesha 10 street, Bandung
E-mail: sofiharaalhazmy@gmail.com

Wono Setia Budhi
Bandung Institute of Technology, Ganesha 10 street, Bandung
E-mail: wono@math.itb.ac.id

Yudi Soeharyadi
Bandung Intsitute of Technology, Ganesha 10 street, Bandung
E-mail: yudish@math.itb.ac.id

Abstract. We will show some properties of Modified Fractional Integral Operators, including the boundedness. There are two modified version that will be discussed. First, doesn't contain logarithm function, and second contains logarithm function.

1. Introduction

M. Idris and H. Gunawan [3] have observed about Bessel-Riesz operator, and showed that the Bessel-Riesz operator is bounded on Morrey space, moreover, they showed the Bessel-Riesz kernel control the Bessel-Riesz operator, see[3]. The Bessel-Riesz operator is defined by

$$I_{\alpha,\gamma}f(x) = \int_{\mathbb{R}^n} \frac{|x-y|^{\alpha-n}}{(1+|x|)^{\gamma}} f(y)dy$$

for every $f \in L^p_{loc}(\mathbb{R}^n)$, which $0 < \alpha < n$ and $\gamma > 0$.

For $1 \leq p < \infty$ and $0 \leq \lambda \leq n$, the Morrey space(classical) is defined by

$$L^{p,\lambda}(\mathbb{R}^n) = \{f \in L^p_{loc}(\mathbb{R}^n) : \|f\|_{L^{p,\lambda}} < \infty\}$$

with $\|f\|_{L^{p,\lambda}} = \sup_{R>0, a\in\mathbb{R}^n} \left(\frac{1}{R^\lambda}\int_{B(a,R)} |f(x)^p dx\right)^{\frac{1}{p}}$. In the second section of this paper, we will generalize the results of M. Idris and H. Gunawan[3], it will be given by Proposition 5.

Let $\gamma > 0$, and $h : \mathbb{R}^+ \cup \{0\} \to \mathbb{R}$ is increasing with $h(0) > 0$, and for some $C > 0$ satisfy the condition

$$h(t) \geq Ct^{\gamma}. \tag{1}$$

Suppose $0 < \alpha < n$, define

$$J_{\alpha,\gamma}f(x) = \int_{\mathbb{R}^n} H_{\alpha,\gamma}(x-y)f(y)dy$$

for every $f \in L^p_{loc}(\mathbb{R}^n)$, with $p \geq 1, H_{\alpha,\gamma}(x) = \frac{|x|^{\alpha-n}}{h(|x|)}, x \in \mathbb{R}^n$. Particularly, for $h(t) = (1+t)^\gamma$, we obtain $J_{\alpha,\gamma}$ is Bessel-Riesz operator.

Another purpose of this research is to observe the boundedness of operator

$$T_{\alpha,\beta}f(x) = \int_{\mathbb{R}^n} G_{\alpha,\beta}(x-y)f(y)dy,$$

$f \in L^p_{loc}(\mathbb{R}^n)$, with $0 < \alpha < n$, $1 \leq \beta < \infty$, and $G_{\alpha,\beta}(x) = \frac{|x|^{\alpha-n}}{\log(\beta+|x|)}, x \in \mathbb{R}^n$. This is interesting, since $\log(\beta + |x|)$ doesn't hold the condition (1), but intuitively it looks quite close with $|x|^\gamma$, for some $0 < \gamma < 1$.

2. Preliminary Results

In this section, we will show the kernel $H_{\alpha,\gamma}$ control the operator $J_{\alpha,\gamma}$ in Lebesgue space and Morrey space (classical). We begin with the following Lemma.

Lemma 1. *Let $0 < \gamma \leq \alpha < n$. If $\frac{n}{n+\gamma-\alpha} < r < \frac{n}{n-\alpha}$ then for every $R > 0$, the series*

$$\left(\sum_{k\in\mathbb{Z}} \frac{(2^kR)^{(\alpha-n)r+n}}{h(2^kR)^r}\right)^{\frac{1}{r}}$$

converge, and we have $H_{\alpha,\gamma} \in L^r$. Moreover, $\|H_{\alpha,\gamma}\|_{L^r} \sim \sum_{k\in\mathbb{Z}} \frac{(2^kR)^{(\alpha-n)r+n}}{h(2^kR)^r}$; it means, there exists C_1 and C_2 positive such that

$$C_1\left(\sum_{k\in\mathbb{Z}} \frac{(2^kR)^{(\alpha-n)r+n}}{h(2^kR)^r}\right)^{\frac{1}{r}} \leq \|H_{\alpha,\gamma}\|_{L^r} \leq C_2\left(\sum_{k\in\mathbb{Z}} \frac{(2^kR)^{(\alpha-n)r+n}}{h(2^kR)^r}\right)^{\frac{1}{r}}.$$

Proof. Let $R > 0$, decompose

$$\sum_{k\in\mathbb{Z}} \frac{(2^kR)^{(\alpha-n)r+n}}{h(2^kR)^r} = \sum_{k=-1}^{-\infty} \frac{(2^kR)^{(\alpha-n)r+n}}{h(2^kR)^r} + \sum_{k=0}^{\infty} \frac{(2^kR)^{(\alpha-n)r+n}}{h(2^kR)^r}.$$

By property of h, the first part of RHS is dominated by $\frac{1}{h(0)}\sum_{k=-1}^{-\infty}(2^kR)^{(\alpha-n)r+n}$, and last part is dominated by $C\sum_{k=-1}^{-\infty}(2^kR)^{(\alpha-n)r+n-\gamma r}$. Both of them will converge simultaneously if $(\alpha-n)r+n > 0$ and $(\alpha-n)r+n-\gamma r < 0$, or $\frac{n}{n+\gamma-\alpha} < r < \frac{n}{n-\alpha}$.

Next, by dyadic decomposition,

$$\begin{aligned}
\|H_{\alpha,\gamma}\|^r_{L^r} &= \sum_{k\in\mathbb{Z}} \int_{2^kR<|x|\leq 2^{k+1}R} \frac{|x|^{(\alpha-n)r}}{h(|x|)^r}dx \\
&\leq \sum_{k\in\mathbb{Z}} \frac{(2^kR)^{(\alpha-n)r}}{h(2^kR)^r} \int_{2^kR<|x|\leq 2^{k+1}R} dx \\
&= C_2 \sum_{k\in\mathbb{Z}} \frac{(2^kR)^{(\alpha-n)r+n}}{h(2^kR)^r}.
\end{aligned}$$

Similarly,

$$\|H_{\alpha,\gamma}\|_{L^r}^r \geq C_1 \sum_{k\in\mathbb{Z}} \frac{(2^{k+1}R)^{(\alpha-n)r+n}}{h(2^{k+1}R)^r} = C_1 \sum_{k\in\mathbb{Z}} \frac{(2^k R)^{(\alpha-n)r+n}}{h(2^k R)^r}.$$

Shows that $\|H_{\alpha,\gamma}\|_{L^r} \sim \sum_{k\in\mathbb{Z}} \frac{(2^k R)^{(\alpha-n)r+n}}{h(2^k R)^r}$ for $\frac{n}{n+\gamma-\alpha} < r < \frac{n}{n-\alpha}$.

□

Now we know $H_{\alpha,\gamma}$ is a member of Lebesgue space, Young's inequality will give the boundedness for $J_{\alpha,\gamma}$ on Lebesgue space.

Theorem 2 (Young's inequality). *Let $1 \leq p, q, r \leq \infty$ and satisfy $\frac{1}{q} + 1 = \frac{1}{p} + \frac{1}{r}$, then*

$$\|g * f\|_{L^q} \leq \|g\|_{L^r}\|f\|_{L^p}$$

for every $f \in L^p$ and $g \in L^r$.

From Lemma 1 dan Young's inequality we get the following direct consquence.

Corollary 3. *Let $0 < \gamma \leq \alpha < n$. If $\frac{n}{n+\gamma-\alpha} < r < \frac{n}{n-\alpha}$ and $1 \leq p, q \leq \infty$ satisfy $\frac{1}{q} + 1 = \frac{1}{p} + \frac{1}{r}$, then*

$$\|J_{\alpha,\gamma} f\|_{L^q} \lesssim \|H_{\alpha,\gamma}\|_{L^r}\|f\|_{L^p}.$$

for every $f \in L^p$

Corrolary 3 didn't only say $J_{\alpha,\gamma}f$ is bounded at Lebesgue space, but also the norm of operator $J_{\alpha,\gamma}$ is dominated by norm of its kernel $H_{\alpha,\gamma}$; that is $\|J_{\alpha,\gamma}\|_{L^p \to L^q} \lesssim \|H_{\alpha,\gamma}\|_{L^r}$. The question is "is the property is also satisfed on Morey space?", the answer is "yes". But Young's inequality is not available on Morrey space. Instead, we will use *Hardy-Littlewood Maximal function M*, which is given by

$$Mf(x) = \sup_{x\in B} \frac{1}{|B|} \int_B |f(y)|dy,$$

$f \in L^p_{loc}(\mathbb{R}^n)$. Also, we will use the fact that M is bounded on Morrey space; that is, for $0 \leq \alpha \leq n$ and $1 \leq p < \infty$ we have $\|Mf\|_{L^{p,\lambda}} \lesssim \|f\|_{L^{p,\lambda}}$ for every $f \in L^{p,\lambda}$. We begin with the following proposition.

Proposition 4. *Let $0 < \gamma \leq \alpha < n$ and $\frac{n}{n+\gamma-\alpha} < s < \frac{n}{n-\alpha}$. Suppose $1 \leq s, p < \infty$ satisfy $\frac{1}{s} + \frac{1}{p} > 1$, and $0 < \lambda < n - \frac{n}{\frac{1}{s}+\frac{1}{p}}$, then*

$$|J_{\alpha,\gamma} f(x)| \lesssim \|H_{\alpha,\gamma}\|_{L^{s,\lambda}} \|f\|_{L^{p,\lambda}}^{\frac{np}{n-\lambda} - \frac{p}{s}} Mf(x)^{1 - \frac{np}{n-\lambda} + \frac{p}{s}}$$

for every $f \in L^{p,\lambda}(\mathbb{R}^n)$.

Proof. It is enough to prove for case $f \not\equiv 0$. Suppose $R > 0$, and x is any element of $\mathbb{R}^n$. Decompose

$$J_{\alpha,\gamma} f(x) = \int_{|x-y|<R} \frac{|x-y|^{\alpha-n}}{h(|x-y|)} f(y)dy + \int_{|x-y|\geq R} \frac{|x-y|^{\alpha-n}}{h(|x-y|)} f(y)dy.$$

Suppose the first part of RHS is I_1 and the last part is I_2. By dyadic decomposition, we get

$$\begin{aligned}
|I_1| &\leq \sum_{k=-1}^{-\infty} \int_{2^kR\leq|x-y|<2^{k+1}R} \frac{|x-y|^{\alpha-n}}{h(|x-y|)}|f(y)|dy \\
&\leq \sum_{k=-1}^{-\infty} \frac{(2^kR)^{\alpha-n}}{h(2^kR)} \int_{2^kR\leq|x-y|<2^{k+1}R} |f(y)|dy \\
&\lesssim Mf(x) \sum_{k=-1}^{-\infty} \frac{(2^kR)^{\alpha-n}}{h(2^kR)} (2^kR)^n \\
&\lesssim Mf(x) \left(\sum_{k=-1}^{-\infty} \frac{(2^kR)^{(\alpha-n)s+n}}{h(2^kR)^s} \right)^{\frac{1}{s}} \left(\sum_{k=-1}^{-\infty} (2^kR)^n \right)^{\frac{1}{s'}} \\
&\lesssim Mf(x) \frac{\left(\int_{|x|\leq R} H_{\alpha.\gamma}{}^s\right)^{\frac{1}{s}}}{R^{\frac{\lambda}{s}}} R^{\frac{\lambda}{s}} R^{\frac{n}{s'}} \\
&\lesssim \|H_{\alpha,\gamma}\|_{L^{s,\lambda}} R^{n-\frac{n-\lambda}{s}} Mf(x).
\end{aligned}$$

On the other hand,

$$\begin{aligned}
|I_2| &\leq \sum_{k=0}^{\infty} \int_{2^kR\leq|x-y|<2^{k+1}R} \frac{|x-y|^{\alpha-n}}{h(|x-y|)}|f(y)|dy \\
&\leq \sum_{k=0}^{\infty} \frac{(2^kR)^{\alpha-n}}{h(2^kR)} \int_{2^kR\leq|x-y|<2^{k+1}R} |f(y)|dy \\
&\leq \sum_{k=0}^{\infty} \frac{(2^kR)^{\alpha-n+\frac{n}{p'}}}{h(2^kR)} \left(\int_{2^kR\leq|x-y|<2^{k+1}R} |f(y)|^p \right)^{\frac{1}{p}} dy \\
&\lesssim \|f\|_{L^{p,\lambda}} \sum_{k=0}^{\infty} \frac{(2^kR)^{\alpha-n+\frac{n}{p'}+\frac{\lambda}{p}}}{h(2^kR)} \\
&\lesssim \|f\|_{L^{p,\lambda}} \sum_{k=0}^{\infty} \frac{(2^kR)^{n-\frac{n-\lambda}{p}}}{(2^kR)^{\frac{n}{s}}} \left(\int_{2^{k-1}R\leq|x|<2^kR} \frac{|x|^{(\alpha-n)s}}{h(|x|)^s} dx \right)^{\frac{1}{s}} \\
&\lesssim \|H_{\alpha,\gamma}\|_{L^{s,\lambda}} \|f\|_{L^{p,\lambda}} \sum_{k=0}^{\infty} (2^kR)^{n-\frac{n-\lambda}{p}-\frac{n-\lambda}{s}} \\
&\lesssim \|H_{\alpha,\gamma}\|_{L^{s,\lambda}} \|f\|_{L^{p,\lambda}} R^{n-\frac{n-\lambda}{p}-\frac{n-\lambda}{s}}.
\end{aligned}$$

The last second part of above this inequlity is convergent, since we have $0 < \lambda < n - \frac{n}{\frac{1}{s}+\frac{1}{p}}$, so $n - \frac{n-\lambda}{p} - \frac{n-\lambda}{p} < 0$. Since $J_{\alpha,\gamma}f(x) = I_1 + I_2$, we have

$$|J_{\alpha,\gamma}f(x)| \lesssim \|H_{\alpha,\gamma}\|_{L^{s,\lambda}} \left(R^{n-\frac{n-\lambda}{s}} Mf(x) + R^{n-\frac{n-\lambda}{p}-\frac{n-\lambda}{s}} \|f\|_{L^{p,\lambda}} \right).$$

By assumption $f \not\equiv 0$, then $Mf(x) \neq 0$ and finite everywhere. Choose $R = \left(\frac{\|f\|_{L^{p,\lambda}}}{Mf(x)}\right)^{\frac{p}{n-\lambda}}$, we have

$$|J_{\alpha,\gamma}f(x)| \lesssim \|H_{\alpha,\gamma}\|_{L^{s,\lambda}} \|f\|_{L^{p,\lambda}}^{\frac{np}{n-\lambda}-\frac{p}{s}} Mf(x)^{1-\frac{np}{n-\lambda}+\frac{p}{s}}$$

□

Let $q = \frac{p}{1-\frac{np}{n-\lambda}+\frac{p}{s}}$, then for every $R > 0$ and $a \in \mathbb{R}^n$ we get

$$\left(\frac{1}{R^\lambda}\int_{B(a,R)} |J_{\alpha,\gamma}f(x)|^q\right)^{\frac{1}{q}} \lesssim \|H_{\alpha,\gamma}\|_{L^{s,\lambda}}\|f\|_{L^{p,\lambda}}^{1-\frac{p}{q}}\left(\frac{1}{R^\lambda}\int_{B(a,R)} Mf(x)^p\right)^{\frac{1}{q}}$$

By taking supremum, and the boundedness of Hardy-Littlewood maximal function M, we have

$$\|J_{\alpha,\gamma}f\|_{L^{q,\lambda}} \lesssim \|H_{\alpha,\gamma}\|_{L^{s,\lambda}}\|f\|_{L^{p,\lambda}}^{1-\frac{p}{q}}\|Mf\|_{L^{p,\lambda}}^{\frac{p}{q}} \lesssim \|H_{\alpha,\gamma}\|_{L^{s,\lambda}}\|f\|_{L^{p,\lambda}}$$

Proposition 5. *Let $0 < \gamma \leq \alpha < n$ and $\frac{n}{n+\gamma-\alpha} < s < \frac{n}{n-\alpha}$. if $1 \leq s, p < \infty$ satisfy $\frac{1}{s}+\frac{1}{p} > 1$, and $0 < \lambda < n - \frac{n}{\frac{1}{s}+\frac{1}{p}}$, then*

$$\|J_{\alpha,\gamma}f\|_{L^{q,\lambda}} \lesssim \|H_{\alpha,\gamma}\|_{L^{s,\lambda}}\|f\|_{L^{p,\lambda}}$$

for every $f \in L^{p,\lambda}$, which $q = \frac{p}{1-\frac{np}{n-\lambda}+\frac{p}{s}}$.

3. Main Results

In this section, we will show the facts that have been obtained about operator $T_{\alpha,\beta}$ on Morrey space. We begin with the result of Adams[1].

Theorem 6 (Adams). *Suppose $1 < p < \frac{n}{\alpha}$ and $0 < \lambda < n - \alpha p$. then*

$$\|I_\alpha f\|_{L^{q,\lambda}} \lesssim \|f\|_{L^{p,\lambda}},$$

for every $f \in L^{p,\lambda}$, which $\frac{1}{q} = \frac{1}{p} - \frac{\alpha}{n-\lambda}$

For case $\beta > 1$, we have $T_{\alpha,\beta} \leq \frac{1}{\log\beta}I_\alpha$. But, by Theorem 6, I_α bounded on Morrey space, as consequence, under the assumptions of Theorem 6 we have $T_{\alpha,\beta}$ bounded on Morrey space.

Corollary 7. *Let $\beta > 1, 1 < p < \frac{n}{\alpha}$ and $0 < \lambda < n - \alpha p$. Then*

$$\|T_{\alpha,\beta}f\|_{L^{q,\lambda}} \lesssim \|f\|_{L^{p,\lambda}},$$

for every $f \in L^{p,\lambda}$, which $\frac{1}{q} = \frac{1}{p} - \frac{\alpha}{n-\lambda}$

Another interesting fact is the kernel $G_{\alpha,1}$ is not a member of Morrey space. For simplicity, notation $G_{\alpha,1}$ is replaced by G_α, and $J_{\alpha,1}$ is replaced by J_α.

Before we prove G_α is not a member of Morrey space, the important thing that should be noted is, for every $\gamma > 0$, $\lim_{R\to\infty} \frac{\log(1+R)}{R^\gamma} = 0$. This shows, for some $C_1 > 0$, $\log(1+R) \leq C_1R^\gamma$ for every R that large enough. For every $0 < \gamma < 1$, $\lim_{R\to 0^+} \frac{\log(1+R)}{R^\gamma} = 0$. This shows for some $C_2 > 0$, $\log(1+R) \leq C_2R^\gamma$ for every R positive that small enough.

Proposition 8. *Let $0 \leq \alpha \leq n$. For every $0 \leq \lambda \leq n$ and $1 \leq p < \infty$, $G_\alpha \notin L^{p,\lambda}$.*

Proof. Let $0 \leq \lambda \leq n$ and $1 \leq p < \infty$. By contradiction, assume that $\|G_\alpha\|_{L^{p,\lambda}} < \infty$. For every $R > 0$,

$$\frac{1}{R^\lambda}\int_{B(0,R)} |G_\alpha(x)|^p dx = \frac{C}{R^\lambda}\int_0^R \frac{r^{(\alpha-n)p+n-1}}{(\log(1+r))^p}dr = \frac{C}{R^\lambda}\int_0^R \frac{r^{\beta-1}}{(\log(1+r))^p}dr, \quad (2)$$

with $\beta = (\alpha - n)p + n$.

IOP Conf. Series: Journal of Physics: Conf. Series **1127** (2019) 012061 doi:10.1088/1742-6596/1127/1/012061

Let $0 < \gamma < 1$, then there exists $C_1 > 0$ and $N > 0$ such that $\log(1+r) \le C_1 r^{\frac{\gamma}{2}}$ for every $r \ge N$. Then, for $R > N$ equation (2) is greater than

$$\frac{C}{R^\lambda}\int_N^R R^{\beta-1-\frac{p\gamma}{2}}dr = \begin{cases} CR^{\beta-\frac{p\gamma}{2}-\lambda} - CNR^{-\lambda} & \beta - \frac{p\gamma}{2} \neq 0 \\ CR^{-\lambda}\log R - CNR^{-\lambda} & \beta - \frac{p\gamma}{2} = 0 \end{cases}.$$

By assumption $\|G_\alpha\|_{L^{p,\lambda}} < \infty$, then $\beta - \frac{p\gamma}{2} - \lambda \le 0$. We have $\beta - p\gamma - \lambda < 0$.

On the other hand, there exists $C_2 > 0$ and $N' > 0$ such that $\log(1+r) \le C_1 r^\gamma$ for every $0 \le r \le N'$. Then, for $R \le N'$ equation (2) is greater than

$$\frac{C}{R^\lambda}\int_0^R R^{\beta-1-p\gamma}dr = \begin{cases} CR^{\beta-p\gamma-\lambda} & \beta - p\gamma \neq 0 \\ -\infty & \beta - p\gamma = 0 \end{cases}.$$

By assumption, should be $\beta - p\gamma - \lambda \ge 0$. Contradicts with $\beta - p\gamma - \lambda < 0$. So, $G_\alpha \notin L^{p,\lambda}$. □

The big question is "is J_α can be bounded on Morrey space?", until now we haven't got the answer yet. However, the ends of this research bring out two possibilities. First, if J_α can't be bounded on Morrey space, we have $J_{\alpha,\beta}$ can be bounded on Morrey space if and only if $\beta > 1$. Second, if J_α can be bounded, we have an example of operator which its kernel is not member of Morrey space, but the operator can be bounded on Morrey space.

Reference

[1] Adams D 1975 A note on Riesz Potential *Duke Math. J.* **42** 768
[2] Chiarenza F and Frasca M 1987 Morrey Spaces and Hardy-Littlewood Maximal Function *Rend. Mat.* **7** 273-279
[3] Idris M, Gunawan H, Lindiarni J and Eridani 2016 The The Boundedness of Bessel-Riesz Operators on Generalized Morrey Spaces *AJMAA* **13** 5

IOP Conf. Series: Journal of Physics: Conf. Series **1127** (2019) 012062 doi:10.1088/1742-6596/1127/1/012062

Some unicyclic Ramsey (P_3, P_n)-minimal graphs

D Rahmadani, E T Baskoro and H Assiyatun

Combinatorial Mathematics Research Group
Faculty of Mathematics and Natural Sciences
Institut Teknologi Bandung (ITB)
Jalan Ganesa 10 Bandung 40132, Indonesia,

E-mail: desi.rahmadani33@students.itb.ac.id, {ebaskoro, hilda}@math.itb.ac.id

Abstract. Let F, G and H be graphs. We write $F \to (G, H)$ if whenever every edge of F is colored by red or blue, then F will contain either a red copy of G or a blue copy of H. The graph F is called Ramsey (G, H)-minimal graph if $F \to (G, H)$ and $F - e \nrightarrow (G, H)$ for every $e \in E(F)$. The class of all Ramsey (G, H)-minimal graphs (up to isomorphism) is denoted by $\mathcal{R}(G, H)$. In this paper, we give explicitly an infinite family of unicyclic graphs in $\mathcal{R}(P_3, P_n)$, for $n = 6$ and 7 constructed from trees in the same class. Moreover, we give a method to construct the unicyclic graphs in $\mathcal{R}(P_3, P_{n+2})$ from existing unicyclic graphs in $\mathcal{R}(P_3, P_n)$, for each $n \geq 6$.

1. Introduction

Let F, G and H be finite undirected graphs without loops and multiple edges. We say $F \to (G, H)$ if any red-blue coloring of F contains a red copy of G or a blue copy of H. The graph F is Ramsey minimal graph for (G, H) if $F \to (G, H)$ and $F - e \nrightarrow (G, H)$ for every $e \in E(F)$. The class of all Ramsey (G, H)-minimal graphs (up to isomorphism) is denoted by $\mathcal{R}(G, H)$. The pair (G, H) will be called *Ramsey-finite* or *Ramsey-infinite* depending upon whether $\mathcal{R}(G, H)$ is finite or infinite.

Some results on Ramsey minimal graphs concerning for the pair paths are obtained. Burr *et al.* [1] proved that the set $\mathcal{R}(P_3, P_3) = \{K_{1,3}, C_{2n+1}\}$ for $n \geq 1$ what implies that the Ramsey set $\mathcal{R}(P_3, P_3)$ is infinite. Then, Burr *et al.* [2] showed that the pair (P_m, P_n), for $3 \leq m \leq n$ is Ramsey-infinite. Yulianti *et al.*[8] determined some infinite families of graphs in $\mathcal{R}(P_3, P_4)$. Faudree and Sheehan [4] determined some trees in $\mathcal{R}(P_3, P_4)$ and $\mathcal{R}(P_3, P_5)$. Rahmadani *et al.*[5] determined some graphs of small order in $\mathcal{R}(P_3, P_6)$. Furthermore, they gave an infinite family of trees in $\mathcal{R}(P_3, P_n)$, for $n \geq 6$ (see [6] and [7]). However, the characterizations of $\mathcal{R}(P_3, P_n)$ for any $n \geq 4$ are still open.

Throughout this paper, we denote a *path* and a *cycle* on n vertices by P_n and C_n, respectively. An acyclic graph, one not containing any cycle, is called a *forest*. A connected forest is called a *tree*. *A unicyclic graph* is a connected graph containing exactly one cycle. A vertex of a graph is said to be *pendant vertex* if its neighborhood contains exactly one vertex. An edge of a graph is said to be *pendant edge* if one of its vertices is a pendant vertex. For other notation and terminology, we refer to [3]. In this paper, we construct explicitly an infinite family of unicyclic graphs in $\mathcal{R}(P_3, P_n)$, for $n = 6$ and 7 constructed from trees in the same class. Moreover, we give a method to construct the unicyclic graphs in $\mathcal{R}(P_3, P_{n+2})$ from existing unicyclic graphs in $\mathcal{R}(P_3, P_n)$, for each $n \geq 6$.

2. Main Results

In this section, we determine an infinite family of unicyclic graphs in $\mathcal{R}(P_3, P_6)$ and $\mathcal{R}(P_3, P_7)$. These unicyclic graphs constructed from trees in $\mathcal{R}(P_3, P_6)$ and $\mathcal{R}(P_3, P_7)$.

Consider trees T_i^6, for any $i \geq 1$ as in Figure 1.

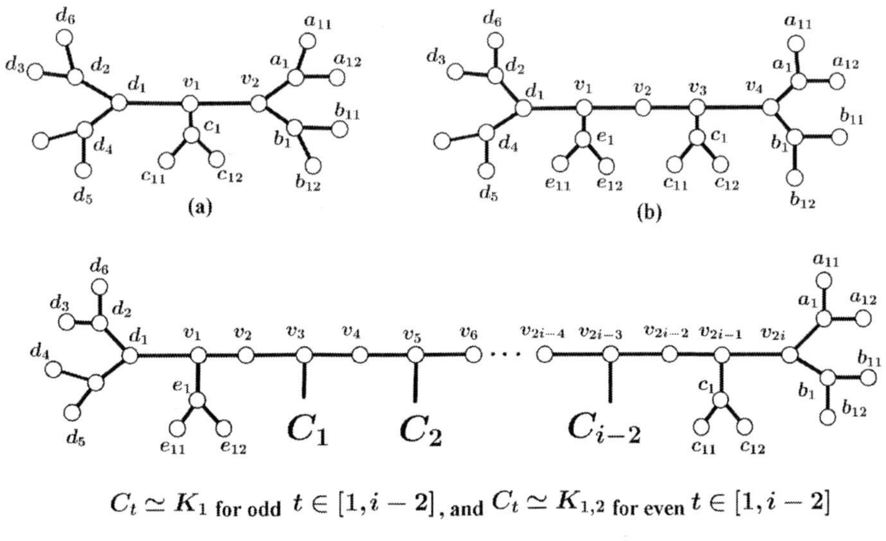

Figure 1. (a). Tree T_1^6 (b). Tree T_2^6 (c). Tree T_i^6, for any $i \geq 3$

Theorem 1. [6] *The trees T_1^6 and T_2^6 are in $\mathcal{R}(P_3, P_6)$.*

Consider the following definition.

Definition 1. *Let k and n be positive integers. The graph U_k^n is a unicyclic graph containing C_4 constructed from T_k^n by attaching two fixed pendant vertices.*

By Definition 1, consider the graphs U_1^6 and U_2^6 as in Figure 2.

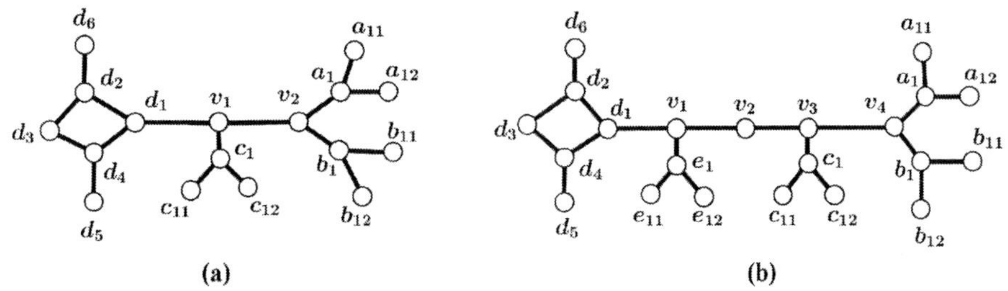

Figure 2. (a) Graph U_1^6 (b) Graph U_2^6

IOP Conf. Series: Journal of Physics: Conf. Series **1127** (2019) 012062 doi:10.1088/1742-6596/1127/1/012062

Theorem 2. *The graph U_1^6 is in $\mathcal{R}(P_3, P_6)$.*

Proof. First, we prove that $U_1^6 \rightarrow (P_3, P_6)$. Consider any red-blue coloring on the edges of U_1^6. Suppose that there is no red P_3. Then, at least two of three edges incident to v_1 must be blue. Observe that each blue edge incident to v_1 can be extended to a blue path of length 2 or 3 where its an end is a pendant vertex. Since the blue paths have an end in v_1 then they form a blue path of length at least 5. Consequently, there will be a blue path P_6 in U_1^6. Thus, $U_1^6 \rightarrow (P_3, P_6)$.

Next, we prove that $U_1^6 - e \nrightarrow (P_3, P_6)$, for any edge $e \in U_1^6$. Let $i, j \in \{1, 2\}$, where $i \neq j$. If $e = a_1a_{1i}$ then color the edges $d_1v_1, b_1v_2, a_1a_{1j}$ by red. Similarly, for edges $e = b_1b_{1i}$. If $e \in \{a_1v_2, v_1v_2, c_1v_1\}$ then color edges d_1v_1 and v_2b_1 by red. If $e = v_2b_1$ then color edges d_1v_1 and v_2a_1 by red. If $e = c_1c_{1i}$ then color the edges $d_1v_1, v_2b_1, c_1c_{1j}$ by red. If $e \in \{d_1v_1, d_1d_2, d_2d_6\}$ then color the edges v_1v_2, d_1d_4 and d_2d_3 by red. If $e \in \{d_1d_4, d_4d_5\}$ then color the edges v_1v_2, d_1d_2 and d_3d_4 by red. If $e = d_2d_3$ then color the edges v_1v_2, d_1d_4 and d_2d_6 by red. If $e = d_3d_4$ then color the edges v_1v_2, d_1d_2 and d_4d_5 by red.

For each this coloring, color the remaining edges by blue. This coloring of $U_1^6 - e$ contains neither red P_3 nor blue P_6. Therefore, $U_1^6 \nrightarrow (P_3, P_6)$, for any $e \in E(U_1^6)$. □

Theorem 3. *The graph U_2^6 is in $\mathcal{R}(P_3, P_6)$.*

Proof. We first prove that $U_2^6 \rightarrow (P_3, P_6)$. Consider any red-blue coloring on the edges of U_2^6. Suppose that there is no red P_3. If two edges incident to v_2 are blue then it is easily to check that there is a blue P_6 in U_2^6. Otherwise, consider two edges adjacent to the red edge. They must be blue. Observe that the blue edges can be extended to a blue path of length 5 where its end vertices are pendant vertices. Hence, $U_2^6 \rightarrow (P_3, P_6)$.

Next, we prove that $U_2^6 - e \nrightarrow (P_3, P_6)$, for every edge $e \in E(U_2^6)$. Let $M_1 = \{v_1v_2, v_3v_4\}$ and $M_2 = \{v_1d_1, v_2v_3, v_4b_1\}$. Let $i, j \in \{1, 2\}$, where $i \neq j$. If $e = a_1a_{1i}$ then color $M2, a_1a_{1j}$ by red. Similarly, for $e = b_1b_{1i}$. If $e = e_1e_{1i}$ then color M_1 and $\{d_2d_3, d_1d_4, e_1e_{1j}\}$ by red. If $e \in \{d_1d_2, d_2d_6, e_1v_1\}$ then color M_1 and $\{d_1d_4, d_2d_3\}$ by red. If $e \in \{d_1d_4, d_4d_5\}$ then color M_1 and $\{d_1d_2, d_3d_4\}$ by red. If $e \in \{d_2d_3, d_1v_1\}$ then color M_1 and $\{d_1d_4, d_2d_6\}$ by red. If $e = d_3d_4$ then color M_1 and $\{d_1d_2, d_4d_5\}$ by red. If $e = c_1c_{1i}$ then color M_2 and c_1c_{1j} by red. If $e \in \{v_1v_2, v_2v_3, v_4a_1\}$ then color v_3v_4, d_1v_1 by red. Similarly, for $e = v_4b_1$. If $e \in \{v_3v_4, c_1v_3\}$ then color M_2 by red.

For each this coloring, color the remaining edges by blue. This coloring of $U_2^6 - e$ contains neither red P_3 nor blue P_6. Therefore $U_2^6 - e \rightarrow (P_3, P_6)$, for every edge $e \in E(U_2^6)$. □

Definition 2. *For $i \geq 3$, the graph U_i^6 is a unicyclic graph containing C_4 and a backbone path P of order $2i$ ($P \equiv v_1v_2 \ldots v_{2i}$). Each of vertices v_{2t+1}, for $t = 1, 2, \ldots, i-2$ is connected to a subtree C_t, where $C_t \simeq K_1$, for odd t and $C_t \simeq K_{1,2}$, for even t as in Figure* 3.

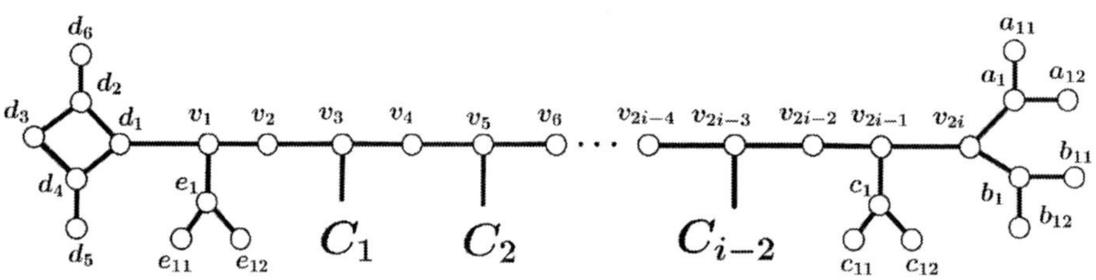

Figure 3. Graph U_i^6 where $C_t \cong K_1$ for odd t and $C_t \cong K_{1,2}$ for even t.

Theorem 4. *For any $i \geq 3$, the graph $U_i^6 \in \mathcal{R}(P_3, P_6)$.*

Proof. Consider the three components of $U_i^6 - \{v_1v_2, v_{2i-1}v_{2i-1}\}$, denoted by H_1 for the component containing v_1; H_2 for the component containing v_{2i-1}; H_3 for the component containing v_2 and v_{2i-2}.

Let μ be any red-blue coloring on the edges of U_i^6. Suppose that there is no red P_3. We first prove that $U_i^6 \to (P_3, P_6)$. Observe that, if one of $\{v_1v_2, v_{2i-2}v_{2i-1}\}$ is red then it is easily to check that there is a blue P_6 in H_1 or H_2. Otherwise, $\mu(v_1v_2) = \mu(v_{2i-2}v_{2i-1}) =$ blue. Then, there exists a blue path of length at least 2 in H_1 and H_2. Next, consider the subtree H_3. Observe that, if there are two blue edges adjacent in a backbone path $v_3, v_4, \ldots, v_{2i-4}, v_{2i-3}$ then there must be a blue P_6 in H_3. Otherwise, we must have a coloring as follows.

$$\mu(v_kv_{k+1}) = \begin{cases} blue, & \text{for odd } i \in [3, 2i-4]; \\ red, & \text{for even } i \in [3, 2i-4]. \end{cases}$$

Since the backbone path $v_3, v_4, \ldots, v_{2i-4}, v_{2i-3}$ has an even length, then $\mu(v_{2i-4}v_{2i-3})$= red. Consequently, there will be a red P_3 in H_3. This contradicts assumption that there is no red P_3. Thus, $U_i^6 \to (P_3, P_6)$.

Next, we prove that $U_i^6 - e \nrightarrow (P_3, P_6)$, for any edge $e \in U_i^6$. Let $M_1 = \{v_kv_{k+1} | \text{for odd } k \in [1, 2i-1]\}$ and $M_2 = \{v_kv_{k+1} |, \text{for even } k \in [1, 2i-2]\}$. Consider the following cases.

- Case 1. e is an edge in H_1.
  If $e = e_1e_{1i}$, for $i, j \in \{1, 2\}$ then color M_1 and e_1e_{1j} by red, for $i \neq j$. If $e \in \{d_1v_1, v_1e_1, d_1d_4, d_4d_5\}$ then color M_1 and d_1d_2, d_3d_4 by red. If $e \in \{d_1d_2, d_2d_6\}$ then color M_1, d_1d_4, d_2d_3 by red. If $e = d_3d_4$ then color M_1, d_1d_2, d_4d_5 by red. If $e = d_2d_3$ then color M_1, d_1d_4, d_2d_6 by red.
- Case 2. e is an edge in H_2.
  Let $j, k \in \{1, 2\}$ and $s, t \in \{a, b\}$. If $e = s_1s_{1j}$ then color M_2, t_1v_{2i}, d_1v_1 and s_1s_{1k} by red, for $j \neq k$ and $s \neq t$. If $e = s_1v_{2i}$ then color M_2, t_1v_{2i}, d_1v_1 by red, for $s \neq t$. If $e = c_1c_{1j}$ then color M_2, a_1v_{2i}, d_1v_1 and c_1c_{1k} by red, for $j \neq k$. If $e \in \{c_1v_{2i-1}, v_{2i-1}v_{2i}\}$ then color a_1v_{2i}, d_1v_1 by red.
- Case 3. e is an edge in $H_3 \cup \{v_1v_2, v_{2i-2}v_{2i-1}\}$. Let $j, k \in [1, 2i-2]$. Let w_j be a vertex in the subtree C_t, for $t \in [1, i-2]$ adjacent to v_j, for some j. Consider the following cases. If $e \in \{v_jv_{j+1}, v_jw_j\}$ then color $e_1v_1, v_{2i-1}v_2i, v_kv_{k+1}$, for any odd $k > j$ and for any even $k < j$ by red. Let x_1 and x_2 be vertices adjacent to w_j (in case the subtree $C_t = K_{1,2}$). If $e \in \{w_jx_p| \text{ for } p, s \in \{1, 2\}\}$ then use the coloring as case 1. In addition, color v_jw_j by blue and color w_jx_s by red, for $p \neq s$.

For each this coloring, color the remaining edges by blue. Then, $U_i^6 - e \nrightarrow (P_3, P_6)$, for any edge $e \in E(U_i^6)$. □

By a similar argument, we construct unicyclic graphs in $\mathcal{R}(P_3, P_7)$ coming from trees in the same class (see [6]). Now, consider the graph U_1^7 and U_2^7 as in Figure 4(a) and (b), respectively.

Theorem 5. *The graph U_1^7 is in $\mathcal{R}(P_3, P_7)$.*

Proof. First, we prove that $U_1^7 \to (P_3, P_7)$. Consider any red-blue coloring on the edges of U_1^7. Suppose that there is no red P_3. It is easy to check that $U_1^7 \supset U_1^6$. Since $U_1^7 \supset U_1^6$ and $U_1^6 \in \mathcal{R}(P_3, P_6)$ then U_1^7 contains a blue path of length at least 5. Observe that U_1^7 contains exactly a blue path of length 5 if one of its end vertices has degree one in U_1^6. However, the vertex has degree three in U_1^7. In addition, at least one of two pendant edges incident to the vertex must be blue then the blue P_6 can be extended to a blue P_7. Otherwise, observe that there exists a blue P_7 in U_1^6. Consequently, U_1^7 also contains a blue P_7. Hence, $U_1^7 \to (P_3, P_7)$.

Next, we prove that $U_1^7 - e \nrightarrow (P_3, P_7)$ for every edge e. If $e \in E(U_1^7) \cap E(U_1^6)$ then use the similar coloring as in $U_1^6 - e$ and color the remaining edges by blue. By this coloring, there is neither red P_3

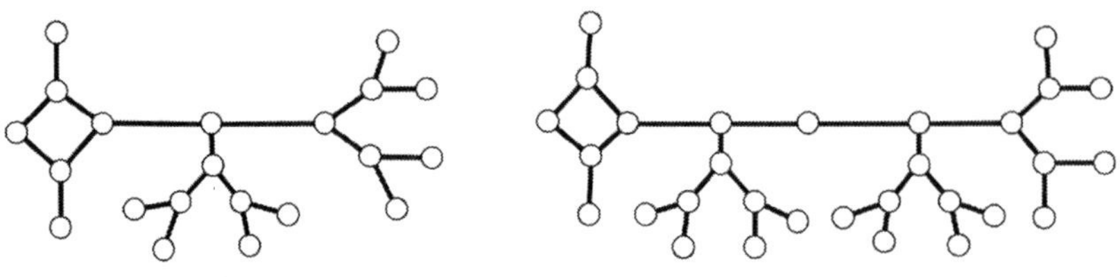

Figure 4. (a) Graph U_1^7 (b) Graph U_2^7

nor blue P_7 in $U_1^7 - e$. If $e \in E(U_1^7) \setminus E(U_1^6)$ then color a pendant edge adjacent to e by red. For the remaining edges, use the similar coloring as in $U_1^6 - e_1$ where e_1 is a pendant edge in U_1^6 adjacent to e in U_1^7 and color e_1 by blue. Consequently, there is neither red P_3 nor blue P_7 in $U_1^7 - e$. Therefore $U_1^7 - e \not\rightarrow (P_3, P_7)$ for any edge $e \in E(U_1^7)$. □

Theorem 6. *The graph U_2^7 is in $\mathcal{R}(P_3, P_7)$.*

Proof. It is easily to show that $U_2^7 \supset U_2^6$. Then, we can verify that graph U_2^7 is an element $\mathcal{R}(P_3, P_7)$. The main idea used in the proof of this theorem is similar to the one in the proof of Theorem 5. □

Definition 3. *For $i \geq 3$, the graph U_i^7 is a unicyclic graph containing C_4 and a backbone path P of order $2i$ ($P \equiv v_1 v_2 \ldots v_{2i}$). Each of vertices v_{2t+1}, for $t = 1, 2, \ldots, i-2$ is connected to a subtree C_t, where $C_t \simeq K_1$ for odd t, and $C_t \simeq J$ for even t as in Figure 5.*

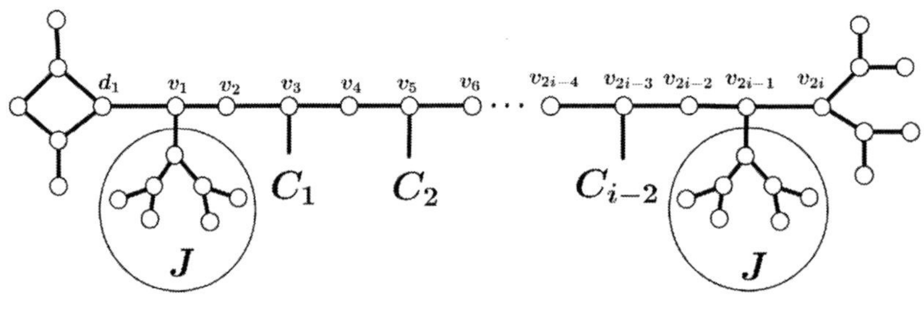

Figure 5. Graph U_i^7 where $C_t \cong K_1$ for odd t and $C_t \cong J$ for even t

Theorem 7. *For any $i \geq 3$, the graph $U_i^7 \in \mathcal{R}(P_3, P_7)$.*

Proof. By Definition 4, we can see that $U_i^7 \supset U_i^6$. Consider any red -blue coloring on the edges of U_i^7. Suppose that there is no red P_3. Consider the subgraph U_i^6 in U_i^7. Since $U_i^6 \in \mathcal{R}(P_3, P_6)$ then there is a blue path of length at least 5 in U_i^7. Let u and v be end vertices of the blue path in U_i^7. Observe that if one of $\{u, v\}$ is a vertex adjacent to a pendant vertex in J then the blue path has length at least 5. If the blue path has length 5, consider that at least one of edges incident to the vertex must be blue. Consequently, the blue path of length 5 can be extended to a blue path of length 6. Otherwise, the blue path has length at least 6. Then, $U_i^7 \rightarrow (P_3, P_7)$.

Next, we prove that $U_i^7 - e \not\rightarrow (P_3, P_7)$, for any edge e. If $e \in E(U_i^7) \cap E(U_i^6)$ then use the similar coloring as in $U_i^6 - e$ and color the remaining edges by blue. By this coloring, there is neither P_3 nor blue P_7 in $U_i^7 - e$. If $e \in E(U_i^7) \setminus E(U_i^6)$ then e is a pendant edge. Color the other pendant edge adjacent to e by red. For the remaining edges, use the similar coloring as in $U_i^6 - e_1$ where e_1 is a pendant edge in U_i^6 having a common vertex with e in U_i^7 and color e_1 by blue. Then, there is neither red P_3 nor blue P_7 in $U_i^7 - e$. Therefore $U_i^7 - e \nrightarrow (P_3, P_7)$, for any edge $e \in E(U_i^7)$. □

Next, the following definition give a method to construct an infinite class of unicyclic graphs in $\mathcal{R}(P_3, P_{n+2})$, for each $n \geq 6$ from existing unicyclic graphs in $\mathcal{R}(P_3, P_{n+2})$.

Definition 4. *For any $n \geq 6$ and $k \geq 1$, the graph U_k^{n+2} is a graph constructed from existing U_k^n by adding two pendant edges to every vertex of degree one on U_k^n.*

By Definition 4, we have the following theorem.

Theorem 8. *For any $n \geq 6$ and $k \geq 1$, if $U_k^n \in (P_3, P_n)$ then $U_k^{n+2} \in (P_3, P_{n+2})$.*

Proof. Assume three components of $U_k^n - \{v_1v_2, v_{2i-2}v_{2i-1}\}$, denoted by H_1 for the component containing v_1; H_2 for the component containing v_{2i-1}; H_3 for the component containing v_2 and v_{2i-2}.

We first prove that $U_k^{n+2} \rightarrow (P_3, P_{n+2})$. Consider any red-blue coloring on edges of U_k^{n+2}. Assume that there is no red P_3. Since $U_k^n \subset U_k^{n+2}$ and $U_k^n \in \mathcal{R}(P_3, P_n)$ then there will be a blue P_n in U_k^{n+2}. Let u and v be end vertices in a blue P_n. Note that if u and v have degree either 1 or 3 in U_k^n then u and v have degree 3 in U_k^{n+2}. In addition, at least one of two edges incident to u and v in U_k^n must be blue, then the blue P_n can be extended into a blue path P_{n+2}.

Otherwise, then at least one of $\{u, v\}$ has degree 2. If exactly one of $\{u, v\}$ has degree 2, say u, then there is a blue P_{n+1} in U_k^{n+2}. Next, consider the other edge incident to u, if the edge is blue then U_k^{n+2} contains a blue P_{n+2}. Otherwise, u must be contained in H_3. By similar argument as Theorem 4, there will a blue P_{n+2} in H_3. Hence, $U_k^{n+2} \rightarrow (P_3, P_{n+2})$.

Next, we prove that $U_k^{n+2} - e \nrightarrow (P_3, P_{n+2})$, for every $e \in E(U_k^n)$. If $e \in E(U_k^{n+2}) \cap E(U_k^n)$ then use a similar coloring as in $U_k^n - e$, for any $e \in E(U_k^n)$ and color the remaining edges by blue. If $e \in E(U_k^{n+2}) \setminus E(U_k^n)$ then e is a pendant edge. Color the other pendant edge adjacent to e by red. For the remaining edges, use the similar coloring as in $U_k^n - e_1$ where e_1 is a pendant edge in U_k^n adjacent to e in U_k^{n+2}. In addition, color e_1 by blue. Each this coloring of $U_k^{n+2} - e$ contains neither red P_3 nor blue P_{n+2}. Therefore, $U_k^{n+2} - e \nrightarrow (P_3, P_{n+2})$, for any e in U_k^n. □

Acknowledgement
This research was supported by research grant "Program Riset dan Inovasi KK-Institut Teknologi Bandung 2018".

3. References

[1] Burr S A, Erdos P, and Lovasz L 1976 On graphs of Ramsey type, *Ars Combin.* **1** 167–190
[2] Burr S A, Erdős P, Faudree R J, Rosseu C C, and Schelp R H 1982 Ramsey-minimal graphs for forests *Studia Scient. Math. Hungar* **15** 265–273
[3] Diestel R 2000 *Graph Theory* 2nd ed (New York: Springer-Verlag)
[4] Faudree R J and Sheehan J 1982 Tree Ramsey-minimal graphs *Congressus Numer* **35** 295–315
[5] Rahmadani D, Baskoro E T and Assiyatun H 2015 On Ramsey minimal graphs for the pair paths *Proc. Comp. Sci.* **74** 15–20
[6] Rahmadani D, Baskoro E T and Assiyatun H 2016 On Ramsey (P_3, P_6)-minimal graphs *AIP Conf. Proc.* 1707, 020016 (2016); doi:10.1063/1.4940817
[7] Rahmadani D, Baskoro E T, Bača M, Assiyatun H and Semaničová-Feňovčíková A (2017) Some infinite families of Ramsey (P_3, P_n)-minimal trees *Proc. Math. Sci.* **127(5)**, 779-786
[8] Yulianti L, Assiyatun H, Uttunggadewa S and Baskoro E T 2010 On Ramsey $(K_{1,2}, P_4)$-minimal graphs *Far East J. Math. Sci.* **40** 23–36

IOP Conf. Series: Journal of Physics: Conf. Series **1127** (2019) 012064 doi:10.1088/1742-6596/1127/1/012064

Explicit Computation of Cheeger Constants for some Classes of Graphs

S Opiyo, Y Soeharyadi and M W Setyabudhi
Department of Mathematics, Institut Teknologi Bandung

samie.opiyo@gmail.com

Abstract. This article deals with the explicit computation of the Cheeger constant and its connection to understanding the properties of some classes of graphs. Cheeger constant is a measure of the connectivity or disconnectivity of graphs and gives the best possible way to cut a graph. More precisely, we deal with the problem of splitting the graph into two large components of approximately equal volumes by making a small cut, which is the idea of Cheeger constant of a graph. We computed the Cheeger constants for simple classes of graphs such as 2-comb graphs, cycle graphs, complete graphs, and cube graphs. We further analyzed the dynamics of Cheeger constants of the graphs under vertex-edge transformation.

1. Introduction

One of the earliest problems in geometry was the isoperimetric problem, which was considered by the ancient Greeks [3]. The basic isoperimetric problem for graphs is essentially removing as little of the graph as possible to separate out a subset of vertices of some desired "size". Isoperimetric problems examine optimal relations between the size of the cut and the sizes of the separated parts. Roughly speaking, isoperimetric problems involving edge-cuts correspond in a natural way to Cheeger constants in spectral geometry. Cheeger constants were studied in the thesis of Cheeger [2], but they can be traced back to Polya and Szego [6]. Related works in isoperimetric/problem are found in [7], [8], [9], [10].

Cheeger constant (sometimes called isoperimetric constant) is the basic idea of cutting a set or graph into two equal volume sub-manifolds. It is defined for n-dimensional Riemannian manifold as well as for a graph. In either case, Cheeger constant measures how well the set/graph can be split into two pieces and its magnitude is a numerical measure of the degree of connectedness of a set/graph. The *Cheeger constant* of the n-manifold M, is defined by:

$$h(M) := \inf_{S_1,S_2} \frac{\text{Area}(M \cap H)}{\min\{vol(S_1), vol(S_2)\}} \tag{1}$$

where $M \cap H$ is any smooth $(n-1)$-dimensional submanifold of M which divides M into two disjoint sub-manifolds S_1, S_2. The infimum may be restricted to all open subsets S of M such that $\partial S \cap M$ (relative boundary) is a smooth sub-manifold of co-dimension 1. The estimation of the Cheeger constant in this case is consistent with the minimization defining the conductance of the neighborhood graph to subsets associated with subsets of M with controlled sliding of the hypersurface [11]. This quantity was introduced by Cheeger [3] in order to bound the eigen-gap of the spectrum of the Laplacian on a manifold. A *Cheeger set* is a subset $S \subseteq M$ such that $h(S; M) = h(M)$. The Cheeger set always exists and it is unique under some conditions on the domain M [4]. Since M is a continuous domain, the area of $M \cap H$ and volume of the subsets of M are easily be determined by integration over the appropriate domains.

Cheeger constant on a compact metric graph, X, is defined for a non-empty open subset $S \subset X$ as.

$$h(X) := \inf_{S \subset X} \frac{|\partial S|}{\min\{vol(S), vol(S^c)\}} \tag{2}$$

where ∂S = set of all edges with one vertex in S and the other in S^c, $vol(S) = \sum_{v \in S} d_v$ is the of degree of vertex v, and $S^c = X - S$ is the complement of S. The quantity $|\partial S|$ denotes the number of points in the boundary and $vol(S) = \int_X 1_S \, dx$ the total length of S. The infimum runs over all open subset $S \subset X$, $S \neq X$ and $S \neq \emptyset$ [5].

Suppose that X is connected and that each vertex is standard, then the corresponding (standard or Kirchhoff) Laplacian (denoted by Δ_X^{std}) has discrete spectrum [5]. Therefore, given a sample $X_1, \ldots, X_n$ in X, it is possible to translate X inside a manifold, M, however, the translation does not share a boundary with M, while its discrete volume and perimeter are left equal to those of X. It is suffices to replace the infimum in the minimization problem by minimum in the Cheeger ratio. The Cheerger ratio for a vertex set S is defined as

$$h(S) = \frac{|\partial S|}{\min\{vol(S), vol(S^c)\}} \tag{3}$$

Let $\Gamma = (V, E)$ be a simple graph and let S be a nonempty subset of vertices V of the graph and S^c the complement of S. The *Cheeger* Constant of $\Gamma = (V, E)$, is defined as

$$h(\Gamma) = \min \frac{|\partial S|}{\min\left(\sum_{v \in S} d_v, \sum_{u \in S^c} d_u\right)} \tag{4}$$

where $\partial S = \{\{u, v\} \in E(\Gamma) | u \in S \text{ and } v \notin S\}$, $vol(S) = \sum_{v \in S} d_v$, and d_v is the degree of the vertex v.

Although Cheeger constant is well-established, it still forms active field of research. The definitions of Cheeger constant given in equations (1), (2) and (4) are generalisation of the subject. In this paper, our focus is establishment of an alternative approach to Cheeger constant of some classes of simple finite graphs such as 2-comb graphs, cycle graphs, complete graphs, and cube graphs. Our results provide simple formulae one would require to compute Cheeger constants of the classes of graphs stated herein.

2. Method

Cheeger constant analysis depends directly on nature of graph someone is dealing with; simple graph, metric graph or a surface. In our approach, we apply first principles to analyse Cheeger constants of some simple classes of graphs. The measure we will use here takes into consideration the degree of a vertex, the subset S of the vertices of a graph Γ, and the partition of the volume of Γ, in term of the sum of the degrees of the vertices in Γ, into the volume of S and S^c respectively. Equation (4) has been used to explicitly compute the Cheeger constant of the graphs followed by analytic approach to obtain formula for each of the classes of the graphs (2-comb, cycle, complete and cube graphs).

A graph $\Gamma = (V, E)$ is called a *comb graph* if it is a tree, all vertices are of degree at most 3, and all the vertices of degree 3 lie on a single simple path. The maximal path which starts and ends in degree 3 vertices is called the *spine* of the comb graph. A path from a degree 3 vertex to a leaf which contains exactly one degree 3 vertex is called a *tooth*. Comb is a graph obtained by joining a single pendant edge to each vertex of a path. We define the *spine length*, denoted by ρ, as the number of edges it contains. To compute Cheeger constant, we partition the given graph into two subsets S and S^c such that $|S| \leq \frac{|V|}{2}$. The Cheeger ratio of the respective subsets are determined from which Cheeger constant obtain by taking the minimum. In a 2-comb graph, Γ_n with n vertices, $\sum_{v \in V} d_v = n(n-1)$. Therefore, the Cheeger constant

$$h(\Gamma_n) = \min \frac{1}{\min(\text{"split" } n(n-1))} \tag{5}$$

IOP Conf. Series: Journal of Physics: Conf. Series **1127** (2019) 012064 doi:10.1088/1742-6596/1127/1/012064

For instance, consider Γ_6 below

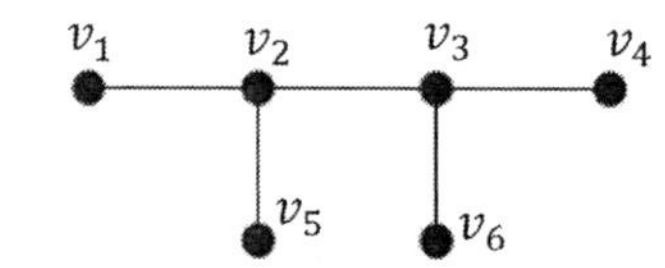

Figure 1. 2-Comb graphs with spine length 3

There are two distinct cuts; $s_1 = \{v_1\}$ with Cheeger ratio $h(S_1) = \frac{1}{\min(1,9)} = 1$ and $s_2 = \{v_1, v_2, v_5\}$ with Cheeger ratio, $h(S_2) = \frac{1}{\min(5,5)} = \frac{1}{5}$. Hence, $h(\Gamma_n) = \min\left(1, \frac{1}{5}\right) = \frac{1}{5}$.

A *cycle graph* C_n, sometimes simply known as an n-cycle, is a graph on n nodes containing a single cycle through all nodes. In a cycle graph, C_n with n vertices, $\sum_{v \in V} d_v = 2n$. Denote the subsets of the graph by S_i, $1 \le i \le n$ and complement S_i by S_i^c. Consider a cycle graph, C_4, below:

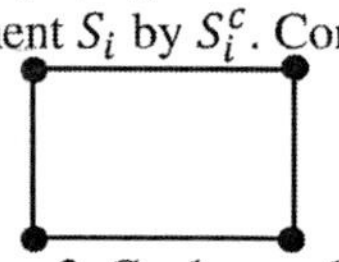

Figure 2. Cycle graphs 4

Here there are also two distinct cuts with Cheeger ratios $h(S_1) = \frac{2}{\min(2,6)} = \frac{2}{2} = 1$ and $h(S_2) = \frac{2}{\min(4,4)} = \frac{2}{4} = \frac{1}{2}$ respectively. Hence, $h(C_4) = \frac{1}{2}$.

A *cubic graph*, C^8,is a graph in which all vertices have degree three. In other words, a cubic graph is a 3-regular graph. Cubic graphs are also called trivalent graphs. The Cheeger constants of cubic graphs were computed in the same way as in the examples above and the result revealed that

$$h(C^8) = \frac{1}{3}. \tag{6}$$

A *complete graph* denoted by K_n is a graph in which each pair of graph vertices is connected by an edge. K_n has n vertices and $\binom{n}{2} = \frac{n(n-1)}{2}$ undirected edges. We computed the Cheeger constants of complete graphs and analysed the results to generate a simple formula as in Lemma 2 here below.

3. Statement of Main Results

The formulae giving exact value of Cheeger constants of cycle graph, complete graph, cubic graph or 2-comb graph are not yet been stated explicitly. In order to determine the Cheeger constant, we used to apply equations (1), (2) or (4), which are sometimes not easy to interpret. Here, we shall give simple formulae of Cheeger constants of cycle graph, complete graph, cubic graph and 2-comb graph, which are very easy to interpret and apply.

3.1 Lemma 1

Let C_n denotes a cycle graph on n vertices. The Cheeger constant of a C_n is given by

$$h(C_n) = \begin{cases} \dfrac{2}{n-1}, & n \text{ odd} \\ \\ \dfrac{2}{n}, & n \text{ even} \end{cases} \tag{7}$$

Proof

Let C_n be a cycle graph with n vertices, then the sum of the degree of the vertices, $\sum_{v \in V} d_v = 2n$. Cut a cycle graph into two disjoint sets, then $|\partial S| = 2$ for all possible cuts.

By definition of Cheeger constant,

$$h(C_n) = \min \frac{2}{\min\left(\sum_{x\in S} d_x, \sum_{y\in S^c} d_y\right)} = \min \frac{2}{\min(\text{"split" } 2n)} = \min \frac{2}{\max_k \min(2k, 2(n-k))}$$

If n even, then $n = 2k$ for some $k \in \mathbb{N}$, then $\max_k \min(2k, 2(n-k))$ is attained for the largest $\min(2k, 2(n-k))$. This occurs when $k = \frac{n}{2}$, hence,

$$h(C_n) = \min \frac{2}{\max_k \min(2k, 2(n-k))} = \min \frac{2}{\min\left(2\left(\frac{n}{2}\right), 2\left(\frac{n}{2}\right)\right)} = \frac{2}{n}$$

Similarly if n odd, then $n = 2r + 1$ for some $r \in \mathbb{N}$, $\max_r \min(2r, 2(n-r))$ is attained for the largest $\min(2r, 2(n-r))$, which occurs when , $r = \frac{\lfloor n \rfloor}{2} = \frac{n-1}{2}$, hence,

$$h(C_n) = \min \frac{2}{\max_r \min(2k, 2(n-k))} = \min \frac{2}{\min\left(2\left(\frac{n-1}{2}\right), 2\left(\frac{n+1}{2}\right)\right)} = \frac{2}{n-1} \qquad \blacksquare$$

3.2 Lemma 2

Let K_n denotes a cycle graph on n vertices. The Cheeger constant of a complete graph, K_n, is given by

$$h(K_n) = \begin{cases} \frac{n+1}{2(n-1)}, & n \; odd \\ \frac{n}{2(n-1)}, & n \; even \end{cases} \tag{8}$$

Proof

Let K_n be a complete graph. Then the degree of each vertex is $n - 1$ and $\sum_{v\in V} d_v = n(n-1)$.
By induction steps, for $n \geq 2$, the Cheeger constant K_n is the minimum of:
Cutting 1 vertex gives,

$$h(S_1) = \frac{n-1}{\min((n-1),\ (n-1)^2)}$$

With 2 vertices cut, we have

$$h(S_2) = \frac{2(n-2)}{\min(2(n-1),\ (n-1)(n-2))}$$

With 3 vertices cut, we have

$$h(S_3) = \frac{3(n-3)}{\min(3(n-1),\ (n-1)(n-3))}$$

$\vdots$

With k steps, $k \in \mathbb{N}$ we have

$$h(S_k) = \frac{k(n-k)}{\min(k(n-1),\ (n-1)(n-k))}$$

We stop when $k = \frac{n}{2}$ for n even. Thus,

$$h(K_n) = \frac{k(n-k)}{\min\left(k(n-1),\ (n-1)(n-k)\right)} = \frac{\frac{n}{2}\left(n - \frac{n}{2}\right)}{\min\left(\frac{n}{2}(n-1),\ \frac{n}{2}(n-1)\right)}$$

$$= \frac{\frac{n}{2}\left(\frac{n}{2}\right)}{\min\left(\frac{n}{2}(n-1),\ \frac{n}{2}(n-1)\right)} = \frac{n}{2(n-1)}.$$

For odd n, we stop when $k = \left\lfloor \frac{n}{2} \right\rfloor$. Here $n = 2k + 1$. Thus,

$$h(K_n) = \frac{\frac{n-1}{2}(n - \frac{n-1}{2})}{\min\left(\frac{n-1}{2}(n-1), (n-1)(n - \frac{n-1}{2}\right)} = \frac{n+1}{2(n-1)} \qquad \blacksquare$$

3.3 Lemma 3
Let C^8denote a cubic graph on 8 vertices. Then the Cheeger constant of C^8 is given by

$$h(C^8) = \frac{1}{3}. \tag{9}$$

Proof
For a cubic graph,

$$\sum_{v \in V} d_v = 3n.$$

Any cubic graph has 8 vertices and is 3-regular. Thus, Cheeger constant of a cubic graph is given by

$$h(C^8) = \min \frac{\frac{n}{2}}{\min\,(\text{"split" } 3n)} = \min \frac{\frac{n}{2}}{\min \frac{3n}{2}} = \frac{1}{3} \qquad \blacksquare$$

3.4 Conjecture 1
Let Γ_n be a 2-comb graph on n vertices. Then the Cheeger constant of Γ_n is given by

$$h(\Gamma_n) = \begin{cases} \frac{1}{\rho}, & n \text{ even} \\ \\ \frac{1}{\rho + 1}, & n \text{ odd} \end{cases} \tag{10}$$

where ρ denotes the spine length of the 2-comg graph with one end fixed to only one vertex while the other end has increasing number of vertices. More works are yet being done on this conjecture in order to provide a stronger result and proof.

4. Conclusion

In this paper, we considered the problem of cutting a graph into two disjoint subsets of approximately equal volumes, followed by analysis of the values of the Cheeger constant of the computation to generalize a simple formulae for each of the graphs named above. The results presented in this article give the easiest way to compute Cheeger constant of the graphs with only minimal efforts as compared to applying the first principles.

Cheeger constant relates directly to the algebraic connectivity of the graph. This signifies that the larger the Cheeger constant, the more difficult is to cut the graph into two disjoint subsets and vice versa. Cheeger constant is not only important in the studying of eigenvalues of Laplacians but also plays very important role in understanding class-structure of data abstracted in terms of graphs as well as graph clustering structures.

5. Further work

Compute Cheeger constant and analyse its relation with graph spectrum properties.

6. Acknowledgement

We greatly express our gratitude to PUPT DPRM Research Grant 2016 for supporting this research. My heartfelt gratitude to my supervisors for their guidance and inspirations.

References

[1] Buser P Cayley graphs and planar isospectral domains, in *Geometry and Analysis on Manifolds* (T. Sunada ed.), Springer Lecture Notes **1339** 1988.

[2] Cheeger, J 1970 A lower bound for the smallest eigenvalue of the Laplacian. in *Problems in Analysis (Papers dedicated to Salomon Bochner, 1969),* Princeton University Press, Princeton, NJ, pp. 195-199.

[3] Chung F R K 1997 Spectral Graph Theory, *CBMS Regional Conference Series in Mathematics* **92**. American Mathematical Society, Providence, Rl.

[4] Caselles V, Cham B A and Novaga M 2010 Some remarks on uniqueness and regularity of Cheegcr sets. *Rend. Sem. Mat. Univ. Padova* **123**, pp. 191-201.

[5] Goulnara A 2009 *Limits of Graphs in Group Theory and Computer Science.* 1st Ed.

[6] Polya G and Szego S Isoperimetric Inequalities in Mathematical Physics, Annals of Math. Studies **27**, Princeton University Press, 1951.

[7] Krejcirik D and Pratelli A The Cheeger constant of curved strips, *Pacific J. Math.* 254(2011)2, pp. 309-333.

[8] Parini E An introduction to the Cheeger problem, *Surv. Math. Appl.* 6(2011), pp. 9-21.

[9] Fridman V and Kawohl B Isoperimetric estimates for the first eigenvalue of the p-Laplace operator and the Cheeger constant, Comment. *Math. Univ. Carolin. 44(2003)4*, pp. 659-667.

[10] Bucur D and Fragala I A Faber-Krahn inequality for the Cheeger constant of N-gons, **204**.

[11] Belkin M and Niyogi P 2001 Lapiacian eigen-maps and spectral techniques for embedding and clustering. In *Advanees in Neural Information Processing Systems,* **Vol. 1**, MIT Press, Cambridge, pp. 585-592.

IOP Conf. Series: Journal of Physics: Conf. Series **1127** (2019) 012065 doi:10.1088/1742-6596/1127/1/012065

Numerical Studying of Soliton in the Korteweg-de Vries (KdV) Equation

Lia Yuliawati[1,2], Wono Setya Budhi[1], Didit Adytia[3]
[1]Department of Mathematics, Institut Teknologi Bandung, Indonesia
[2]STKIP Sebelas April Sumedang, Indonesia
[3]School of Computing, Telkom University, Bandung, Indonesia

E-mail: liaylwt@gmail.com

Abstract. In this paper, we use a numerical approach for finding soliton of the Korteweg-de Vries (KdV) equation in infinite dimensionalization. The traveling wave hypothesis is used to extract the analytic soliton solution of KdV equation. Barbera (1993) shown that KdV equation is a solution of Hamiltonian energy optimization in the level set of the momentum. We numerically solve the optimization problem by using steepest descent method and adding the assumption to guarantee the constraint will be fulfilled. From this method, the dynamic system of the problem is obtained and finite difference implementation is used for solving the dynamic system. Based on the method and the hypothesis, the soliton is obtained.

1. Introduction

A soliton is a certain kind of wave which travels with permanent form and with the constant speed in certain material. Solitons can be found in a variety of materials and are used for many different purposes. The first observation of a wave with similar characteristics to those of a soliton was made in 1834 by John Scott Russell [1]. This was the beginning of a whole new field of study to which scientists and mathematicians over the years have contributed extensively. In general, a soliton is obtained as the solution of a nonlinear partial differential equation (PDE). Today there are a number of equations known to have soliton solutions. The most famous of these is the Korteweg-de Vries equation (KdV equation), which is our main focal point.

In this paper, we interest to find the soliton solution of KdV equation by using numerical approach. The KdV equation is a completely integrable Hamiltonian system which can be solved explicitly. Thus, some analytical solutions of it are found, and for appropriate initial conditions, their existence and uniqueness have been shown by Gardner, Grrne, and Kruskal [2]. By using variational approach, Barbera [3] shown that the KdV equation is solution of the Hamiltonian energy optimization in the level set of momentum. Here, we solve this optimization problem by using steepest descent method. The constraint of the problem will be guaranteed by adding the first order differential equation with the exponential decay function as the solution. When the steepest descent method is used, we have the dynamical system with step as a time. We use finite different method to solve it. Some simulation will be shown for solving the Hamiltonian maximization in the level set of momentum and the others for solving the momentum minimization in the level set of Hamiltonian. The speed of convergence and the stability of the solutions will be compared for both problem.

2. Analytic solution of KdV

We know that Korteweg-de Vries equation describe surface water waves above a flat bottom. Suppose that the elevation surface waves denoted by $\eta(\mathbf{x}, t)$, with $\mathbf{x}$ is horizontal space and t is time. KdV equation for surface waves above a flat bottom has the form [4]

$$\partial_t \eta = -c\partial_x \left[\eta + \frac{1}{6}h^2\partial_x^2\eta + \frac{3}{4h}\eta^2 \right], \tag{1}$$

with $c = \sqrt{gh}$, g is gravity and h is the depth.

Let us to consider the special family of solitary waves called soliton. This wave travel with undisturbed shape and at constant speed. Their profile decays at infinity and consists of a single positive hump. Based on Brauer [5], we suppose that the KdV equation has a traveling solution with the form

$$\eta(x, t) = S(x - \lambda t), \tag{2}$$

where the function S represent the shape of soliton and λ is the constant speed. By substituting (2) in (1), we have the following equation

$$-\lambda S' = -c\left[S + \frac{1}{6}h^2 S'' + \frac{3}{4h}S^2 \right]'. \tag{3}$$

Integrating once in (3) and assume that the function S has the zero value at infinity. Then we can rewritten (3) as follow

$$\lambda S = c\left[S + \frac{1}{6}h^2 S'' + \frac{3}{4h}S^2 \right] \tag{4}$$

or

$$\frac{1}{6}ch^2 S'' + (c - \lambda)S + \frac{3c}{4h}S^2 = 0 \tag{5}$$

We can check that this equation has the homoclinic orbit that related with the soliton profile if the value of λ is greater than the value of c.

Let us to consider the explicit expression for soliton profile with the form

$$S(\xi) = \frac{a}{\cosh^2(b\xi)} \tag{6}$$

By substituting this expression in (5), then we have the relation between a, b and λ as follow

$$\lambda = c\left[1 + \frac{a}{2h}\right] \text{ dengan } b = \frac{1}{2h}\sqrt{\frac{3a}{h}}.$$

Therefore, we have the analytic solution for KdV in the following expression

$$\eta(\mathbf{x}, t) = \frac{a}{\cosh^2\left(\sqrt{\frac{3a}{4h^3}}\left(\mathbf{x} - c\left[1 + \frac{a}{2h}\right]t\right)\right)} \tag{7}$$

3. Numerical Solution of KdV

The Hamiltonian of KdV [4] has the following functional

$$\mathcal{H}(\eta) = c\int \left[-\frac{1}{12}h^2(\partial_x\eta)^2 + \frac{1}{2}\eta^2 + \frac{1}{4h}\eta^3 \right] dx \tag{8}$$

and the momentum of KdV [4] as follow

$$\mathcal{M}(\eta) = \frac{1}{2}\int \eta^2 dx \tag{9}$$

The KdV soliton can be expressed as maximization problem of Hamiltonian in the level set of momentum (see [3]).

$$\max_{\eta}\{\mathcal{H}(\eta)|\mathcal{M}(\eta)=\gamma\}, \tag{10}$$

or

$$\min_{\eta}\{\mathcal{M}(\eta)|\mathcal{H}(\eta)=\beta\}. \tag{11}$$

Here, we will solve numerically the problem (10) and (11) by using steepest descent method. First, we will solve the problem (10). By using steepest descent method, we have the dynamic as follow

$$\partial_\tau \eta = \delta_\eta \mathcal{H} - \lambda \delta_\eta \mathcal{M}, \tag{12}$$

where τ is the "time" for iteration. To guarantee that the momentum will converge to the value of β, we assume that the momentum satisfy the exponential decay function or the solution of the first order differential equation in the form [3]

$$\partial_\tau(\mathcal{M}-\gamma) = -(\mathcal{M}-\gamma). \tag{13}$$

By using the chain rule in the left hand side of (13), we can rewrite (13) as follow

$$\langle \delta_\eta \mathcal{M}, \partial_\tau \eta\rangle = -(\mathcal{M}-\gamma). \tag{14}$$

By substituting (12) in (14), we have the expression for λ in the form

$$\lambda = \frac{\langle \delta_\eta \mathcal{M}, \delta_\eta \mathcal{H}\rangle + (\mathcal{M}-\gamma)}{\langle \delta_\eta \mathcal{M}, \delta_\eta \mathcal{M}\rangle}. \tag{15}$$

Thus the problem (10) is equivalent with the problem as follow

$$\partial_\tau \eta = \delta_\eta \mathcal{H} - \frac{\langle \delta_\eta \mathcal{M}, \delta_\eta \mathcal{H}\rangle + (\mathcal{M}-\gamma)}{\langle \delta_\eta \mathcal{M}, \delta_\eta \mathcal{M}\rangle}\delta_\eta \mathcal{M}. \tag{16}$$

Similarly, for the problem (11) is equivalent with

$$\partial_\tau \eta = \delta_\eta \mathcal{M} - \frac{\langle \delta_\eta \mathcal{M}, \delta_\eta \mathcal{H}\rangle - (\mathcal{H}-\beta)}{\langle \delta_\eta \mathcal{H}, \delta_\eta \mathcal{H}\rangle}\delta_\eta \mathcal{H}. \tag{17}$$

Let us to consider for $\eta(\xi) = \frac{a}{\cosh^2(b\xi)}$. We can explicitly find the expression of momentum in the form

$$\mathcal{M}(\eta) = \frac{1}{2}\int \left[\frac{a}{\cosh^2(b\xi)}\right]^2 d\xi = \frac{2a^2}{3b}.$$

For the momentum given, we can found the analytic solution $\eta(\xi) = \frac{a}{\cosh^2(b\xi)}$ with

$$a = \frac{3(\gamma/4)^{2/3}}{h} \text{ and } b = \frac{\sqrt{3a}}{2h\sqrt{h}}$$

Suppose we have the initial value of the problem (16) is $\eta_0(x) = A\exp(-B(x-\mu)^2)$ with $A = 2.8, B = 0.01$, and $\mu = 0$. For $0 \le \tau \le 60$, we can see the snapshots of the solution in Figure 2. In this figure, we consider that the numeric solution is close to the analytic solution. On other case, we use the same initial value, but for the problem (17). Then the solution is close to analytic solution.

IOP Conf. Series: Journal of Physics: Conf. Series **1127** (2019) 012065 doi:10.1088/1742-6596/1127/1/012065

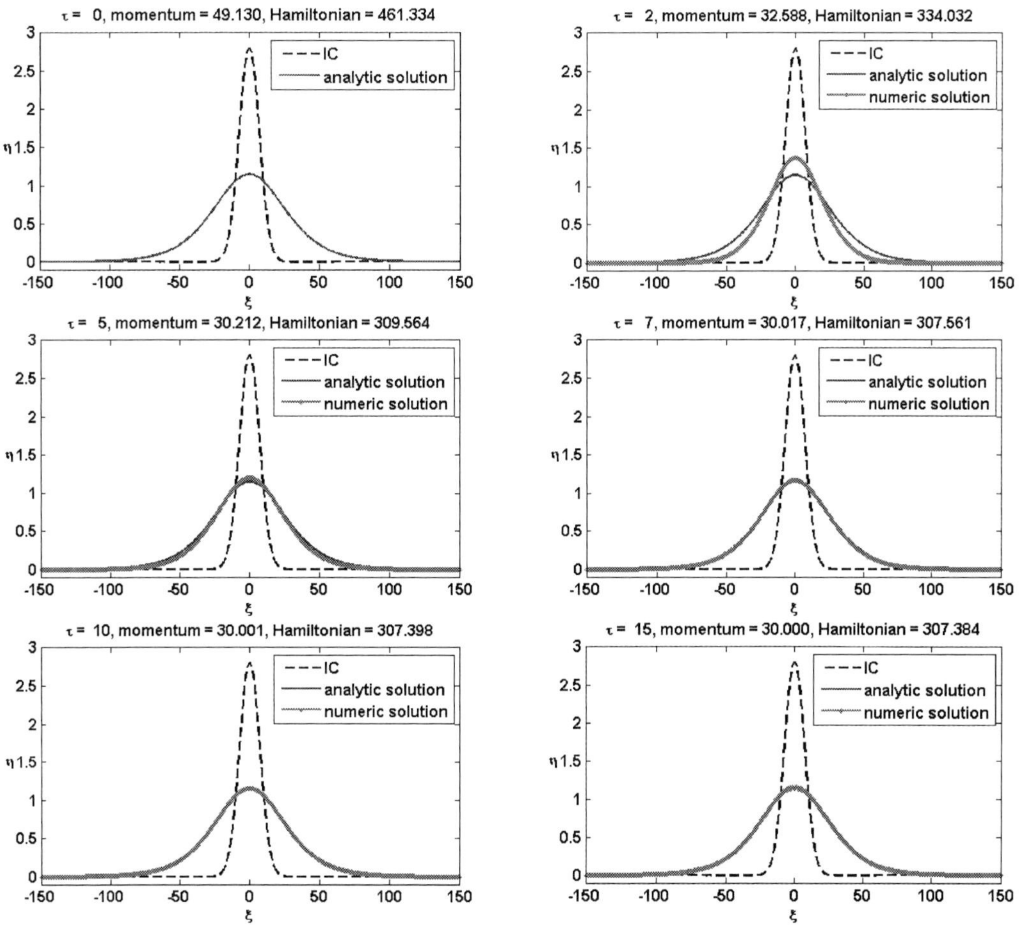

Figure 1. Numerical solution of the problem $\max_{\eta}\{\mathcal{H}(\eta)|\mathcal{M}(\eta) = 30\}$ with the initial value $\eta_0(x) = A\exp(-B(x-\mu)^2)$ and the values $A = 2.8, B = 0.01, \mu = 0$.

4. Conclusion

The KdV equation can be represented as optimization problem of Hamiltonian in the level set of momentum. In case maximization of Hamiltonian in the level set of the momentum, we add the assumption that the momentum is the exponential decay function. Therefore, for minimization of momentum in the level set of the Hamiltonian, the assumption is that the Hamiltonian is the exponential decay function. By using steepest descent method, and using finite difference method in both of the problem, we have the result that the solutions is converge to analytic soliton solution of KdV.

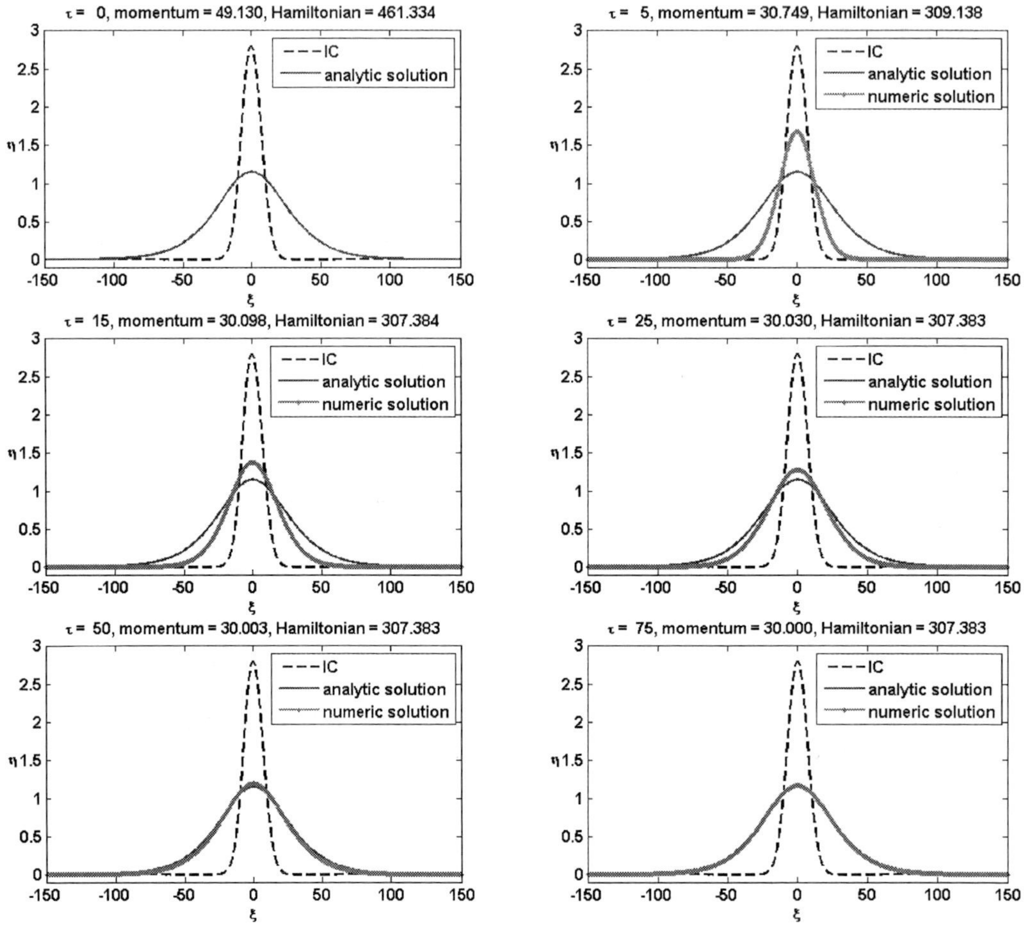

Figure 2. Numerical solution of the problem $\min_\eta\{\mathcal{M}(\eta)|\mathcal{H}(\eta) = 307.383\}$ with the initial value $\eta_0(x) = A\exp(-B(x-\mu)^2)$ and the values $A = 2.8, B = 0.01, \mu = 0$.

5. Acknowledgments

The authors wish to thank Prof. E. van Groesen from LabMath-Indonesia for giving the problem. We also thank to Peri Turnip for helping and discussing in LabMath-Indonesia. Moreover the first author feels very proud for the internship in LabMath-Indonesia. The research of the first author is supported by Research Grant "Riset KK Analisis dan Geometri ITB" and PuPT.

References

[1] Anna Bcklund and Daniel Weston, Analytical and Numerical Study of Soliton Collisions, SA104X Degree Project in Engineering Physics, First Level Department of Theoretical Physics Royal Institute of Technology (KTH), 2010.

[2] Gardner, C. S., and Marikawa, G. K. (1965). The effect of temperature of the width of a small amplitude solitary wave in a collision free plasma. Communications on Pure and Applied Mathematics, 18, 3549.

[3] Barbera W. van de Fliert, Variational Formulation for Coherent Vortex Structures. Dissertation, University of Twente, 1993.

[4] E. van Groesen and Jaap Molenaar, Continuum Modeling in the Physical Sciences, SIAM 2007.

[5] Klaus Brauer, The Korteweg-de Vries Equation: History, exact Solutions, and graphical Representation, University of Osnabrck/Germany, revision in 2014.

Spectral method for vorticity-streamfunction equations, with application to Rayleigh-Benard convection

M A Ginting and S R Pudjaprasetya

Industrial & Financial Mathematics Research Group, Faculty of Mathematics and Natural Sciences, Institut Teknologi Bandung, Jalan Ganesha 10 Bandung, 40514 Indonesia

E-mail: mariaaginting@gmail.com

Abstract. The spectral method with an explicit time integration scheme is implemented to solve the vorticity-streamfunction equations. In spectral method, functions are approximated with finite sums of periodic sine and cosine functions, which makes this method suitable for periodic problems. Numerical experiments for several cases are discussed. In Rayleigh-Benard convection problem, a horizontal layer of incompressible fluid is kept within high and low temperatures along the bottom and top boundaries, respectively. Triggered by a periodic initial condition, we simulate the occurrence of a regular pattern of convection cells, known as Benard cells.

Keywords: spectral method, vorticity-streamfunction equation, Rayleigh-Benard convection

1. Introduction

Transfer of heat in a medium can occur by means of three possible mechanisms, conduction, convection and radiation. Convective heat transfer is the transfer of heat from one place to another by the movement of fluid molecules due to temperature difference in the fluid. If one region of the fluid is heated, it will become less dense compare to its surroundings, and due to buoyancy, give rise to transport of fluid molecules. This is the underlying physical mechanism of heat convection process. Rayleigh-Benard's convection is a simple convection system that occurs in a layer of fluid heated from below (Hepworth 2014 [6]). In this phenomenon, the fluid develops a regular pattern of convection, known as Benard cells.

In Rayleigh-Benard convection, heat transfer in a fluid between two rigid plates with different fixed temperature is observed. This problem was firstly initiated by Henri Benard in 1900. He made an experiment on the instability of fluid in a thin layer that was heated from below involving the surface tension and thermo-capillary convection (Sandberg 2011 [7]). In 1916 Lord Rayleigh [9], who was inspired by Benard's experiment, derived the theoritical demands for temperature gradient convection. He predicted that instability occurs when the temperature gradient is large enough, and that the onset of the convection is governed by Rayleigh number, R_a, that exceed a certain critical value. Further work in this area is recorded by Chandrasekhar 1961 [5], which is also be the foremost text on linear theory of Rayleigh-Benard convection.

In Peyret 2002 [3], the Rayleigh-Benard problem was solved by using Fourier - Chebyshev approximation. In this paper, we simulate the similar phenomenon using the spectral method. In this method, the solution is represented as a truncated Fourier series expansion, therefore this method is appropriate for periodic problem. Derivation of the governing equations for Rayleigh-Benard convection is presented in Sections 2 and 3. Then we describe the numerical implementation and the results of linear part in Sections 4 and 5.

2. Vorticity-streamfunction equations

Vorticity-streamfunction equations are derived from the Navier-Stokes equations for incompressible flow [4]. It consists of the conservation of momentum and mass

$$\partial_t \mathbf{V} + \mathbf{V}\cdot\nabla\mathbf{V} + \nabla p - \nu\nabla^2\mathbf{V} + \frac{\mathbf{V}}{\tau} = \mathbf{f} \tag{1}$$

$$\nabla\cdot\mathbf{V} = 0. \tag{2}$$

Let $\mathbf{V} = (u, v)$ describes the velocity vector of the fluid, p as the pressure (divided by constant density ρ), ν as the kinematic viscosity, τ as the time constant for linear frictional damping in the 2D flow, and $\mathbf{f}$ as a forcing term. To obtain the vorticity-streamfunction equations, we apply curl operator on the momentum equation so that we have

$$\partial_t \omega + \mathbf{V}\cdot\nabla\omega - \nu\nabla^2\omega + \frac{\omega}{\tau} = F \tag{3}$$

where ω is the vorticity defined as $\omega \equiv |\nabla\times\mathbf{V}| = (|\partial_x v - \partial_y u|)\mathbf{k}$ with $F = |\nabla\times\mathbf{fk}|$ is the forcing term, and $\mathbf{k}$ the unit vector normal to the (x, y) plane of the flow. Vorticity is related to the interaction of local swirls or eddies in the fluid.

Considering the conservation of mass in equation (2), there exist a scalar function $\Psi(x, y)$ which is called streamfunction so that the velocity vector can be written as $\mathbf{V} = \nabla\times(\psi\mathbf{k})$. Hence, the conservation of mass can be satisfied automatically since div curl is a zero operator. So the fluid velocity and the vorticity are related to the stream function by

$$u = \partial_y\psi, \qquad v = -\partial_x\psi, \qquad \nabla^2\psi + \omega = 0. \tag{4}$$

3. Rayleigh-Benard convection problem

Rayleigh-Benard is a thermal convection problem that occurs when a layer of fluid is heated from below. As response to the non-uniform temperature distribution, the fluid develops a regular pattern of convection known as Benard cells. To accommodate the non-uniform temperature, the following energy conservation equation is taken

$$\partial_t T + \mathbf{V}\cdot\nabla T - \kappa_T\nabla^2 T = 0, \tag{5}$$

with T is the fluid temperature, and κ_T thermal diffusivity. Fluid density depends on temperature according to $\rho = \rho_0(1 - \alpha_T(T - T_0))$ with ρ_0 is the reference fluid density associated with the reference temperature T_0, and α_T is the thermal expansion coefficient. Here, fluid motion is affected by the buoyant force $\mathbf{f} = \frac{\rho}{\rho_0}\mathbf{g}$, with $\mathbf{g}$ the gravitational vector field. Therefore, the forcing term in the vorticity equation is $F = |\nabla\times\mathbf{f}| = g\alpha_T\partial_x T$. If we neglect the frictional damping, the governing equations written in full are as follows

$$\partial_t\omega + \mathbf{V}\cdot\nabla\omega - \nu\nabla^2\omega = g\alpha_T\partial_x T, \tag{6}$$

$$\nabla^2\psi + \omega = 0, \tag{7}$$

$$\partial_t T + \mathbf{V}\cdot\nabla T - \kappa_T\nabla^2 T = 0, \tag{8}$$

In this problem, three independent variables are involved, those are vorticity ω, stream function ψ, and temperature T, all of them are functions of dependent variables x, y, t. The governing equations (6-8) hold in a 2D domain with $-H/2 < y < 0$ and $0 < x < L$, with H and L are typical height and length, respectively. The Rayleigh-Benard problem takes place in a long channel, and therefore periodicity in x-direction can be assumed. Further, there is no fluid flow across the upper and lower boundary, which leads to the following boundary conditions

$$\begin{array}{llll} \psi = 0, & \partial_y \psi = 0 & \text{along} & y = -\frac{H}{2} \\ \psi = 0, & \partial_y \psi = 0 & \text{along} & y = 0. \end{array} \tag{9}$$

The upper and lower temperatures are kept constant, to yield the following boundary conditions

$$\begin{array}{lll} T = T_1 & \text{at} & y = -\frac{H}{2}, \\ T = T_2 & \text{at} & y = 0, \end{array} \tag{10}$$

with $T_1 - T_2$ is positive, since in this case the fluid is heated from below.

3.1. Normalized variables

Our computations will be conducted to the normalized equations, and therefore we first introduce the following normalized variables

$$X = x\frac{2\pi}{L}, \quad Y = y\frac{2\pi}{H}, \quad \tilde{t} = t/t_0, \text{ with } t_0 = \frac{H^2}{4\pi^2 \kappa_T}, \tag{11}$$

$$\mathbf{W} = \mathbf{V}\left(\frac{2\pi}{H}\right) t_0, \quad \Omega = \omega t_o, \quad \Psi = \psi/\kappa_T, \quad \theta = \frac{T - T_2}{T_1 - T_2}. \tag{12}$$

The governing equations (6)-(8) written in dimensionless variables are

$$\partial_{\tilde{t}} \theta + \mathbf{W} \cdot \nabla \theta - \nabla^2 \theta = 0, \tag{13}$$

$$\partial_{\tilde{t}} \Omega + \mathbf{W} \cdot \nabla \Omega - P_r \nabla^2 \Omega = \frac{a}{8\pi^3} P_r R_a \partial_X \theta, \tag{14}$$

$$\nabla^2 \Psi + \Omega = 0. \tag{15}$$

The above equations hold in a domain $0 < X < 2\pi$ and $-\pi < Y < 0$. The boundary conditions (9),(10) are now read

$$\begin{array}{llll} \theta = 1, & \Psi = 0, & \partial_Y \Psi = 0, & \text{along } Y = -\pi, \\ \theta = 0, & \Psi = 0, & \partial_Y \Psi = 0, & \text{along } Y = 0. \end{array} \tag{16}$$

The velocity vector $\mathbf{W} = (U, V)$ is related to the stream function by

$$U = \partial_Y \Psi, \; V = -a \partial_X \Psi, \tag{17}$$

with $a = H/L$. In the above equations, gradient, Laplace operator, and the vorticity in the transformed variables are

$$\nabla = (a\partial_X, \partial_Y), \; \nabla^2 = a^2 \partial_{XX} + \partial_{YY}, \; \Omega = a\partial_X V - \partial_Y U. \tag{18}$$

We note that in the normalized formulation, there is only two non-dimensional numbers, those are the Prandtl number P_r and Rayleigh number R_a defined as

$$P_r = \frac{\nu}{\kappa_T}, \; R_a = \frac{g\alpha_T (T_1 - T_2) H^3}{\kappa_T \nu}. \tag{19}$$

In the next section, these normalized equations will be solved numerically using spectral method, and in this article we restrict to the linear model.

4. Implementation

The Rayleigh-Benard problem that we consider has Dirichlet boundary conditions of different temperatures along the top and bottom boundaries. In order to implement the spectral method, the problem should be formulated in periodic form, so here we extend the $y-$interval to $[-\pi, \pi]$. Hence, equations (13)-(15) is solved in a domain $0 < X < 2\pi$ and $-\pi < Y < \pi$ with boundary conditions

$$\begin{array}{llll} \theta = 1, & \Psi = 0, & \partial_Y \Psi = 0, & \text{along } Y = \pm\pi, \\ \theta = 0, & \Psi = 0, & \partial_Y \Psi = 0, & \text{along } Y = 0. \end{array} \tag{20}$$

Consider the linear part of equation (13) and (14)

$$\partial_{\bar{t}}\theta - \nabla^2\theta = 0, \tag{21}$$

$$\partial_{\bar{t}}\Omega - P_r\nabla^2\Omega = \frac{a}{8\pi^3}P_r R_a \partial_X \theta. \tag{22}$$

Apply Fourier transform to the above equations to get the following ordinary differential equations

$$\partial_{\bar{t}}\hat{\theta}_{k,l} = -\left(a^2k^2 + l^2\right)\hat{\theta}_{k,l}, \tag{23}$$

$$\partial_{\bar{t}}\hat{\Omega}_{k,l} = -P_r\left(a^2k^2 + l^2\right)\hat{\Omega}_{k,l} + \frac{a}{8\pi^3}P_r R_a ik\hat{\theta}_{k,l}. \tag{24}$$

The explicit time integration of Crank-Nicholson scheme is implemented to yield

$$\hat{\theta}^{n+1}_{k,l} = \frac{1 - \frac{\Delta t}{2}\left(a^2k^2 + l^2\right)}{1 + \frac{\Delta t}{2}\left(a^2k^2 + l^2\right)}\hat{\theta}^n_{k,l}, \tag{25}$$

$$\hat{\Omega}^{n+1}_{k,l} = \frac{\left(1 - \frac{\Delta t}{2}P_r\left(a^2k^2 + l^2\right)\right)\hat{\Omega}^n_{k,l} + \frac{\Delta t}{16\pi^3}aP_r R_a ik\left(\hat{\theta}^{n+1}_{k,l} + \hat{\theta}^n_{k,l}\right)}{1 + \frac{\Delta t}{2}P_r\left(a^2k^2 + l^2\right)}. \tag{26}$$

Streamfunction can be computed by applying Fourier transform to equation (15) so that $\hat{\Psi} = \hat{\Omega}/(a^2k^2 + l^2)$. The streamfunction is needed to calculate the velocity $\mathbf{W}$ and the nonlinear terms that is not investigated in this paper.

5. Results

In this section the numerical results for the linear part of Rayleigh-Benard convection problem is presented. There are three cases of simulation with different initial conditions. For the first one, we set the initial condition for temperature $\theta(X, Y, 0) = \cos(X)$ and vorticity $\Omega(X, Y, 0) = 0$. We use $P_r = 0.71$, $R_a = 6 \times 10^3$, and $\Delta t = 10^{-3}$ as the parameters. The simulation shows that the temperature becomes $\theta = |Y|/\pi$ which is the steady state solution, and the vorticity develops a regular pattern of convection cells known as Benard cells. These results are shown in Figure 1 and 2.

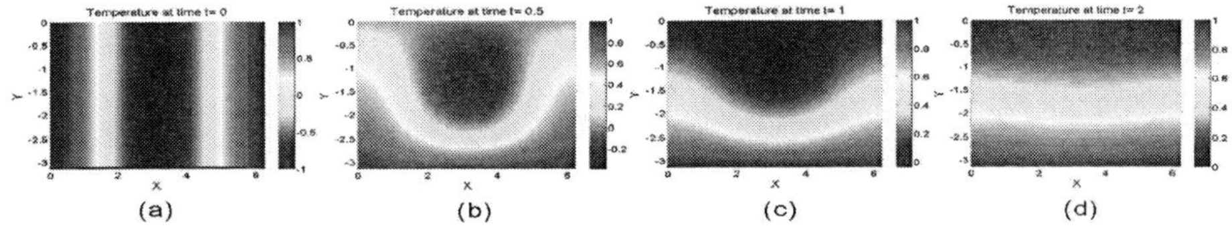

Figure 1. Temperature with initial condition $\theta(X, Y, 0) = \cos(X)$ and $\Omega(X, Y, 0) = 0$ (a) at $t = 0$ (b) at $t = 0.5$ (c) at $t = 1$ (d) at $t = 2$.

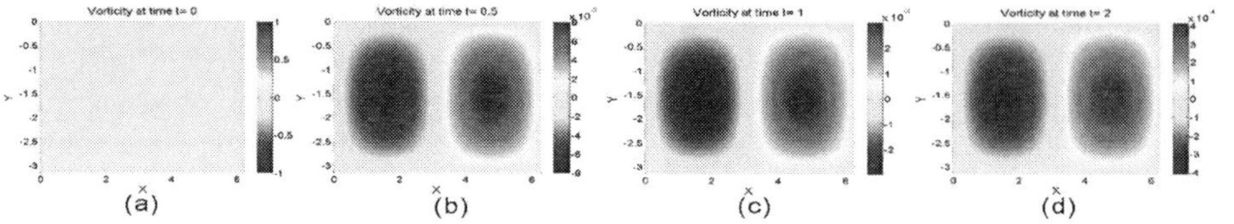

Figure 2. Vorticity with initial condition $\theta(X, Y, 0) = \cos(X)$ and $\Omega(X, Y, 0) = 0$ (a) at $t = 0$ (b) at $t = 0.5$ (c) at $t = 1$ (d) at $t = 2$.

In the next case, we set the initial condition of the temperature $\theta(X, Y, 0) = \cos(4x)$ and the vorticity $\Omega(X, Y, 0) = 0$. Figure 3 and 4 present the result of this case. The temperature has the same behavior as the previous case, which becomes the steady solution. And the regular pattern of convection cells appear in the vorticity field.

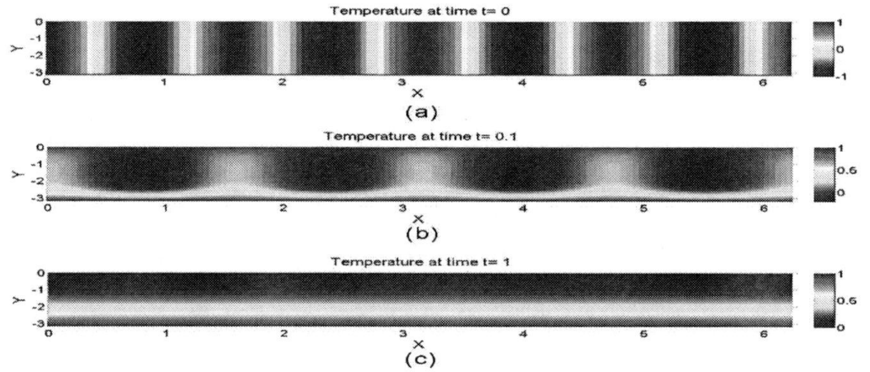

Figure 3. Temperature with initial condition $\theta(X, Y, 0) = \cos(4X)$ and $\Omega(X, Y, 0) = 0$ (a) at $t = 0$ (b) at $t = 0.1$ (c) at $t = 1$.

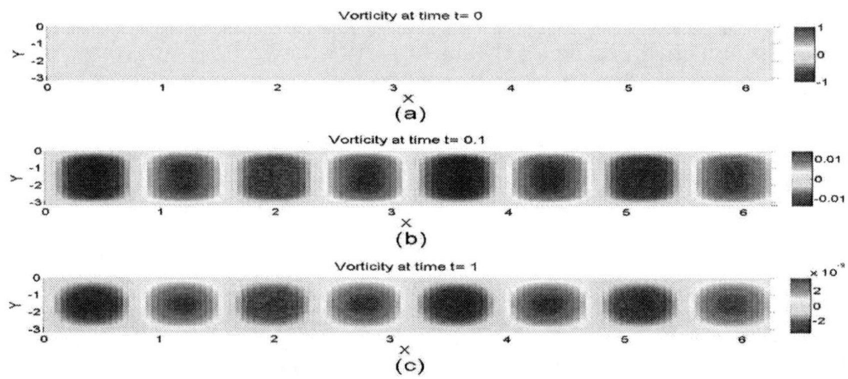

Figure 4. Vorticity with initial condition $\theta(X, Y, 0) = \cos(4X)$ and $\Omega(X, Y, 0) = 0$ (a) at $t = 0$ (b) at $t = 0.1$ (c) at $t = 1$.

In the previous simulations, the spectral method has been succesfully applied to simulate the Rayleigh-Benard convection problems. It was shown that due to high temperature as bottom boundary, fluid temperature will eventually change to a steady temperature, whereas the vorticity develop Benard cells.

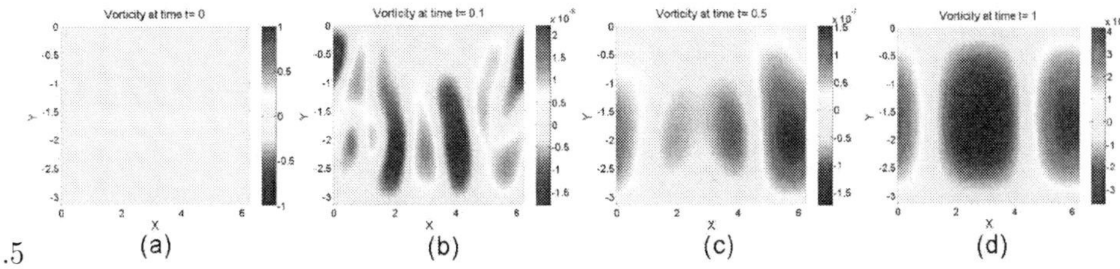

Figure 5. Vorticity with initial condition $\theta(X,Y,0) = |Y|/\pi + 10^{-3}$ and $\Omega(X,Y,0) = 0$ (a) at $t = 0$ (b) at $t = 0.1$ (c) at $t = 0.5$ (d) at $t = 1$.

Finally, in the last simulation, we start with an initial temperature, which is a perturbation of the steady temperature. To be explicit, we set the initial condition for the temperature $\theta(X,Y,0) = |Y|/\pi + 10^{-3}$ and vorticity $\Omega(X,Y,0) = 0$, whereas parameters are the same as the previous simulations. Result for the vorticity is shown in Figure 5. At the early stage of simulation, an irregular pattern of convection cells appear, but then as time progresses, it developes into a regular pattern. In this case, we also have fluid temperature that tends to the steady temperature $\theta = |Y|/\pi$. But the figure is not given here because the perturbation is too small to be visible in temperature plots.

6. Conclusions

The spectral method with an explicit time integration scheme has been formulated to solve the vorticity stream function equations. For the linearized model, the resulting scheme has been used to simulate various thermal convection problems. Steady state solution was simulated, also the development of Benard cells in a layer of horizontal fluid heated from below. Moreover, simulation result shows that perturbation of steady temperature also induces a regular pattern.

7. Acknowledgement

The authors acknowledge the financial support of research grant Riset KK ITB 2017.

References

[1] Cushman-Roisin B and Beckers J-M 2007 *Introduction to Geophysical Gluid Dynamics Physical and Numerical Aspects* (Academic Press)
[2] Mattheij R M M, Rienstra S W and Boonkkamp J H M 2005 *Partial Differential Equations* (Philadelphia: Society for Industrial and Applied Mathematics)
[3] Peyret R 2002 *Spectral Methods for Incompressible Viscous Flow* (New York: Springer)
[4] Storey B D 2012 *Hands-on Research: Modeling 2D Turbulence*
[5] Chandrasekhar S 1961 *Hydrodynamic and Hydromagnetic Stability* (Oxford: Clarendon Press)
[6] Hepworth B J 2014 *Nonlinear Two-dimensional Rayleigh-Benard Convection*
[7] Sandberg M, Berg N and Johnsonn G 2011 *Rayleigh-Benard Convection*
[8] Gelfgat A Y 1999 Different Modes of Rayleigh-Benard Instability in Two- and Three-Dimensional Rectangular Enclosures *Journal of Computational Physics* **156** 300324
[9] Rayleigh L 1916 On convection currents in a horizontal layer of fluid, when the higher temperature is on the under side *Philosophical Magazine* **32** 529546
[10] Majda A J and Bertozzi A L 2002 *Vorticity and Incompressible Flow* (Cambridge: Cambridge University Press)

IOP Conf. Series: Journal of Physics: Conf. Series **1127** (2019) 012067 doi:10.1088/1742-6596/1127/1/012067

Dissipativity of Nonlinear Networked Control Systems Modeled by Markovian Jump System

A Najmurrokhman[1], A S Rohman[2] and P H Rusmin[2]

[1]Dept. of Electrical Engineering, Universitas Jenderal Achmad Yani, Cimahi 40533, Indonesia
[2]School of Electrical Engineering and Informatics, Institut Teknologi Bandung, Bandung 40132, Indonesia

E-mail: asepnajmu@yahoo.com

Abstract. Networked control system (NCS) is a control system in which the control loop is closed by a shared communication networked. There are some parameters that arise when using a network in an NCS such as packet dropout, time delay, limited bandwidth, and so forth. Due to the stochastic property of such parameters, the entire system under consideration realizes the Markovian jump system. This paper considers a dissipativity notion for a class of nonlinear networked control systems. Dissipativity refers to the existence of a supply rate function dealing with the system such that the system has a dissipative property. The main result in this paper is a solvability condition for achieving dissipativity of a class of nonlinear networked control systems modeled by Markovian jump system. The solvability condition is shown by the nonlinear matrix inequalities (NLMI).

1. Introduction

A networked control system (NCS) is a control system in which a control loop is closed via a shared communication network [1]. Compared to the conventional point-to-point system connection, the use of an NCS has advantages of reduced system wiring, ease of system diagnosis and maintenance, and increased system agility [2]. Since sensors, actuators, and controllers are connected via a shared band-limited communication network, there are some parameters which appear such as network-induced delays, packet dropout, bandwidth constraints, etc. The existence of these parameters will degrade performance of NCS or even cause instability [3]. Due to the widespread of the shared networks, NCS has attracted many researchers especially for large systems with distributed structure such as the distributed energy system, traffic control systems, water distribution in a city area, irrigation systems, etc. Though the theoretical foundation of NCS has been improved considerably during the last decade, many issues should be resolved before the all the advantages of NCS can be harvested.

A typical NCS is shown in Fig 1. In that illustration, a personal computer equipped with the data acquisition card enables the plant to send out the sensor data and receive the control data through the Ethernet LAN. A dedicated controller receives the sensor data, computes the control law for a certain purpose, and then sends out the control data through the same LAN. Because of the existence of the network parameter, a designed controller should be capable of overcoming that parameter. To cope with such situation, the network parameter has to be explicitly incorporated into a controller design using a certain model, such as Markov process, so that the system under consideration renders a Markovian jump system.

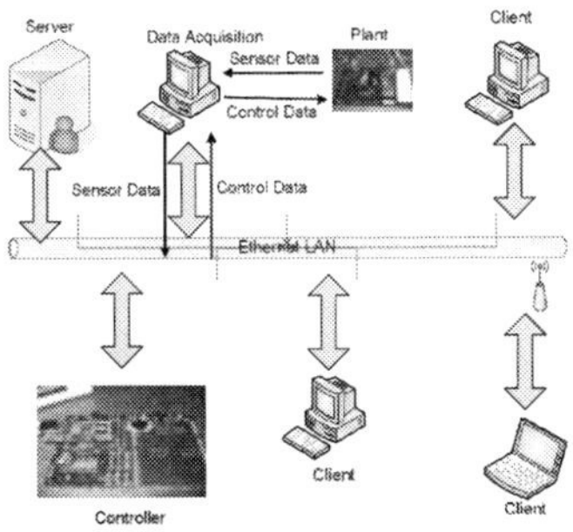

Fig. 1 A typical NCS

The Markovian jump systems (MJS) has attracted increasing attention in the recent literature, see e.g [4-9] for linear systems and [10-11] for nonlinear systems. Such systems are those having transition between models determined by a Markov chain. MJS is considered to be appropriate to model plants whose structures are subject to random abrupt changes due to component failures or repairs, sudden environmental changes, abrupt variations of the operating point of a nonlinear plant, changing subsystem interconnections, and so on [12]. There is a quite deep literature for a class of this system [5] and previous results relevant to this class of system in NCS modeling was reported in [4,6-8,10-13] for name a few. Whenever employing MJS in NCS modeling, the transition probabilities in the jumping process determine the overall system behavior [9-10].

In this paper, we address the dissipativity condition in terms of the nonlinear matrix inequalities (NLMI) in analysis of a class of nonlinear networked control system modeled by Markovian jump system. The dissipative concept used to analyze and design the control system was initially developed by Willems [14]. This concept concerns the analysis and design of control systems that use input-output properties based on the energy-related description [15]. Practically speaking, a system has a dissipative property if it always dissipates energy. When the dissipativity is assured, the stability problem can be solved. Meanwhile, the dissipative performance is a generalization of the performance measure, such as finite gain (H_∞) and passivity [14]. Thus, the development of a dissipative framework will generalize the existing ones, including H_∞ control as well as passivity-based design.

Using a mathematical abstraction of the notions of physical power and energy, researchers have developed a stability analysis and designed a controller for various applications in dissipative system frameworks. Recent studies concerning the dissipative approach that are in line with our interest are reported in [16-20]. Aliyu has proposed a dissipativity analysis for the nonlinear Markovian jump system [16]. Zhang *et al.* have considered the dissipative problem of a class of stochastic hybrid systems and focused on the analysis of whether a stochastic hybrid system with time delay is stochastically asymptotically stable and strictly dissipative [17]. Mahmoud *et al.* have worked on a robust dissipative control problem applied to a class of hybrid multi-rate control models with time-delay and a switching controller [18]. Zhang *et al.* derived a linear state feedback controller in a reliable dissipative control problem for a class of stochastic hybrid systems [19]. Wang *et al.* addressed the dissipative problem for uncertain time-delay NCS [20]. The aforementioned papers consider the linear model of the networked control systems under consideration. So far, to the best of our knowledge, a dissipative control design for a nonlinear NCS modeled by the MJS has not been investigated yet. Therefore, the development of such a framework would extend the existing linear ones.

For convenience, we adopt the following notations in this paper. Capital letters denote matrices and small letters denote vectors. For real matrices or vectors, $(.)^T$ indicates their transpose. For a real square matrix P, we write $P > 0$ (resp., $P < 0$) if $P = P^T$ is symmetric and positive (resp. negative) definite. Matrices and vectors, if their dimensions are not explicitly stated, are assumed to be compatible for algebraic operations. The symbol (*) within a matrix represents the symmetric term of the matrix. $\Re^n$ denotes the space of all n-dimensional real vectors. The symbol $E[X]$ denotes the mathematical expectation of X. 0 denotes the zero scalar as well as zero matrices with appropriate dimensions. $Pr(x)$ means the probability of a random variable x. The set of integer numbers is denoted by $\mathbb{Z}$, i.e. $\mathbb{Z} = \{1, \ldots, N\}$.

2. Preliminaries

Dissipativity theory of nonlinear systems has become one of the major approaches to the study of complex systems [21]. Based on the consideration of abstract energy that may have certain physical meaning for a particular system, dissipativity has a meaning that the increase in energy storage in the systems is no more than the supplied energy from outside the system. Dissipativity is characterized by storage functions and supply rates, which represent the energy stored inside the system and energy supplied from outside the system, respectively. By using the notion of energy used in the system, dissipative systems means that the amount of energy that can be extracted from the system is equal to the energy supplied to the system with or without energy dissipation occurs.

Brogliato *et al.* summarizes dissipativity notion for discrete system in the definition as follows [15]:

Definition 1. Consider the following class of discrete system in the form of affine representation

$$\Sigma_d \begin{cases} x(k+1) = f(x(k)) + g(x(k))u(k) \\ y(k) = h(x(k)) + l(x(k))u(k) \end{cases} \tag{1}$$

where $x(k)$ denote state variable of the system, $u(k)$ is control input, and $y(k)$ is an output. The functions f, g, h, and l are smooth mappings. The system Σ_d *is said to be dissipative* with respect to the *supply rate* $W(u,y)$ if for all admissible $u(k)$ there exists a *storage function* $V(x) \geq 0$ such that the following dissipation inequality holds

$$V(x(k+1)) - V(x(k)) \leq W(y(k), u(k)); \forall u \in \mathbf{U}, \forall y \in \mathbf{Y}, \forall k \in \mathbb{Z} \tag{2}$$

Inequality (2) could be said a step-wise dissipativity. By monotonicity assumption, we can get easily the following dissipation inequality

$$V(x(k+1)) - V(x(0)) \leq \sum_{i=0}^{k} W(y(i), u(i))\,; \forall u \in \mathbf{U}, \forall y \in \mathbf{Y}, \forall i, k \in \mathbb{Z} \tag{3}$$

Gupta in [22] classifies two notions of dissipativity, i.e dissipative and strictly dissipative based on the inequality sign in (2) or (3) respectively. A system is called dissipative if the inequality (2) or (3) satisfy. Meanwhile, the system is strictly dissipative if the mathematical operators in inequality (2) or (3) turns into the strict inequality.

Now, consider the Markovian jump nonlinear systems in the following discrete-time form

$$\begin{aligned} x(k+1) &= f(x(k), r(k)) + g(x(k), r(k))u(k) \\ y(k) &= h(x(k), r(k)) + l((x(k), r(k))u(k) \end{aligned} \tag{4}$$

where $x(k) \in \mathbf{X} \subset \Re^n$ is state vector with initial condition x_0, $u(k) \in \mathbf{U} \subset \Re^m$ and $y(k) \in \mathbf{Y} \subset \Re^p$ are system's input and output respectively. The functions f, g, h, and l are the smooth and time-varying nonlinear mappings in the form of $f: \mathbf{X} \times \mathbb{Z} \rightarrow \Re^n$, $g: \mathbf{X} \times \mathbb{Z} \rightarrow \Re^{n \times m}$, $h: \mathbf{X} \times \mathbb{Z} \rightarrow \Re^p$, and $l: \mathbf{X} \times \mathbb{Z} \rightarrow \Re^{p \times m}$. All the mappings are functions of a discrete-time Markov chain taking values in a finite set $\mathbb{Z} = \{1, \ldots, N\}$. The Markov chain has transition probabilities $p_{ij} = Pr(r(k+1) = j | r(k) = i)$ which satisfy $p_{ij} > 0$ and $\sum_{j=1}^{N} p_{ij} = 1$ for each $i \in \mathbb{Z}$. The initial conditions are given by specifying $r(0)$ and $x(0)$. Furthermore, dissipativity corresponds to the existence of a supply rate and a storage function as stated in the following definitions.

Definition 2. Consider a function $W : \Re^m \times \Re^p \to \Re$ associated with a system of the form (4). This function is called as the *supply rate*, if for any $u \in \mathbf{U} \subset \Re^m$ and $y \in \mathbf{Y} \subset \Re^p$, the system has the following property

$$E\left[\sum_{k=0}^{T} \left|W\left(y(k), u(k)\right)\right|\right] < \infty, T = 0,1,\cdots \quad (5)$$

Definition 3. A system of the form (4) with supply rate $W(y(k),u(k))$ is *said to be locally dissipative in the stochastic sense*, if there exists a nonnegative continuous function $V(x(k),r(k)) : \mathbf{X} \times \mathbb{Z} \to \Re^+$, called the storage function, such that for all $k \geq 0$

$$E\left[V\left(x(k+1), r(k+1)\right) - V\left(x(k), r(k)\right)\right] < E\left[\sum_{k=0}^{T} \left|W\left(y(k), u(k)\right)\right|\right], \forall x \in \mathbf{X}, \forall u \in \mathbf{U} \quad (6)$$

Concerning to the supply rate, if the form of a supply rate is as the followings

$$W\left(y(k), u(k)\right) = \begin{pmatrix} y(k) \\ u(k) \end{pmatrix}^T \begin{pmatrix} Q & S \\ S^T & R \end{pmatrix} \begin{pmatrix} y(k) \\ u(k) \end{pmatrix} \quad (7)$$

where Q, S, and R are suitable dimension with $y(k)$ and $u(k)$, then system (4) is *said to be locally* QSR-*disipative* in the stochastic sense.

3. Main results

Proposition 1 below characterizes a necessary and sufficient condition for Markovian jump nonlinear system as in (4) to be locally stochastis QSR-dissipative.

Proposition 1. A system of the form (4) is said to be locally stochastic QSR-dissipative if and only if there exists a nonnegative storage function $V(x(k),r(k)) : \mathbf{X} \times \mathbb{Z} \to \Re^+$ with $V(\mathbf{0},r(k)) = 0$, the supply rate (7), and two real functions $\rho(x(k),r(k)) : \mathbf{X} \times \mathbb{Z} \to \Re^+$, $\varepsilon(x(k),r(k)) : \mathbf{X} \times \mathbb{Z} \to \Re^+$, such that three conditions below hold for all $i, j \in \mathbb{Z}$

$$\sum_{j=1}^{N} p_{ij} V\left(f\left(x(k), i\right), j\right) - V\left(x(k), i\right) = h^T\left(x(k), i\right) Q h\left(x(k), i\right) - \rho^T\left(x(k), i\right) \rho\left(x(k), i\right) \quad (8)$$

$$\frac{1}{2}\sum_{j=1}^{N} p_{ij} \left.\frac{\partial V(z, j)}{\partial z}\right|_{z=f(x(k),i)} g\left(x(k), i\right) = h^T\left(x(k), i\right) Q l\left(x(k), i\right) + h^T\left(x(k), i\right) S - \rho^T\left(x(k), i\right) \varepsilon\left(x(k), i\right) \quad (9)$$

$$\frac{1}{2}\sum_{j=1}^{N} p_{ij} g^T\left(x(k), i\right) \left.\frac{\partial^2 V(z, j)}{\partial z^2}\right|_{z=f(x(k),i)} g\left(x(k), i\right) = l^T\left(x(k), i\right) Q l\left(x(k), i\right) + 2 l^T\left(x(k), i\right) S + R - \varepsilon^T\left(x(k), i\right) \varepsilon\left(x(k), i\right) \quad (10)$$

Proof: See [23].

Proposition 1 states the necessary and sufficient conditions for the system represented in Equation (4) to be stochastically dissipative with respect to the supply rate (7). Based on that proposition, dissipativity is characterized by three elements, namely the existence of a positive definite function that can twice differentiable property, the real function $\rho(x(k),r(k))$, and the real function $\varepsilon(x(k),r(k))$. As reported by Wang *et al.*, there is no specific method to determine the real function in order to conclude dissipativity of the underlying system [23]. However, if a class of the nonlinear form are affine, then the dependence on the two real functions can be eliminated.

To be precise, consider the following affine Markovian jump nonlinear system

$$\begin{aligned} x(k+1) &= A(x(k),r(k))x(k) + B(x(k),r(k))u(k) \\ y(k) &= C(x(k),r(k))x(k) + D(x(k),r(k))u(k) \end{aligned} \tag{11}$$

and the quadratic storage function $V(x(k),r(k)) = x(k)^T P(x(k),r(k))x(k)$. Assume that matrix $P(x(k),r(k))$ in the storage function only depends on the state of the Markov chain. Then, theorem 1 below characterizes the dissipativity in the stochastic sense for affine Markovian jump nonlinear system (11).

Theorem 1. Consider the affine Markovian jump nonlinear system (11). The system is locally stochastic QSR-dissipative if and only if there exists some positive definite matrices P_i such that the following state dependent LMI's are satisfied for all $i, j \in \mathbb{Z}, x \in \mathbf{X}$

$$\begin{pmatrix} A_i^T(x)P_{pi}A_i(x) - P_i - C_i^T(x)QC_i(x) & * \\ B_i^T(x)P_{pi}A_i(x) - (QD_i(x)+S)^T C_i(x) & B_i^T(x)P_{pi}B_i(x) - (R + S^T D_i(x) + D_i^T(x)S + D_i^T(x)QD_i(x)) \end{pmatrix} < 0$$

$$i, j \in \mathbb{Z}, x \in \mathbf{X} \tag{12}$$

with A_i, B_i, C_i, and D_i denote system matrices in (11) when the plant mode is in mode $i \in \mathbb{Z}$, i.e $r(k) = i$. P_{pi} are matrices produced from P_i after multiplication by the transition probability of the associated Markov chain and given in the following form

$$P_{pi} = \sum_{j=1}^{N} p_{ij} P_j, \forall i, j \in \aleph . \tag{13}$$

Proof: By using Proposition 1, substitution the storage function $V(x(k),r(k)) = x(k)^T P(x(k),r(k))x(k)$ and its derivative into (8)-(10) renders the following state dependent LMI

$$\begin{aligned} &\begin{pmatrix} A_i^T(x)P_{pi}A_i(x) - P_i - C_i^T(x)QC_i(x) & * \\ B_i^T(x)P_{pi}A_i(x) - (QD_i(x)+S)^T C_i(x) & B_i^T(x)P_{pi}B_i(x) - (R + S^T D_i(x) + D_i^T(x)S + D_i^T(x)QD_i(x)) \end{pmatrix} \\ &= \begin{pmatrix} -\rho^T\rho & * \\ -\varepsilon^T\rho & -\varepsilon^T\varepsilon \end{pmatrix} \\ &= -(\rho \quad \varepsilon)^T (\rho \quad \varepsilon) \\ &< 0 \end{aligned} \tag{14}$$

Theorem 1 characterizes the dissipativity of affine Markovian jump nonlinear system based on matrices of its state space representation. With Theorem 1, QSR-dissipativity of the system (11) can be obtained by finding some positive definite matrices P_i for all mode in its Markov chain which satisfy a state dependent LMI (14). By Schur complement, the state dependent LMI (14) can be written in the 4 x 4 matrix blocks as follows

$$\begin{pmatrix} P_i & * & * & * \\ SC_i(x) & R + S^T D_i(x) + D_i^T(x) & * & * \\ A_i(x) & B_i(x) & P_{pi}^{-1} & * \\ -QC_i(x) & -QD_i(x) & 0 & -Q \end{pmatrix} > 0 \qquad \forall i \in \mathbb{Z}, x \in \mathbf{X} \tag{15}$$

4. Concluding remarks

The paper describes on dissipativity of affine nonlinear networked control systems via Markovian jump system approach. Dissipativity of such system is assured by the existence of positive definite matrices P_i which satisfy the solvability condition in terms of nonlinear matrix inequalities.

References

[1] Gupta R A, Chow M Y 2010 Networked Control Systems : Overview and Research Trends *IEEE Trans. on Industrial Electronics* 57 (7) 2527-2535

[2] Zhang W, Branicky M S, Phillips S M 2001 Stability of Networked Control Systems *IEEE Control Systems Magazine* 21 (2) 84-99
[3] Heemels W P M H, Teel A R, van de Wouw N, Nesic D Networked Control Systems with Communication Constraints: Tradeoffs between Transmission Intervals, Delays and Performance, *IEEE Trans. Autom. Control* 55 (8) 1781-1796
[4] Najmurrokhman A, Riyanto B, Rohman A S, Hendrawan 2013 Dissipative Controller Design for Networked Control Systems via the Markovian Jump System Approach *Journal of Engineering and Technological Sciences* 45 (1) 25-46
[5] Costa O L V, Fragoso M D, Marques R P 2005 *Discrete-Time Markov Jump Linear Systems*, Springer-Verlag, London, UK
[6] Li L, Ugrinovskii V A 2007 On Necessary and Sufficient Conditions for H_∞ Output Feedback Control of Markov Jump Linear Systems *IEEE Trans. on Autom. Control* 52 (7) 1287-1292
[7] Fioravanti A R, Goncalves A P C, Geromel J C 2008 H_2 and H_∞ Filtering of Discrete-time Markov Jump Linear Systems through Linear Matrix Inequalities *Proc. of the 17th World Congress The International Federation of Automatic Control* Seoul, Korea, July 6-11, pp. 2681-2688
[8] Geromel J C, Goncalves A P, Fioravanti A R 2009 *Dynamic Output Feedback Control of Discrete-Time Markov Jump Linear Systems through Linear Matrix Inequalities* SIAM J. Control Optim. 48 (2) 573-593
[9] Wang J, Liu C, Yang H 2012 Stability of a class of networked control systems with Markovian characterization *Appl. Math. Model.* 36 (7) 3168-3175
[10] Wang Z, Huang L, Yang X 2011 $H\infty$ performance for a class of uncertain stochastic nonlinear Markovian jump systems with time-varying delay via adaptive control method *Appl. Math. Model.* 35 (4) 1983-1993.
[11] Aliyu M D S, Boukas E K 2000 Robust H-infinity Control for Markovian Jump Nonlinear Systems *Journal of Mathematical Control and Information* 17 295-308.
[12] Aberkane S, Ponsart J C, Sauter D 2008 Output-Feedback H_2/H_∞ Control of A Class of Networked Fault Tolerant Control Systems *Asian J. Control* 10 (1) 34-44
[13] Shi Y, Yu B 2009 Output Feedback Stabilization of Networked Control Systems with Random Delays Modeled by Markov Chains *IEEE Trans. on Autom. Control* 54 (7) 1668-1674
[14] Willems J C 1972 Dissipative dynamical systems Part I : General Theory, Part II : Linear Systems with quadratic supply rates, *Arch. Ration. Mech. Analysis* 45 321-393
[15] Brogliato B, Lozano R, Maschke B, Egeland O 2007 *Dissipative Systems Analysis and Control* Springer-Verlag London
[16] Aliyu M D S 1999 Dissipativity and Stability of Nonlinear Jump Systems, *Proc. of American Control Conference,* pp. 795-799
[17] Zhang H., Yan H. & Chen Q 2010 Stability and Dissipative Analysis for A Class of Stochastic System with Time-Delay *Journal of the Franklin Institute* 347 882-893
[18] Mahmoud MS, Nounou H N, Xia Y 2010 Robust Dissipative Control for Internet-Based Switching Systems *Journal of the Franklin Institute* 347 (1) 154-172
[19] Zhang H, Guan Z H, Feng G 2008 Reliable Dissipative Control for Stochastic Impulsive Systems *Automatica* 44 1004-1010
[20] Wang J, Han Q L, Yang F 2011 Robust Dissipative Control for Networked Control Systems with Multiple Packet Dropout, *Proc. of the 18th IFAC World Congress* pp. 78-83
[21] Zhao J, Hill D J 2008 Dissipativity Theory for Switched Systems *IEEE Trans. on Autom. Control* 53 (4) 941-953
[22] Gupta S 1996 Robust Stabilization of Uncertain Systems Based on Energy Dissipation Concepts, *NASA Contractor Paper* 4713
[23] Wang Y, Gupta V, Antsaklis P J 2013 Stochastic Passivity of Discrete-Time Markovian Jump Nonlinear Systems, *Proc. of American Control Conference.*

IOP Conf. Series: Journal of Physics: Conf. Series **1127** (2019) 012068 doi:10.1088/1742-6596/1127/1/012068

Characterization of split distinguished triangles

Dellavitha Nasution

Algebra Research Group, Faculty of Maths and Natural Sciences, Institut Teknologi Bandung, Indonesia

E-mail: dellavitha@math.itb.ac.id

Abstract. Let A be a finite dimensional $\mathbf{k}$-algebra over an algebraically closed field $\mathbf{k}$. The aim of this paper is to show an equivalent condition on parameterizations of complexes of projective A-modules for existence of short exact sequences of complexes of projective A-modules. Moreover, we show an equivalent condition for split distinguished triangle $U \to W \to V \to U[1]$ related to the difference between dimension of $\mathrm{Hom}(U[i] \oplus V[i], X)$ and dimension of $\mathrm{Hom}(W[i], X)$ for every X object in triangulated category.

1. Introduction

Representation theory has been successfully studied parametrization of modules over $\mathbf{k}$-algebras. One can thereby regard left A-modules structure on vector space $\mathbf{k}^d$ are nothing but sequences of $d \times d$-matrices which satisfy some relations. When view as elements of affine space $\mathbb{A}^N$, that sequences of matrices are roots of some polynomials simultaneously. That variety is named Module Variety which denote by mod_d^A. To this variety one can define an action of general linear group $GL_d(\mathbf{k})$ by conjugation whose orbits corresponding to isomorphism classes of d-dimensional left A-modules.

The framework of A-modules was extended to objects of derived category of A-modules. Jensen, Su, and Zimmermann [2] defined a topological space parameterizing right bounded complexes of projective A-modules $\mathrm{comproj}_{\mathbf{d}}^A$ with fixed dimension array $\mathbf{d}$. The topological group $G_{\mathbf{d}} \cong \prod_{i\in\mathbb{Z}} Aut_A\left(\bigoplus_{j=1}^{l} P_j^{d_{ij}}\right)$ acting on it so that orbits correspond to quasi-isomorphism classes of right bounded complexes of projective A-modules in the category of complexes.

In module category, Zwara [4] showed that the existence of short exact sequence $0 \to U \to V \to W \to 0$ if $\begin{pmatrix} u & z \\ 0 & v \end{pmatrix}$ is a block matrix. In this paper, we generalize it in the frame of derived category. Moreover, we show an equivalent condition for split distinguished triangle $U \to W \to V \to U[1]$ related to the difference between dimension of $\mathrm{Hom}(U[i] \oplus V[i], X)$ and dimension of $\mathrm{Hom}(W[i], X)$ for every X object in triangulated category.

2. Basic Definitions

Let A be a finite dimensional algebra over an algebraically closed field $\mathbf{k}$. Let mod_d^A denote the affine variety of d-dimensional left A-modules. To this variety one can define an action of general linear group $GL_d(\mathbf{k})$ by conjugation whose orbits corresponding to isomorphism classes of d-dimensional left A-modules.

Let $P_1, \ldots, P_l$ be a complete set of representatives of isomorphism classes of indecomposable projective A-modules. For an element $\underline{d} = (d_1, \ldots, d_l) \in \mathbb{N}^l$ define $\alpha(\underline{d}) = \sum_{i=1}^{l} d_i\ dim_{\mathbf{k}} P_1$ the dimension of $\bigoplus_{i=1}^{l} P_j^{d_{ij}} \in mod_{\alpha(\underline{d})}^A$. Now for a matrix $\mathbf{d} = (d_{ij})$ with l columns and coefficients in $\mathbb{N}$ defined $\underline{d}_i$ to be i-th row of $\mathbf{d}$ and defined $\mathrm{comp}_{\mathbf{d}}^A$ to be the subset of

$$\left(\prod_{i\in\mathbb{Z}} mod_{\alpha(\underline{d}_i)}^A\right) \times \left(\prod_{i\in\mathbb{Z}} Hom_{\mathbf{k}}\left(\mathbf{k}^{\alpha(\underline{d}_i)}, \mathbf{k}^{\alpha(\underline{d}_{i-1})}\right)\right)$$

consisting elements $M_\bullet = (M_i, \partial_i)_{i\in\mathbb{Z}}$ with the properties ∂_i is an A-homomorphism as a map from M_i to M_{i-1} and $\partial_{i-1}\partial_i = 0$. Such matrix $\mathbf{d}$ is called a dimension array.

Jensen, Su and Zimmermann [2] define a topological space

$$\mathrm{comproj}_{\mathbf{d}}^A := \pi_M^{-1}\left(\prod_{i\in\mathbb{Z}} \bigoplus_{j=1}^{l} P_j^{d_{ij}}\right)$$

with $\pi_M : comp_{\mathbf{d}}^A \longrightarrow \prod_{i\in\mathbb{Z}} mod_{\alpha(\underline{d}_i)}^A$ the natural projection. That topological space parameterize objects in bounded derived category $\mathbf{D}^b(A)$. They use the fact that $\mathbf{D}^b(A) \cong \mathbf{K}^{-,b}(A-proj)$. Concerning conventions for derived categories we shall follow [3].

Theorem 1. Let $M_\bullet$ and $N_\bullet$ be two right bounded complexes with same dimension array $\mathbf{d}$. Any short exact sequence $0 \to N_\bullet \to M_\bullet \oplus Z_\bullet \to Z_\bullet \to 0$ gives arise a distinguished triangle $N_\bullet \to M_\bullet \oplus Z_\bullet \to Z_\bullet \to N[1]_\bullet$ in derived category $\mathbf{D}^b(A)$.

Proof. See [1].

3. Results

The following theorems characterize existence of short exact sequences of complexes of projective A-modules in $\mathrm{comproj}_{\mathbf{d}}^A$. Let $\mathbf{d}, \mathbf{d}', \mathbf{d}''$ be dimension arrays and let $\partial_\bullet^U, \partial_\bullet^V, \partial_\bullet^W$ be points in $\mathrm{comproj}_{\mathbf{d}'}^A, \mathrm{comproj}_{\mathbf{d}''}^A, \mathrm{comproj}_{\mathbf{d}}^A$ corresponding to complexes $U_\bullet, V_\bullet, W_\bullet$. The following theorems hold.

Theorem 2. Let $\partial_\bullet^U \in \mathrm{comproj}_{\mathbf{d}'}^A(\mathbf{k})$ and $\partial_\bullet^V \in \mathrm{comproj}_{\mathbf{d}''}^A(\mathbf{k})$. Define $\mathbf{d} = \mathbf{d}' + \mathbf{d}''$. Then $\partial_\bullet^W \in \mathrm{comproj}_{\mathbf{d}}^A(\mathbf{k})$ with

$$\partial_i^W = \begin{bmatrix} \partial_i^U & \nu_i \\ 0 & \partial_i^V \end{bmatrix}$$

for some ν_i in $Hom_A\left(\mathbf{k}^{\alpha(\underline{d}_i'')}, \mathbf{k}^{\alpha(\underline{d}_{i-1}')}\right)$ if and only if there exists a short exact sequence Σ of complexes of A-modules

$$0 \to U_\bullet \xrightarrow{\mathfrak{i}} W_\bullet \xrightarrow{\mathfrak{p}} V_\bullet \to 0,$$

where $\mathfrak{i}$ is a natural injection and $\mathfrak{p}$ is a natural projection. Moreover, the short exact sequence Σ is split if and only if there exist t_i in $Hom_A\left(\mathbf{k}^{\alpha(\underline{d}_i'')}, \mathbf{k}^{\alpha(\underline{d}_i')}\right)$, $i \in \mathbb{Z}$, such that $\nu_i = t_{i-1}\partial_i^V - \partial_i^U t_i$.

Proof. For all $i \in \mathbb{Z}$, let $W_i := U_i \oplus V_i$ and $\partial_i^W = \begin{bmatrix} \partial_i^U & \nu_i \\ 0 & \partial_i^V \end{bmatrix}$. By introducing the injection $\mathfrak{i}$ of U_i to the first factor of W_i and the projection $\mathfrak{p}$ of the second factor of W_i to V_i, we get a short exact sequence of complexes Σ as follows

$$\Sigma : \qquad 0 \to U_\bullet \xrightarrow{\mathfrak{i}} W_\bullet \xrightarrow{\mathfrak{p}} V_\bullet \to 0.$$

(i) The verification that $\mathfrak{i}$ is a morphism of complexes:

$$\begin{array}{ccc} x & \longrightarrow & \partial_i^U(x) \\ \downarrow & & \downarrow \\ \begin{bmatrix} x \\ 0 \end{bmatrix} & \longrightarrow & \begin{bmatrix} \partial_i^U & \alpha_i \\ 0 & \partial_i^V \end{bmatrix}\begin{bmatrix} x \\ 0 \end{bmatrix} = \begin{bmatrix} \partial_i^U(x) \\ 0 \end{bmatrix} \end{array}$$

(ii) The verification that $\mathfrak{p}$ is a morphism of complexes

$$\begin{array}{ccc} \begin{bmatrix} x \\ y \end{bmatrix} & \longrightarrow & \begin{bmatrix} \partial_i^U & \alpha_i \\ 0 & \partial_i^V \end{bmatrix}\begin{bmatrix} x \\ y \end{bmatrix} = \begin{bmatrix} \partial_i^U(x) + \alpha_i(y) \\ \partial_i^V(y) \end{bmatrix} \\ \downarrow & & \downarrow \\ y & \longrightarrow & \partial_i^V(y) \end{array}$$

Conversely suppose that we have an exact sequence Σ. Since $\mathfrak{i}$ and $\mathfrak{p}$ are morphisms of complexes, for $u \in U_i$ we obtain

$$\mathfrak{p}_{i-1}\partial_i^W \begin{bmatrix} u \\ 0 \end{bmatrix} = \partial_i^V \mathfrak{p}_i \begin{bmatrix} u \\ 0 \end{bmatrix} = 0.$$

Thus

$$\partial_i^W \begin{bmatrix} u \\ 0 \end{bmatrix} = \begin{bmatrix} \psi_i(u) \\ 0 \end{bmatrix}$$

for some A-module homomorphism $\psi_i : U_i \to U_{i-1}$. On the other side

$$\mathfrak{i}_{i-1}\partial_i^U(u) = \partial_i^W \mathfrak{i}_i(u) = \partial_i^W \begin{bmatrix} u \\ 0 \end{bmatrix} = \begin{bmatrix} \psi_i(u) \\ 0 \end{bmatrix}.$$

Then $\partial_i^U(u) = \psi_i(u)$ for every $u \in U_i$. Hence $\psi_i = \partial_i^U$. This yields $\partial_i^W \begin{bmatrix} u \\ 0 \end{bmatrix} = \begin{bmatrix} \partial_i^U(u) \\ 0 \end{bmatrix}$ for every $u \in U_i$. For every $v \in V_i$

$$\mathfrak{p}_{i-1}\partial_i^W \begin{bmatrix} 0 \\ v \end{bmatrix} = \partial_i^V \mathfrak{p}_i \begin{bmatrix} 0 \\ v \end{bmatrix} = \partial_i^V(v).$$

Hence $\partial_i^W \begin{bmatrix} 0 \\ v \end{bmatrix} = \begin{bmatrix} u_{i-1} \\ \partial_i^V(v) \end{bmatrix}$ for some u_{i-1} in U_{i-1}. It determines an A-module homomorphism $\nu_i : V_i \to U_{i-1}$, given by $v \mapsto u_{i-1,k}$ for all $v \in V_i$. Then we obtain

$$\partial_i^W = \begin{bmatrix} \partial_i^U & \nu_i \\ 0 & \partial_i^V \end{bmatrix}.$$

Now we are going to see when the short exact sequence Σ splits. Since $\mathfrak{p} : W_\bullet \to V_\bullet$ just projects onto the second factor of $W_\bullet$, such sections are precisely the chain map $V_\bullet \to W_\bullet$ of the form

$$t = \left(\begin{bmatrix} t_i \\ id_{V_i} \end{bmatrix} \right)_{i \in \mathbb{Z}}, t_i : V_i \to U_i \text{ an } A\text{-module homomorphism.}$$

The stipulation that family t is a chain map of complexes gives the equality

$$\begin{bmatrix} \partial_i^U & \alpha_i \\ 0 & \partial_i^V \end{bmatrix} \begin{bmatrix} t_i \\ id_{V_i} \end{bmatrix} = \begin{bmatrix} t_{i-1} \\ id_{V_{i-1}} \end{bmatrix} \partial_i^V$$
$$\begin{bmatrix} \partial_i^U t_i + \alpha_i \\ \partial_i^V \end{bmatrix} = \begin{bmatrix} t_{i-1} \partial_i^V \\ \partial_i^V \end{bmatrix}$$

so $\nu_i = t_{i-1}\partial_i^V - \partial_i^U t_i$. $\blacksquare$

The dual of the above theorem is as follows.

Theorem 3. $\partial_\bullet^{\mathcal{W}'} = \left(\begin{bmatrix} \partial_i^{\mathcal{U}} & 0 \\ \mu_i & \partial_i^{\mathcal{V}} \end{bmatrix} \right)_{i \in \mathbb{Z}} \in \text{comproj}_{\mathbf{d}}^A(\mathbf{k})$ for some μ_i in $Hom_A\left(\mathbf{k}^{\alpha(\underline{d}'_i)}, \mathbf{k}^{\alpha(\underline{d}''_{i-1})}\right)$ if and only if there exists an exact sequence of complexes Σ' of the form

$$0 \to V_\bullet \xrightarrow{\mathfrak{i}} W'_\bullet \xrightarrow{\mathfrak{p}} U_\bullet \to 0.$$

where $\mathfrak{i}$ is a natural injection and $\mathfrak{p}$ is a natural projection. Moreover Σ' is split if and only if $\mu_i = t'_{i-1}\partial_i^u - \partial_i^v t'_i$ for some $t'_i \in Hom_A\left(\mathbf{k}^{\alpha(\underline{d'}_i)}, \mathbf{k}^{\alpha(\underline{d''}_i)}\right)$.

From Theorem 1, Theorem 2, and Theorem 3 we get the following corollaries.

Corollary 4. Let $\partial_\bullet^{\mathcal{U}} \in \text{comproj}_{\mathbf{d}'}^A(\mathbf{k})$ and $\partial_\bullet^{\mathcal{V}} \in \text{comproj}_{\mathbf{d}''}^A(\mathbf{k})$. Define $\mathbf{d} = \mathbf{d}' + \mathbf{d}''$. Then $\partial_\bullet^{\mathcal{W}} \in \text{comproj}_{\mathbf{d}}^A(\mathbf{k})$ with

$$\partial_i^{\mathcal{W}} = \begin{bmatrix} \partial_i^{\mathcal{U}} & \nu_i \\ 0 & \partial_i^{\mathcal{V}} \end{bmatrix}$$

for some ν_i in $Hom_A\left(\mathbf{k}^{\alpha(\underline{d}''_i)}, \mathbf{k}^{\alpha(\underline{d}'_{i-1})}\right)$ if and only if there exists a distinguished triangle

$$U_\bullet \xrightarrow{\mathfrak{i}} W_\bullet \xrightarrow{\mathfrak{p}} V_\bullet \to U[1]_\bullet$$

in derived category $\mathbf{D}^b(A)$.

Corollary 5. $\partial_\bullet^{\mathcal{W}'} = \left(\begin{bmatrix} \partial_i^{\mathcal{U}} & 0 \\ \mu_i & \partial_i^{\mathcal{V}} \end{bmatrix} \right)_{i \in \mathbb{Z}} \in \text{comproj}_{\mathbf{d}}^A(\mathbf{k})$ for some μ_i in $Hom_A\left(\mathbf{k}^{\alpha(\underline{d}'_i)}, \mathbf{k}^{\alpha(\underline{d}''_{i-1})}\right)$ if and only if there exists a distinguished triangle

$$V_\bullet \xrightarrow{\mathfrak{i}} W'_\bullet \xrightarrow{\mathfrak{p}} U_\bullet \to V[1]_\bullet$$

in derived category $\mathbf{D}^b(A)$.

Let $\mathfrak{T}$ be a triangulated category. Assume that for every Y and Z objects in $\mathfrak{T}$, $Hom_{\mathfrak{T}}(Y, Z)$ is a finite dimensional $\mathbf{k}$-vector space. Let

$$\Gamma : U \xrightarrow{f} W \xrightarrow{g} V \xrightarrow{h} U[1]$$

be a triangle in the triangulated category $\mathfrak{T}$. We define an additive function ϑ_Γ from all objects of $\mathfrak{T}$ to the set of sequences with coefficients in integers,

$$\vartheta_\Gamma(X) = (\dim_{\mathbf{k}} Hom(V[i] \oplus U[i], X) - \dim_{\mathbf{k}} Hom(W[i], X))_{i\in\mathbb{Z}}$$

for any object X in $\mathfrak{T}$.

The following theorem shows an equivalent condition for split distinguished triangle $U \to W \to V \to U[1]$ related to the difference between dimension of $\mathrm{Hom}(U[i] \oplus V[i], X)$ and dimension of $\mathrm{Hom}(W[i], X)$ for every X object in triangulated category.

Theorem 6. Let $\Gamma : U \xrightarrow{f} W \xrightarrow{g} V \xrightarrow{h} T(U)$ be a distinguished triangle in the triangulated category $\mathfrak{T}$. Then $\vartheta_\Gamma(X)$ is a sequence with coefficients in $\mathbb{N} \cup \{0\}$ for any object X in $\mathfrak{T}$. Moreover, the distinguished triangle Γ is split if and only if $\vartheta_\Gamma(X) = 0$ for any object X in $\mathfrak{T}$.

Proof. By applying the functor $Hom_{\mathfrak{T}}(-, X)$ to the distinguished triangle Γ, we obtain the exact sequence

$$\cdots \to Hom_{\mathfrak{T}}(T(U), X) \xrightarrow{h^*} Hom_{\mathfrak{T}}(V, X) \xrightarrow{g^*} Hom_{\mathfrak{T}}(W, X) \xrightarrow{f^*} Hom_{\mathfrak{T}}(U, X) \to \cdots$$

Without lost of generality, we will see only the 0-th position of the sequence $\vartheta_\Gamma(X)$

$$\begin{aligned} &\dim_{\mathbf{k}} Hom_{\mathfrak{T}}(V \oplus U, X) \\ &= \dim_{\mathbf{k}} Hom_{\mathfrak{T}}(V, X) + \dim_{\mathbf{k}} Hom_{\mathfrak{T}}(U, X) \\ &= \mathrm{Rank} g^* + \mathrm{Null} g^* + \dim_{\mathbf{k}} Hom_{\mathfrak{T}}(U, X) \\ &= \mathrm{Null} f^* + \mathrm{Null} g^* + \dim_{\mathbf{k}} Hom_{\mathfrak{T}}(U, X) \\ &= \dim_{\mathbf{k}} Hom_{\mathfrak{T}}(W, X) - \mathrm{Rank} f^* + \mathrm{Null} g^* + \dim_{\mathbf{k}} Hom_{\mathfrak{T}}(U, X) \\ &= \dim_{\mathbf{k}} Hom_{\mathfrak{T}}(W, X) + \mathrm{Null} g^* + \varepsilon, \quad \varepsilon \geq 0. \end{aligned}$$

with f^* defined by $f^*(\alpha) = \alpha \circ f$, $\alpha \in Hom_{\mathfrak{T}}(W, X)$, similarly with g^* and h^*. Hence the entries of $\vartheta_\Gamma(X)$ are in $\mathbb{N} \cup \{0\}$.

Moreover if we let Γ be split then $h = 0$, so $\left(T^i(h)\right)^* = 0$ for any object X in $\mathfrak{T}$ and any $i \in \mathbb{Z}$. It yields $\left(T^i(f)\right)^*$ is an epimorphism and in particular $\left(T^i(g)\right)^*$ is a monomorphism so Null $\left(T^i(g)\right)^* = 0$. Thus

$$dim_{\mathbf{k}}(T^i(V) \oplus T^i(U), X) = \dim_{\mathbf{k}}(T^i(W), X)$$

for all $i \in \mathbb{Z}$, so $\vartheta_\Gamma(X) = (0, \ldots, 0)$.

Conversely if $\vartheta_\Gamma(X) = (0, \ldots, 0)$ then

$$\dim_{\mathbf{k}} Hom_{\mathfrak{T}}(T^i(U), X) - \mathrm{Rank}\ \left(T^i(f)\right)^* + \mathrm{Null}\ \left(T^i(g)\right)^* = 0.$$

It implies $\dim_{\mathbf{k}} Hom_{\mathfrak{T}}(T^i(U), X) = \mathrm{Rank}\ \left(T^i(f)\right)^*$ and $\mathrm{Null}\ \left(T^i(g)\right)^* = 0$, since $\mathrm{Rank}\ \left(T^i(f)\right)^* \leq \dim_{\mathbf{k}}(T^i(U), X)$ and $\mathrm{Null}\ \left(T^i(g)\right)^* \geq 0$, for every $i \in \mathbb{Z}$. Thus $\mathrm{Ker}\ \left(T^i(g)\right)^* = 0$ for every $i \in \mathbb{Z}$. This condition implies $\left(T^i(h)\right)^* = 0$ for every $i \in \mathbb{Z}$. In particular, if we put $X = T^{i+1}(U)$ and evaluate the identity morphism $id_{T^{i+1}(U)} \in Hom_{\mathfrak{T}}(T^{i+1}(U), T^{i+1}(U))$,

$$0 = \left(T^i(h)\right)^* \left(id_{T^{i+1}(U)}\right) = id_{T^{i+1}(U)} \circ T^i(h) = T^i(h)$$

so that $h = 0$. Hence the distinguished triangle splits. ∎

References

[1] Holm T and Jorgensen P 2010 *Triangulated Categories (London Mathematical Society Lecture Note Series, Book 375)*, ed Holm T, Jrgensen P and Rouquier R (Cambridge: Cambridge Univ. Press) p 33
[2] Jensen BT, Su X and Zimmermann A 2005 Degenerations for derived categories *J.Pure Appl. Algebr.* **198** 281-95
[3] Zimmermann A 2014 *Representation Theory: A Homological Algebra Point of View* (Switzerland: Springer) p 259
[4] Zwara G 2000 Degenerations of finite dimensional modules are given by extensions *Compos. Math.* **121** 205-18

Modelling of Islamic Calendar System based on Moon Phase

R Wahyuni, G M Issaac, G Rachmaputri, I Muchtadi-Alamsyah

Department of Mathematics, Faculty of Mathematics and Natural Science, Institut Teknologi Bandung, Indonesia

risant.96@gmail.com, gardhani.issaac@s.itb.ac.id, gantina.r@math.itb.ac.id, ntan@math.itb.ac.id

Abstract. The determination of the Islamic calendar system still relies on the Gregorian calendar, so until now, there is no Islamic calendar system prevailing in the international scale. By making function based on synodic period of the Moon, Wada Date (Julian Date version of the Islamic calendar) can be made, and in the end, the Islamic calendar which based purely on the position of the Moon can be reconstructed. The purpose of this modelling is to create an Islamic calendar that has a higher accuracy than the Islamic calendar that we use today. The method that we chose here is the study of literature with the resources that come from articles, books, and internet. As for the analysis of its own data, we chose the numerical analysis method to approach the results of synodic period of the Moon data. From this analysis, we obtained a fact that the graph of synodic period of the Moon turned out to resemble a specific function. Based on the graph from original data, we selected two functions to approach this graph, and we named it the Fourier function and the Moon Phase function. Furthermore, these two functions will be compared to determine which of these functions will be selected. The function that has been selected will be used to make Wada Date which will be used to search for the length of each month in the Islamic calendar. If we have got the length of each month, the last process is to map out the dates for the month to a matrix that would produce an Islamic calendar.

1. Introduction

The **Islamic calendar** is a lunar calendar, a calendar which is based on cycles of the phase of the Moon, consisting of 12 months in a year of 354 or 355 days. The twelve months in Islamic calendar are Muharram, Safar, Rabiul awal, Rabiul akhir, Jumadil awal, Jumadil akhir, Rajab, Sya'ban, Ramadhan, Syawal, Dzulkaidah, and Dzulhijjah.[6] Each month in the Islamic calendar begins at the time of the monthly "conjunction", called **Ijtima'**, when the Moon is located on a straight line between the Earth and the Sun. So, the length of each month has either 29 or 30 days, usually in no discernible order. Traditionally, the first day of each month is the day (beginning at sunset) of the first sighting of the **hilal** (crescent moon) shortly after sunset. If the hilal is not observed immediately after the 29th day of a month, then the day that begins at that sunset is the 30th.

On the other hand, the **Gregorian calendar** is a solar calendar whose dates indicate the position of Earth on its revolution around the Sun. The length of each calendar month in Gregorian calendar has always remain the same in each year, except in February which length can be 28 or 29 days depends on the year's position in the leap year cycle.[5]

In the world of astronomy, we also know the term of **Julian Date (JD)**. Julian Date is the continuous count of days since the beginning of the Julian Period used primarily by astronomers. JD starts from the number 0 on 1st of January 4713 BC. The number of JD is incremented by 1 unit per 1 day, thus continuously.[4] Always keep in mind that the numbers of JD always starts at 12:00 pm. It is very important to understand Julian Date because one of its functions is to determine the time of the new moon.

Until today, the determination of the Islamic calendar system still relies on the Gregorian calendar, yet it is clear that both of these calendars have different reference base. Moreover, unlike the Gregorian calendar, there is no Islamic calendar system that applicable on international scale. This causes the difference in the Islamic calendar applied in various regions of the world. By utilizing the length of synodic period of the Moon data in a few years[2], we can look for a function to create Wada Date (Julian Date version of the Islamic calendar). The purpose is to reconstruct the Islamic calendar system which purely based on the position of the Moon. Furthermore, this Islamic calendar system is expected to replace the past Islamic calendar system that less accurate. It lead us to our goal, so we can determine any important dates in the Islamic calendar more effectively, such as determining the first of Syawal and the first of Ramadhan.

2. The construction of synodic period of the Moon function

The data that we used as reference to construct the function is the list of synodic period of the Moon data that began in September 1924 AD to July 2005 AD in the Gregorian calendar, which has the total amount of 1000 data. Here is the result of the plot of the data, with the x-axis as months in Islamic calendar and the y-axis as the length of each month.

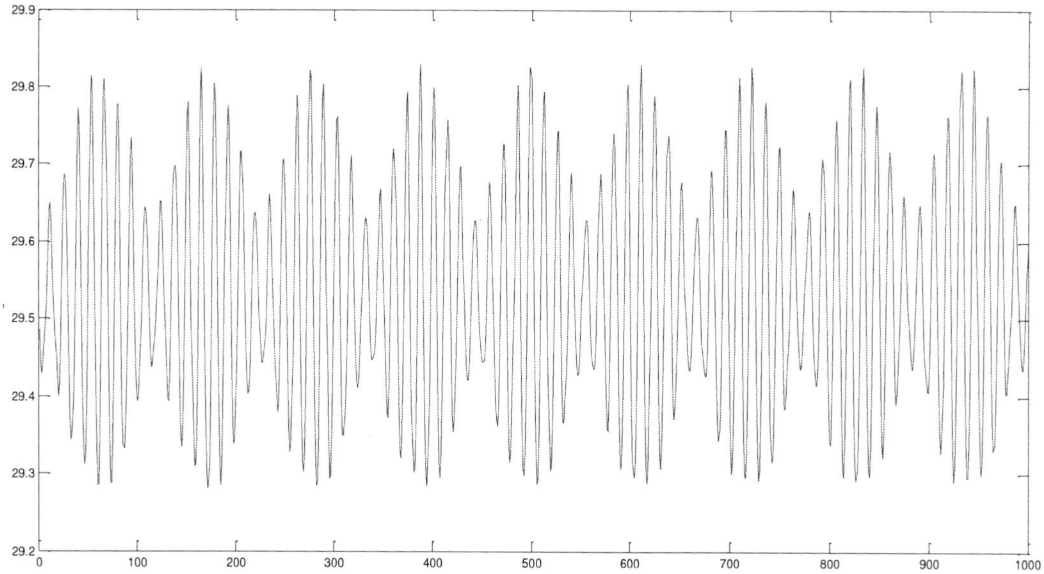

Figure 1. Synodic period of the Moon starting from September 1924 AD - July 2005 AD in Gregorian calendar.

If we observed graph of synodic period of the Moon above, it appears that it resembles a function in the form of a Fourier series or the sum of sin function. So, we used the **cftool** menu in MATLAB to compare the results of curve fitting using a Fourier function and the sum of sin function. But before doing curve fitting using cftool, the midpoint of the original data is calculated in advance to be transformed to the point $y = 0$. The goal is to make the graphics becoming more similar with the sin function that has the midpoint in the $y = 0$.

From the results of curve fitting which approaching the Fourier function, we obtained a function with $R^2 = 0{,}8088$. The function is:

$$f(x) = 29{,}52776 + \sum_{i=1}^{8} (a_i \cos 0{,}06439ix + b_i \sin 0{,}06439ix)$$

with the value of each coefficient are in Table 1.

Table 1. The value of the coefficients on the $f(x)$ function.

i	a_i	b_i
1	0,0002284	−0,00007051
2	0,0002969	0,0001431
3	0,0001086	0,0001913
4	0,0001976	−0,00003377
5	0,000672	0,0002167
6	0,0005739	0,001125
7	0,08028	−0,1623
8	−0,009149	0,002935

On the other hand, from the results of curve fitting which approaching sum of sin function, we obtained a function with $R^2 = 0{,}9917$. The function is:

$$g(x) = 29{,}551977 + \sum_{i=1}^{4} a_i \sin(b_i x + c_i)$$

with the value of each coefficient are in Table 2.

Table 2. The value of the coefficients on the $g(x)$ function.

i	a_i	b_i	c_i
1	0,1819	0,4506	−3,537
2	0,08662	0,508	−0,8411
3	0,00001028	0,4511	2,109
4	0,02179	0,000301	−1,9

Since the $g(x)$ function has a greater value of R^2 than the $f(x)$ function, then we selected the $g(x)$ function as the Moon phase function that will be used later to make Wada Date (WD).

3. Assumptions used in this modelling

Before we create Wada Date, we should notice that the data that we used is the synodic period of the Moon in September 1924 AD or coincide with the month of Safar 1343 H in the Islamic calendar. Meanwhile, we want to make the starting point of Wada Date (WD 1) at the start of the Islamic calendar days in accordance with the history, which falls on the 9th of Dzulhijjah 10 H or coincide with the 6th of March 632 M in the Gregorian calendar.[1,6] Therefore, the point x in the $g(x)$ function needs to be shifted back so that $x = 1$ is the month of Dzulhijjah in the year 10 H. Due to the difference between the month of Dzulhijjah 10 H with Safar 1343 H is 1332 years 2 months, then the point x need to be moved back 15986 units. Thus, the Moon phase function becomes:

$$g(x) = 29{,}551977 + \sum_{i=1}^{4} a_i \sin(b_i(x - 15986) + c_i)$$

with the values of each coefficient as listed in Table 2

Besides that, based on Danjon Limit, the hilal will be visible if the angular separation (center to center) between Sun and Moon is 7°.[3] It means, according to Danjon, at least the hilal will be visible 14 hours after Ijtima'. Since the number of days in each month in the Islamic calendar depends on the sighting of the hilal shortly after sunset, then it is assumed that the Ijtima' event occurred on the 29th of each month in the Islamic calendar. If 14 hours after Ijtima' falls before 6:00 pm, then the next day is the 1st day of next month. But if not, then the next day will be the 30th day of that month.

4. The construction of Wada Date (Julian Date version of the Islamic calendar)

We have known that Wada Date will be made to have the same rules with the Julian Date that has been made in advance by previous mathematicians.[4] So the construction of Wada Date can be done by utilizing the Julian Date through the logic as follows:

- WD (1) = 9th of Dzulhijjah 10 H = 6th of March 632 AD.
- To construct Wada Date, we utilized the formula of Julian Date that already existed to build the formula: WD (day, month, year in AD) = JD (day, month, year in AD) - (JDX-1). With JDX is the Julian Date of WD (1).

The final goal of this modelling is to reconstruct the Islamic calendar to the more accurate version and depends just to the movement of the Moon only. So, to determine the number of days in each month, it is necessary to find the time difference between two ijtima' point that adjacent to each other. The way that we can use to find it is to convert each of the ijtima' point to the Wada Date that we have build first, so that we can calculate the time difference. The algorithm logic to convert each Ijtima' point to its Wada Date is as follows:

- We assumed that the first Ijtima' was set at 29th of Dzulhijjah 10 H.
- Based on the Moon phase function that has been selected, the function can be interpreted as follows: $g(x)$ = the function of time difference between the xth Ijtima with the $(x-1)^{th}$ Ijtima'.
- From $g(x)$ above, we searched its cumulative function, $G(x) = \sum_{i=1}^{x} g(i)$.
- Thus, we obtained Wada Date as follows:

$$WD'(x) = WD \text{ in } x\text{th Ijtima}' = -8.5 + \mathrm{G(x)} + \mathrm{TD}.$$

The number 8,5 obtained from the initial assumption that WD (1) is 9th of Dzulhijjah 10 H at 12:00 pm. While TD is the time difference between time of the 0-Ijtima' point in the data with time of the 0-Ijtima' point in the function.

Next, we will look for a function to convert Wada Date to the Islamic calendar, so that when we input the xth Wada Date then we will know if it falls on which day, month, and year in the Islamic calendar. The algortihm logic of these functions are as follows:

- Input x (x^{th} Wada Date that will be converted to Islamic calendar).
- Find the biggest ${n_1}^{th}$ Ijtima' and smallest ${n_2}^{th}$ Ijtima' so that $WD'(n_1) < x \leq WD'(n_2)$.
- Find the date of x in Islamic calendar by:

$$\text{z = date in the Islamic calendar} = 29 - (\lfloor WD'(n_2) \rfloor - \lfloor x \rfloor).$$

- Find the month of x in Islamic calendar by:

$$\text{l = month in the Islamic calendar} = mod(n_2, 12) - 1.$$

- Find the year of x in Islamic calendar by:

$$\text{h = year in the Islamic calendar} = \left\lfloor \frac{n_2+11}{12} \right\rfloor + 10.$$

The last stage is to implement the algorithm logic to reconstruct the Islamic calendar as follows:

- Build matrix A := ['Monday ' 'Tuesday', .., 'Sunday'].
- Build matrix B := 6 × 7 sized matrix.
- Map each 29th in every month of the following ways:

$$g = floor(mod\ (WD'(n_1) + 4), 7)$$

and then map: $29 \mapsto B(1, g)$.

- Find the length of each month by:

$$t = \lfloor WD'(n_2) \rfloor - \lfloor WD'(n_1) \rfloor.$$

- Map the 30th of the same month or the 1st of next month by:

  If $t = 30$ then
    If the g + 1 = 8 ⇒ B (3,1) = 30.
    Otherwise B (1, g + 1) = 30.
  If $t = 1$,
    If the g + 1 = 8 ⇒ B (3,1) = 1.
    Otherwise B (1, g + 1) = 1.
- Fill matrix B after the number 29 or 30 with the numbers 1 to 29.
- Combining matrices A and B.

5. The error generated in the Islamic calendar

As previously described above that the value of R^2 at the Moon phase function created is 0,9917. It means that there is still an error between the length of each month, or in other words the number of days, in the Islamic calendar created with the original length of each month in the Islamic calendar. However, an error on this monthly basis is not an error that directly visible on the Islamic calendar created. Errors that occur on each month in the Islamic calendar are the sum of all errors in previous months. For example, suppose if we want to know the result of error from the Moon phase function when Ramadhan 1400 H. Then the error on this month in the Islamic calendar is the result of the sum of all errors in each month before Ramadhan 1400 H.

After the program created in MATLAB to calculate the error that will be inflicted on a monthly basis, it is known that the function $g(x)$ only give a maximum error of ± 0,21 days, or approximately 5 hours per month. It is caused by the result of the plot of errors in each month that resembles the sin function which has a fixed intervals.

References

[1] Ammi N B 2015 *Sejarah Penetapan Kalender Hijriyah.* https://konsultasisyariah.com/14956-sejarah-penetapan-kalender-hijriah.html
[2] Fred E 2014 *PHASES OF THE MOON: 1901 TO 2000.* http://astropixels.com/ephemeris/phasescat/phases1901.html
[3] Jim M 2009 *Danjon Limit.* https://the-moon.wikispaces.com/Danjon+Limit
[4] Muhammad N 2009 *Kalender Julian, Kalender Gregorian dan Julian Day.* http://www.eramuslim.com/peradaban/ilmu-hisab/kalender-julian-kalender-gregorian-dan-julian-day.htm
[5] Wikipedia 2016 *Gregorian calendar.* https://en.wikipedia.org/wiki/Gregorian_calendar
[6] Wikipedia 2016 *Islamic calendar.* https://en.wikipedia.org/wiki/Islamic_calendar

IOP Conf. Series: Journal of Physics: Conf. Series **1127** (2019) 012070 doi:10.1088/1742-6596/1127/1/012070

Mathematical model and numerical simulation of dual-channel multiphase bulk arrival queuing system with heterogeneous service time

Atina Ahdika

Department of Statistics, Faculty of Mathematics and Natural Sciences, Universitas Islam Indonesia, Jalan Kaliurang Km 14.5 Sleman, Yogyakarta 55584, Indonesia

E-mail: atina.a@uii.ac.id

Abstract. There are many queue models have been developed by many researchers. Many previous studies have developed queue model for large system. This paper constructs a queue model for small system which consists of dual server with multiphase in one system. Each server has different service time and the customer comes in batch (bulk arrival). The model has been derived based on the concept of birth-death process and solved using global balance condition to obtain performance measurements of the system. Numerical simulation is carried out to get a better understanding about the correlation between each variable involved in the system. The result shows that the more dense the movement of the customers in the system, the more people and the longer time spent by a customer in the system.

1. Introduction

Queue model is a mathematical model of waiting lines which is constructed so that the time length spent in the line can be predicted. Many events in the real life can be modelled using queue model such as problems related to industrial manufactures, network systems, traffic systems, biological systems, etc. Researchers have developed many models which gave an optimal solution to minimize the number of queues and the length time of the queue in a system by adding the server or by accelerating the service rate in each server. Li and Stanford [1] developed a multi-class multi-server queuing model with heterogeneous servers under the accumulating priority queuing discipline. Sikdar and Gupta [2] developed a finite-buffer batch arrival and batch service queue with single and multiple vacations, they also derived various performance measures of the model. Ekpenyong and Udoh [3] extended and improved on the performance measure of the single-server single-queue system with multiple phases under the conditions of FCFS, infinite population source, Poisson arrival, and Erlang service time. Ghimire *et al.* [4] constructed a single server bulk queueing model with fixed batch size $'b'$ and customers arrive to the system with Poisson rate λ and served exponentially with the rate μ. Maurya [5] studied a mathematical model of two-state batch arrivals and batch service Markovian queue with server vacation model under transient state. Sultan *et al.* [6] extended a Monte Carlo study to perform analysis of a multi-server bulk arrival queuing system with two-modes server breakdown. Sultan [7] introduced an easily applicable algorithm to solve problems involving bulk-arrival queues with a breakdown of one of the heterogeneous servers in case of steady state.

IOP Conf. Series: Journal of Physics: Conf. Series **1127** (2019) 012070 doi:10.1088/1742-6596/1127/1/012070

Most of the models discussed before are the model for a large queue system. While for small queue system is still rarely discussed. Whereas this small queue system can be helpful for small office or home-based industries. The models for large queue system have fairly high complexity and many conditions that must be met by the system. Departing from these matter, to simplify the model so that the model can be implemented easier for small queue system, then comes the idea to build a mathematical model and numerical simulation to the small queue system which is a combination of many cases. The model is a dual-channel multiphase bulk arrival queuing system with heterogeneous service time and infinite customer capacity. The arrival of the batch in this model is assumed as Poisson distributed and the service rate is exponentially distributed.

2. System construction and mathematical model

Queuing model built in this study is a model with two-channel services, k phases of service that must be passed by every arrival, and the customer comes in the form of groups or large quantities. The construction of the model is described in Figure 1

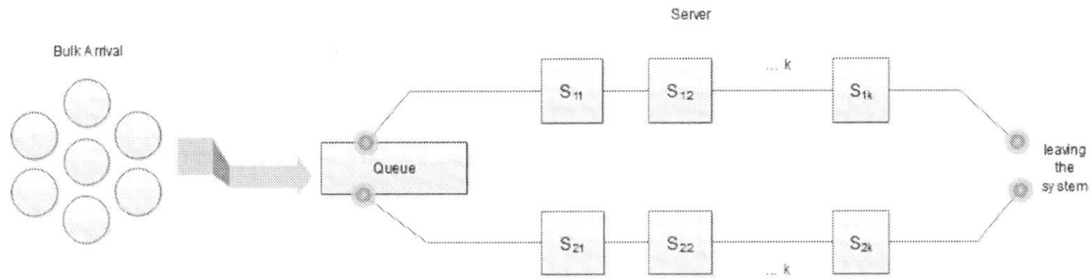

Figure 1. Construction of the Model

The system shows that each arrival comes in batch, however customer leaves the system individually. The queue model can be met in home-based industry, service process in district office, new student registration system, or in the airport with two arrival gates where the arrival can comes in a group. The service rate in each channel is assumed to be different, so that each individual leaves the system in accordance with the minimum service rate. However, if only left one arrival, then the arrival will go to the channel with the fastest service rate. Suppose random variables represent the service rate $X_i \sim EXP(\mu_i)$, $i = 1, 2$. To determine which channel has faster service rate, we defined a new random variable $Y = min(X_1, X_2) \sim EXP(\mu_1 + \mu_2)$. The construction is developed from the smaller size of batch (bulk), notated by b. Figure 2 shows the process diagram of the constructed model.

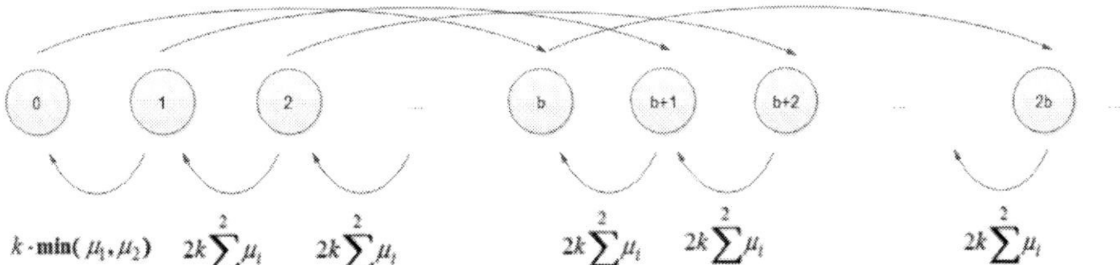

Figure 2. Process Diagram of the Constructed Model

λ is the arrival rate, k is the number of phases, $2k\sum_{i=1}^{2}\mu_i$ is total service rate of all phases, and $k\cdot min(\mu_1,\mu_2)$ is total service rate of the faster channel if there is only one customer left. Probability function of the model is obtained by solving the global balance equation as follows:

$$k\cdot min(\mu_1,\mu_2)P_1 - \lambda P_0 = 0 \tag{1}$$

$$2k\sum_{i=1}^{2}\mu_i P_2 - [\lambda + k\cdot min(\mu_1,\mu_2)]\,P_1 = 0 \tag{2}$$

$$2k\sum_{i=1}^{2}\mu_i P_{n+1} - \left[\lambda + 2k\sum_{i=1}^{2}\mu_i\right]P_n = 0, \qquad n = 2,3,\ldots,b-1 \tag{3}$$

$$\lambda P_{n-b} + 2k\sum_{i=1}^{2}\mu_i P_{n+1} - \left[\lambda + 2k\sum_{i=1}^{2}\mu_i\right]P_n = 0, \qquad n \geq b \tag{4}$$

Eq.(1) to (4) are used to obtain the probability generating function (pgf) of N (the number of customers in the system) that is $G(z) = \sum_{n=0}^{\infty} z^n P_n$, and we obtain

$$G(z) = \frac{2k\sum_{i=1}^{2}\mu_i(zP_1 + P_0)}{2k\sum_{i=1}^{2}\mu_i - \lambda z\sum_{j=0}^{b-1} z^j} \tag{5}$$

Eq.(5) is an implicit function in form of P_0 and P_1. The equation can be changed into explicit form by using the correlation between P_0 and P_1 from eq.(1)

$$P_1 = \frac{\lambda}{k\cdot min(\mu_1,\mu_2)}P_0 \tag{6}$$

The value of P_0 is obtained by substituting eq.(6) into eq.(5) at $z = 1$, where $G(1) = \sum_{n=0}^{\infty} 1^n P_n = \sum_{n=0}^{\infty} P_n = 1$ and we obtain

$$P_0 = 1 - \frac{\rho}{2} - P_1 = \frac{k\cdot min(\mu_1,\mu_2)[2-\rho]}{2[\lambda + k\cdot min(\mu_1,\mu_2)]} \tag{7}$$

where $\rho = \lambda b/k\sum_{i=1}^{2}\mu_i$ represents the traffic intensity of the queue model for the case of bulk arrival and heterogeneous service rate. Then substitute eq.(7) into eq.(5) and obtained the pgf

$$G(z) = \frac{\frac{k\sum_{i=1}^{2}\mu_i[2-\rho]}{[\lambda + k\cdot min(\mu_1,\mu_2)]}[z\lambda + k\cdot min(\mu_1,\mu_2)]}{2k\sum_{i=1}^{2}\mu_i - \lambda\sum_{j=1}^{b-1} z^j} \tag{8}$$

Probability of a certain amount of customers in the system can be calculated by differentiating $G(z)$ n times at $z = 0$ and divided it by $n!$, ie

$$P(N = n) = \frac{1}{n!}\cdot\frac{d^{(n)}}{dz}G(z)\mid_{z=0} \tag{9}$$

3. Performance measurements

3.1. Average Number of Customers in the System

The first performance measure is the average number of customers in the system, notated by L_S. L_S is obtained by calculating the expectation of N. However, we have probability generating function (pgf) instead of probability density function (pdf) of N, therefore the expectation value is obtained by differentiating pgf in eq. (8) one time at $z = 1$.

$$L_S = \sum_{n=0}^{\infty} nP_n = \frac{d}{dz}G(z) \mid_{z=1} = \frac{\rho k \sum_{i=1}^{2} \mu_i}{\rho k \sum_{i=1}^{2} \mu_i + kb \cdot min(\mu_1, \mu_2)} + \frac{\rho(b+1)}{2[2-\rho]} \tag{10}$$

3.2. Average Amount of Time a Customer Spent in the System

The average amount of time a customer spent in the system is notated by W_S. The smaller the value of W_S, the more efficient the existing queue system. The value of W_S is obtained by dividing average number of customer in the system with the total arrival rate.

$$W_S = \frac{L_S}{\lambda b} = \frac{1}{\rho k \sum_{i=1}^{2} \mu_i + kb \cdot min(\mu_1, \mu_2)} + \frac{(b+1)}{2k \sum_{i=1}^{2} \mu_i [2-\rho]} \tag{11}$$

3.3. Average Amount of Time a Customer Spent Waiting in the Queue

The average amount of time a customer spent waiting in the queue is notated by W_Q. W_Q is calculated only when a customer waiting in the queue not including when they are being served. The value of W_Q is obtained by subtracting the value of W_S with the expected value of the service time $E(S) = 1/k \sum_{i=1}^{2} \mu_i$, thus

$$W_Q = W_S - E(S) = \frac{1}{\rho k \sum_{i=1}^{2} \mu_i + kb \cdot min(\mu_1, \mu_2)} + \frac{(b-3+2\rho)}{2k \sum_{i=1}^{2} \mu_i [2-\rho]} \tag{12}$$

3.4. Average Number of Customers Waiting in the Queue

The last measure is the average number of customer waiting in the queue, notated by L_Q. Similar with W_Q, L_Q is only calculated based on the number of customers waiting in the queue. The value of L_Q is obtained by multiplying the average amount of time a customer waiting in the queue with the total arrival rate.

$$L_Q = \lambda b W_Q = \frac{\rho k \sum_{i=1}^{2} \mu_i}{\rho k \sum_{i=1}^{2} \mu_i + kb \cdot min(\mu_1, \mu_2)} + \frac{\rho(b-3+2\rho)}{2[2-\rho]} \tag{13}$$

4. Numerical simulation

In this section, numerical simulation is conducted to explain correlation among the performance measurements so that it can be used as a consideration in determining the efficient system. The system called an efficient system if the value of traffic intensity is less than one, $\rho < 1$, i.e. $\rho = \lambda b / k \sum_{i=1}^{2} \mu_i < 1$ so that $\lambda b < k \sum_{i=1}^{2} \mu_i$.

IOP Conf. Series: Journal of Physics: Conf. Series **1127** (2019) 012070 doi:10.1088/1742-6596/1127/1/012070

In other words, when the arrival rate is less than the service rate, then there will be no solid queue in the system. The relation between ρ and the performance measurements for different size of arrival b is discussed in this section. The simulation was conducted by setting the value of ρ from 0.1 to 1 and generated the value of n, μ_1, μ_2, and k for various value of b. The value of λ is representated by the value of ρ. Table 1 shows the variables generated in the simulation.

Table 1. Generated Variables of the Simulation

Simulation	ρ	b	n	k	μ_1	μ_2
1	$0.1, 0.2, \ldots, 0.9, 1$	$5, 10, 15, 20, 25$	50	5	7	8
2		$15, 30, 45, 60, 75$	200	10	32	28
3		$10, 20, 30, 40, 50$	100	5	10	15
4		$25, 50, 75, 100, 125$	300	10	30	25

The result of the simulations (are only shown for simulation 1 and 2 because the rest provide the same behaviour), the correlation between ρ, b, L_S, L_Q, W_S, and W_Q are shown by Figure 3 and 4.

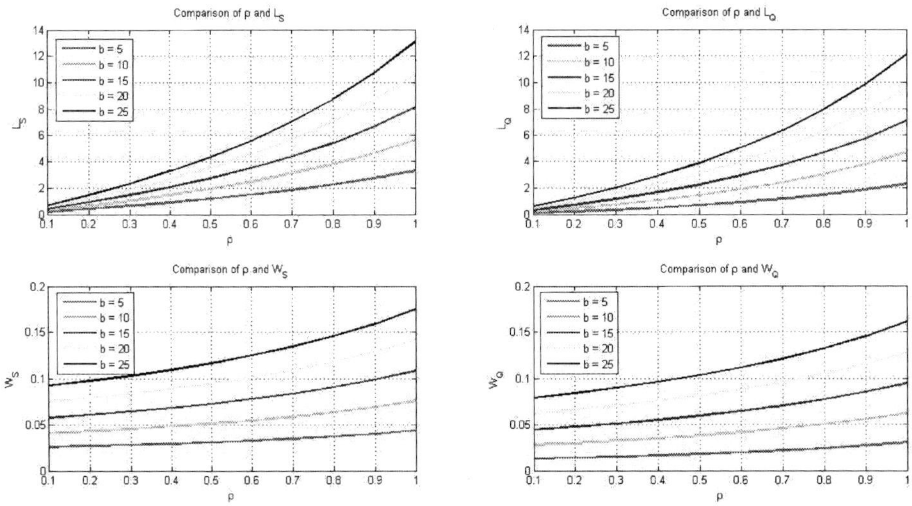

Figure 3. Correlation of ρ and Performance Measurements of the Model for Simulation 1

Figure 3 and 4 show that the larger the value of ρ, the larger the value of the performance measurements L_S, L_Q, W_S, and W_Q for various value of b, n, k, μ_1, and μ_2. It is indicated that the more dense the movement of customers in the queue, the more people who are in the system or in the queue and also the longer the time spent by a customer in the system or in the queue. In addition, the larger the size of arrival b, the greater the values of L_S, L_Q, W_S, and W_Q. From the simulations, it can also be seen that Simulation 2 has the shortest time spent by a customer in the system and in the queue. It is happened because Simulation 2 has the fastest service rate among others. Based on the results of the numerical simulation, it is concluded that the model can be implemented for small queue system and the system can shortened the queue by accelerating the service rate of the servers.

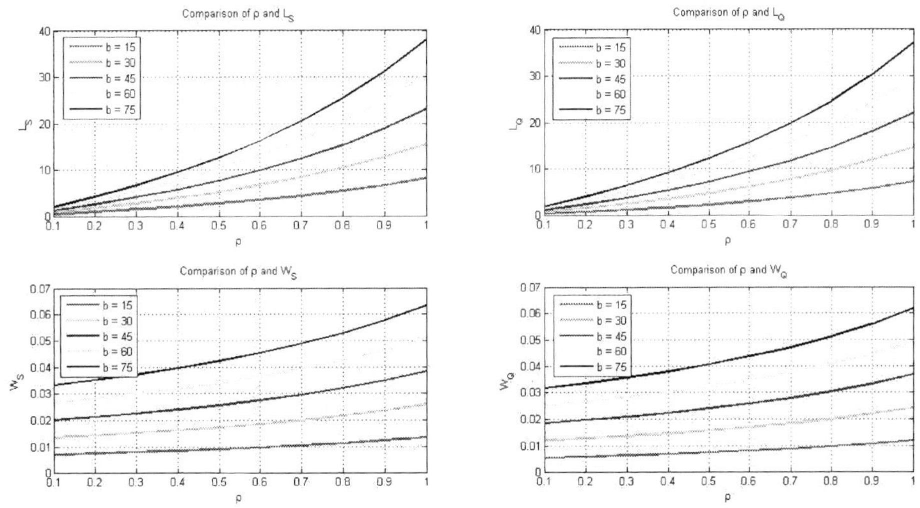

Figure 4. Correlation of ρ and Performance Measurements of the Model for Simulation 2

5. Conclusion

The construction of a dual-channel multiphase bulk arrival queuing system with heterogeneous service time has been developed. It shows that the mathematical model in form of explicit pgf can be derived. Performance measurements of the model have also been obtained. The correlation between variables involved in the model with the performance measurements has been shown by the numerical simulation. Overall, it can be concluded that the traffic intensity ρ is directly proportional with the performance measurements and the size of the arrival b. The model which has been constructed can be implemented to small queue system.

6. References

[1] Na Li and David A. Stanford, 2016, Multi-server Accumulating Priority Queues with Heterogeneous Servers, *Europian Journal of Operational Research 252* 866-878.

[2] K. Sikdar and U.C. Gupta, 2008, On the Batch Arrival Batch Service Queue with Finite Buffer Under Server's Vacation: $M^X/G^Y/1/N$ Queue, *Computers and Mathematics with Application 56* 2861-2873.

[3] Emmanuel John Ekpenyong and Nse Sunday Udoh, 2011, Analysis of Multi-Server Single Queue System with Multiple Phases, *Pak.j.stat.oper.res. Vol VII No.2* 305-314.

[4] Sushil Ghimire, R.P. Ghimire, and Gyan Bahadur Thapa, 2014, Mathematical Models of $M^b/M/1$ Bulk Arrival Queueing System, *Journal of the Institute of Engineering Vol. 10 No. 1* 184-191.

[5] Vishwa Nath Maurya, 2014, Mathematical Modelling and Performance Analysis of Single Server Two-State Batch Arrivals and Batch Service Markovian Queue with Multiple Vacations, *American Journal of Modeling and Optimization Vol. 2 No. 2* 39-46.

[6] A.M. Sultan, N.A. Hassan, and N.M. Elhamy, 2005, Computational Analysis of A Multi-Server Bulk Arrival with Two Modes Server Breakdown, *Mathematical and Computational Applications Vol. 10 No. 2* 249-259.

[7] Ahmed M.M. Sultan, Multi-Channel Bi-Level Heterogeneous Servers Bulk Arrival Queueing System with Erlangian Service Time, *Mathematical and Computational Applications Vo. 12 No. 2* 97-105 (2007).

AUTHOR INDEX

AUTHOR INDEX